# Lecture Notes in Physics

Springer
*Berlin*
*Heidelberg*
*New York*
*Barcelona*
*Hong Kong*
*London*
*Milan*
*Paris*
*Tokyo*

**Physics and Astronomy**

ONLINE LIBRARY

http://www.springer.de/phys/

## Editorial Policy

The series *Lecture Notes in Physics* (LNP), founded in 1969, reports new developments in physics research and teaching -- quickly, informally but with a high quality. Manuscripts to be considered for publication are topical volumes consisting of a limited number of contributions, carefully edited and closely related to each other. Each contribution should contain at least partly original and previously unpublished material, be written in a clear, pedagogical style and aimed at a broader readership, especially graduate students and nonspecialist researchers wishing to familiarize themselves with the topic concerned. For this reason, traditional proceedings cannot be considered for this series though volumes to appear in this series are often based on material presented at conferences, workshops and schools (in exceptional cases the original papers and/or those not included in the printed book may be added on an accompanying CD ROM, together with the abstracts of posters and other material suitable for publication, e.g. large tables, colour pictures, program codes, etc.).

## Acceptance

A project can only be accepted tentatively for publication, by both the editorial board and the publisher, following thorough examination of the material submitted. The book proposal sent to the publisher should consist at least of a preliminary table of contents outlining the structure of the book together with abstracts of all contributions to be included.
Final acceptance is issued by the series editor in charge, in consultation with the publisher, only after receiving the complete manuscript. Final acceptance, possibly requiring minor corrections, usually follows the tentative acceptance unless the final manuscript differs significantly from expectations (project outline). In particular, the series editors are entitled to reject individual contributions if they do not meet the high quality standards of this series. The final manuscript must be camera-ready, and should include both an informative introduction and a sufficiently detailed subject index.

## Contractual Aspects

Publication in LNP is free of charge. There is no formal contract, no royalties are paid, and no bulk orders are required, although special discounts are offered in this case. The volume editors receive jointly 30 free copies for their personal use and are entitled, as are the contributing authors, to purchase Springer books at a reduced rate. The publisher secures the copyright for each volume. As a rule, no reprints of individual contributions can be supplied.

## Manuscript Submission

The manuscript in its final and approved version must be submitted in camera-ready form. The corresponding electronic source files are also required for the production process, in particular the online version. Technical assistance in compiling the final manuscript can be provided by the publisher's production editor(s), especially with regard to the publisher's own Latex macro package which has been specially designed for this series.

## Online Version/ LNP Homepage

LNP homepage (list of available titles, aims and scope, editorial contacts etc.):
http://www.springer.de/phys/books/lnpp/

LNP online (abstracts, full-texts, subscriptions etc.):
http://link.springer.de/series/lnpp/

K. Porsezian   V.C. Kuriakose   (Eds.)

# Optical Solitons

Theoretical and Experimental Challenges

 Springer

**Editors**

Dr. K. Porsezian
Raman School of Physics
Pondicherry University
Kalapet
Pondicherry 605014, India

Dr. V.C. Kuriakose
Department of Physics
Cochin University
of Science and Technology
Kochi 682022, India

*Cover Picture*: (see Figure 1 of the contribution of M. Lakshmanan in this volume)

Cataloging-in-Publication Data applied for

A catalog record for this book is available from the Library of Congress.

Bibliographic information published by Die Deutsche Bibliothek

Die Deutsche Bibliothek lists this publication in the Deutsche Nationalbibliografie;
detailed bibliographic data is available in the Internet at http://dnb.ddb.de

ISSN 0075-8450
ISBN 3-540-00155-7 Springer-Verlag Berlin Heidelberg New York

Springer-Verlag Berlin Heidelberg New York
a member of BertelsmannSpringer Science+Business Media GmbH

http://www.springer.de

© Springer-Verlag Berlin Heidelberg 2003
Printed in Germany

Typesetting: Camera-ready by the authors/editor
Camera-data conversion by Steingraeber Satztechnik GmbH Heidelberg
Cover design: *design & production*, Heidelberg

Printed on acid-free paper
SPIN: 10900154      57/3141/du - 5 4 3 2 1 0

# Preface

In the 21st Century, information technology will be essential for realizing a worldwide communication network. Optical transmission using a short optical pulse train is a fundamental technology for achieving a high-speed and long distance global network. Pulse propagation in optical fiber communication systems is affected by the fiber nonlinearity even at relatively low power levels. One example of the action of nonlinearity is the formation of solitons in anomalous dispersion fibers. Among many optical transmission formats, an optical soliton, which is created by balancing the anomalous group velocity dispersion with the fiber nonlinearity, called the self-phase modulation, offers great potential for realizing an advanced optical transmission system, since the soliton pulse can maintain its waveform over long distances. An ideal soliton can exist only in a lossless fiber with constant dispersion. Interest in optical solitons has grown steadily in recent years. The field has considerable potential for technological applications, and it presents many exciting research problems both from a fundamental and an applied point of view. Now, about a quarter of a century after the first theoretical predictions of optical solitons, the industrial application of the optical soliton concept is close to becoming reality in the booming field of modern telecommunications, where the demand for high-speed data transmission and routing is constantly growing. However, all real systems have losses and varying dispersion, which severely degrades the quality of the optical soliton pulse. Recent progress in optical soliton technology has overcome these difficulties by introducing the idea of dispersion compensation. Various technologies have arisen in the race to move more bits of data faster and farther, with increasing reliability and decreasing costs.

Solitons are one among many exciting new technologies emerging in optical networking and they are poised to benefit the commercial ultra-long-haul all-optical multi-terabit networks spanning distances up to many millions of kilometers. It could well become one of the fundamental technologies in the current communication revolution. Solitons are localized nonlinear waves having stable properties which allow them to propagate very long distances with little change. New optical devices are in various stages of development: Soliton information processing is the most advanced one. In the last decade, many distinguished physicists and mathematicians have contributed to the relevant research. However, to a large extent the various communities involved – engineers, physicists and mathematicians – still operate within their own boundaries. One of the purposes of this volume is to feature articles that bring together different perspectives, with the hope of stimulating future interaction between researchers with different backgrounds. Reflecting the importance of optical solitons in future technologies, in this book, we include articles written by eminent scientists in the field. To reach a diversified audience, a pool of internationally renowned scientists has generously contributed articles treating the different aspects of soliton engineering that they have mastered. The editors wish to thank them warmly for devoting their valuable time to the preparation of these articles. We believe that we have covered a majority emerging areas of optical solitons and their applications. We are happy to bring out this volume, a compilation of articles from experts, and hope that the reader will come to appreciate the importance of optical solitons and other related optical devices in future communication systems. We acknowledge with thanks the editorial help of Dr.Angela Lahee, Editor, Springer-Verlag.

We are very grateful to Prof. M. Lakshmanan for the help and encouragement we received from him in bringing out this volume. Finally, our thanks go to Dr.T. Ramesh Babu, R. Ganapathy, K. Senthil Nathan, Minu Joy, M.N. Vinoj, C. Ramesh Kumar, C.D. Ravikumar and D. Vijay for their help in the process of editing of the book.

Pondicherry and Kochi                                          K. Porsezian
October 2002                                              V.C. Kuriakose

# Table of Contents

# List of Contributors

**M.J. Ablowitz**
Department of Applied Mathematics
University of Colorado at Boulder
Boulder, CO 80309-0526, USA
Mark.Ablowitz@colorado.edu

**A.B. Aceves**
Department of Mathematics and
Statistics
University of New Mexico
Albuquerque, NM 87131
aceves@math.unm.edu

**N.N. Akhmediev**
Optical Sciences Centre
Research School of Physical Sciences
and Engineering
Institute of Advanced Studies
Australian National University
Canberra, ACT 0200, Australia
nna124@rsphy1.anu.edu.au

**A. Ankiewicz**
Optical Sciences Centre
Research School of Physical Sciences
and Engineering
Institute of Advanced Studies
Australian National University
Canberra, ACT 0200, Australia
ana124@rsphysse.anu.edu.au

**K.J. Blow**
Aston University
Aston Triangle
Birmingham, B4 7ET,UK
k.j.blow@aston.ac.uk

**C. Conti**
INFM-RM3, Via della
Vasca Navale 84, 00146 Roma, Italy
c.conti@ele.uniroma3.it

**J.I. Dijkhuis**
Atom Optics and Ultrafast Dynamics
Debye Institute, University of Utrecht
P.O. Box 80 000, 3508 TA
Utrecht, Netherlands
J.I.Dijkhuis@phys.uu.nl

**P.T. Dinda**
Laboratoire de Physique
de l'Université de Bourgogne
UMR CNRS No. 5027, Av.A. Savary
B.P. 47 870, 21078 Dijon Cédex,
France
Patrice.Tchofo-Dinda
@u-bourgogne.fr

**J.M. Dudley**
Laboratoire d'Optique P.M. Duffieux
Université de Franche-Comté, 25030
Besançon, France
john.dudley@univ-fcomte.fr

**M. Haelterman**
Service d'Optique et Acoustique
Université Libre de Bruxelles, B-1050
Bruxelles, Belgium
Marc.Haelterman@ulb.ac.be

**A. Hasegawa**
Himeji Dokkyo University
and Soliton Communications
43, 19-1 Awataguchi Sanjobocho
Higashiyama, Japan
soliton@mbox.kyoto-inet.or.jp

**J. Herrmann**
Max Born Institute for Nonlinear
Optics and Short Pulse Spectroscopy
Max-Born-Str. 2a, D-12489 Berlin
Germany
jherrman@mbi-berlin.de

**T. Hirooka**
Department of Applied Mathematics
University of Colorado at Boulder
Boulder, CO 80309-0526, USA
hirooka@riec.tohoku.ac.jp

**A. Husakou**
Max Born Institute for Nonlinear
Optics and Short Pulse Spectroscopy
Max-Born-Str. 2a, D-12489 Berlin
Germany
gusakov@mbi-berlin.de

**T. Inoue**
Graduate School of Engineering Osaka
University
2-1 Yamada-oka, Suita, Osaka
565-0871, Japan
inoue@comf5.comm.eng.
osaka-u.ac.jp

**V.P. Kalosha**
Max Born Institute for Nonlinear
Optics and Short Pulse Spectroscopy
Max-Born-Str. 2a, D-12489 Berlin
Germany
kalosha@mbi-berlin.de

**T. Kanna**
Centre for Nonlinear Dynamics
Department of Physics
Bharathidasan University

Tiruchirapalli 620 024, India
tkans@yahoo.com

**V.C. Kuriakose**
Department of Physics
Cochin University of Science and
Technology, Kochi-682 022, India
vck@cusat.ac.in

**A. Labruyere**
Laboratoire de physique
de l'Université de Bourgogne
UMR CNRS No. 5027, Av.A. Savary
B.P. 47 870, 21078 Dijon
Cédex, France
lbruyere@u-bourgogne.fr

**M. Lakshmanan**
Centre for Nonlinear Dynamics
Department of Physics
Bharathidasan University
Tiruchirapalli 620 024, India
lakshman25@satyam.net.in

**S.B. Leble**
Faculty of Applied Physics and
Mathematics, Technical University
of Gdansk ul G Narutowicza, 11/12
80-952 Gdansk-Wrzeszcz, Poland
lleble@mif.pg.gda.pl

**A. Maruta**
Graduate School of Engineering
Osaka University
2-1 Yamada-oka, Suita
Osaka, 565-0871, Japan
maruta@comm.eng.osaka-u.ac.jp

**G. Millot**
Laboratoire de Physique
de l'Université de Bourgogne
21078 Dijon, France
Guy.Millot@u-bourgogne.fr

**C. Montes**
Centre National
de la Recherche Scientifique
Laboratoire de Physique
de la Matière Condensée
Université de Nice - Sophia Antipolis
Parc Valrose
06108 Nice Cedex 2, France
montes@unice.fr

**O.L. Muskens**
Atom Optics and Ultrafast Dynamics
Debye Institute, University of Utrecht
P.O. Box 80 000, 3508 TA
Utrecht, Netherlands
O.L.Muskens@phys.uu.nl

**K. Nakkeeran**
Laboratoire de physique
de l'Université de Bourgogne
UMR CNRS No. 5027, Av.A. Savary
B.P. 47 870, 21078 Dijon
Cédex, France
ennaks@polyu.edu.hk

**Y. Nonaka**
Graduate School of Engineering
Osaka University
2-1 Yamada-oka, Suita
Osaka, 565-0871, Japan
nonaka@comf5.comm.eng.
osaka-u.ac.jp

**S. Pitois**
Laboratoire de Physique
de l'Université de Bourgogne
21078 Dijon, France
Stephane.Pitois@u-bourgogne.fr

**K. Porsezian**
Raman School of Physics
Pondicherry University
Pondicherry-605 014, India
ponzsol@yahoo.com

**J.M. Soto-Crespo**
Instituto de Optica, Consejo Superior
de Investigaciones Cientificas,
Serrano 121
28006 Madrid, Spain
iodsc09@io.cfmac.csic.es

**V.B. Taranenko**
Physikalisch-Technische
Bundesanstalt
38116 Braunschweig
Germany
Victor.Taranenko@ptb.de

**K. Thyagarajan**
Physics Department
Indian Institute of Technology
New Delhi 110016, India
ktrajan@physics.iitd.ernet.in

**G.E. Town**
Department of Electronics
Macquarie University
NSW 2109, Australia
gtown@ics.mq.edu.au

**S. Trillo**
University of Ferrara, Via Saragat 1
44100 Ferrara, Italy
strillo@fub.it

**C.O. Weiss**
Physikalisch-Technische
Bundesanstalt
38116 Braunschweig, Germany
Carl.Weiss@ptb.de

**Y. Yoshika**
Graduate School of Engineering
Osaka University
2-1 Yamada-oka, Suita
Osaka, 565-0871, Japan
yoshika@comf5.comm.eng.
osaka-u.ac.jp

# Introduction

K. Porsezian[1] and V.C. Kuriakose[2]

[1] Raman School of Physics, Pondicherry University, Pondicherry-605 014, India
[2] Department of Physics, Cochin University of Science and Technology,
Kochi-682 022, India

Today we are on the verge of another industrial revolution, the information revolution. The present day earth is becoming more digitized and light is playing an indispensable role in keeping the communication lines open. Because of the Internet, information flows across the continents as easily as it flows across an office. The ever-increasing Internet traffic will soon exceed today's performance limit of Terabits per second per fiber. The rapid increase in network traffic demands reliability, more transmission capacity, good performance, rapid transmission with less transmission loss, etc. The above traits are easily achievable with fibers. To offer high-bandwidth services ranging from home-based PCs to large business and research organizations, telecommunication companies worldwide are using light waves travelling within optical fibers as the dominant transmission system. Other examples include: services such as database queries and updated shopping, video-on-demand, remote education, video conferencing and web-based courses.

In addition to the above, optical technology is all set to replacing electronics in the very near future, and finds applications in computer memories (CD ROM), sound systems (compact disc), video systems (laser disc) etc. An immense amount of research is in progress on optical computers. It needs to interface electronics with the optical devices. Researchers are working towards all-optical devices and systems with the integrated optical technology using optical switches. For example, the intensity-induced switches generate undesirable effects like pulse reshaping and breaking, which degrade the performance of the device. Moreover, optical switches do not produce any noise as there are no moving parts, like in mechanical switches. This leads to a great reduction in the noise contribution to the system as a whole.

## 1 Importance of Optical Fiber Communications

After the invention of the laser in 1960, people attempted to use it for communication purposes because of its coherent radiation. Researchers struggled to find a suitable medium for communications for more than five years. Fortunately in 1966, the fiber medium emerged as the best choice to transmit the optical signals. The fiber is selected as a medium, because it has a peculiar property of confining an electromagnetic field in the plane perpendicular to the axis of the fiber core. The physical principle behind such devices is the principle of total internal reflection, which was first demonstrated by John Tyndall in 1870. The first optical fibers were made from glass. The fibers are capable of guiding the light in a similar way as electrons are guided in copper cables. Though the number of bits sent per second increased, the fiber loss was a serious problem.

In traditional electrical communications, the message is sent to distant places through copper cables, after superposing them over a carrier wave. The process of superposition is called modulation. During modulation, the amplitude, the frequency or the phase of the carrier changes in accordance with the message and these modulations are respectively called Amplitude Modulation (AM), Frequency Modulation (FM) and Phase Modulation (PM). The carrier here is one among the known electromagnetic spectrum. At the receiving end the message is extracted from the carrier by a demodulation process. The medium of transmission was later extended to atmosphere in which the carrier is either radio waves or microwaves. The satellite communication system operates in the microwave region. With the merits of fibers like low cost, easy installation, signal security, life more than 25 years, accommodation of more channels, low transmission loss, no electromagnetic interference, abundant availability of raw material etc. The decade-old battle between the fiber medium and the satellite systems for international communications points to the fiber system being the clear economic and technological winner. This is because the satellite communication system is costly, life is limited to only five to ten years, handling the system requires more technical and professional people and it has lower channel handling capacity. With the astonishing development of lasers, various telecommunication companies across the world have started using light as the ascendant transmission system. In fiber systems, the transmission of signals is through dielectric media called waveguides. The medium is of hair thin glass fibers that guide the light signals over long distances. As the carrier here is light, this type of communication is called Optical Fiber Communication (OFC).

In both wired and wireless communications the amount of information transmitted can be increased by increasing the range of the frequency of the carrier called bandwidth. OFC system is an excellent communication system compared to the other media such as copper or atmosphere. They offer low-loss transmission over a wide range of frequency of about 50 THz. This range is several times more than the bandwidth available in copper cables or any other transmission medium. Because of this property, this system allows signals to be transmitted over very long distances at higher speeds before they need to be amplified. Unlike the electrical communications, in optical systems the transmission of information is in an optical format, which is carried out not by AM, FM or PM of the carrier but by varying the intensity of the optical power.

The information to be sent is converted into bits (zero or one) using an Analog-to-Digital Converter (ADC). The swift developments in communication technology keep on increasing this bit rate. Non-return to zero (NRZ) is a binary modulation with square pulses in which the signal is OFF for a '0' bit and ON for a '1' bit. Return to Zero (RZ) is the same, except that the one pulse is shorter than the bit time. As the signal travels over a long distance, the energy of the pulse will decrease with time. This energy loss may be due to linear and nonlinear effects of the fiber. This can be compensated by periodic amplification using optical amplifiers, also called as repeaters. The distance between the repeaters depends on the amount of energy loss of the pulse. The bit-rate product measures the transmission capacity of the optical fiber links.

Since the implementation of OFC at the beginning of 1980s transmission capacity has experienced a 10-fold increase every four years. Transmission of information over long distances using fibers is limited by intrinsic properties such as loss, dispersion, nonlinearity and amplifier induced noise. The fiber loss and dispersion can be minimized by selecting the pulse that propagates through the fiber with a wavelength of 1550nm and by using dispersion - shifted fibers. Operating in this wavelength region gives a fiber bandwidth of approximately 20,000 GHz. This corresponds to a fiber capacity of about four million TV channels, with a fiber attenuation coefficient of about 0.18 dB/km. This fiber loss can still be limited by using periodic amplifiers. The dispersion loss can be reduced by a dispersion compensation scheme, which requires the wavelength to be at 1330nm for almost zeroing dispersion.

A major accomplishment in the development of OFC was the invention of Erbium Doped Fiber Amplifiers (EDFA) in 1987. Cost reduction is possible by replacing the conventional electronic repeaters by EDFA. The fibers used in EDFA are doped with the rare earth element erbium, which acts as the gain medium for amplifying the input optical signal. These are transparent to bit rates and different modulation techniques. As there are no active and passive components in these amplifiers, as in the case of existing optical amplifiers, the reduction in noise is possible. EDFA also allows the OFC system to be upgraded in bit rates at a later date by amplifying signals at many wavelengths simultaneously. This is another way of increasing the system capacity other than increasing the bit rate. Keeping the bit rate constant and using more than one wavelength, the process called Wavelength Division Multiplexing (WDM) is possible. It finds applications ranging from local area-network (LAN) to the ultra-long intercontinental links.

## 2    Birth of Optical Solitons

The extensive use of fibers for communication is mainly because they are nonlinear materials. The refractive index of the fiber depends nonlinearly on the intensity of the pulse travelling through it. The nonlinear nature of the fiber requires the optical source to insert a very high-energy pulse for propagation. In fibers, if the intensity of a pulse changes on its travel, since it depends on the refractive index of the fiber; the phase of the pulse also changes with time. This change of phase with time leads to the generation of additional frequencies on either side of the pulse. The intensity of the pulse itself modulates the phase of the pulse. This circumstance is called as Self Phase Modulation (SPM). This effect leads to a distortion of the pulse and the signal becomes unstable.

As we know, no optical pulse is perfectly monochromatic, as it excites a spectrum of frequencies, each spectral component travelling at different speeds called group velocity. An optical signal becomes increasingly distorted as it travels along a fiber resulting in chromatic dispersion. As this dispersion is a function of wavelength of the pulse travelling, this is known as group velocity dispersion (GVD). Since the medium is dispersive, the pulse width will spread in the time domain with increasing distance along the fiber called normal dispersion, having positive GVD value. The reverse will takes place in the case of anomalous dispersion, having negative GVD value. Thus if nonlinearity alone is introduced then the high intensity pulse may break into short waves as time passes, whereas if the dispersion alone is allowed in a fiber then the broadening of pulse will take place. So, if the nonlinearity is exactly balanced with the dispersion effect, the pulse will propagate without any loss in energy. Such a pulse is called an optical soliton. This phenomena was theoretically predicted in optical fiber by Hasegawa and Tappert in 1974. The concept of soliton was first introduced in another context by Zabusky and Kruskal in 1964.

The above-discussed effects are in an anomalous regime, the respective solitons are called bright solitons having negative GVD parameter. These solitons have hyperbolic secant shape for their intensity profile. In the normal dispersion regime, the GVD parameter is positive and the resulting solitons are called dark solitons. The intensity profile in this case contains a dip in its shape with uniform background having hyperbolic tangent shape. Both bright and dark solitons are robust to perturbations of their parameters like the width and the shape of the intensity profile. The force of attraction between dark solitons is always repulsive whereas in bright solitons it is attractive up to some relative phase difference and there after repulsive force comes in to play.

Existing linear systems of communication suffer from chromatic dispersion, which has to be managed in some way. Typically 10 Gbits/s systems require periodic dispersion compensating fiber. As bit rates increase to 40 Gbits/s and beyond, and as distance increases up to 10,000 km, the effects of dispersion become severe. In addition, as bit rates increase, pulses get smaller and their instantaneous power gets higher (i.e. the energy has to be squeezed into a shorter pulse), thus producing nonlinearities. Moreover in linear systems, higher powers are required to combat noise as distances increase. Thus, it becomes increasingly difficult to manage dispersion and to limit nonlinearities in high-bit-rate long distance communications. Nonlinear or soliton systems change the game by accepting, and in fact using the nonlinearity to combat dispersion, solving two problems at once.

As we have already discussed,the effects due to nonlinearity and dispersion are 'destructive' in OFC but useful in Optical Soliton Fiber Communication (OSFC). The soliton type optical pulses are highly stable, their transmission rate is more than 100 times better than that in the best linear system and they are not affected by the imperfections in the fiber geometry or structure. Solitons can be propagated without any distortion if the nonlinear characteristics (like amplitude, intensity of the pulse depending on velocity) and dispersion characteristics (frequency depending on velocity) of the media are balanced

Solitons can also be multiplexed at several wavelengths, without the interaction between channels usually suffered in NRZ systems. Nowadays, most of the communication systems use RZ format (e.g. Trans Oceanic Transmission (TOT) 10 Gbits/s per channel) for information transfer in dispersion-managed fibers. This format is the only stable form for pulse propagation through the fiber in the presence of fiber nonlinearity and dispersion in all-optical transmission lines with minimum loss. In dispersion-managed fibers, a large pulse width is allowed, pulse height is reduced and nonlinear interaction between adjacent pulses as well as among different wavelength channels are believed to be reduced. This issue is the focus of much recent soliton research and few articles in this volume mainly discuss this important problem.

Apart from the usage of solitons in communication, they also find application in the construction of optical switches. The advantage of using solitons is that they do not change their shapes even on interaction with other pulses. In these switches, the propagation of one optical pulse affects the other, the 'signal' pulse by the 'control' pulse. Here it behaves like the control pulse opening a gate for a signal pulse, so as to allow it to pass through. Photonic logic gates operate on this principle. So logic gates can also be realized using solitons.

Another very important application of optical solitons is pulse compression. The production of Ultra Short Pulses (USP) is the new technology. USP are needed in many branches of science, which have been using nanosecond and picosecond pulses. Now the focus of attention is to generate femtosecond pulses, which are only few wavelengths of the light wave. Optical solitons help in the generation of USP.

The following important advantages of soliton systems will influence the preferred choice for future communication systems. First, solitons are unaffected by an effect called polarization mode dispersion due to the imperfection in the circular symmetry of fiber, which leads to a small and variable difference between the propagation constants of orthogonal polarized modes. This dispersion becomes a major problem over long distances and at high data rates. Secondly, solitons are well matched with all-optical processing techniques. Our long-term goal is to create networks in which all of the key high-speed functions, including routing, demultiplexing and switching are performed in the optical domain. So, the signals need not be converted into an electrical form on the way. Most of the devices and techniques designed for these tasks work only with well-separated optical pulses, which are particularly effective with solitons. If the solitons are controlled properly they can be more robust than NRZ pulses. Schemes have been devised that can not only provide control over the temporal positions of the solitons, but also remove noise added by amplifiers. Such schemes would allow the separations between amplifiers to be many times greater than that in the schemes that are used with NRZ pulses. This is probably the strongest economic argument for using solitons, since fewer amplifiers would be needed for achieving the same data capacity.

Solitons, if used, would replace the traditional NRZ and RZ modulations, which are used in almost all commercial terrestrial WDM systems. Although nonlinearities exist in NRZ and RZ systems there are undesirable effects limiting the performance of the system. Typically the design of a conventional WDM system involves an effort to increase the power as much as possible (to counteract attenuation and noise) without introducing too much nonlinearity. Thus NRZ and RZ systems are often called linear systems. Recent advancements in soliton communication with 3.2 Tbits/s have been demonstrated. But the total capacity is limited by the available bandwidth of EDFA.

## 3   Present Book

In this book, Hasegawa, father of optical solitons, presents the basic aspects of optical solitons and their applications in optical fiber communication. In particular, he has given the basic mathematical ideas required to understand the concept of optical soliton. In addition, he has also discussed the recent experimental results on optical solitons . Thyagarajan explains linear and nonlinear propagation effects in optical fiber. In this article, the different linear and nonlinear effects in fiber and amplifiers and their characteristics are discussed in detail. Leble explains the nonlinear waves in optical wave-guides and soliton theory applications. He reviews the recent results in guided propagation of electromagnetic waves and supports the theoretical results through experimental verifications. He discusses the different theoretical models and establishes their integrability and soliton properties.

Akhmediev and Ankiewicz presents a detailed analysis of integrable, Hamiltonian and dissipative systems. Considering the complex cubic-quintic Ginzburg-Landau equation, the above aspects of this equation are explained in detail. Blow explains the split step operator technique and its role in optical communication studies. He presents how this technique gives rise to real systems applications of the principles. Lakshmanan and Kanna discuss the soliton in coupled NLS equations and explain the inelastic interaction of soliton and their consequences in optical computing and other modern optical devices.

Aceves presents an overview of mathematical modeling in fiber and waveguide structure. In this article, he discusses the possible formation of soliton-like pulses in fibre, waveguide and fiber arrays. Trillo and Conti review the basic properties of gap soliton propagating along one-dimensional short-period gratings and by considering the Bragg grating with Kerr nonlinearity, they discuss other soliton bearing structures.

Dinda et al discuss the effects of stimulated Raman scattering on ultra short pulse propagation in optical fiber systems. They present a general theory for the soliton self frequency shift which applies to any pulse whose spectral bandwidth lies within the third order telecommunication window and discuss the disastrous impact of soliton self frequency shift in communication systems.

Ablowitz and Hirooka present an analytical framework that describes the propagation and interaction of quasi-linear optical pulses in strongly dispersion managed transmission systems. Introducing the multiple scale analysis, dispersion managed nonlinear Schrodinger (NLS) equation is derived. And investigate the pulse evolution and elucidate the role of nonlinearity in quasi-linear transmission systems using the dispersion managed NLS equation and also analyzed the intra-channel pulse interactions. Maruta et al discuss stable bi-soliton in a dispersion managed fiber system. In addition, they propose novel error preventable line-coding schemes in which binary data are assigned to bi-soliton and single dispersion managed soliton. Using the above scheme, impairments arising from intra-channel interaction can be drastically reduced compared with the conventional scheme for the same bit rate.

Town et al review the optical fiber soliton lasers with gain medium in an optical fiber amplifier. They have given an extensive review of the various aspects in the construction of soliton lasers. Husakou et al discuss the nonlinear optical phenomena with ultra-broadband radiation in photonic crystal fibers and hollow wave guides without applying the slowly varying envelope approximation and Taylor expansion for the refractive index and studied the generation of supercontinua and extremely short pulses. A new method for the generation of extremely short optical pulses by higher order stimulated Raman scattering is also investigated.

Millot et al discuss the experimental study of modulational instability (MI) and vector soliton in optical fibers. They present an overview of some of the most important effects of MI in normally depressive hi-bi fibers. They show that an induced process of MI may be exploited for the generation of THz train of vector soliton and reports the experimental observation of Bragg MI induced by a dynamic grating obtained through cross phase modulation with a beating wave in a hi-bi fiber. Montes presents the self-structuration of three-wave dissipative soliton in quadratic media with absorption losses. Considering both nondegenerate and degenerate CW-pumped backward optical parametric oscillators, he discusses the self-structuration of three-wave dissipative soliton.

Taranenko and Weiss demonstrate experimentally and numerically the existence of spatial solitons in multiple-quantum-well semiconductor microresonators driven by an external coherent optical field. They also discuss different manipulation of such solitons. Muskens and Dijkhuis present the recent development on propagation of picosecond acoustic wave packets created by nJ femtosecond optical pulses in a lead molybdate single crystal, employing the Brillouin scattering technique as a local probe of acoustic strain.

From the recent theoretical and experimental investigations on this exciting topic it is clear that the soliton is going to play an indispensable role in future communication systems as well as in photonic technology in general. Thus, the present book contains a collection of reviews and papers that highlight the recent theoretical and experimental developments in this challenging field. Reviews on the basic concepts on optical solitons and fiber optics are included so that a beginner in this field will find this book very useful.

# Optical Soliton Theory
# and Its Applications in Communication

A. Hasegawa

Soliton Communications, www.solitoncomm.com

**Abstract.** The concept of optical soliton was born in 1973 and was experimentally demonstrated in 1980. It took another two decades for its application to practical systems. During this period, the research on optical solitons has induced the development of various new theories and stimulated discoveries of new forms of solitons. The history of optical soliton research presents an interesting case study in which an abstract mathematical concept has been transferred into practical public interests and opened new fields of vigorous scientific activities. The lecture summarizes theory of optical solitons and their applications for optical communications.

## 1 Introduction

The optical soliton in fibers is probably the best studied form of a soliton because of its remarkable behavior that agrees well with theoretical predictions and its potential as optical information carrier. The optical soliton discussed in this chapter, as well as in most articles in this book, is a soliton of the envelop of light waves whose fundamental properties are described by the nonlinear Schrödinger equation (NLSE). The NLSE contains the lowest order group velocity dispersion (or diffraction) and the cubic nonlinearity and the soliton is formed by the balance of these terms. The NLSE is integrable by the method of inverse scattering transform (IST) as most solitons are. For the review purpose of the lecture series, I first present essential feature of the IST and introduce soliton concept in Sect.2. Modern optical communication utilizes optical fibers as medium of transmission of information. The information is carried by modulation on light waves in fibers. In Sect.3, I discuss electrical properties of fiber dielectric media that influence the propagation of information in fibers. I show that group velocity dispersion (or the group delay) and the cubic nonlinearity of fiber dielectric (or the Kerr effect) plays the dominant role in the deformation of the information in fibers where the loss is compensated by optical amplifiers. In Sect.4, I derive the master equation to describe information transfer in fiber, which in an ideal limit becomes NLSE. In Sect.5, I introduce Lagrangian method of solving NLSE with perturbations by taking advantage of the fact that the equation admits the soliton solution. Knowing that soliton is the unique solution of NLSE, one can obtain the evolution of information in fibers by studying only the soliton parameters, the amplitude, frequency, position and the phase. While an infinite dimensional analysis is required when a non-soliton format of information is used, the use of the soliton format reduces the problem to finite dimensional analysis. In addition to the fact that the soliton is the only stable format of light wave envelope in fibers, that the finite dimensional treatments suffice in most cases is an important merit of the use of solitons as information carrier. For communication applications, optical soliton is modified so that it is more immune to external perturbations by applying proper variation of the fiber dispersion profile. Solitons that are created in such fibers are often called dispersion-managed solitons (DMS). I present an example of DMS and present their properties in Sect.5. In the final section, (Sect.6), I introduce some of the most recent experimental results of single channel and multi-channel DMS transmission. Here, I show examples of error free transmission of a single channel 40Gb/s over 10,000 km as well as multi-channel 1.4Tb/s (42.7 Gb/s x 32 channel) over 6,000 km.

## 2    Properties of Optical Solitons as a Result of IST

In an ideal case of loss less and homogeneous media, the optical soliton is described by the equation called the nonlinear Schrödinger equation (NLSE). The NLSE is the expression for evolution of the normalized Fourier amplitude q of the light wave electric field in a medium with cubic (Kerr) nonlinearity and (group) dispersion or diffraction. In a case of spatial (Z) evolution, the NLSE is given by

$$i\frac{\partial q}{\partial Z} \pm \frac{1}{2}\frac{\partial^2 q}{\partial T^2} + |q|^2 q = 0, \tag{1}$$

Here, $T$ represents time and $\pm$ signs correspond to anomalous ( $k'' < 0$ is the wave number and the prime indicates the derivative with respect to the angular frequency) and normal ($k'' > 0$) dispersions respectively.

The NLSE given by Eq.(1) is known to be integrable [1]. The basic idea of solving the NLSE is to recognize that the equation can be expressed as the compatibility condition of the following two *linear* evolution equations for a wave function, $\psi(T, Z; \xi)$

$$L(Z)\psi = \xi\psi, \tag{2}$$

$$\frac{\partial \psi}{\partial Z} = M(Z)\psi, \tag{3}$$

with a subsidiary condition that the eigenvalue $\xi$ is invariant under the $Z$ evolution,

$$\frac{d\xi}{dZ} = 0, \tag{4}$$

where $L$ and $M$ are differential operators in the $T$ - derivatives and involve $q$ of Eq.(1). The pair $(L\ M)$ is called a Lax pair of the integral system [2]. The compatibility condition under the requirement that the eigenvalue is invariant in $Z$ (or stationary) becomes,

$$\frac{\partial L}{\partial Z} = ML - LM = [M, L]. \tag{5}$$

Equation (5) is derived by taking the $Z$ derivative of Eq.(2) and using Eqs.(3) and (4). Equation (5) is called the Lax equation and gives the operator representation of an integrable system . The Lax pair for NLSE with anomalous dispersion is given by [1],

$$L = \begin{pmatrix} i\frac{\partial}{\partial T} & q \\ -q^* & -i\frac{\partial}{\partial T} \end{pmatrix},$$  (6)

$$M = \begin{pmatrix} i\frac{\partial^2}{\partial T^2} + \frac{i}{2}|q|^2 & q\frac{\partial}{\partial T} + \frac{1}{2}\frac{\partial q}{\partial T} \\ -q^*\frac{\partial}{\partial T} - \frac{1}{2}\frac{\partial q^*}{\partial T} & -i\frac{\partial^2}{\partial T^2} - \frac{i}{2}|q|^2 \end{pmatrix}.$$  (7)

It is tedious but straightforward to check that the Lax equation (5) with $L$ and $M$ given by Eq.(6) and Eq.(7) gives the NLSE. Now, what is the significance of this? If we consider $q$ as a parameter representing the potential, the eigenvalue equation (2) is a linear equation for the eigen function $\psi$ whose eigen value is invariant of $Z$. We are interested in obtaining the $Z$ evolution of $q$. Solving for the potential $q$ for a given eigenvalue (and scattering data) is known as inverse scattering problem. It is called inverse because normal eigenvalue problem is to solve for eigenvalue for a given potential. The solution of inverse scattering problem is well established in quantum mechanics and the potential is given by a linear integral equation involving the eigenvalue and the scattering data. The evolution of $q$ in $Z$ is given by the evolution of the eigen function given by Eq.(3).

The soliton solution is characterized by discrete eigenvalue(s) of Eq. (2). The complex eigen values of Eq.(2), which are invariant by definition, give soliton parameters $\eta$ (designating the amplitude and the inverse of the pulse width) and $\kappa$ (designating the frequency shift and the velocity) through,

$$\xi_n = \frac{\kappa_n + i\eta_n}{2}, (n = 1, 2)$$  (8)

and the one-soliton solution is given by

$$q(T, Z) = \eta \operatorname{sech}\left[\eta\left(T + \kappa Z - T_0\right)\right] \exp[-i\kappa T + \frac{i}{2}(\eta^2 - \kappa^2)Z + i\sigma)].$$  (9)

Since $\xi$ is invariant, the soliton parameters are determined from the initial pulse form, $q(0, T)$ with $q$ replaced by $q(0, T)$ in Eq.(2).

The evolution of $q(Z, T)$ is obtained by means of the inverse scattering transform (IST). However, for practical purposes derivation of the eigenvalue from $q(0, T)$ provides the most critical information on solitons that emerge from $q(0, T)$. For example, if we approximate the input pulse shape as

$$q(0, T) = A \operatorname{sech} T,$$  (10)

the eigenvalues of Eq.(2) are obtained analytically and the number of eigenvalues $N$ is given by [3]

$$A - \frac{1}{2} < N \le A + \frac{1}{2},$$  (11)

where the corresponding eigenvalues are imaginary and given by

$$\zeta_n = i\frac{\eta_n}{2} = i\left(A - n + \frac{1}{2}\right), \quad n = 1, 2, \ldots, N.$$  (12)

If $A$ is exactly equal to $N$, the solution can be obtained in terms of $N$ solitons and their amplitudes are then given by

$$\eta_n = 2\,(N - n) + 1 = 1, 3, 5, \ldots, (2N - 1)\,. \tag{13}$$

If $A$ is not an integer, the initial pulse evolves to $N$ numbers of solitons and dispersive wave.

We note here that in this particular example of the initial pulse given by Eq.(10), all the eigenvalues are purely imaginary. Consequently, $\kappa_n = 0$, and the velocities (in the frame of reference of the group velocity) of all the solitons are 0; as a results $N$ numbers of solitons propagate at the same speed and the resultant pulse has oscillatory structure due to phase interference among different solitons.

## 3    Dielectric Properties of Optical Fibers

When an electric field is applied, dielectric material polarizes and polarization current is induced. This current, like the current in a condenser, is proportional to the time variation of the electric field. The effect of polarization is expressed by the use of electric displacement vector, $\mathbf{D}$, as

$$\mathbf{D} = \varepsilon_0 \mathbf{E} + \mathbf{P}. \tag{14}$$

Here, $\varepsilon_0 (= 8.854 \times 10^{-12} \mathrm{F/m})$ is the dielectric constant of vacuum and $\mathbf{P}$ represents the polarization and is given by

$$\mathbf{P} = -en\mathbf{x}(\mathbf{E}). \tag{15}$$

In this expression, $n$ is the density of electrons that participate to the polarization, $\mathbf{x}$ represents the displacement of electron position in a dielectric molecule induced by the electric field $\mathbf{E}$.

The displacement $\mathbf{x}$ is given classically by the response of electrons described by the equation of motion in the presence of ionic potential and the electric field. If the ionic potential is approximated by parabolic in $\mathbf{x}$, the response becomes linear and is proportional to $\mathbf{E}$. However, the actual potential deviates from parabolic due to the influence of neighboring ions. This deviation appears as a reduction of the potential field in proportion to $x^4$. This results in the nonlinear response that is proportional to the cubic power of the electric field. The displacement as the response to the electric field may be obtained by means of the Fourier transform with the nonlinear term as perturbation. Substituting the result into Eq. (15) and (14) gives the relation between the Fourier amplitude of the displacement in terms of that of the electric field,

$$\overline{\mathbf{D}}(\omega) = \varepsilon_0 \varepsilon \left( \omega, |\overline{E}|^2 \right) \overline{E}(\omega), \tag{16}$$

where $\varepsilon$ is the relative permittivity constant of the material and the bar indicates the Fourier amplitude.

It is convenient to introduce the index of refraction, $n\left(\omega, |\overline{E}|^2\right)$ as the ratio of the speed of light in vacuum, $c$ and the phase velocity of the light wave in the dielectric medium, $\omega/k = c/\sqrt{\varepsilon}$,

$$n\left(\omega, |\overline{E}|^2\right) = \frac{ck}{\omega} = \sqrt{\varepsilon} \equiv n_0(\omega) + n_2(\omega) |\overline{E}|^2, \tag{17}$$

where $n_2$ is the Kerr coefficient defined as the incremental change of the index of refraction with respect to the electric field intensity $|\overline{E}|^2$. For glass, the nonlinear coefficient $n_2$ is extremely small and has a value of $10^{-22} (\text{m/V})^2$. Equation (17) indicates that the wave number $k$ given by

$$k = \frac{m}{c} n = \frac{m}{c} \left[n_0(\omega) + n_2(\omega) |\overline{E}|^2\right], \tag{18}$$

becomes a function of frequency and electric field intensity. The former presents the group velocity dispersion (GVD) and the latter the self phase modulation (SPM) of light waves in fibers. Most high-speed transmission systems at present use an all-optical scheme with loss compensated for by periodic optical amplifications. When the fiber loss is effectively eliminated by optical amplifiers, the major limitations come from these effects. Let us first discuss the effect of the GVD. The GVD originates from the combination of the wave-guide property and the material property of the fiber. In the presence of the GVD, information carried by a different frequency component of electric field propagates at different speed thus arrives at the different time. The relative delay $\Delta t_D$ of arrival time of information at frequencies $\omega_1$, and $\omega_2$ at distance $z$ is given by

$$\Delta t_D = \frac{z}{v_g(\omega_1)} - \frac{z}{v_g(\omega_2)} = \frac{(\partial v_g/\partial \omega)(\omega_2 - \omega_1)}{v_g^2}. \tag{19}$$

If we use

$$v_g = \frac{\partial \omega}{\partial k} = \frac{1}{(\partial k/\partial \omega)} = \frac{1}{k'},$$
$$\frac{\partial v_g}{\partial \omega} = \frac{\partial}{\partial \omega}\left(\frac{1}{k'}\right) = -\frac{k''}{(k')^2} = -k'' v_g^2, \tag{20}$$

Eq.(19) becomes

$$\Delta t_D = k''(\omega_1 - \omega_2)z, \tag{21}$$

or in terms of the wavelength difference $\Delta\lambda \equiv \lambda_1(2\pi c/\omega_1) - \lambda_2(2\pi c/\omega_2)$,

$$\Delta t_D = -k'' \frac{2\pi c}{\lambda^2} \Delta\lambda z. \tag{22}$$

Technically $\Delta t_D$ expressed in the unit of picosecond (ps), with $\Delta\lambda$ in nanometer (nm) and the distance $z$ in kilometer (km) is called the group delay $D(\text{ps/nm/km})$. The dispersion shifter fiber (DSF) has a value of $D$ in the range of -1 to 1 ps/nm/km at the wavelength of 1550 nm where the fiber loss is minimum, while the standard single mode fiber (SMF) has $D$ of 16 ps/nm/km at this wavelength.

Currently fibers with various value of $D$ are available with the help of proper design of a fiber. From Eq.(22), $k''$ may be expressed in terms of $D$ as,

$$k'' = -\frac{\lambda^2}{2\pi c} D. \tag{23}$$

Equation (21) shows that the difference of arrival time of information is proportional to the group dispersion, $k''$, the difference of the frequency components $\omega_1 - \omega_2$ and the distance of probation $z$. We note that if $k'' < 0$ (called anomalous dispersion regime), higher frequency component of information arrives earlier and for normal dispersion, $k'' > 0$ (normal dispersion regime), the other way around. If the information at different frequency components arrives at different time, the information may be lost. The problem becomes more series if the amount of information is large so that $\omega_1 - \omega_2$ is large. If we consider one pulse with the pulse width given by $t_0$, in the presence of a group delay, two different frequency components of the pulse arrive at different time, thus the pulse will be broadened. The distance $z_0$ at which the pulse width becomes twice as that of the initial pulse is called the dispersion distance. The dispersion distance may be obtained from Eq.(21) by replacing $\omega_1 - \omega_2$ by $1/t_0$ and $t_D$ by $t_0$ as,

$$z_0 = \frac{t_0^2}{|k''|} = \frac{t_0^2}{D} \frac{2\pi c}{\lambda^2}. \tag{24}$$

For an example of $D = 1\text{ps/nm/km}$, $t_0 = 10\text{ps}$, $\lambda = 1550\text{nm}$, the dispersion distance become approximately 100km.

Let us now consider the effect of SPM. In the presence of the Kerr effect, the wave number is given by Eq.(18). This equation indicates that the Kerr effect induces a nonlinear phase shift $\Delta\Phi_N$ through the nonlinear part of the wave number $k_N$ given by

$$\Delta\Phi_N = k_N z = \frac{\omega}{c} n_2 |\overline{E}|^2 z = \frac{2\pi z n_2 |\overline{E}|^2}{\lambda}. \tag{25}$$

Here $n_0$ is the linear index of refraction and $n_2$ is the Kerr coefficient having a value $\approx 10^{-22} (m/V)^2$. For a light wave with a peak power of 1 mW, $|\overline{E}|$ becomes about $10^5 \text{V/m}$ in a typical fiber. Thus $n_2 |\overline{E}|^2$ has a value of $10^{-12}$. Therefore, even if $n_2 |\overline{E}|^2 \approx 10^{-12}$ since $z/\lambda \approx 10^{12}$ for $z = 10^2 km$ and $\lambda \approx 1500nm$, $\Delta\Phi_N$ between the high intensity portion and the low intensity portion can become of order unity over this distance of propagation. This indicates that the phase information is lost over a distance of $10^2$ km if the light wave power is as low as a milliwatt. In addition it creates a mixture of information in amplitude and phase. This means that information transfer by means of coherent modulation is not appropriate for a light wave in fibers for a propagation distance beyond $10^2$ km. Similarly, the phase sensitive duo-binary format faces loss of information when adjacent pulses overlap during the transmission.

# 4    Master Equation for Information Transfer in Optical Fibers

The information carried by the light wave in fibers is expressed by the modulation $\overline{E}(z,t)$ on the light wave electric field E(z,t) , where,

$$E\left(z,t\right) = \frac{1}{2}\left[\overline{E}\left(z,t\right)e^{i(k_0 z - \omega_0 t)} + \text{c.c}\right]. \tag{26}$$

We now derive the equation, which describes evolution $\overline{E}$ along the direction $z$ of the propagation of information. The most convenient way to derive the envelope equation is to Taylor expand the wave number $k(\omega, |\overline{E}|^2)$ , around the carrier frequency $\omega_0$ and the electric field intensity $|\overline{E}|^2$,

$$k - k_0 = k'\left(\omega_0\right)\left(\omega - \omega_0\right) + \frac{k''\left(\omega_0\right)}{2}\left(\omega - \omega_0\right)^2 + \frac{\partial k}{\partial |\overline{E}|^2}|\overline{E}|^2, \tag{27}$$

and to replace $k - k_0$ by the operator $i\partial/\partial t$ and $\omega - \omega_0$ by $-i\partial/\partial t$ , and to operate on the electric field envelope, $\overline{E}(z,t)$. The resulting equation reads,

$$i\left(\frac{\partial \overline{E}}{\partial z} + k'\frac{\partial \overline{E}}{\partial t}\right) - \frac{k''}{2}\frac{\partial^2 \overline{E}}{\partial t^2} + \frac{\partial k}{\partial |\overline{E}|^2}|\overline{E}|^2\overline{E} = 0. \tag{28}$$

As shown in Sect.3, the wave number in this expression is given by (18). Thus $k'$, $k''$, $\partial k/\partial |\overline{E}|^2$ in Eq.(28) are given approximately by

$$k' \approx \frac{n_0\left(\omega_0\right)}{c}$$

$$k'' \approx \frac{2}{c}\frac{\partial n_0}{\partial \omega} \tag{29}$$

$$\frac{\partial k}{\partial |\overline{E}|^2} \approx \frac{\omega_0}{c}n_2.$$

We note that to obtain $k''$ in this expression, we should go back to Eq.(18) and take the second derivative of $k$ with respect to $\omega$. It is often convenient to study the evolution of $E$ in the coordinate moving at the group velocity $\tau = t - k'z$ . Then the envelope equation becomes,

$$i\frac{\partial \overline{E}}{\partial z} - \frac{k''}{2}\frac{\partial^2 \overline{E}}{\partial \tau^2} + \frac{\omega_0 n_2}{c}|\overline{E}|^2\overline{E} = 0. \tag{30}$$

For a light wave envelope in a fiber, the coefficients of this equation depend on the fiber geometry and modal structure of the guided light wave. In particular, for a SMF, $k'' = 0$ occurs at $\lambda = 1.3\mu$m which is determined primarily by the glass property itself, while $k''$ becomes zero at $\lambda = 1.55\mu$m for a DSF because of the wave-guide property.

For a guided wave in a fiber is modified by the wave-guide dispersion which depends on the modal structure in the fiber [4]. In particular, for a weekly-guided mode, the wave number k is given by the eigen function $\phi(x_\perp)$ for the wave-guide mode,

$$k^2 = \frac{(\omega/c)^2 \int |\nabla_\perp \phi|^2 n_0^2 d\mathbf{S} - \int |\nabla_\perp^2 \phi|^2 d\mathbf{S}}{\int |\nabla_\perp \phi|^2 d\mathbf{S}}. \tag{31}$$

Here $n_0$ is the linear refractive index, which is in general a function of the transverse coordinates $x_\perp$ and frequency $\omega$. The integration $\int d\mathbf{S}$ is evaluated across the cross section of the fiber and $\phi$ is normalized such that

$$\int |\nabla_\perp \phi|^2 d\mathbf{S} = A_{eff} E_0^2, \tag{32}$$

where $E_0$ is the peak intensity of the light electric field, and $A_{eff}$ is the (effective) cross section of the fiber. In addition, since the light intensity varies across the fiber, $n_2$ in Eq.(30) is reduced by the factor $g$ given by

$$g = \frac{\omega}{kcA_{eff}E_0^4} \int n_0 |\nabla_\perp \phi|^2 d\mathbf{S} \approx \frac{1}{2}. \tag{33}$$

A linear wave packet deforms due to the group velocity dispersion $k''$. For a light wave pulse with a scale size of $t_0$, the deformation takes place at the dispersion distance given Eq.(24). Thus it is convenient to introduce the distance $Z$ normalized by $z_0$ and time $T$ normalized by $t_0$. Equation (28) reduces to the NLSE,

$$\frac{\partial q}{\partial Z} = \frac{i}{2} \frac{\partial^2 q}{\partial T^2} + i |q|^2 q. \tag{34}$$

Here $q$ is the normalized amplitude given by

$$q = \sqrt{\frac{\omega_0 n_2 g z_0}{c}} \overline{E}. \tag{35}$$

Equation (34) is the master equation that describes the evolution of information propagation in fibers. We recall that $|q|^2$ represents the self-induced phase shift, which is of order unity for a mW level of light wave power with dispersion distance of a few hundred kilometers.

In deriving Eq.(34), $k'' < 0$ (anomalous dispersion) is assumed. For a normal dispersion $k'' > 0$, the coefficient of the first term on rhs of Eq.(34) becomes negative. Equation (34) is the master equation first derived by the Hasegawa and Tappert [5] and is now widely used, with proper modification, in the design of light wave transmission systems. In a practical system, the fiber dispersion $k''$ often varies in $Z$. In addition, fiber has amplifiers with gain G(Z) and loss with loss rate $\gamma$. Then Eq.(34) should be modified to [6],

$$\frac{\partial q}{\partial Z} = \frac{i}{2} d(Z) \frac{\partial^2 q}{\partial T^2} + i |q|^2 q + [G(Z) - \gamma] q. \tag{36}$$

Here $d(z)$ is the group dispersion normalized by its average value. However, if the variation of $d$ is small compared with the average and if the amplifier distance is much shorter than the dispersion distance, one can show that Eq.(36) can be reduced the ideal NLSE [7].

## 5    Lagrangian Method and Soliton Perturbation Theory

As was shown in Eq.(9), a soliton can be described by the four parameters, $\eta, \kappa, T_0$ and $\sigma$. This means that one can study the transmission properties by following the behavior of these limited number of parameters in a soliton transmission system instead of infinite numbers of parameter required in non-soliton systems. In an ideal loss less transmission line of a constant dispersion, these four parameters are conserved exactly from the inverse scattering theorem of the NLSE [1]. However, in the presence of various perturbations these parameters evolve in $Z$.

If we represent the perturbation by $iR[q.q']$ the perturbed nonlinear Schrödinger equation becomes

$$i\frac{\partial q}{\partial Z} + \frac{1}{2}\frac{\partial^2 q}{\partial T^2} + |q|^2 q = iR[q, q^*]. \tag{37}$$

The evolution equations for the four parameters may be obtained by perturbation method of conserved quantities of the nonlinear Schrödinger equation, such as energy and momentum, perturbed inverse scattering transform, or the Lagrangian method. For practical cases, all of these methods give the same result. Since the Lagrangian method is also applicable to nonintegrable cases, such as the dispersion-managed system introduced in Sect.6, we present this method here. We first note that the NLSE can be derived from the variation of the Lagrangian density, $\mathbf{L}(q, q^*)$ ,

$$\delta \int_{-\infty}^{\infty} \int_{-\infty}^{\infty} \mathbf{L} dZ dT = 0, \tag{38}$$

with

$$\mathbf{L} = \frac{i}{2}\left(q^*\frac{\partial q}{\partial Z} - q\frac{\partial q^*}{\partial Z}\right) + \frac{1}{2}\left(|q|^4 - \left|\frac{\partial q}{\partial T}\right|^2\right). \tag{39}$$

The nonlinear Schrödinger equation (1) can be expressed by means of the functional derivative of $\mathbf{L}$ with respect to $q^*$ ,

$$\frac{\delta \mathbf{L}(q, q^*)}{\delta q^*} = 0. \tag{40}$$

Thus the perturbed nonlinear Schrödinger equation (37) may be written as

$$\frac{\delta \mathbf{L}(q, q^*)}{\delta q^*} - iR[q, q^*] = 0. \tag{41}$$

Evolution equations for soliton parameters, $\eta, \kappa, T_0$ and $\sigma$ can be obtained from the variation of the finite dimensional Lagrangian $L\left(\eta, \kappa, T_0, \mathrm{d}\eta/\mathrm{d}T, \mathrm{d}\kappa/\mathrm{d}Z, \mathrm{d}T_0/\mathrm{d}Z, \mathrm{d}\sigma/\mathrm{d}Z\right)$. In the absence of the perturbation, the evolution equation is obtained from

$$\delta \int_{-\infty}^{\infty} L \mathrm{d}Z = 0. \tag{42}$$

If we use the soliton solution expressed by the four parameters in the form,

$$q\left(Z, T\right) = \eta\left(Z\right) \operatorname{sech}\left[\eta\left(Z\right)\left(T - T_0\left(Z\right)\right)\right] \exp\left[-i\kappa\left(Z\right)T + i\sigma\left(Z\right)\right], \tag{43}$$

the Lagrangian $L(r_j, \mathbf{\acute{Y}}_j)$ becomes,

$$L = \int_{-\infty}^{\infty} \mathbf{L}\mathrm{d}T = -2\eta\left(\kappa\frac{\mathrm{d}T_0}{\mathrm{d}Z} + \frac{\mathrm{d}\sigma}{\mathrm{d}Z}\right) + \frac{1}{3}\eta^3 - \eta k^2. \tag{44}$$

where

$$r_j = \eta, \ \kappa, \ T_0, \ \sigma \tag{45}$$

and

$$\mathbf{\acute{Y}}_j = \frac{\mathrm{d}\eta}{\mathrm{d}Z}, \frac{\mathrm{d}\kappa}{\mathrm{d}Z}, \frac{\mathrm{d}T_0}{\mathrm{d}Z}, \frac{\mathrm{d}\sigma}{\mathrm{d}Z} \tag{46}$$

Equation (42) naturally gives the Euler Lagrange equation of the form,

$$\frac{\mathrm{d}}{\mathrm{d}Z}\left(\frac{\partial L}{\partial \mathbf{\acute{Y}}_j}\right) - \frac{\partial L}{\partial r_j} = 0. \tag{47}$$

For example if we take $r_j = \eta$, Eq.(47) reads

$$\frac{\mathrm{d}}{\mathrm{d}Z}\left(\frac{\partial L}{\partial \eta}\right) = 0 = \frac{\partial L}{\partial \eta} = -2\left(\kappa\frac{\mathrm{d}T_0}{\mathrm{d}Z} + \frac{\mathrm{d}\sigma}{\mathrm{d}Z}\right) + \frac{1}{3}\eta^2 - k^2,$$

or

$$\frac{\mathrm{d}\sigma}{\mathrm{d}Z} = -\kappa\frac{\mathrm{d}T_0}{\mathrm{d}Z} + \frac{1}{2}\left(\eta^2 - k^2\right). \tag{48}$$

Similarly by taking $r_j = \kappa$,

$$\frac{\partial L}{\partial \kappa} = -2\eta\left(\kappa + \frac{\mathrm{d}T_0}{\mathrm{d}Z}\right) = 0,$$

or,

$$\frac{\mathrm{d}T_0}{\mathrm{d}Z} = -\kappa. \tag{49}$$

We note that in terms of unnormalized time position shift $\triangle t_0$, Eq.(49) gives,

$$\frac{\triangle t_0}{\mathrm{d}Z} = \triangle \omega_0 k'', \tag{50}$$

which corresponds to Eq.(21). In addition by taking $r = \sigma$ and $T_0$, we have

$$\frac{\partial L}{\partial \delta} = \frac{\mathrm{d}}{\mathrm{d}Z}\left(\frac{\partial L}{\partial\left(\mathrm{d}\delta/\mathrm{d}Z\right)}\right) = -2\frac{\mathrm{d}\eta}{\mathrm{d}Z} = 0,$$

or,

$$\eta = \text{const.} \tag{51}$$

and

$$\frac{\partial L}{\partial T_0} = \frac{\mathrm{d}}{\mathrm{d}Z}\left(\frac{\partial L}{\partial\left(\mathrm{d}T_0/\mathrm{d}Z\right)}\right) = -2\left(\kappa\frac{\mathrm{d}\eta}{\mathrm{d}Z} + \eta\frac{\mathrm{d}\kappa}{\mathrm{d}Z}\right) = 0,$$

or

$$\kappa = \text{const.} \tag{52}$$

Equations (51) and (52) confirm the soliton property of invariant eigen values, while Eqs.(48) and (49) reproduce the soliton solution shown in Eq.(9) if $T_0$ and $\sigma$ are given by their initial values.

In the presence of perturbation the variational equation is modified in accordance with Eq.(41). In order to accommodate the modification in the evolution equation (sometimes called the dynamical equation), we write the derivatives of Lagrangian $L$ in Eq.(47) in the form of a chain-rule,

$$\frac{\partial L}{\partial r_j} = \int_{-\infty}^{\infty}\left(\frac{\delta \mathbf{L}}{\delta q_0\left(T\right)}\frac{\partial q_0\left(T\right)}{\partial r_j} + \frac{\delta \mathbf{L}}{\delta q_0^*\left(T\right)}\frac{\partial q_0^*\left(T\right)}{\partial r_j}\right)dT. \tag{53}$$

Then with the help of Eq.(41) we can derive the Lagrange equation of motion with the perturbation in the following form,

$$\frac{\partial L}{\partial r_j} - \frac{\mathrm{d}}{\mathrm{d}Z}\left(\frac{\partial L}{\partial \acute{\mathbf{Y}}_j}\right) = \int_{-\infty}^{\infty}\left(R^*\frac{\partial q}{\partial r_j} + R\frac{\partial q^*}{\partial r_j}\right)dT, \tag{54}$$

$$r_j = \eta,\ \kappa,\ T_0, \text{and } \sigma. \tag{55}$$

For example, soliton parameters and are modified according to the following equations,

$$\frac{\mathrm{d}\eta}{\mathrm{d}Z} = \int_{-\infty}^{\infty}\text{Re}\left\{R\left[q_0, q_0^*\right]e^{-i\varphi}\right\}\text{sech}\tau\mathrm{d}\tau, \tag{56}$$

$$\frac{\mathrm{d}\kappa}{\mathrm{d}Z} = -\int_{-\infty}^{\infty}\text{Im}\left\{R\left[q_0, q_0^*\right]e^{-i\varphi}\right\}\text{sech}\tau\tanh\tau\mathrm{d}\tau, \tag{57}$$

$$\frac{\mathrm{d}T_0}{\mathrm{d}Z} = -\kappa + \frac{1}{\eta^2}\int_{-\infty}^{\infty}\text{Re}\left\{R\left[q_0, q_0^*\right]e^{-i\varphi}\right\}\tau\text{sech}\tau\mathrm{d}\tau + \varphi\int_{-\infty}^{\infty}\text{Im}\left\{q_1 e^{-i\varphi}\right\}\text{sech}\tau\tanh\tau\mathrm{d}\tau, \tag{58}$$

$$\frac{d\sigma}{dZ} = -\kappa \frac{dT_0}{dZ} + \frac{1}{2} \left( k^2 + \eta^2 \right) + \frac{1}{\eta} \int\limits_{-\infty}^{\infty} \text{Im} \left\{ R \left[ q_0, q_0^* \right] e^{-i\varphi} \right\} \tau \, \text{sech} \tau \left( 1 - \tau \tanh \tau \right) d\tau$$

$$+ \frac{\kappa}{\eta^2} \int\limits_{-\infty}^{\infty} \text{Re} \left\{ R \left[ q_0, q_0^* \right] e^{-i\varphi} \right\} \tau \, \text{sech} \tau \, d\tau + \eta \int\limits_{-\infty}^{\infty} \text{Re} \left\{ q_1 e^{-i\varphi} \right\} \text{sech} \tau \, d\tau,$$

$$\tag{59}$$

where

$$\phi = -\kappa T + \sigma. \tag{60}$$

These equations indicate that the invariant properties of soliton parameters are generally lost in the presence of perturbation. In particular the frequency modulation inducted through Eq.(57) results in soliton position jitter through Eq.(58). This effect presents a serious problem of timing jitter when solitons are used as information carrier. Unlike other transmission methods, where deformation of wave forms leads to error, in a soliton transmission system, timing jitter is the major contribution to the error since the pulse shape itself is robust. However, the problem in soliton systems is manageable since control of a limited number of parameters is sufficient.

## 6    Dispersion Managed Solitons

An optical pulse can propagate free of distortion if the fiber response is completely linear and integrated dispersion and the loss compensated by gain over the entire span is zero even if the local value of dispersion is finite and fiber has a finite loss. Similarly, the timing jitter of solitons can also be eliminated if the average dispersion is zero since the jitter is induced by the group dispersion as shown in Eq.(50). In case of the soliton transmission, one can not arbitrarily reduce the dispersion since the soliton intensity is proportional to the dispersion and reduction of the soliton intensity leads to a wider pulse and also reduces the signal to noise ratio. By recognizing this, Suzuki et al [8] reduced the integrated dispersion periodically by connecting dispersion compensating fiber (DCF) and succeeded transmission of 10 Gb/s soliton signal over a cross pacific distance without soliton control. The result was puzzling for most soliton experts since abrupt change of dispersion was expected to destroy the soliton property. One year later Smith et al [9] came up with an idea of a new soliton that can be transmitted in a fiber having a periodic variation of dispersion as shown in Fig. 1. So let us now study the pulse propagation in such a fiber.

We first note that If the fiber is linear, the pulse that starts at point $a$ can recover the original shape completely at point $e$ if the dispersion in anomalous region $d_0$ at $a \leq Z \leq b$ and $d \leq Z < e$ is opposite from that in normal dispersion region, $-d_0$, at $b \leq Z < d$, provided $\overline{ab} + \overline{de} = \overline{bd}$. However if the dispersion map is not symmetric, the pulse at $Z = e$ cannot recover the original shape.

Let us consider what happens if we take into account of the fiber nonlinearity. As was discussed in Sect.3, the major nonlinearity of a fiber originates from the Kerr effect. This produces self-induced phase shift and induces chirp (frequency

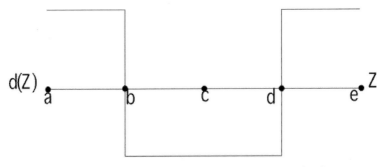

**Fig. 1.** A canonical dispersion map for dispersion managed soliton systems

changes in time) in the pulse. The fiber dispersion induces also a chirp. In an anomalous (normal) dispersion region, the direction of the chirp due to nonlinearity is opposite from (same as ) that due to the dispersion. Remember that a soliton is produced if the self-induced chirp due to nonlinearity cancels with that due to the anomalous dispersion. Let us now consider what happens if the local dispersion is much larger than that which allows a soliton solution for the given pulse intensity. Then the pulse acquires certain amount of chirp at the end of the anomalous dispersion section $Z = b$ of Fig. 3. However this amount of chirp is reversed as the pulse propagates through the normal dispersion section and may be completely reversed at point $Z = d$. Then the amount of the chirp can again become zero at the end of periodicity at $Z = e$. In fact, it can be shown that for a proper choice of the initial pulse width and pulse intensity, one can construct a nonlinear pulse that can recover the initial pulse shape at each end of the periodic dispersion map for a wide range of the value of the average dispersion $\langle d \rangle$. This includes the case $\langle d \rangle$ and even in case $\langle d \rangle$ is slightly negative (average normal dispersion) [9]. Furthermore it can be shown that even if the pulse does not recover its original shape at the end of one period, it recovers the original shape after several spans of the periodic map with a proper choice of the initial pulse width and the intensity. Such nonlinear stationary pulses are produced by a balance between properly averaged dispersion, weighted by spectral change, and nonlinearity and behave quite similar to the ideal soliton in terms of their stability and dynamic range of nonlinearity and dispersion. Consequently, they are often called dispersion-managed solitons (DMS). A DMS is attractive because, by taking a map having zero or close to zero average dispersion, their timing jitter induced by frequency modulation can be made to zero or close to zero, since the rate of change of the time position is proportional to the amount of the dispersion.

Let us now analyze the behavior of a nonlinear pulse in a fiber having a periodic variation of the dispersion as shown in Fig. 1 using the Lagrangian method introduced in Sect.5.

The envelope equation for properly normalized electric field of optical field in a fiber with the group velocity dispersion variation, $d(Z)$ in the direction of propagation $Z$ satisfies

$$i\frac{\partial q}{\partial Z} + \frac{\mathrm{d}(Z)}{2}\frac{\partial^2 q}{\partial T^2} + \alpha |q|^2 q = -i\Gamma q + iG(Z)q, \tag{61}$$

As before, $T$ is the time normalized by the pulse width $t_0$, $Z$ is the distance normalized by the dispersion distance $\left(t_0^2/\left|k''\right|\right)$, $\Gamma$ is the loss rate per dispersion distance and $G(Z)$ is the amplifier gain. Equation (6.1) may be reduced to a Hamiltonian structure by introducing a reduced amplitude $u$

$$u = q/a, \tag{62}$$

where

$$\mathrm{d}a/\mathrm{d}Z = [-\Gamma + G(Z)]\,a, \tag{63}$$

and

$$i\frac{\partial u}{\partial Z} + \frac{\mathrm{d}(Z)}{2}\frac{\partial^2 u}{\partial T^2} + \alpha a^2(Z)|u|^2 u = 0. \tag{64}$$

Equation (64) is not generally integrable because of inhomogeneous coefficients $a(Z)$ and $d(Z)$. However, it can be derived by variation of the Lagrangian density defined by,

$$\mathbf{L}(T, Z) = \frac{i}{2}(u_Z u^* - u_Z^* u) + \frac{\alpha a^2(Z)}{2}|u|^4 - \frac{\mathrm{d}(Z)}{2}|u_T|^2. \tag{65}$$

Now, a DMS normally requires the average $d(Z)$, $\langle d(Z)\rangle$, much smaller than the local $|d(Z)|$ ;

$$\langle d(Z)\rangle = \frac{1}{L_p}\int_0^{L_p} d(Z)\,\mathrm{d}z \ll |d(Z)|, \tag{66}$$

where $L_p$ is a periodic length of the dispersion map. If $\langle d(Z)\rangle = 0$ and in the absence of nonlinearity, Eq.(64) has an exact periodic solution given by a Gaussian with frequency chirp. Thus, let us introduce a Gaussian ansatz as the local (constant d) solution of Eq.(64),

$$\begin{aligned}
u(T, Z) = {} & \sqrt{p(Z)}\exp\left[-p(Z)^2\{T - T_0(Z)\}^2/2\right] \\
& \times \exp\left[iC(Z)\{T - T_0(Z)\}^2/2 - i\kappa(Z)T + i\theta_0(Z)\right],
\end{aligned} \tag{67}$$

where $p$, $C$ and $\kappa$, represent the inverse of the pulse width, chirp coefficient and frequency shift and $T_0$ and $\theta_0$ represent the soliton position and phase respectively. Then the evolution of these parameters are obtained by the Euler-Lagrange equation of motion, Eq.(47),

$$\frac{\mathrm{d}p}{\mathrm{d}Z} = -pCd, \tag{68}$$

$$\frac{\mathrm{d}C}{\mathrm{d}Z} = -C^2 d - \frac{\alpha a^2 p^3}{\sqrt{2}} + 2dp^4, \tag{69}$$

$$\frac{d\kappa}{dZ} = 0, \tag{70}$$

$$\frac{dT_0}{dZ} = -\kappa d, \tag{71}$$

and

$$\frac{d\theta_0}{dZ} = \frac{\kappa^2 d}{2} - dp^3 + \frac{5\alpha a^2}{4\sqrt{2}}. \tag{72}$$

If $d$ is piecewise constant as shown in Fig. 1, Eqs.(68) and (69) may be integrated to give,

$$\frac{C^2}{2} = C_0 p^2 - \frac{p^4}{2} + p^2 \int \frac{\alpha a^2 p^2}{2\sqrt{2}d} dp, \tag{73}$$

where $C_0$ is a constant and plays a role of the Hamiltonian.

To demonstrate the nature of the solution $p(Z)$ and $C(Z)$, let us take a simple example of a loss less fiber $a(Z) = a_0 (= \text{constant})$. Then Eq.(73) reduces to

$$\frac{C^2}{2} = C_0 p^2 - \frac{p^4}{2} - A_0 p^3, \tag{74}$$

where

$$A_0 = \frac{\alpha a_0^2}{2\sqrt{2}d_0}, \tag{75}$$

represents the strength of nonlinearity. One can obtain a periodic solution with the periodicity $L_p$ by a proper choice of initial conditions $p(0)$ and $C(0)$ so that $p(L_p) = p(0)$ and $C(L_p) = C(0)$. For a linear pulse with $A_0 = 0$, the periodic solution for f is a Gaussian with a periodically varying chirp parameter $C(Z)$ and the trajectories in p-C plane at $d = d_0$ and $d = -d_0$ completely overlays themselves. However the nonlinearity produces a gap in these trajectories as shown in Fig. 2, because of the frequency chirp produced by the self-induced phase shift.

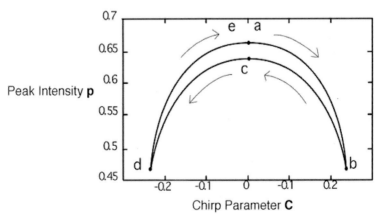

**Fig. 2.** Phase space trajectories of $p$ and $C$ for the dispersion map of Fig. 3

The peak value of $p$ at point $c$ (normal dispersion regime) is less than that at anomalous dispersion regime because of the nonlinearity induced self phase shift. We note here that the effective dispersion for a pulse whose spectral width changes in proportion to $p$ as assumed in Eq.(67) is not simply given by $d$ but by that weighted by $p^2$. This provides $\langle p^2 d \rangle > 0$ even if $\langle d \rangle = 0$ and the balance between the weighted average $\langle d \rangle = 0$ and the nonlinearity may produce a stationary nonlinear pulse. This situation is analogous to an ideal soliton solution, which is constructed by a balance of (constant) dispersion and nonlinearity. However, in the dispersion-managed case, the above argument indicates that the nonlinear stationary solution is possible even for $\langle d \rangle < 0$ or $\langle d \rangle \geq 0$, provided that $\langle p^2 d \rangle > 0$ and proper nonlinearity exists. A DMS is created in this manner. $\tilde{a}$ An ideal soliton solution for a fiber with $d = d_0 =$const, can be constructed for an arbitrary value of $d_0$ by a proper choice of the amplitude, while the linear stationary solution exists only for $d_0 = 0$. Similarly for a dispersion managed case, a nonlinear stationary periodic pulse can be constructed for an arbitrary value of $\langle d \rangle$ by a choice of initial amplitude and chirp, while the linear stationary pulse is possible only for $\langle d \rangle = 0$. This allows DMS to have much larger tolerance in the fiber dispersion. We here note again that, if the system is linear, the trajectory in $p - c$ plane shown in Fig. 2 returns at the original point $d$ only when the average dispersion $\langle d \rangle$ is exactly zero. We further note that if $p(0)$ and/or $C(0)$ is not chosen so that after one period $p(L)$ and/or $C(L)$ does not return to the original value, it was numerically confirmed that $p(nL)$ and/or $C(nL)$ returns to a limit area in $p - C$ plane, where $n = 1, 2, \ldots$. In other words Eqs.(68) and (69) in general have doubly periodic solutions.

In the design of a dispersion managed soliton system, one important parameter is $S$ that is defined as the ratio of dispersion distance given by the local value of the dispersion in one period to the length of the map period. If $S$ is much smaller than unity, the system is close to a case of uniform dispersion and the DMS becomes similar to the ideal soliton with little oscillations in the pulse shape. However, in this case, the long tail of the sech$t$ pulse tends to introduce nonlinear interactions between adjacent pulses unless the pulse separation is made sufficiently large. While if $S$ is much larger than unity, the pulse shape approached to a Gaussian having much shorter tail. But the pulse width oscillates by a large factor leading to also undesirable inter-pulse interactions. It was found by numerical simulations that there exists an optimum value of $S$ that provides the best transmission capacity and the value is given about 1.65 for a simple map shown in Fig. 1 for single channel transmission. For a wavelength division multiplexed (WDM), the optimum value of $S$ is somewhat larger than the single channel case since inter-channel nonlinear interaction should be suppressed. This may be achieved by reducing the pulse intensity by means of stretching the pulse by a choice of a relatively large $S$ value.

# 7  Some Recent Experimental Results of Optical Soliton Transmission

Dispersion managed solitons are now experimented both in a single channel ultra-high speed transmissions and in wavelength division multiplexed (WDM) transmissions. Here two representative experimental results are introduced, one, 40 Gb/s single channel DMS transmission over 10,000km by Morita et al [10] and the other, 1.4 Tb/s (42.7 Gb/s x 32 channel) over 6,000 km by Sugahara et al [11].

The most remarkable single channel experiment that the author is aware of is the one done by Morita et al of KDD group [10]. Figure 3 shows the experimental set up.

The transmission experiments were conducted in a 140 km recirculating loop. In the 20 Gb/s transmitter, a 20 Gb/s optical soliton data stream was produced by optically time-division-multiplexing (OTDM) 10 Gb/s RZ data pulses, which were generated with a distributed feed back laser diode (DFB-LD), sinusoid ally-driven electro absorption (EA) modulators and two intensity modulators operated at 10 Gb/s with a $2^{15} - 1$ pseudorandom binary sequence. The signal

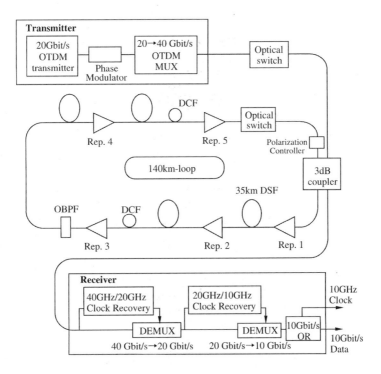

**Fig. 3.** Experimental setup for 40 Gb/s soliton transmission experiment by Morita et al [10]

wavelength was 1552.8 nm and obtained pulse width was about 9 ps. To improve the transmission performance, input phase modulation was applied to the output signal from the 20 Gb/s transmitter. Then 40 Gb/s signal was generated by OTDM 20 Gb/s signals in 2 ways regarding the state of polarization. The state of polarization of adjacent OTDM channels was set to be parallel or orthogonal. The 140 km recirculating loop consisted of 4 spans of dispersion-shifted fiber (DSF), 5 EDFA repeaters pumped at 980 nm and an optical band pass filter (OBPF) with 6 nm bandwidth. At the signal wavelength, the average dispersion of DSF was 0.29 ps/nm/km and the system average dispersion was reduced to 0.028 ps/nm/km by compensating for the most of the accumulated dispersion after every two DSF spans. Fig. 4 shows this dispersion map schematically. The average span length of the DSF was 35.7 km. The repeater output power was set to about +4 dBm.

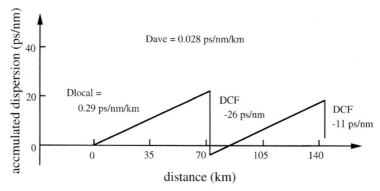

**Fig. 4.** Dispersion map for the 40 Gb/s transmission [10]

In the receiver, the transmitted 40 Gb/s signals were optically time-division-demultiplexed with optical gates generated by sinusoidally driven polarization insensitive EA modulators in two stages; 40 Gb/s to 20 Gb/s and 20 Gb/s to 10 Gb/s. The bit error rate (BER) for the demultiplexed 10 Gb/s signals was measured. Fig. 5 shows the average BER for the four OTDM channels as a function of transmission distance in the cases with the state of polarization of the adjacent OTDM channels parallel (○) and orthogonal (●). In these experiments, the condition of the initial phase modulation was optimized. In addition, in the single-polarization experiment, a low-speed polarization scrambler was used to reduce polarization hole-burning effects of the EDFA repeaters. As shown in Fig. 5 by setting the state of the polarization orthogonal transmission performance was greatly improved and transmission distance for BER of was extended from 8,600 km to 10,200 km. The signal waveforms were measured using a high-speed photo detector with and without a polarizer. Fig. 6 shows the measured eye diagrams of the 40 Gb/s signals and polarization-division-demultiplexed 20 Gb/s signals before and after 10,000 km transmission. The preserved polarization orthogonality reduced the soliton-soliton interaction effectively through the

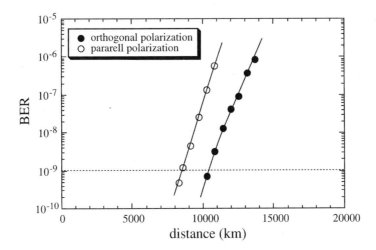

40 Gbit/s single-channel transmission
- Dlocal:0.3 ps/nm/km, Dave:0.03 ps/nm/km
- DCF : 2span(-26,-11 ps/nm/km), filter: 5nm

**Fig. 5.** Bit error rate measurements after 10,000 km transmission of the 40 Gb/s soliton signals by Morita et al [10]. ○ for parallel polarization and ● for orthogonal polarization between adjacent pulses

transmission and made possible to transmit 40 Gb/s data over 10,000 km. This experimental result clearly shows effectiveness of the dispersion-managed soliton in TDM soliton transmissions.

We now introduce a WDM experiment by Sugahara et al [11]. The experiment is an extension of 20 Gb/s based WDM experiment of Fukuchi et al [12]. In order to increase the transmission rate, Sugahara et al made three improvements. One was to introduce Raman amplifications. Since a Raman amplifications using the transmission fiber itself provides a distributed gain, they add less integrated noise than a system with EDFA. This allows to use less signal power and can reduce cross talks among different channels. Two was to use a doubly periodic map with the shorter periodicity approximately one half of the earlier experiment. This allowed to keep approximately the same value of S as the earlier experiment yet could keep the local averaged dispersion relatively large to reduce cross talks originating from a long collision distance, yet by making integrated dispersion low by an introduction of DCF after every four spans, they could keep the power per channel relatively low. Three was to use the forward error corrector (FEC) to improve the error free transmission distance.

The experimental setup is shown in Fig. 7. The transmitter comprised 32 DFB-LDs equally spaced at 100 GHz intervals ranging from 1539 nm to 1563 nm. Even and odd channels were combined and modulated separately into 42. Gb/ carrier suppressed return to zero (CS-RZ) signals by two LiNbO3 intensity

input

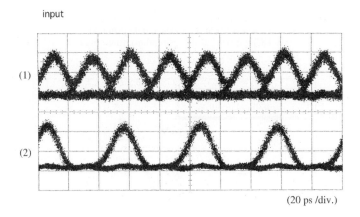

(20 ps /div.)

after 10000 km transmission

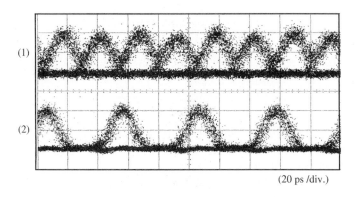

(20 ps /div.)

**Fig. 6.** Measured eye diagram of soliton pulses after 10,000 km of transmission. The picture is taking by overlaying a sequence of arrived pulses in the bit frame. The opening of the eye indicates error free transmission [10]

modulators. The first one was for CS-RZ modulation driven by a 21.3 GHz sinusoidal wave and the second one was for data modulation driven by a true $2^{23}$ -1 pseudorandom binary sequence. The even and odd channels were then combined using a polarization beam coupler with orthogonal polarization between adjacent wavelengths. The transmission line was composed of four quadruple-hybrid span as shown in Fig. 8(b). The total loop length was 308.8 km The dispersion, dispersion slope, and effective core area were +20 ps/nm/km, +0.06 ps/nm2/km, and 106 μm² for low loss pure silica core fiber (PSCF) and -63 ps/nm/km -0.19 ps/nm2/km, and 19 μm² for dispersion compensated fiber (DCF).The average dispersion in each span was 0.8 ps/nm/km At the end of each span, Raman pump unit provided the total power of 450 m to compensate for the 11.5 dB span loss.

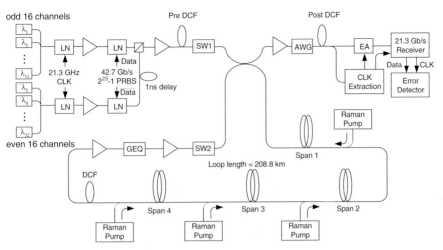

**Fig. 7.** Experimental setup of 32 ch × 42.7 Gb/s transmission experiment by Sugahara et al [11]

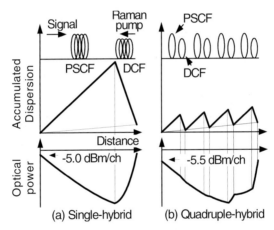

**Fig. 8.** Dispersion map (b) used in the experiment [11]

The authors used 1424 nm, 1437 nm and 1465 nm wavelength LDs as Raman pump sources. The fiber input signal power was optimized to be -5.3 dBm/ch. In the receiver, the measured channel selected by an array wave-guide (AWG) was optically demultiplexed to 21.3 Gb/s signals and then received by a 21.3 Gb/s receiver The transmission performance was evaluated by the bit-error-rate (BER), and was converted to the Q-value.

With the help of FEC, they could achieve error free transmission of the entire channels over the distance of 6,050 km using this scheme as shown in Fig. 9. It is expected that the error free distance without the use of FEC would have been around 3 to 4,000 km. It is remarkable that the authors could double the bit-

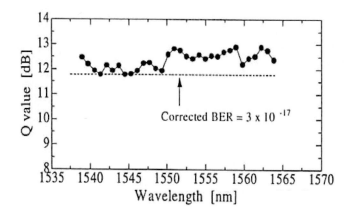

**Fig. 9.** Q values measured at the distance of 6,050 km of transmission for the entire channels [11]

rate per channel without degrading the transmission quality by the use of proper dispersion map and Raman amplifications.

## 8    Concluding Remarks

After a brief introduction of the theory of optical soliton as the solution of the nonlinear Schrödinger equation, historical evolution of the soliton concept for application to communications is presented. It is shown that by a proper management of the fiber dispersion, transmission quality can be improved significantly. The Lagrangian method is introduced as a way of analyzing perturbation on soliton parameters as well as studying the dispersion managed solitons. Finally, by introducing some of the most recent results of transmission experiments, practical potential of soliton based ultra-high bit rate transmission has been demonstrated.

## References

1. V. E. Zakharov, A. B. Shabat: Sov. Phys. JETP **34**, 62 (1972)
2. P. D. Lax: Comm. Pare Appl. Math. **21**, 467 (1968)
3. J. Satsuma, N. Yajima: Suppl. Prog. Theor. Phys. **55**, 284 (1974)
4. Y. Kodama, A. Hasegawa: IEEE J. Quantum Electron. **23**, 510 (1987)
5. A. Hasegawa, F. D. Tappert: Appl.Phys. Lett. **23**, 142 (1973)
6. A. Hasegawa: Opt. Lett. **23**, 3302 (1983)
7. A. Hasegawa, Y. Kodama:Opt. Lett. **16**, 1385 (1991)
8. M. Suzuki, I. Morita, N. Edagawa, S. Yamamoto, H. Toga, S. Akiba: Electron. Lett. **31**, 2027 (1995)
9. N. J. Smith, N. J. Doran, F. M. Knox, W. Forysiak: Electron. Lett.**32**, 54 (1996)

10. I. Morita, K. Tanaka, N. Edagawa, M. Suzuki: 1998 European Conference on Optical Communication (ECOC 98) Volume 3, 47-52, Madrid Spain (1998)
11. H. Sugahara, K. Fukuchi, A. Tanaka, Y. Inada, T. Ono: 2002 Optical Fiber Conference (OFC 2002) Pos Deadline Paper, FC6-1, Anaheim CA (2002)
12. K. Fukuchi, M. Kakui, A. Sasaki, T. Ito, Y. Inada, T. Tsuzaki, T. Shitomi, K. Fujii, S. Shikii, H. Sugahara, A. Hasegawa: 1999 European Conference on Optical Communication (ECOC 99) Post Deadline Paper PD2-10, , Nice, France (1999)

# Linear and Nonlinear Propagation Effects in Optical Fibers

K. Thyagarajan

Physics Department, Indian Institute of Technology
New Delhi 110016,
India

**Abstract.** The development of low loss optical fibers, compact and efficient semiconductor lasers operating at room temperature, optical detectors and optical amplifiers have truly revolutionized the field of telecommunication and have provided us with a communication system capable of carrying enormous amount of information over intercontinental distances. When information carrying light pulses propagate through an optical fiber, they suffer from attenuation, temporal broadening and they even interact with each other through nonlinear effects in the fiber. These effects which tend to distort the signals need to be overcome to achieve high speed communication over long distances. In this chapter, we will give a brief outline of the various linear and nonlinear propagation effects in optical fibers and their impact on optical fiber communication systems. Important components such as optical fiber amplifiers and dispersion compensators, which are playing a very important role in the fiber optic revolution, will also be discussed.

## 1 Introduction

Light waves have frequencies that are orders of magnitude higher than radio waves or microwaves and thus have much higher capacities to carry information. The development of low loss optical fibers, compact and efficient semiconductor lasers operating at room temperatures, optical detectors and optical amplifiers have truly revolutionized the field of telecommunication and have provided us with a communication system capable of carrying enormous amount of information over intercontinental distances. For communication using an optical fiber, the given information is first digitized and then sent in the form of pulses of light through an optical fiber. When these light pulses propagate through the fiber, they suffer from attenuation, temporal broadening and they even interact with each other through nonlinear effects in the fiber. These effects which tend to distort the signals need to be overcome to achieve high speed communication over long distances. Over the years, a number of scientific and technological developments have been taking place to achieve this objective. These include the development of optical amplifiers which amplify the optical signals in the optical domain itself without any need for conversion to electrical signals and new fiber designs which overcome/reduce the nonlinear and temporal broadening effects, dispersion compensation schemes etc. In this chapter, we will give a brief outline of the various linear and nonlinear propagation effects in optical fibers and discuss their significance in an optical communication system. Further details on many of the topics may be found in numerous texts that are available in literature [1-6].

## 2    The Optical Fiber

The simplest optical fiber is a cylindrical structure consisting of a central core of doped silica ($SiO_2$)surrounded by a concentric cladding of pure silica (see Fig. 1). Such a fiber is referred to as a step index fiber.

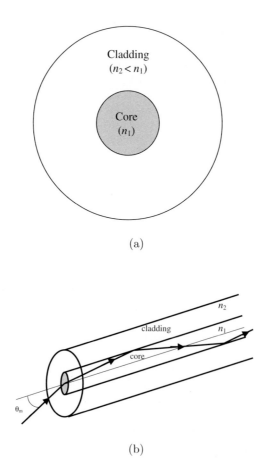

(a)

(b)

**Fig. 1.** (a) and (b) A step index optical fiber consists of a cylindrical structure made of a central core of refractive index $n_1$ surrounded by a concentric cladding of refractive index $n_2$. Light guidance takes place through the phenomenon of total internal reflection

The refractive index of the core ($n_1$) is slightly greater than that of the cladding ($n_2$). In the simplest picture, the phenomenon of light guidance by the core can be thought of occurring by the process of total internal reflection. All those light rays that are incident at the core-cladding interface at an angle greater than the critical angle given by

$$\phi_c = \sin^{-1}(n_2/n_1) \tag{1}$$

will get total internally reflected and will be guided by the core. If the corresponding maximum angle of incidence at the entrance face of the fiber is $\theta_m$, then we define the numerical aperture (NA) of the fiber as

$$NA = \sin\theta_m = \sqrt{(n_1^2 - n_2^2)}. \tag{2}$$

Telecommunication optical fibers have typically NA = 0.2 which corresponds to a maximum angle of acceptance of about $11.5^0$. There are mainly two fiber types: multimode fibers and single mode fibers. Multimode fibers are characterized by core diameters of 50 μm and cladding diameters of 125 μm while single mode fibers have typically core diameters of 9 to 10 μm and cladding diameters of 125 μm. There are two main varieties of multimode fibers namely step index and graded index fibers. Step index fibers are characterized by a homogeneous core of constant refractive index while graded index fibers have an inhomogeneous core in which the refractive index decreases in an almost parabolic fashion from the center of the core to the core-cladding interface.

## 3    Modes in an Optical Fiber

In order to model light propagation through optical fibers, we need to solve Maxwell's equations in the core and cladding of the optical fiber and apply appropriate boundary conditions at the core cladding interface. It is convenient to analyze light propagation through optical fibers using the concept of modes. The modes of an optical fiber are those transverse field distributions of electric and magnetic fields that propagate through the optical fiber with definite phase and group velocities and without any change in their polarization state or in their transverse field distribution. Thus, for example, the $x$-component of the electric field of a mode will have a spatial and temporal dependence of the form

$$E(x, y, z, t) = A\psi(x, y)e^{i(\omega t - \beta z)} \tag{3}$$

where $\omega$ represents the angular frequency and $\beta$ is referred to as the propagation constant of the mode. The axis of the fiber is assumed to be along the $z$-axis. In Eq. (3), $A$ is a normalization constant and $\psi(x, y)$ represents the transverse electric field distribution of the mode. For step index fibers, complete vector solutions of the Maxwell's equations can be obtained. Since for telecommunication fibers, the core-cladding index difference is usually very small ($\sim$ 0.004 to 0.01), we can make the scalar approximation and solve for the scalar modes of the optical fiber. In this approximation we neglect the vector nature of the modes and assume the modes to be linearly polarized; this leads to what are referred to as linearly polarized (LP) modes of the fiber. The scalar wave equation is given by

$$\nabla^2 \Psi = \varepsilon_0 \mu_0 n^2 \frac{\partial^2 \Psi}{\partial t^2} \qquad (4)$$

where $n(x, y)$ represents the transverse refractive index distribution of the optical fiber and $\Psi$ represents a transverse component of the electric field ($E_x$ or $E_y$) of the mode. For step index fibers the refractive index profile is given by

$$\begin{aligned} n(r) &= n_1; \ r < a \quad \text{(core)} \\ &= n_2; \ r > a \quad \text{(cladding)} \end{aligned} \qquad (5)$$

where $r(= \sqrt{(x^2 + y^2)})$ represents the cylindrical radial coordinate. There are two classes of modes, namely guided modes and radiation modes. We define the effective index of a mode by

$$n_{\text{eff}} = \frac{\beta}{k_0} \qquad (6)$$

such that the phase velocity of the mode is given by $c/n_{\text{eff}}$. Guided modes are characterized by $n_1 > n_{\text{eff}} > n_2$ and for these modes, the power is mainly confined to the core of the optical fiber. The fields of guided modes are oscillatory in the core and decay in the cladding. They form a discrete set of solutions and thus any given fiber will support a finite number of guided modes. On the other hand, radiation modes are characterized by $n_{\text{eff}} < n_2$ and their fields are oscillatory even in the cladding. Unlike the guided modes, they form a continuum of solutions. Guided and radiation modes form a complete set of solutions in the sense that any arbitrary field distribution in the optical fiber can be expressed as a linear combination of the discrete guided modes and the continuum radiation modes:

$$\Psi(x, y, z) = \sum a_j \psi_j(x, y) e^{-i\beta_j z} + \int a(\beta) \psi(x, y, \beta) e^{-i\beta z} d\beta \qquad (7)$$

with $|a_j|^2$ being proportional to the power carried by the $j^{th}$ guided mode and $|a(\beta)|^2 d\beta$ being proportional to the power carried by radiation modes with propagation constants lying between $\beta$ and $\beta + d\beta$. The constants $a_j$ and $a(\beta)$ can be determined from the incident field distribution at $z = 0$. For a step index profile given by Eq.(5), the field distributions of the guided LP modes are given by [2]

$$\psi(r, \phi) = \begin{cases} \frac{A}{J_l(U)} J_l\left(\frac{Ur}{a}\right) & r < a \\ \frac{A}{K_l(W)} K_l\left(\frac{Wr}{a}\right) & r > a \end{cases} \begin{bmatrix} \cos l\phi \\ \sin l\phi \end{bmatrix} \qquad (8)$$

where $l = 0, 1, 2$, and

$$U = a\left(k_0^2 n_1^2 - \beta^2\right)^{1/2}; W = a\left(\beta^2 - k_0^2 n_2^2\right)^{1/2} \qquad (9)$$

$$V^2 = U^2 + W^2 = k_0^2 a^2 \left(n_1^2 - n_2^2\right)$$

In Eq. (8) $J_n(x)$ and $K_n(x)$ represent Bessel functions and modified Bessel functions respectively. Applying the boundary conditions of continuity of field $\psi$ and its derivative at the core cladding interface, we obtain the following eigenvalue equation which specifies the allowed values of the propagation constant $\beta$:

$$U\frac{J_{l-1}(U)}{J_l(U)} = -W\frac{K_{l-1}(W)}{K_l(W)} \tag{10}$$

For a given value of $l$, the above equation will have a finite number of solutions and the $m^{th}$ solution is referred to as the $LP_{lm}$ mode. For modes with $l = 0$, the eigenvalue equation reduces to

$$U\frac{J_1(U)}{J_0(U)} = W\frac{K_1(W)}{K_0(W)} \tag{11}$$

For step index fibers with

$$V < 2.405 \tag{12}$$

there is only one solution to the eigenvalue equation. This mode is referred to as the $LP_{01}$ mode and such fibers are called as *single mode fibers*. Typical single mode fibers operating at 1310 nm have a core radius of about 8 to 10 μm and an index difference of about 0.004. Although the mode field profile of step index fibers can be described by Bessel functions, the field profiles are very close to being Gaussian and as such, Gaussian approximation is a widely used approximation to describe the fundamental mode distribution in such fibers.

## 4    Typical Fiber Optic Communication System

Information to be sent through a fiber optic communication system is first coded into a binary sequence of electrical pulses which then modulate a laser beam to produce a sequence of ones and zeroes which are represented by the presence and absence of pulses (see Fig. 2). The rate at which information gets transmitted is defined in terms of a quantity referred to as the bit rate which defines the number of pulses being sent per second. Typical bit rates of transmission today are 2.5 Gbits/s ($= 2.5 \times 10^9$ bits/s) and 10 Gbits/s. Light pulses propagating through an optical fiber suffer from three effects: 1. *Attenuation*: The power carried by light pulses propagating through the fiber continuously decreases as it

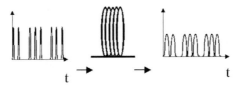

**Fig. 2.** In an optical fiber communication system, information is sent in the form of pulses of light. The pulses suffer from attenuation, pulse dispersion and nonlinear effects as they propagate through the optical fiber

propagates along the fiber. Some of the main mechanisms responsible for this include Rayleigh scattering, absorption by impurities, mainly water, waveguide imperfections such as bends, etc. and intrinsic infrared and ultraviolet absorption. 2. *Pulse dispersion*: The temporal width of the light pulse increases as it propagates through the fiber. This leads to an overlapping of adjacent pulses resulting in loss of resolution and hence increased errors in deciphering the information. 3. *Nonlinear effects*: Even moderate powers in the light pulses, leads to a significant intensity of the pulses due to confinement in a small cross sectional area. Due to this confinement over long distances, the nonlinearity in the optical fiber leads to effects such as self phase modulation, cross phase modulation and four wave mixing. These effects modify the propagation of the pulses through the fiber and also lead to cross talk among the various wavelength channels that are simultaneously propagating through the fiber. The main objective of any fiber optic communication system is to achieve increased bit rates of transmission and to be able to transmit information carrying signals over very long distances without the need for any regeneration.

## 5   Attenuation

Figure 3 shows a typical spectral dependence of loss for a silica fiber. Also shown are the fundamental losses due to Rayleigh scattering, ultraviolet and infrared absorptions and due to waveguide imperfections. The peaks in the attenuation curve are mainly due to the presence of water in the fiber. Losses in optical fiber

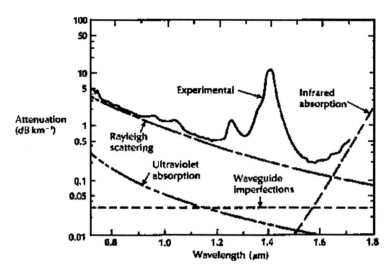

**Fig. 3.** Typical attenuation spectrum of an optical fiber. Notice windows of low loss transmission around 1310 nm and 1550 nm. The presence of water impurity leads to a peak in the attenuation curve around 1390 nm. [After Ref. 7]

are specified in terms of the unit dB/km. It is defined as

$$\alpha(\text{dB/km}) = \frac{10}{L(\text{km})} \log\left(\frac{P_{in}}{P_{out}}\right) \tag{13}$$

Here $P_{in}$ and $P_{out}$ are the input and output powers corresponding to an optical fiber of length $L$ (km). Thus if a launched power of 1 mW reduces to 0.5 mW after propagation through 15 km of the fiber, then the loss coefficient of the fiber corresponds to 0.2 dB/km. There are mainly two low loss windows of silica optical fibers, namely the 1310 nm window with typical losses of about 0.4 dB/km and the lowest loss window of 1550 nm with losses of about 0.25 dB/km. Most of the current systems operate in the 1550 nm window where the losses are a minimum and where, as we will see later, erbium doped fiber amplifiers operate. Although Fig. 3 shows a typical loss curve, current commercially available fibers have almost no water impurity and consequently, the two peaks are almost absent. Fibers having a low loss window over the entire band of 1250 nm to 1650 nm which is approximately 50 THz of bandwidth are now commercially available. This opens up the possibility of sending multiple optical signals at different wavelengths through a fiber leading to what is referred to as wavelength division multiplexing (WDM). Attenuation leads to drop in optical power of the pulses. At the receiver the optical pulses are received by a photodetector and converted to electrical signals for further processing. The process of detection adds noise to the converted electrical signal. There are three main noise components at the detection process: shot noise, thermal noise and dark current noise. Shot noise is important at higher signal levels while thermal noise is dominant for weak signals. Thus the output electrical pulse sequence is noisy. Since the receiver has to detect the absence or presence of the pulses, if the signal levels are very low, noise added to the signal can lead to errors in the detection. Error in detection is defined in terms of a quantity referred to as Bit Error Rate (BER):

$$\text{BER} = \frac{(\text{number of bits read erroneously in time } \tau)}{(\text{total number of bits received during time } \tau)}$$

Typical bit error rates required are $10^{-9}$ to $10^{-12}$ which correspond to one bit error out of $10^9$ bits and $10^{12}$ bits respectively. For example, if a 2.5 Gbits/s system has a BER of $10^{-9}$, then on an average, the number of errors in one second is about 2.5. In order to keep the BER below a certain level, a minimum signal level is required. The conventional technique to send signals over long distances is to electrically regenerate the optical signal when the signal levels become low enough. This process is referred to as regeneration. In a regenerator, the incoming optical signals are first converted to electrical signals which are then processed in the electrical domain. The processed electrical signals are then used to modulate a laser to regenerate the optical pulse stream. The regenerator removes all added noise from the pulse and thus the output from the regenerator is almost identical to the pulse stream that was sent into the link. Since in the regenerator, all processing is performed after conversion of the optical signal into the electrical domain, any required upgradation in the bit rate would need a replacement of the regenerator. Also, for a WDM system in which multiple light wavelengths each carrying independent signals are simultaneously sent through an optical fiber, the incoming signals would have to be first demultiplexed into individual wavelengths and then each wavelength would have to be processed by a separate regenerator after which the signals need to be demultiplexes for onward transmission. This makes such a solution very expensive.

# 6    Optical Amplifiers

If the maximum transmission distance in a link is determined by attenuation, then it is clear that a simple amplification of the signal would be able to increase the link length. Optical amplifiers are devices that amplify the incoming optical signals in the optical domain itself without any conversion to the electrical domain, and have truly revolutionized fiber optic communications. Because the signals are processed in the optical domain itself, optically amplified communication systems can be easily upgraded. Due to the large gain bandwidth of optical amplifiers, multiple wavelength channels can be simultaneously amplified. This has led to a revolution in fiber optic communication by employing the technique of wavelength division multiplexing (WDM).

## 6.1    Erbium Doped Fiber Amplifier(EDFA)

The most important optical amplifier is the *Erbium Doped Fiber Amplifier* (EDFA) [2,8,9]. Optical amplification by EDFA is based on the process of stimulated emission, which is the basic principle behind laser operation. In fact a laser without any optical feedback is just an optical amplifier. Figure 4 shows two levels of an atomic system: the ground level with energy $E_1$ and an excited level with energy $E_2$. Under thermal equilibrium, most of the atoms are in the ground level. Thus if light corresponding to an appropriate frequency ($\nu = (E_2 - E_1)/h$; $h$ being the Planck's constant) falls on this collection of atoms then it will result in a greater number of absorptions than stimulated emissions and the light beam will suffer from attenuation. On the other hand, if the number of atoms in the upper level could be made more than in the lower level, then an incident light beam at the appropriate frequency could induce more stimulated emissions than absorptions, thus leading to optical amplification. This is the basic principle behind optical amplification by EDFA. Figure 5 shows the three lowest lying energy levels of Erbium ion in silica matrix. A pump laser at 980 nm excites the erbium ions from the ground state to the level marked $E_3$. The level $E_3$ is a short-lived level and the ions jump down to the level marked $E_2$ after a few microseconds. The lifetime of level $E_2$ is much larger and is about 12 milliseconds. Hence ions brought to level $E_2$ stay there for a long time. Thus by pumping hard enough, the population of ions in the level $E_2$ can be made larger than the population of level $E_1$ and thus achieve population inversion between levels $E_1$ and $E_2$. In such

## Attenuation

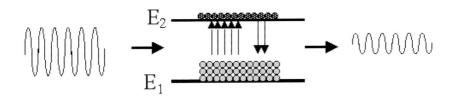

## Amplification

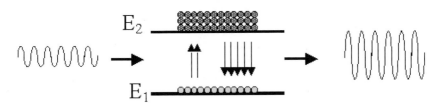

**Fig. 4.** Population inversion between two energy levels can be used for optical amplification of light waves at a frequency corresponding to the energy difference $E_2 - E_1$

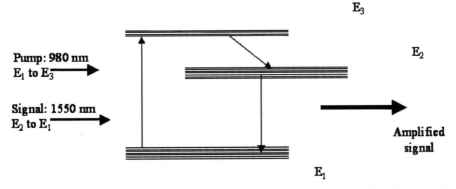

**Fig. 5.** The low lying energy levels of erbium ion in silica matrix. Incident light at 980 nm leads to the excitation of erbium ions from the ground state to the $E_3$ state from which they make a transition to the long lived level $E_2$. Population inversion between levels $E_2$ and $E_1$ leads to amplification of signals in the 1550 nm window

a situation, if a light beam at a frequency corresponding to the energy difference $(E_2 - E_1)$ falls on the collection, it will get amplified. For erbium ions, the energy difference $(E_2 - E_1)$ falls in the 1550 nm band and thus it is an ideal amplifier for signals in the 1550 nm window. In the case of erbium ions in silica matrix, the energy levels are not sharp levels but are broadened. Hence the system is capable of amplifying optical signals over a band of wavelengths. Figure 6 is a schematic of an EDFA which consists of a short piece ($\sim$ 20 m in length) of erbium doped fiber (EDF), a single mode fiber doped with erbium in the core and which is pumped by a 980 nm pump laser through a wavelength division multiplexing (WDM) coupler. The WDM coupler multiplexes light of wavelength 980 nm and 1550 nm from two different input arms to a single output arm. The 980 nm pump light is absorbed by the erbium ions to create population inversion between levels $E_2$ and $E_1$. Thus, incoming signals in the 1550 nm wavelength region get amplified as they propagate through the population inverted doped fiber. Figure 7 shows a typical measured gain spectrum of an EDFA for an input pump power of 45 mW and an input signal power of -20 dBm (= 10 $\mu$W). As can be seen in the figure, EDFAs can provide amplifications of greater than 20 dB over the entire band of 40 nm from 1525 nm to about 1565 nm. This wavelength band is referred to as the C band (conventional band) and is the most common wavelength band of operation. With proper amplifier optimization, EDFAs can also amplify signals in the wavelength range of 1570 to 1610 nm; this band of wavelengths is referred to as the L-band (long wavelength band). The C-band and L-band amplifiers together can be used to simultaneously amplify 160 wavelength channels. Such systems are indeed commercially available now. It can be seen from Fig. 7 that although EDFAs can provide gains over an entire band of 40 nm, the gain is not flat, i. e., the gain depends on the signal wavelength. Thus if multiple wavelength signals with same power are input into the amplifier, then their output powers will be different. In a communication system employing a chain of amplifiers, a differential signal gain among the various signal wavelengths (channels) from each amplifier will result in a significant difference in signal power levels and hence in the signal to noise ratio (SNR) among the various channels. In fact, signals for which the gain in the amplifier is greater than the loss suffered in the link, will keep on increasing in power level while those channels for which the amplifier gain is less than the loss suffered will keep on reducing in power. The former channels will finally saturate the amplifiers and will also lead to increased nonlinear effects in the link while the latter will have reduced SNR leading to increased errors in detection. Thus such a differential amplifier gain is not desirable in a communication system and it is very important to have gain flattened amplifiers.

## Gain Flattening of EDFAs

There are basically two main techniques for gain flattening: one uses external wavelength filters while the other one relies on modifying the amplifying fiber properties to flatten the gain. Figure 8 shows the principle behind gain flattening using external filters. In this, the output of the amplifier is passed through a

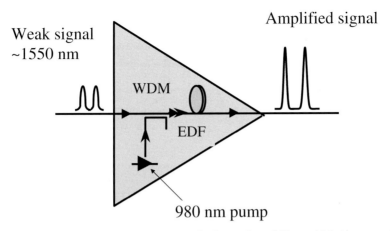

**Fig. 6.** An EDFA consists of a short piece of erbium doped fiber which is pumped by a 980 nm laser diode through a WDM coupler. Signals around the 1550 nm wavelength get amplified as they propagate through the pumped doped fiber

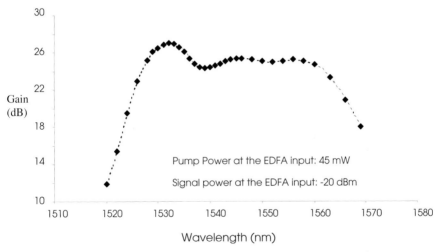

**Fig. 7.** A typical measured gain spectrum of an EDFA

special wavelength filter whose transmission characteristic is exactly the inverse of the gain spectrum of the amplifier. Thus channels which have experienced greater gain in the amplifier will suffer greater transmission loss while channels which experience smaller gain will suffer smaller loss. By appropriately tailoring the filter transmission profile it is possible to flatten the gain spectrum of the amplifier. Filters with specific transmission profiles can be designed and fabricated using various techniques. These include thin film interference filters and filters based on long period fiber gratings (LPG). Typical gain flatness of better than 1 dB can be achieved and gain flattened EDFAs are commercially available.

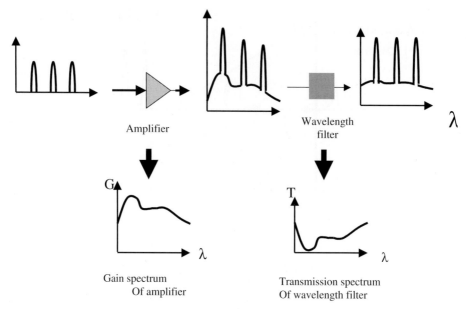

**Fig. 8.** The principle of gain flattening of EDFA using an external wavelength filter

## Gain Flattening Using Long Period Gratings (LPG)

Long period gratings are periodic perturbations formed on an optical fiber which induce phase matched coupling among the core mode and cladding modes [10-14]. Cladding modes are modes which are guided by the cladding-air interface of the optical fiber. Since the cladding diameters are usually much larger than wavelength of light and the index difference between the cladding and air is also very large, the cladding usually supports an extremely large number of guided modes. If $\beta_{co}$ and $\beta_{cl}$ are the propagation constants of the core mode and a cladding mode respectively, then a periodic perturbation with a period $\Lambda$ satisfying the following equation can efficiently couple power among the two modes, provided of course that the coupling coefficient describing their coupling is non-zero:

$$\Lambda = \frac{2\pi}{(\beta_{co} - \beta_{cl})} \tag{14}$$

For typical fiber parameters, the required period $\Lambda$ is in the range of a few hundred micrometers and hence the name long period gratings. These gratings are different from fiber Bragg gratings (short period gratings) in which the periodic perturbation has a period of about 0.5 μm and a specific wavelength suffers a strong reflection while the other wavelengths get transmitted. The perturbation required to form long period gratings are usually obtained by exposing the fiber to ultraviolet light of wavelength of 244 nm. Thus when an optical fiber is exposed to an uv interference pattern with the required period, then due to photosensitivity of the optical fiber, a permanently induced refractive index grating

with the same period is formed in the core of the fiber [12]. This periodic refractive index modulation acts as a perturbation and leads to coupling among
the various modes of the optical fiber. To achieve coupling between the core
mode and a particular cladding mode, a grating with a specific period given
by Eq. (14) needs to be formed inside the fiber. Since the effective indices are
wavelength dependent, coupling of power from the core to the cladding takes
place at a particular wavelength. This coupled light gets lost from the fiber and
this acts as a bandstop filter for the incident light. By appropriately choosing
the period, length and the refractive index modulation of the gratings, one can
make an effective gain flattening filter. Long period gratings can also be formed
within the fiber by locally heating the fiber over a small region and repeating
this operation with the required period of the grating. Such a heating operation
can be easily provided by a fiber fusion splice machine [15]. Figure 9 shows the
transmission spectrum from a single mode fiber in which an LPG with a period
of 480   μm has been fabricated using a fiber splice machine. Two prominent
loss peaks are visible in the spectrum. These are due to the coupling of power
from the mode guided in the core to different cladding modes. The value of peak
loss, the width of the loss peak and the position of the loss peak are determined
by the period of the grating, the value of refractive index perturbation and the
length of the grating. By choosing appropriate values of these parameters, it is
possible to achieve LPGs so as to realize a wavelength filter that can flatten the
gain of an EDFA. Figure 10 shows the amplified spontaneous emission outputs
from an erbium doped fiber, with and without a long period grating filter. The
doped fiber is pumped by a high power 980 nm laser. The figure shows clearly
that using LPGs we can indeed achieve gain flattening of EDFAs.

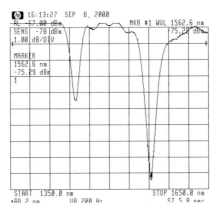

**Fig. 9.** The transmission spectrum of a LPG with a period of 480 mm fabricated using
a fusion splice machine on a standard single mode fiber [After Ref. 15]

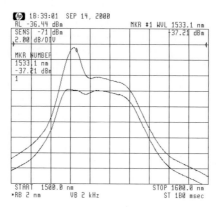

**Fig. 10.** Flattening of the ASE spectrum from an EDF pumped by a 980 nm laser, using an LPG of appropriate characteristics

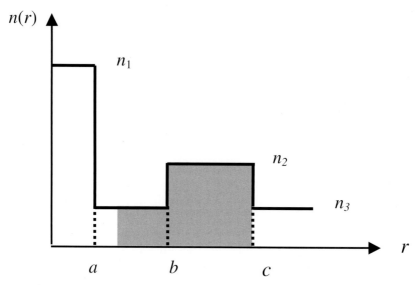

**Fig. 11.** A schematic of a refractive index profile of an erbium doped fiber which exhibits intrinsic gain flattening

## Intrinsically Flat Gain Spectrum

We note that the gain of the EDFA is not flat due to the spectral dependence of the absorption and emission cross sections and also due to the variation of the modal overlap between the pump, signal and the erbium doped region of the fiber. Thus it is in principle possible to flatten the gain of the amplifier by appropriately choosing the transverse refractive index profile of the fiber and the doping profile of the fiber to achieve flatter gain. Figure 11 shows a schematic of a refractive index profile distribution and the corresponding erbium doped

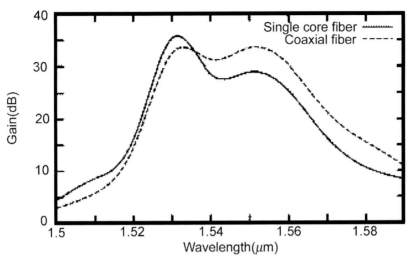

**Fig. 12.** Comparison of the gain spectrum of a EDFA with a conventional erbium doped fiber and the proposed erbium doped fiber [After Ref. 16]

region which can provide gain flattening by appropriately optimizing the various parameters. Figure 12 shows the comparison of the gain profile of an EDFA with a conventional fiber and the gain profile of an optimized EDFA with the proposed design [16]. As is evident, much flatter gain profiles can be achieved using proper optimization of the refractive index profile and the doping profile of an EDF.

## 6.2    Fiber Raman Amplifier (FRA)

If we send an intense light beam at a frequency $\nu_p$ through an optical fiber, at the output of the fiber, in addition to the power at $\nu_p$, we observe light exiting at higher wavelengths also. The new wavelengths get generated via the process of *stimulated Raman scattering* [1]. Figure 13 shows a typical output spectrum from a single mode fiber in which intense pump light beam at 1450 nm has been coupled. As evident, a continuous spectrum of radiation starting from the pump wavelength to about 150 nm above are generated. There is a peak in the emission at around 1550 nm, which is about 100 nm higher than the pump wavelength. This Raman scattered spectrum is characteristic of silica optical fiber. In fact in Raman scattering, light at an incident frequency $\nu_p$ gets scattered to light at other frequencies with $\nu_s < \nu_p$; the difference in energy being given away to the silica lattice. This is referred to as Stokes emission. Raman scattering in silica occurs over a frequency band of approximately 40 THz with a peak at about a frequency downshifted by approximately 13 THz from the pump frequency. Thus for any pump frequency of $\nu_p$, we can observe peak Raman scattering at a frequency approximately $\nu_p - 13$ (THz). For $\nu_p$ corresponding to 1450 nm, the peak occurs at about 1550 nm. If a signal at 1550 nm is sent along with a strong

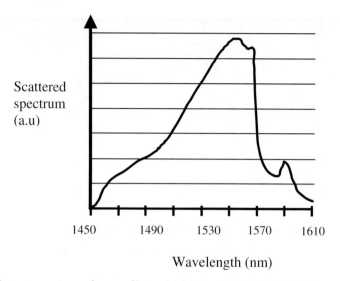

**Fig. 13.** Output spectrum from a fiber which is pumped by a strong pump laser at 1450 nm. Raman scattering produces the scattered spectrum lying between the pump wavelength and 1610 nm. The peak of the emission lies at a frequency which is about 13 THz lower than the pump frequency

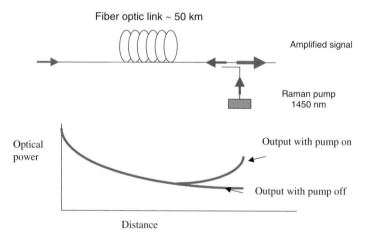

**Fig. 14.** In a Raman amplifier, an intense pump which is upshifted in frequency by about 13 THz from the signal frequency is coupled in the opposite direction to the signal. The signal gets amplified as it propagates through the fiber by the process of stimulated Raman scattering

pump at 1450 nm, then the signal at 1550 nm can get amplified through the process of stimulated Raman scattering. In fact any wavelength starting from 1450 nm to about 1600 nm can experience gains of different amounts; the gain spectrum will be approximately corresponding to the curve shown in Fig. 13.

Since Raman scattering is found in all media, FRAs do not require any special fiber. Even the link fiber that is used for transmission of information can be used for achieving Raman amplification. Figure 14 shows a typical FRA which uses the link fiber for amplification. For achieving amplification at around 1550 nm, a pump beam at 1450 nm is coupled into the exit end of the fiber and propagating in the opposite direction to the signal. The signal experiences gain via stimulated Raman scattering as it propagates through the fiber and encounters the pump wave. Thus the output from the fiber end will be higher in the presence of the pump than in the absence of the pump. Thus the effective loss suffered by the signal is lower in the presence of the pump. The on-off gain of Raman amplifiers (defined with respect to the output power from the fiber in the presence of and in the absence of the pump power) can be about 10 dB. Raman gain G is described by the following equation [1,17]:

$$G = \exp\left(\frac{g_r P_{po} L_{eff}}{A_{eff} K} - \alpha_s L\right) \tag{15}$$

where $g_r$ represents the Raman gain coefficient (peak value of    $6 \times 10^{-14}$ m/W at 1550 nm for silica fibers), $P_{p0}$ is the incident pump power, $\alpha_s$ is the signal attenuation coefficient. $L_{eff}$ is the effective length defined by Eq. (39) with $\alpha$ replaced by $\alpha_p$, the attenuation coefficient of the fiber at the pump wavelength and $A_{eff}$ is the effective mode area given approximately by Eq. (37) with $\psi$ representing the mode field profile at the pump or signal wavelength (here assumed to be approximately equal). $K$ represents a polarization factor which has a value 1 for parallel pump and signal polarizations and is equal to 2 for random polarization states and is equal to infinity when signal and pump are perpendicularly polarized. The on-off gain coefficient is given by:

$$G_{on-off} = \exp\left(\frac{C_R P_{p0} L_{eff}}{K}\right)$$

$$C_R = \frac{g_r}{A_{eff}} \tag{16}$$

where $C_R$ represents the Raman gain efficiency and depends on the effective area of the fiber. Since $C_R$ depends on the effective area of the fiber, it has different values for different fiber designs. Typical peak values of $C_R$ are 0.7 $W^{-1}$ $km^{-1}$ for Corning NZ-DSF fiber, about 0.4 $W^{-1}$ $km^{-1}$ for standard single mode fibers and as high as 6.5 $W^{-1}$ $km^{-1}$ for special highly nonlinear fibers. Dispersion compensating fibers (which will be discussed in Sec. 8) have much smaller mode area and hence possess a much higher Raman gain efficiency ($\sim 3$ $W^{-1}$ $km^{-1}$). Thus such fibers can indeed be used a discrete Raman amplifiers in a link. Raman amplifiers have many advantages as compared to EDFAs:

• They can be used at any signal wavelength by appropriately choosing the wavelength of the pump upshifted in frequency by about 13 THz.
• They have a broad gain spectrum. The gain spectrum can be flattened and broadened by simultaneously using multiple pump wavelengths at appropriate wavelengths

• The link fiber itself can be used as the amplifying medium
• They have lower noise figure than EDFAs.
• For the same span distance, with Raman amplifiers lower launch power levels can be used which leads to reduced nonlinear effects
• They can be used to increase span distance or increase bit rate or number of channels for a given link.

Although distributed Raman amplifiers using the fiber link are most common, discrete Raman amplifiers with additional functions such as dispersion compensation (using the dispersion compensating fiber as the Raman amplifying medium) or to provide dynamic channel equalization are becoming attractive. indeed when WDM channels propagate simultaneously through an optical fiber, SRS leads to transfer of power from lower wavelength channels to higher wavelength channels leading to Raman induced cross talk. Using discrete amplifiers, simultaneous amplification and channel power equalization is possible [18]. Raman amplifiers require high pump powers of 300 to 500 mW and such high power lasers have started to become available only recently [19,20]. Hybrid amplifiers that combine Raman amplifiers and EDFAs are being studied for application to high speed (40 Gbits/s) and ultralong haul communication systems. With the availability of high power pumps, Raman amplifiers have started to enter real systems.

## 6.3   Noise in Amplifiers

In this section we will briefly discuss noise characteristics of EDFAs. As discussed earlier, in an EDFA population inversion between two energy levels of erbium ion leads to optical amplification by the process of stimulated emission. Erbium ions occupying the upper energy level can also make spontaneous transitions to the ground state and emit radiation. This radiation appears over the entire fluorescent band of emission of erbium ions and travels both in the forward as well as in the backward direction along the fiber. The spontaneous emission generated at any point along the fiber can also get amplified just like the signal as it propagates through the population inverted fiber. The resulting radiation is called amplified spontaneous emission (ASE). This ASE is the basic mechanism leading to noise in the optical amplifier [8]. ASE appearing in wavelength region not coincident with the signal can be filtered using an optical filter. On the other hand, the ASE that appears in the signal wavelength region cannot be separated and constitutes the minimum added noise from the amplifier. If $P_{in}$ represents the signal input power (at frequency $\nu$) into the amplifier and $G$ represents the gain of the amplifier then the output signal power is given by $G\,P_{in}$. Along with this amplified signal, there is also ASE power which can be shown to be given by [8]

$$P_{ASE} = 2n_{sp}(G-1)h\nu B_0 \tag{17}$$

where $B_o$ is the optical bandwidth over which the ASE power is being measured (which must be at least equal to the optical bandwidth of the signal)

$$n_{sp} = \frac{N_2}{(N_2 - N_1)} \tag{18}$$

Here $N_2$ and $N_1$ represent the population densities in the upper and lower amplifier energy levels of erbium in the fiber. Minimum value for $n_{sp}$ corresponds to a completely inverted amplifier for which $N_1 = 0$ and thus $n_{sp} = 1$. As a typical example, we have $n_{sp} = 2$, $G = 100$ (20 dB), $\lambda = 1550$ nm, $B_o = 12.5$ GHz (= 0.1 nm) which gives $P_{ASE} = 0.6$ μ W (= - 32 dBm). We can define the optical signal to noise ratio (OSNR) as the ratio of the output signal power to the ASE power:

$$\text{OSNR} = \frac{P_{out}}{P_{ASE}} = \frac{G P_{in}}{2 n_{sp}(G - 1) h\nu B_o} \tag{19}$$

where $P_{in}$ is the average power input into the amplifier (which is about half of the peak power in the bit stream, assuming equal probability of ones and zeroes). For large gains $G \gg 1$ and assuming $B_0 = 12.5$ GHz, for a wavelength of 1550 nm, we obtain $\text{OSNR(dB)} \approx P_{in}(\text{dBm}) + 58 - F$ where $F(\text{dB}) = 10\log(2n_{sp})$ is the noise figure of the amplifier (for large gains). For $n_{sp} = 2$ and $P_{in} = -30$ dBm, we obtain an OSNR of 22 dB. In system designs, typically one looks for an OSNR of greater than 20 dB for the detection to have low bit error rates (BER). For a fiber optic system consisting of $N$ amplifiers and $(N - 1)$ fiber spans, each of the amplifiers adds its own noise contribution to the OSNR at the end of the link. If each amplifier is assumed to be identical and to exactly compensate for the loss of the fiber link preceding it, then the OSNR at the end of the link is given by OSNR (dB) $= P_{in}(\text{dBm}) + 58 - F - 10\log N$. Thus the OSNR will keep falling as more and more amplifiers are added to the link and at some point in the link when the OSNR falls below a certain value, the signal would need to be regenerated. In the case of Raman amplifiers, since it is distributed over the link fiber, part of the noise power generated by the amplification process gets attenuated as it propagates along the fiber unlike in an EDFA where the noise is generated over a very short length of doped fiber after propagation through the link fiber. Thus the effective noise generated in Raman amplification is smaller than in the case of EDFA and improvements in OSNR of about 5 dB are possible. Thus there is a maximum number of amplifiers that can be placed in a link beyond which the signal needs to be regenerated.

# 7   Pulse Dispersion

In this section, we will discuss about pulse dispersion and how it limits the bit rate in a communication system. We will restrict our discussion to single mode fibers, which are the principal types of fibers used in today's long distance communication systems. Fiber optic communication system uses digital transmission in which the information to be sent is first digitized and coded in terms of 1s and

0s. In the optical domain, each 1 is represented by the presence of a light pulse while each 0 is represented by the absence of a light pulse. Each speech signal, which is an analog signal, is digitized into a stream of pulses at the rate of 64000 bits per second or 64 kbits/s. This represents the capacity required to send one speech signal. Similarly in order to send 15000 speech signals simultaneously through an optical fiber, we would need to send optical pulses at the rate of

$$15000 \times 64000 = 0.96 \times 10^9 = 0.96 \text{Gbits/s.}$$

Now, in order to retrieve the information at the end of the link, it is necessary that the individual optical pulses be resolvable. When optical pulses travel through an optical fiber, in general, they broaden in time; this is referred to as pulse dispersion. In multimode fibers, this is primarily caused by the different times taken by the different modes in propagating through the optical fiber. This is referred to as *intermodal dispersion*. For typical step index optical fibers, this dispersion is about 50 ns/km. Thus over a distance of 50 km, the pulse dispersion would be about 2.5 $\mu$s. If we assume that one can allow for a pulse dispersion equal to the pulse duration, then over 50 km, the minimum separation between the pulses would be 2.5 $\mu$s which corresponds to a maximum bit rate of $1/(2.5 \times 10^{-6})$ = 400 kbits/s, which is indeed a very small bit rate. Pulse dispersion in multimode fibers can be significantly reduced by using graded index fibers. In particular, for a fiber having a parabolic refractive index profile in the core, in which the refractive index decreases quadratically from the axis, the pulse dispersion caused due to different travel times of various modes reduces to 0.25 ns/km. For the same 50 km link, this would allow for a pulse rate of 80 Mbits/s. The problem of intermodal dispersion in multimode fibers can be completely eliminated by using single mode fibers. As discussed earlier, single mode fibers support only a single mode of propagation and are characterized by $V < 2.405$. Single mode fibers also suffer from pulse dispersion. Different wavelength components present in the input pulse travel at different velocities resulting in pulse dispersion. Apart from this, pulses can also suffer from polarization mode dispersion (PMD) which is caused due to the slightly different times taken by two orthogonally polarized components of the incident light beam. PMD becomes important only for bit rates of 10 Gbits/s and higher. In order to understand pulse dispersion in a single mode fiber, we consider a Gaussian input pulse described by

$$\psi(z = 0, t) = C e^{-t^2/\tau_0^2} e^{i\omega_0 t} \tag{20}$$

where $\tau_0$ is the input pulse width, $\omega_0$ is the central frequency of the light wave and $C$ is a constant. The frequency spectrum of such a pulse can be obtained by taking a Fourier transform of Eq. (20) to obtain

$$A(\omega) = \frac{C\tau_0}{2\sqrt{\pi}} e^{-\tau_0^2(\omega-\omega_0)^2/4} \tag{21}$$

If $\beta(\omega)$ represents the frequency dependent propagation constant of the mode, then each frequency component of the incident pulse suffers a phase shift of

$\beta(\omega)$ $z$ after propagating through a distance $z$ in the fiber. Thus the output pulse can be written as

$$\psi(z,t) = \int_{-\infty}^{+\infty} A(\omega)e^{i[\omega t - \beta(\omega)z]}d\omega \tag{22}$$

Since the frequency spectrum given by Eq. (21) is usually very sharply peaked, we make a Taylor series expansion of $\beta(\omega)$ around $\omega_0$:

$$\beta(\omega) = \beta(\omega_0) + \left.\frac{d\beta}{d\omega}\right|_{\omega_0}(\omega - \omega_0) + \frac{1}{2}\left.\frac{d^2\beta}{d\omega^2}\right|_{\omega_0}(\omega - \omega_0)^2 + \dots \tag{23}$$

Substituting the expansion given by Eq. (23) in Eq. (22) and integrating, we obtain the following expression for the output pulse [2]:

$$\Psi(z,t) = \frac{C}{(1+\sigma^2)^{1/4}}\exp[-\frac{(t-z/v_g)^2}{\tau^2(z)}]\exp[i(\Phi(z,t) - \beta(\omega_0)z)] \tag{24}$$

where

$$\Phi(z,t) = \omega_0 t + \kappa(t - \frac{z}{v_g})^2 - \frac{1}{2}\tan^{-1}(\sigma) \tag{25}$$

$$\kappa = \frac{\sigma}{(1+\sigma^2)\tau_0^2}; \quad \sigma = \frac{2\alpha z}{\tau_0^2}; \quad \frac{1}{v_g} = \left.\frac{d\beta}{d\omega}\right|_{\omega_0} \tag{26}$$

$$\tau^2(z) = \tau_0^2(1+\sigma^2); \quad \alpha = \left.\frac{d^2\beta}{d\omega^2}\right|_{\omega_0} = -\frac{\lambda_0^2}{2\pi c}D \tag{27}$$

We notice from Eq. (24) that as the pulse propagates, it gets broadened in time; the pulse width at any value of $z$ is given by $\tau(z)$. We also notice that the phase of the pulse is no more proportional to time t but varies quadratically with time t. This implies that the instantaneous frequency of the pulse varies with time and such a pulse is referred to as a chirped pulse. The temporal broadening and chirping of the pulse are determined by the value of the dispersion coefficient $D$ (usually measured in units of ps/km-nm, i.e., the dispersion suffered in picoseconds per kilometer of propagation length per nanometer of spectral width of the source) which in turn depends on the variation of $\beta$ with frequency. The dependence of $\beta$ on frequency or wavelength can be approximately split into two parts:

●That due to the dependence of $\beta$ on wavelength via the dependence of refractive indices of the core and cladding on the wavelength. This is referred to as material dispersion.

● That due to the explicit dependence of $\beta$ on wavelength. This is referred to as waveguide dispersion. The total dispersion is approximately given by the algebraic sum of material and waveguide dispersions. Figure 15 shows the spectral variation of material, waveguide and total dispersions for a standard single mode fiber made of silica with $\triangle = 0.0027$, where $\triangle = \frac{n_1 - n_2}{n_1}$ and $a = 4.1$ $\mu$m. It can be seen from the figure that material dispersion in silica passes through a zero

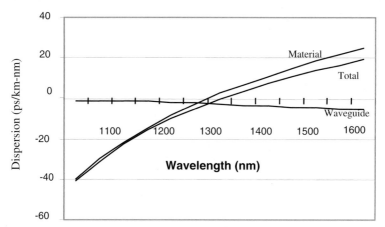

**Fig. 15.** Typical spectral variation of material, waveguide and total dispersion in a typical single mode fiber. Such fibers are referred to as conventional single mode fibers or standard single mode fibers and have zero dispersion around 1310 nm

around 1270 nm and changes sign at this wavelength. However, the waveguide dispersion contribution is negative over the wavelength range plotted. Thus the total dispersion (sum of material and waveguide dispersions) passes through zero around 1310 nm. Hence such fibers exhibit zero dispersion in the 1310 nm wavelength window and are referred to as conventional single mode fibers or standard single mode fibers. Note that such fibers posses about +17 ps/km-nm dispersion value around 1550 nm. Since waveguide dispersion component depends on

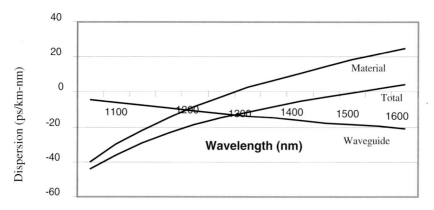

**Fig. 16.** Typical spectral variation of material, waveguide and total dispersion in dispersion shifted fibers with zero dispersion around 1550 nm

the fiber refractive index profile, by appropriately choosing the fiber parameters, the zero dispersion wavelength can be shifted to any wavelength above 1270 nm. Figure 16 shows the material, waveguide and total dispersion of a fiber with zero dispersion in the 1550 nm wavelength window. The fiber parameters are $\triangle = 0.0076$ and $a = 2.3$ $\mu$m. Such fibers are referred to as dispersion shifted fibers (DSF). From Eq. (25), the instantaneous frequency within the pulse can be evaluated and is given by

$$\omega(t) = \omega_0 + 2\kappa \left( t - \frac{z}{v_g} \right) \tag{28}$$

In the normal dispersion region (e. g., for $\lambda_0 < 1310$ nm in standard SMF), $D$ is negative and $\kappa$ is positive. Thus for such a case, the instantaneous frequency within the dispersed pulse increases with time. Thus the leading edge of the pulse is red shifted while the trailing edge is blue shifted. On the other hand, in the anomalous dispersion region, $D$ is positive and hence $\kappa$ is negative. Thus the instantaneous frequency decreases with time. This leads to a leading edge which is blue shifted and a trailing edge which is red shifted. Figure 17 shows the chirping in the normal and anomalous dispersion regions of propagation.

## 8    Dispersion Compensation

Conventional single mode fibers (CSF) have a zero dispersion wavelength of 1310 nm and there is more than 70 million kilometers of such a fiber lying underground in the world. We have seen that EDFAs operate in the 1550 nm window where CSFs have about +17 ps/km-nm of dispersion. Also as we will see, to minimize nonlinear effects, the fiber should have a finite non zero dispersion value. In the presence of dispersion, the optical pulses will suffer from broadening and thus before they completely overlap, the pulses have to be processed. Similar to the concept of optical amplification where the attenuation in the fiber is compensated in the optical domain itself without any conversion to the electrical domain, it is possible to compensate optically for the dispersion in the optical pulses using the concept of dispersion compensation. We saw earlier that when pulses at 1550 nm propagate through a CSF, the leading edge of the pulse is blue shifted while the trailing edge is red shifted. If such dispersed pulses are propagated through an appropriate length of a fiber having $D < 0$ (in which longer wavelengths travel faster than shorter wavelengths), then such optical pulses can be decompressed to their original width. Such a fiber is called a *dispersion compensating fiber* (DCF). The zero dispersion wavelength of such a fiber is above the operating wavelength. Such fibers are commercially available and can be used to typically compensate for 80 kilometers of propagation through CSF. Note that adding such a fiber in the link for dispersion compensation also adds to the loss budget of the link and usually needs additional optical amplifiers to compensate for the additional losses. At the same time since the effective area of DCFs are much smaller than conventional single mode fibers, the Raman gain efficiency of such fibers is very high (see Eq. (16)). Thus by pumping a DCF with a high

Unchirped
Input pulse

Chirped &
broadened
Output pulse

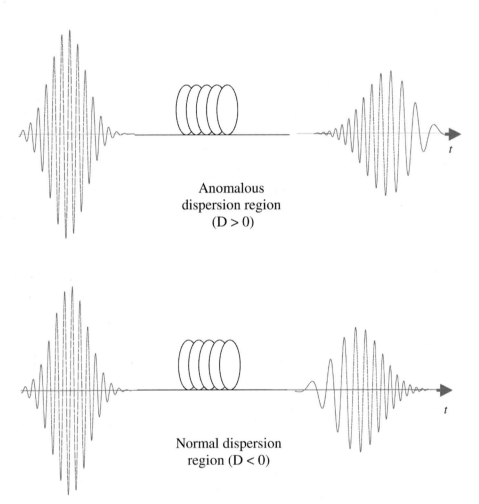

**Fig. 17.** Typical chirping caused in the normal and anomalous dispersion regions of an optical fiber. Notice that the chirp is of opposite sign in the two cases

power Raman pump, simultaneous dispersion compensation and amplification is possible. Let $L_s$ and $L_c$ represent the lengths of the link fiber and the dispersion compensating fiber and their dispersion coefficients be given by $D_s(\lambda)$ and $D_c(\lambda)$ respectively. For complete dispersion compensation at a specific wavelength $\lambda_n$, we must have $D_s(\lambda_n) \, L_s + D_c(\lambda_n) \, L_c = 0$ and hence the length of the DCF must be given by

$$L_c = -\frac{D_s(\lambda_n)}{D_c(\lambda_n)}L_s$$

In a WDM system there are a number of wavelengths propagating simultaneously through the link. Hence the residual dispersion at a nearby wavelength $\lambda_n + \Delta\lambda$ will be

$$\text{Dispersion} = D_s(\lambda_n + \Delta\lambda)L_s + D_c(\lambda_n + \Delta\lambda)L_c$$

$$= D'_s\,L_s\Delta\lambda\left(1 - \frac{\frac{D'_c}{D_c}}{\frac{D'_s}{D_s}}\right)$$

where primes denote differentiation with respect to wavelength and we have made a Taylor series expansion about $\lambda_n$ and retained terms upto order $\Delta\lambda$. Thus for the dispersion compensation over a band of wavelengths the relative dispersion slopes (RDS) of the two fibers must be equal:

$$(\text{RDS})_{\text{DCF}} = \frac{D'_c}{D_c} = (\text{RDS})_{\text{SMF}} = \frac{D'_s}{D_s}$$

Not meeting the condition of the dispersion slope compensation as given by the above equation, may lead to accumulation of unacceptable levels of dispersion in many channels in the WDM system. Thus design of DCFs for broadband dispersion compensation has to take this factor into account. For high bit rates, compensation of dispersion even in the systems employing NZ-DSF fibers become important. DCFs with dispersion slope compensation for broadband transmission using conventional SMFs as well as NZ-DCFs are now available. There has been a lot of work in the realization of DCFs with high negative dispersion coefficients by modifying the refractive index profile of the fiber. A dual core fiber was recently proposed by Thyagarajan et al in Ref.[21] which was shown to have greater than -5000 ps/km-nm of dispersion. Figure 18(a) shows the refractive index profile of a fabricated fiber [22,23] having such a profile and Fig.18 (b) shows the corresponding measured and simulated dispersion coefficient showing the possibilities of realizing extremely large dispersion coefficients by tailoring the refractive index profile. By an appropriate choice of the fiber parameters such a fiber design can also be made to compensate for a number of wavelengths in a band [23].

## 9  Nonlinear Effects in Optical Fibers

Consider a light beam having a power of 100 mW propagating through an optical fiber having an effective mode area of 50 $\mu m^2$. The corresponding optical intensity is $2 \times 10^9$ W/m$^2$. At such high intensities, the nonlinear effects in optical fibers start to influence the propagation of the light beam and can significantly affect the capacity of a WDM optical fiber communication system [24]. The most important nonlinear effects that affect optical fiber communication systems include

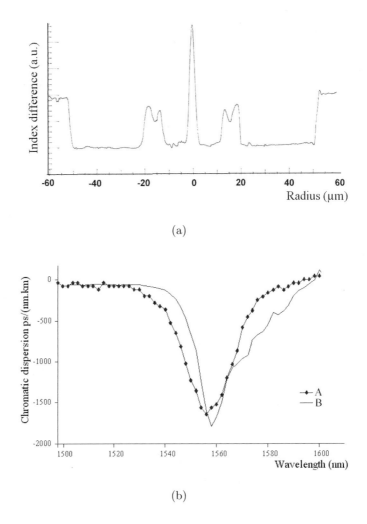

(a)

(b)

**Fig. 18.** (a) Refractive index profile of a fiber possessing a very high negative dispersion coefficient and (b) the simulated and measured dispersion as a function of wavelength showing the possibility of achieving extremely large dispersion coefficients. [After Refs. 21 and 22].

- Self phase modulation (SPM)
- Cross phase modulation (XPM)
- Four wave mixing (FWM).

Stimulated Raman scattering (SRS) and Stimulate Brillouin scattering (SBS) are also important nonlinear phenomena and in Sec. 6.2 we have seen how SRS can be used for optical amplification [25]. In this section, we will discuss mainly SPM, XPM and FWM which affect pulse propagation through optical fibers.

## 9.1   Self Phase Modulation (SPM)

The lowest order nonlinearity present in an optical fiber is the third order non-linearity. Thus for such a fiber, the polarization generated consists of a linear and a nonlinear term:

$$P = \varepsilon_0 \chi E + \varepsilon_0 \chi^{(3)} E^3 \tag{29}$$

where $\chi$ and $\chi^{(3)}$ represent respectively the linear and third order susceptibility of the medium (silica) and $E$ represents the electric field of the propagating light wave/pulse. If we assume the incident electric field to be given by

$$E = E_0 \cos(\omega t - \beta z) \tag{30}$$

where $\beta$ is the propagation constant, then substituting in Eq. (29), we obtain the following expression for the induced polarization at frequency $\omega$:

$$P = \varepsilon_0 \left( \chi + \frac{3}{4} \chi^{(3)} E_0^2 \right) E_0 \cos(\omega t - \beta z) \tag{31}$$

Now, the intensity of the propagating light wave is given by

$$I = \frac{1}{2} c \varepsilon_0 n_0 E_0^2 \tag{32}$$

Substituting Eq. (32) in Eq. (31) and using the fact that polarization and the refractive index are related through

$$P = \varepsilon_0 \left( n^2 - 1 \right) E \tag{33}$$

we obtain the following expression for the refractive index of the medium in the presence of nonlinearity:

$$n = n_0 + n_2 I \tag{34}$$

where $n_2$ is the nonlinear coefficient given by

$$n_2 = \frac{3}{4} \frac{\chi^{(3)}}{c \varepsilon_0 n_0^2} \tag{35}$$

and we have assumed the second term in Eq. (34) to be very small in comparison to $n_0$. Equation (34) gives the expression for the intensity dependent refractive index of the medium due to the third order nonlinearity. It is this intensity dependent refractive index that gives rise to self phase modulation (SPM). In the case of an optical fiber, the light wave propagates in the form of a mode having a specific field distribution in the transverse plane of the fiber. For example, the fundamental mode is approximately Gaussian in the transverse distribution. Thus in optical fibers, it is more convenient to express the propagation in terms of modal power rather than intensity which is dependent on the transverse coordinate. If $A_{eff}$ is the effective cross sectional area of the mode, then $I = P/A_{eff}$, where $P$ is the power carried by the optical beam. Due to intensity dependent refractive index, the propagation constant of a mode can be written as

$$\beta_{NL} = \beta + \gamma P \tag{36}$$

where

$$\gamma = \frac{k_0 n_2}{A_{eff}}; \quad A_{eff} = 2\pi \frac{\left(\int \psi^2(r) r dr\right)^2}{\int \psi^4(r) r dr} \tag{37}$$

represent the nonlinear coefficient and the nonlinear mode effective area respectively and $k_0$ represents the free space propagation constant of the optical wave and $\beta$ is the propagation constant of the mode at low powers. If we assume the mode to be described by a Gaussian function, then $A_{eff} = \pi w_0^2$ where $w_0$ is the Gaussian mode spot size. Note that the nonlinear coefficient $\gamma$ of the fiber depends on the effective area of the mode. The larger the effective mode area, the smaller are the nonlinear effects. If $\alpha$ represents the attenuation coefficient of the optical fiber, then the power propagating through the fiber decreases exponentially as $P(z) = P_0 e^{-\alpha z}$ where $P_0$ is the input power. In such a case, the phase shift suffered by an optical beam in propagating through a length $L$ of the optical fiber is given by

$$\Phi = \int_0^L \beta_{NL} dz = \beta L + \gamma P_0 L_{eff} \tag{38}$$

where

$$L_{eff} = \frac{\left(1 - e^{-\alpha L}\right)}{\alpha} \tag{39}$$

is called the effective length of the fiber. If $\alpha L \gg 1$ then $L_{eff} \sim 1/\alpha$ and if $\alpha L \ll 1$ then $L_{eff} \sim L$. The effective length gives the length of the optical fiber wherein most of the nonlinear phase shift has accumulated. For single mode fibers operating at 1550 nm, $\alpha \sim 0.25$ dB/km ($= 5.8 \times 10^{-5}$ m$^{-1}$) and thus $L_{eff} \sim L$ for $L \ll 17$ km and $L_{eff} \sim 17$ km for $L \gg 17$ km. Since the propagation constant $\beta_{NL}$ of the mode depends on the power carried by the mode, the phase $\Phi$ of the emergent wave depends on its power and hence this is referred to as *self phase modulation* (SPM). Consider a Gaussian input pulse at a center frequency of $\omega_0$ and having an electric field given by

$$E = E_0 e^{-t^2/\tau_0^2} e^{i\omega_0 t} \tag{40}$$

incident on an optical fiber of length $L$. In the presence of only SPM (i.e., no dispersion), the output electric field would be (neglecting attenuation)

$$E = E_0 e^{-t^2/\tau_0^2} e^{i(\omega_0 t - \beta L - \gamma P L)} \tag{41}$$

where $P = |E|^2$ is the power carried by the pulse. Since $P$ is a function of time, the phase of the output pulse is no more a linear function of time. Thus the

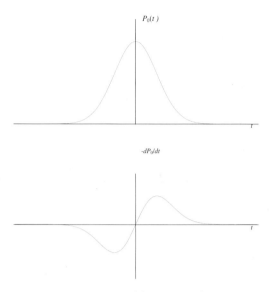

**Fig. 19.** Temporal variation of $P(t)$ and $-dP/dt$ for a Gaussian pulse

output pulse is chirped and the instantaneous frequency of the output pulse is given by

$$\omega(t) = \frac{\mathrm{d}}{\mathrm{d}t}\left(\omega_0 t - \gamma P L\right) = \omega_0 - \gamma L \frac{\mathrm{d}P}{\mathrm{d}t} \tag{42}$$

Figure 19 shows the temporal variation of $P(t)$ and $-dP/dt$ for a Gaussian pulse. The leading edge of the pulse corresponds to the left of the peak of the pulse while the trailing edge corresponds to the right of the peak. Thus in the presence of SPM, the leading edge gets downshifted in frequency while the trailing edge gets upshifted in frequency. The frequency at the center of the pulse remains unchanged from $\omega_0$. Figure 20 shows an input unchirped and the output chirped pulse generated due to SPM. The chirping due to nonlinearity without any corresponding increase in pulse width leads to increased spectral broadening of the pulse. This spectral broadening coupled with the dispersion in the fiber leads

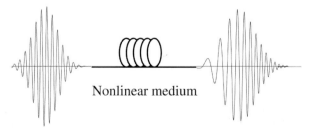

Nonlinear medium

**Fig. 20.** Due to intensity dependent refractive index of the medium, an input unchirped pulse gets chirped as it propagates through a nonlinear medium

to modified dispersive propagation of the pulse. We saw earlier that in the normal dispersion region the chirping due to dispersion is to downshift the leading edge and upshift the trailing edge of the pulse. This is of the same sign as that due to SPM. Thus in the normal dispersion regime (wavelength less than the zero dispersion wavelength) the chirping due to dispersion and nonlinearity add. Thus at high powers, where the nonlinear effects are not negligible, the pulse will suffer additional dispersion as compared to the dispersion of the same pulse at low powers. On the other hand, in the anomalous dispersion region (wavelength greater than the zero dispersion wavelength), the chirping due to dispersion is opposite to that due to nonlinearity and thus in this wavelength region, nonlinearity and dispersion induced chirpings can partially or even totally cancel each other. When total cancellation takes place, the pulse neither broadens in time nor in its spectrum and such a pulse is called a *soliton*. Such solitons can hence be used for dispersionless propagation of pulses to realize very high bit rate systems. Assuming only second order dispersion and $\chi^{(3)}$ nonlinearity, the amplitude $A(z,t)$ of the electric field of an incident pulse can be shown to satisfy the following equation [1]:

$$\frac{\partial A}{\partial z} = i\frac{\beta_2}{2}\frac{\partial^2 A}{\partial T^2} - i\gamma|A|^2 A \tag{43}$$

where $T = t - z/v_g$. Equation (43) is referred to as the nonlinear Shrödinger equation and describes the propagation of a pulse in a medium possessing second order dispersion and $\chi^{(3)}$ nonlinearity. The solution of the above equation give us solitons which are described mathematically by:

$$A(t,z) = A_0\text{sech}(\sigma T)e^{-igz} \tag{44}$$

where

$$g = -\frac{1}{2}\beta_2\sigma^2; \quad \beta_2 = \frac{d^2\beta}{d\omega^2}; \quad A_0^2 = -\frac{\beta_2}{\gamma}\sigma^2 \tag{45}$$

The peak power required to form a soliton is related to the pulse width and is given by

$$P_0 = |A_0|^2 \approx 1.55\frac{\lambda_0^2|D|}{\pi c\gamma\tau_f^2} \tag{46}$$

where $\tau_f$ is the full width at half maximum (FWHM) of the soliton pulse and is given by

$$\tau_f = \frac{2}{\sigma}\ln(1+\sqrt{2}) \approx \frac{1.7627}{\sigma}. \tag{47}$$

As a numerical example, we calculate the soliton power required to form a soliton with $\tau_f = 100$ ps, in a fiber with $D = 2$ ps/km-nm at a wavelength of 1550 nm. Assuming $A_{eff} = 75$ μm$^2$, $n_2 = 3.2 \times 10^{-20}$ m$^2$/W, we obtain $\gamma \sim 1.73 \times 10^{-3}$ m$^{-1}$ W$^{-1}$. Substituting these values in Eq. (46), we find that the peak power required is about 0.5 mW. A heuristic derivation of the power required to form a soliton by cancellation of chirping due to dispersion and nonlinearity can be found in Ref.[2]. Even if the cancellation between dispersive and nonlinear chirping is not perfect, the nonlinear effects in an optical fiber lead to reduced pulse broadening in the anomalous dispersion region. Thus the net dispersion suffered by the pulse decreases as the power increases. This fact needs to be kept in mind while designing dispersion compensation schemes.

## 9.2   Cross Phase Modulation (XPM)

Consider the simultaneous launching of two or more different light beams of different wavelengths into an optical fiber. In such a case, each individual light wave will lead to a change in refractive index of the fiber due to the intensity dependent refractive index. This change in refractive index of the fiber then affects the phase of the other light beam(s); this results in what is referred to as cross phase modulation (XPM). To study XPM, we assume simultaneous propagation of two waves at two different frequencies through the fiber. If $\omega_1$ and $\omega_2$ represent the two frequencies, then one obtains for the variation of the amplitude $A_1$ of the frequency $\omega_1$ as [1]

$$\frac{dA_1}{dz} = -i\gamma \left( \widetilde{P_1} + 2\widetilde{P_2} \right) A_1 \tag{48}$$

where $\widetilde{P_1}$ and $\widetilde{P_2}$ represent the powers at frequencies $\omega_1$ and $\omega_2$ respectively. The first term in Eq.(48) represents SPM while the second term corresponds to XPM. If the powers are assumed to attenuate at the same rate, i.e.,

$$\widetilde{P_1} = P_1 e^{-\alpha z}, \widetilde{P_2} = P_2 e^{-\alpha z} \tag{49}$$

then the solution of Eq.(48) is

$$A_1(L) = A_1(0)e^{-i\gamma(P_1 + 2P_2)L_{eff}} \tag{50}$$

where, as before, $L_{eff}$ represents the effective length of the fiber. From Eq.(50) it is apparent that the phase of signal at frequency $\omega_1$ is modified by the power at another frequency $\omega_2$. This is referred to as XPM. Note also that XPM is twice as effective as SPM. Similar to the case of SPM, we can now write for the instantaneous frequency of the signal at frequency $\omega_1$ in the presence of XPM as [cf. Eq. (42)]

$$w(t) = \omega_1 - 2\gamma L_{eff}\frac{dP_2}{dt} \tag{51}$$

Hence the part of the signal at $\omega_1$ that is influenced by the leading edge of the signal at $\omega_2$ will be down shifted in frequency (since in the leading edge $dP_2/dt > 0$) and the part overlapping with the trailing edge will be up shifted in frequency (since $dP_2/dt < 0$). This leads to a frequency chirping of the signal at $\omega_1$ just like in the case of SPM. Conventional detectors detect the intensity variation of the signal and hence are not affected by phase variations caused by XPM. However, these phase variations get converted to intensity variations due to dispersive effects in the fiber and these intensity variations can cause further bit errors in a fiber optic communication system. If the two light waves are pulses, then XPM leads to chirping of the pulses. In the presence of finite dispersion (i.e., operation away from the zero dispersion wavelength), the two pulses will move with different velocities and thus the pulses will walk off from each other. In case the pulses enter the fiber together, then due to walk off each pulse will see only the trailing or the leading edge of the other pulse which will lead to chirping. On the other hand, if the 'collision' is complete, i. e., if the pulses start separately and walk through each other and again separate as they propagate through the fiber, then there would be no XPM induced chirping since the pulses would have interacted with both the leading and the trailing edge of the other pulse. In an actual system, this cancellation will not be perfect since the pulses suffer attenuation and this leads to reduced nonlinear interaction as the pulses walk through each other. We can define a parameter called the walk off length $L_{wo}$ which defines the length required for one pulse to walk off from the other:

$$L_{wo} = \frac{\triangle\tau}{D\triangle\lambda} \tag{52}$$

where $D$ is the dispersion coefficient and $\triangle\lambda$ is the wavelength spacing between the interfering channels. For return to zero (RZ) pulses $\triangle\tau$ corresponds to the pulse width while for non-return to zero (NRZ) pulses, $\triangle\tau$ corresponds to the rise time or fall time of the pulse. Closely spaced channels will interact over longer fiber lengths and hence will suffer greater XPM effects. Also larger dispersion will reduce the walk off length and hence the XPM effects. In an actual fiber the power carried by the pulses also decreases with propagation and we have earlier defined a characteristic length, $L_{eff}$ for this purpose. Now, if $L_{wo} \ll L_{eff}$, then over the length of interaction of the pulses, the power levels do not change appreciably and the magnitude of the XPM effects will be proportional to the wavelength spacing between the interfering channels. On the other hand, if $L_{wo} \gg L_{eff}$, then the interaction lengths are now determined by the losses and the XPM effects become almost independent of the channel spacing. These conclusions are consistent with experimental observations [26]. XPM induced intensity interference can be studied by simultaneously propagating an intensity modulated pump signal and a continuous wave (cw) probe signal at a different wavelength. The intensity modulated pump signal will induce phase modulation on the cw probe signal and the dispersion of the medium will convert the phase modulation to intensity modulation of the probe. Thus the magnitude of the intensity fluctuation of the probe signal serves as an estimate of the XPM induced interference. Figure 21 shows the variation of the RMS value of probe intensity modulation with the wavelength separation between the intensity modulated pump signal and the probe. The experiment has been performed over four amplified spans of 80 km of standard single mode fiber (SMF) and non-zero dispersion shifted fiber (NZDSF). The large dispersion in SMF has been compensated using dispersion compensating chirped gratings. The probe modulation in the case of SMF decreases approximately linearly with $1/\triangle\lambda$ for all $\triangle\lambda$s, while for the NZ-DSF, for small $\triangle\lambda$s, the modulation is independent of $\triangle\lambda$. This is consistent with the earlier discussion in terms of $L_{wo}$ and $L_{eff}$.

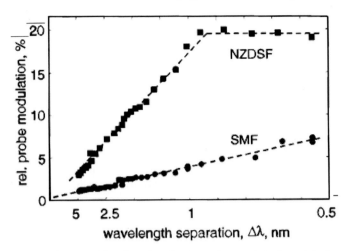

**Fig. 21.** Variation of the RMS value of probe intensity modulation with the wavelength separation between the intensity modulated signal and the probe [After Ref.[27]]

### 9.3   Four Wave Mixing (FWM)

Consider the incidence of three waves at three frequencies $\omega_2$, $\omega_3$ and $\omega_4$ into an optical fiber. In the presence of the three waves, the third order nonlinearity in the fiber leads to the generation of a nonlinear polarization at a frequency $\omega_1$ given by

$$\omega_1 = \omega_3 + \omega_4 - \omega_2 \tag{53}$$

This nonlinear polarization can, under some circumstances, lead to the generation of electromagnetic waves at the new frequency $\omega_1$. This phenomenon is referred to as four wave mixing (FWM). In a dense wavelength division multiplexed (DWDM) system, FWM leads to cross talk among different wavelength channels of the system. In order to study FWM, we write for the electric field of each wave as

$$E_i = \frac{1}{2}[A_i\psi_i(x,y)e^{i(\omega_i t - \beta_i z)} + c.c.]; \quad i = 1,2,3,4 \tag{54}$$

where $A_i$ represents the amplitude, $\psi_i(x,y)$ the transverse filed distribution and $\beta_i$, their propagation constants and c.c represents complex conjugate. The total electric field at any point in the fiber is given by

$$E = E_1 + E_2 + E_3 + E_4 \tag{55}$$

Substituting for the total electric field in the expression for the nonlinear polarization given by Eq.(29), we obtain the following expression for the nonlinear polarization at the frequency $\omega_1$:

$$P_{NL}^{\omega_1} = \frac{1}{2}[p_{nl}e^{i(\omega_1 t - \beta_1 z)} + c.c.] \tag{56}$$

where

$$p_{nl} = \frac{3\varepsilon_0}{2} \chi^{(3)} A_2^* A_3 A_4 \psi_2 \psi_3 \psi_4 e^{-i\triangle\beta z} \tag{57}$$

$$\triangle\beta = \beta_3 + \beta_4 - \beta_2 - \beta_1$$

From the expression for the nonlinear polarization it can be seen that, in general, the velocity of the nonlinear polarization is not equal to the velocity of the electromagnetic wave that it is generating. For efficient generation of the electromagnetic wave at the frequency $\omega_1$, there must be phase matching, i. e., the velocity of the nonlinear polarization (which acts as the source of the radiation at frequency $\omega_1$) and the electromagnetic wave (at frequency $\omega_1$) must be equal. From Eq. (56) we see that the velocity of the nonlinear polarization is given by

$$v_{np} = \frac{\omega_1}{\beta_3 + \beta_4 - \beta_2} = \frac{\omega_1}{\triangle\beta + \beta_1} \tag{58}$$

which will be equal to the velocity of the generated electromagnetic wave at $\omega_1$ only when $\triangle\beta = 0$. This is the phase matching condition required for efficient four wave mixing:

$$\triangle\beta = \beta_3 + \beta_4 - \beta_2 - \beta_1 = 0 \tag{59}$$

Thus efficient FWM will take place when Eq. (59) is satisfied. Substituting the expression for $P_{NL}^{(\omega_1)}$ in the wave equation for $\omega_1$ and making the slowly varying approximation (in a manner similar to that employed in the case of SPM and XPM), we obtain the following equation for $A_1(z)$ [1]:

$$\frac{dA_1}{dz} = -2i\gamma A_2^* A_3 A_4 e^{-i\triangle\beta z} \tag{60}$$

where $\gamma$ is defined by Eq.(37) with $k_0 = \omega/c$, $\omega$ representing the average frequency of the four interacting waves, and $A_{eff}$ is the average effective area of the modes. Assuming that all frequencies have the same attenuation coefficient $\alpha$ and neglecting depletion due to conversion, the power generated at the frequency $\omega_1$ due to mixing of the other three frequencies can be shown to be

$$P_1(L) = 4\gamma^2 P_2 P_3 P_4 L_{eff}^2 \eta e^{-\alpha L} \tag{61}$$

where

$$\eta = \frac{\alpha^2}{\alpha^2 + \triangle\beta^2} \left[ 1 + \frac{4e^{-\alpha L} \sin^2 \frac{\triangle\beta L}{2}}{(1 - e^{-\alpha L})^2} \right] \tag{62}$$

As evident from Eq. (61), maximum conversion occurs when $\eta = 1$, i. e. $\triangle\beta = 0$. Now,

$$\triangle\beta = \beta(\omega_3) + \beta(\omega_4) - \beta(\omega_2) - \beta(\omega_1) \tag{63}$$

Assuming the frequencies to be closely and equally spaced (i.e., $\omega_1 = \omega_2 - \triangle\omega$, $\omega_3 = \omega_2 - 2\triangle\omega$, $\omega_4 = \omega_2 + \triangle\omega$) and making a Taylor series expansion of all $\beta$s about a frequency $\omega_2$, we get

$$\triangle\beta = -\frac{4\pi D\lambda^2}{c}(\triangle\nu)^2 \tag{64}$$

where $\triangle\omega = 2\pi\triangle\nu$ represents the frequency spacing between adjacent channels. Thus when the channels lie around the zero dispersion wavelength of the fiber, $D = 0$ and we have phase matching and thus an efficient FWM. If one wishes to reduce FWM, then one must operate away from the zero dispersion wavelength. This has led to the development of non-zero dispersion shifted fibers (NZ-DSF), wherein a finite but small dispersion ($\sim 2$ to $8$ ps/km-nm) is introduced to reduce FWM effects in an actual fiber optic communication system. On the other hand, if a strong FWM is desired for an application such as all-optical signal processing or wavelength conversion, then the interacting wavelengths must lie close to the zero dispersion wavelength. If we assume all wavelengths to carry the same power $P_{in}$, then under phase matching, the ratio of the power generated due to FWM to that exiting from the fiber is given by

$$R = \frac{P_g}{P_{out}} = \frac{P_1(L)}{P_{in}e^{-\alpha L}} = 4\gamma^2 P_{in}^2 L_{eff}^2 \tag{65}$$

Thus using $\gamma = 1.73 \times 10^{-3}$ m$^{-1}$W$^{-1}$, $L_{eff} = 20$ km, we obtain

$$R = 4.8 \times 10^{-3} P_{in}^2 \text{(mW)} \tag{66}$$

Thus if in each channel we have an input power of 1 mW, then the FWM generated output will be about 0.5 % of the power exiting in the channel. This gives us the level of cross talk among the channels created due to FWM. Figure 22 shows the output spectrum measured at the output of a 25 km long dispersion shifted fiber ($D = -0.2$ ps/km-nm at the central channel) when three 3 mW wavelengths are launched simultaneously. Notice the generation nine new frequencies with different amplitudes (with a maximum peak ratio of 1% to the input signals) due to FWM. Newly generated waves will interfere with power present in those channels and lead to cross talk. Figure 23 shows the ratio of

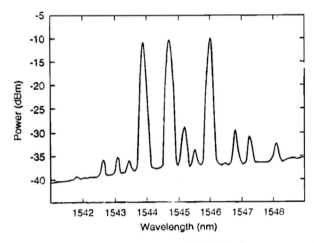

**Fig. 22.** Generation of new frequencies because of FWM when waves at three frequencies are incident in the fiber [After Ref. 28]

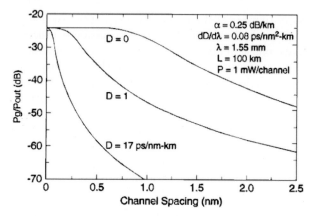

**Fig. 23.** Ratio of FWM generated power to the output power as a function of channel spacing Dl for different dispersion coefficients [After 28]

generated power to the output as a function of channel spacing $\triangle\lambda$ for different dispersion coefficients. It can be seen that by choosing a non zero value of dispersion, the four wave mixing efficiency can be significantly reduced. Larger the dispersion coefficient, smaller can be the channel spacing for the same cross talk. Although FWM leads to cross talk among different wavelength channels in an optical fiber communication system, it can be used for various optical processing functions such as wavelength conversion, high speed time division multiplexing, pulse compression etc [29,30]. For such applications, there is a concerted worldwide effort to develop highly nonlinear fibers with much smaller mode areas and higher nonlinear coefficients. Some of the very novel fibers that have been developed recently include holey fibers, photonic bandgap fibers or photonic crystal fibers which are very interesting since they posses extremely small mode effective areas ($\sim 2.5\mu m^2$ at 1550nm) and can be designed to have zero dispersion even in the visible region of the spectrum [31,32]. This is expected to revolutionize nonlinear fiber optics by providing new geometries to achieve highly efficient nonlinear optical processing at lower powers.

## 10   Conclusions

Recent developments in erbium doped fiber amplifiers and Raman fiber amplifiers and dispersion compensation techniques have helped increase distance between regenerator sites as well as in the realization of very high capacity optical fiber communication systems using dense wavelength division multiplexing (DWDM) techniques. Increased channel numbers and increased powers in the system have given rise to increased nonlinear effects in the system. Solitons use the nonlinearity to overcome dispersive effects and ultra long soliton systems have been demonstrated. Other nonlinear effects such as FWM have been controlled by developing new fibers such as non zero dispersion shifted fibers. With very high intensities existing within the fibers and extremely long interaction lengths provided by optical fibers, nonlinear effects can be observed even at moderate power levels. Some of these effects are being currently investigated for applications in all optical signal processing. New types of fibers namely holey fibers and photonic crystal fibers are being developed which are expected to further revolutionize optical fiber communication and signal processing.

# References

1. G.P.Agrawal: *Nonlinear fiber optics*, (Academic Press, New York, 1989)
2. J. L. Auguste, R. Jindal, J. M. Blondy, M. Clapeau, J. Marcou, B. Dussardier, G. Monnom, D. B. Ostrowsky, B. P. Pal and K. Thyagarajan:Electron. Lett. **36**, 1689 (2000)
3. J. L. Auguste, J. M. Blondy, J. Maury, J. Marcou, B. Dussardier, G. Monnom, R. Jindal, K. Thyagarajan and B. P. Pal: Opt. Fiber Tech.**8**, 89 (2002)
4. O. Aso, M. Tadakuma and S. Namiki:Furukawa Review **19**, 63 (2000)
5. P. C. Becker, N. A. Olsson and J. R. Simpson: *Erbium doped fiber amplifiers: Fundamentals and technology*, (Academic Press, San Diego, 1999)
6. A. R. Chraplyvy:J. Lightwave Tech. **8**, 1548 (1990)
7. E. Desurvire: *Erbium doped fiber amplifiers*, (John Wiley, New York, 1994)
8. T. Erdogan: J. Opt. Soc. Am.B **14**,1760 (1997)
9. T. Erdogan:J. Lightwave Tech. **15**, 1277 (1997)
10. A. Ghatak and K. Thyagarajan:*Introduction to fiber optics*, (Cambridge University Press, UK, 1998)
11. E. Iannone, F. Matera and M. Settembre:*Nonlinear optical communication networks*, (John Wiley, New York, 1998)
12. M. Islam and M. Nietubyc:Raman reaches for ultralong haul DWDM, WDM solutions (2002)
13. R. Kashyap:*Fiber Bragg gratings*, (Academic Press, SanDiego, 1999)
14. J. L. Auguste, R. Jindal, J. M. Blondy, M. Clapeau, J. Marcou, B. Dussardier, G. Monnom, D. B. Ostrowsky, B. P. Pal and K. Thyagarajan:Electron. Lett. **36**, 1689 (2000)
15. L. Kazovsky, S. Benedetto and A. Willner:*Optical fiber communication systems*, (Artech House, Boston, 1996)
16. T. Miya, Y. Terunama, T. Hosaka and T. Miyashita:Electron Lett. **15** 106 (1979)
17. D. K. Mynbaev and L. L. Schiener: *Fiber communications technology*, (Pearson Education, Asia, 2001)
18. S. Namiki and Y. Emori:IEEE J. Selected Topics in Quant. Electron. **7**, 3(2001)
19. P. Palai, M. N. Satyanarayana, Mini Das, K. Thyagarajan and B. P. Pal:Opt. Comm. **193**, 181 (2001)
20. P. Petropoulos, T. M. Monro, W. Belardi, K. Furusawa, J. H. Lee and D. J. Richardson:Opt. Lett. **26**, 1233 (2001)
21. R. Ramaswami and K. N. Sivarajan: *Optical networks: A practical perspective*, (Morgan Kaufmann Harcourt Asia, Singapore, 2001)
22. J. K. Ranka and R. S. Windeler:Opt. Photon. News **August Issue**, 20(2000)
23. L. Rapp:IEEE Photon Tech.Letts **9**, 1592 (1997)
24. M. Saruwatari:IEEE J. Selected Topics in Quant. Electron. **6**, 1363 (2000)
25. M. Shtaif, M. Eiselt and L. D. Garrett:IEEE Photon. Tech. Letts. **12**, 88 (2000)
26. R. G. Smith:App.Opt. **11**, 2489 (1972)

27. R. W. Tkach, A. R. Chraplyvy, F. Forghieri, A. H. Gnauck and R. M. Derosier:J. Lightwave Tech. **13**, 841(1995)

28. O. Aso, M. Tadakuma and S. Namiki:Furukawa Review **19**, 63(2000)

29. K. Thyagarajan and J. Kaur:Opt. Comm. **183**, 407 (2000)

30. K. Thyagarajan, R. K. Varshney, P. Palai, A. Ghatak and I. C. Goyal:Photon. Tech. Letts. **8**, 1510 (1996)

31. A. M. Vengsarkar, P. J. Lemaire, J. B. Judkins, V. Bhatia, T. Erdogan and J. E. Sipe:J. Lightwave Tech. **14**, 58 (1996)

32. A. M. Vengsarkar, J. R. Pedrazzani, J. B. Judkins, P. E. Lemaire, N. S. Bergans and C. R. Davidson:Opt. Lett. **21**, 336 (1996)

# Nonlinear Waves in Optical Waveguides and Soliton Theory Applications

S.B. Leble

[1] Faculty of Applied Physics and Mathematics
Technical University of Gdańsk,
ul. G.Narutowicza, 11/12 80-952, Gdańsk-Wrzeszcz, Poland
email: leble@mif.pg.gda.pl

[2] Kaliningrad State University, Theoretical Physics Department,
Al.Nevsky st.,14,
236041 Kaliningrad, Russia

**Abstract.** Starting with dielectric slab as a waveguide, we discuss the formal aspects of a derivation of model (soliton equations) reducing the description of the three-dimensional electromagnetic wave. The link to the novel experiments in planar dielectric guides is shown. The derivation we consider as an asymptotic in a small parameter that embed the soliton equation into a general physical model. The resulting system is coupled NS (c NS). Then we go to the nonlinear resonance description; N-wave interactions. Starting from general theory and integration by dressing method arising from Darboux transformation (DT) techniques. Going to linear resonance, we study N-level Maxwell-Bloch (MB) equations with rescaling. Integrability and solutions and perturbation theory of MB equations is treated again via DT approach. The solution of the Manakov system (the cNS with equal nonlinear constants) is integrated by the same Zakharov-Shabat (ZS) problem. The case of non reduced MB equation integrability is discussed in the context of the general quantum Liouville-von Neumann (LvN) evolution equation as associated ZS problem.

## 1 Introduction

The development of telecommunications presents a wide field for applications of recent achievements of nonlinear physics. May be the most impressive is the use of soliton effects in optical fibers [1]. The striking example of such use in an experimental realization is a high rate (upto 100Gb/sec) transmission on huge (up $10^5$ km) distances [1]. The very important development of the conventional NS soliton propagation technique give the so-called dispersion managed (DM) soliton notion. The notion is introduced via non-autonomous NS models, it is based on the NS equation with the coordinate dependent coefficients (dispersion constant). Technically the dependence corresponds to incorporation of sections of some compensating fibers into the standard periodic transmission line. We send the reader to recent lectures from Les Houches School [2,3]. We however will touch some actual problems of the soliton propagation, e.g. four-wave mixing, (see the same lectures), [4] via integrable N-wave system with account of wave asynchronism phenomena on the basis of our recent results [5].

Next important technology is based on efficient all-optical logic gates use [6]. It seems to be connected with solitonic behavior of light beams in the direction transversal to the propagation axis, the form of interaction is very similar (compare, e.g.,[7]) to two-soliton-like solutions of Manakov system [8]. Such solutions are discussed in Sect.5.1.

Investigations of laser emission interaction with resonant media sufficiently promoted an understanding and mathematical description of this phenomena [9]. From mathematical point of view this wonderful stability of self-induced transparency (SIT) pulses is the corollary of the complete integrability of Maxwell-Bloch (MB) equations that describe SIT phenomenon in two-level medium approximation with nondegenerate levels [9], recently developed for the case of two-photon processes [10]. The SIT pulses correspond to soliton solutions of the equations and their interaction properties reflect the integrability of evolution equations to model this phenomena. There is the growing interest to other solutions of the MB systems (e.g. periodic ones) that may also give the basis for description of such phenomena as pulse amplification in (e,g. Erbium-) doped fibers. The possibility of the equations generalizations is outlined below (look also at [17,16,18])

In this paper we would review, develop and illustrate our recent results in the theory of guide propagation of electromagnetic waves. Ten years after the publication of the book [19] were very abundant in theoretic ideas as well as in experimental results or even in technical achievements.

The general theory of such complicated phenomena as waveguide propagation in a quantum medium connects at least three ingredients. The first one uses the translation symmetry along the propagation axis $(x)$, it is

i) classical theory of guided modes based on some appropriate Sturm-Liuville spectral problem in transverse coordinates, with rather complicated matching conditions on boundaries,

ii) the second one is the averaged quantum evolution of density matrix of an atomic system model, it gives thermodynamics connection (material equations) between field components and polarization;

iii) last one deals with the final averaging procedure by means of statistical physics description; it accounts spectral broadening and related effects.

The solution of the first problem is based on boundary conditions choice and may be on additional simplifications that allow to split the general system of Maxwell equations in both adjacent media. The matching procedure may simultaneously account properties of both media solutions that leads generally to integral dispersion law [19]. The solution defines the modes notion and a spectrum as the basis of the whole propagation problem.

The second problem stands for quantum evolution described by Liouville-von Neumann equation (LvN) with essential nonlinearity incorporation at least via perturbation theory. There are two contributions (resonant and non-resonant) described in the Sects. 4 and 5. We would also mention the development of the "integrable" approach by such model LvN equations [11,12,14,15].

The joint quantum - classical (unified) N-state Maxwell-Bloch model is integrable in the reduced case (fixed direction of propagation) and if either special conditions of transition probability or degeneration [17,20,21] are fulfilled. In the Sect.4 the papers [22,23] are compiled and illustrated. Important reversed relaxation processes are accounted namely inside the statistical averaging procedure. The result depends on the distribution function choice and introduce typical times such as an inverse level width and a time of coherence $T^*$.

There are efforts to solve the MB equations in its original, non-reduced theory (e.g. - numeric ones [24]). It would allow to study processes of nonlinear reflection starting from [25]. We describe here (Sect.4.3) one of attempts to account both directions of waves propagation [26], suggesting an approach in the spirit of Landau-Ginsburg models, originally proposed in field theory for phase transitions description. The appropriate LvN equation exhibit Euler top-type evolution $i\dot{\rho} = [S, \rho]$ that found to be integrable with abundant set of explicit solutions [11].

In the Sect.2.2 a review of relevant recent experiments is presented. We start from the paper [27] published in 1995 and only mention some historical publications sending the reader to the reviews [1–3]. The observation of effects in connection with the NS system [27] gave rise to the growing interest for further experimental efforts and attempts to find new integrable cases of the systems [28,29]. A discussion of the observations results is presented in Sects. 3.2, 4.2, 5.1. We describe the eight-parameter soliton-like solution of the integrable coupled NS = Manakov equation and illustrate the result with the special emphasis of possible application in a gate (dielectric slab)for all-optical computers. We also write down a solution that stands for interaction of a soliton with a periodic solution: this one candidates to a pumping mechanism in the fiber connection lines.

# 2    Planar Dielectric Waveguides

## 2.1    Dielectric Slab as a Waveguide

Let us begin from the general theory development giving details of the modification of the results from the book [19], (Sect.3.4). Let a slab of a dielectric span the interval $z \in [-h, h]$. We shall name it the "planar waveguide" in the direct correspondence with the mentioned publications (e.g. [30], see also Sect.2.2)). Assume the linear isotropic (averaged) part of a dielectric constant is denoted $\varepsilon$ inside the interval and the constant is equal 1 outside.

In the paper [30] the Manakov soliton observation was declared with the direct demonstration of output laser beams profiles. The geometry of the experiments implies the nonlinear diffraction (soliton evolution) in the direction (say - $x$) that is orthogonal to the both direction of propagation $z$ and the transverse coordinate of a slab $y$. Hence, developing the results of [19] within the weak nonlinearity realm we first summarize the generalities in a way to fit the choice of notations and the reference frame.

Maxwell equations

$$\mathbf{D}_t = c \operatorname{rot} \mathbf{H}, \tag{1}$$

$$\mathbf{B}_t = -c \operatorname{rot} \mathbf{E}, \tag{2}$$

$$\operatorname{div}\mathbf{D} = 0. \tag{3}$$

are chosen to account for the averaged constants of the dielectric layer $\varepsilon$ and $\mu$ separating the nonlinearity and anisotropy by the definition of the polarization vector $\mathbf{P}$. Let us restrict ourselves with the following form of the permittivity, namely

$$\mathbf{D} = \varepsilon\mathbf{E} + 4\pi\mathbf{P}, \tag{4}$$

where

$$P_i = \chi_{ik}^0 E_k + \chi_{ik}^1 |\mathbf{E}|^2 E_k + \ldots,$$

The repetition of indices imply summation, the only Kerr nonlinearity is shown. Combining the Maxwell equations (1,2), after the D'Alambert operator definition $\Box = \frac{\mu\varepsilon}{c^2}\frac{\partial^2}{\partial t^2} - \triangle$, one arrives at the set of the basic wave equations

$$\Box\mathbf{E} = -4\pi\left[\frac{\mu}{c^2}\mathbf{P}_{tt} + \nabla(\nabla\mathbf{P})\right]. \tag{5}$$

The algorithm of the coupled NS system derivation may be illustrated as follows. We would account the geometry and physics of the experiment of [30]. The two parallel beams of TE field ($E_e||110$) and TM mode ($E_m||110$) inputs are chosen orthogonal, provided by the expressions

$$E_x = E_e(z, x, t)Y(y)\exp i(kz - \omega t) + \text{c.c}, \tag{6}$$

$$E_y = E_m(z, x, t)Y(y)\exp[i(kz - \omega t)] + \text{c.c.} \tag{7}$$

These are the simplest one-mode wavetrains which however account the 2+1 nature of evolution along the both axis of propagation and the orthogonal one. For example, the action of the operator quabla on the TE wave gives

$$\exp[-i(kz - \omega t)]\Box E_x = \exp[-i(kz - \omega t)]\Box[E_e(z, x, t)Y(y)\exp i(kz - \omega t) + \text{c.c.}]$$
$$= (-\omega^2/c_0^2 + k^2 + \alpha^2)YE_e$$
$$+ \left[\frac{-i\omega}{c_0^2}\frac{\partial}{\partial t} - ik\frac{\partial}{\partial z} + \triangle_\perp + \frac{1}{c_0^2}\frac{\partial^2}{\partial t^2}\right]YE_e. \tag{8}$$

Relation of scales in $x$ and $z$ directions depend on the pulse duration and laser beam cross-section and define the linear part of a final model equation. One easily recognize propagation and dispersion operators in Eq.(8).

Let us plug the $\nabla\mathbf{P} = -\varepsilon\nabla\mathbf{E}$ from Eqs.(3), (4) into the equation (5). If the amplitude functions depend weakly on the variables, the substitution of $P_i$, Eqs.(6), (7), (8) and similar for $E_y$ into Eq.(5), after account of the dispersion relation $\omega^2 = c_0^2(k^2 + \alpha^2)$, $c_0 = c/\sqrt{\mu\varepsilon}$ and multiplication by $\exp[-i(kz - \omega t)]$, gives, for the first ($x$) component

$$\left[\frac{-2i\omega}{c_0^2}\frac{\partial}{\partial t} - 2ik\frac{\partial}{\partial z} - \triangle_\perp + \frac{1}{c_0^2}\frac{\partial^2}{\partial t^2}\right]YE_e = -4\pi\left[\frac{\mu}{c^2}\mathbf{P}_{1tt} - \varepsilon\nabla_1(\nabla\mathbf{E})\right].$$

This equation is the 2+1 evolution of a boundary TE pulse due to nonlinearity, diffraction along x, and dispersion along $z$. After transformations to the moving frame, for a relatively long pulse, i.e. if $\lambda_x \ll \lambda_z$, $\lambda_{x,z}$ are the scales along x and z correspondingly, we arrive at

$$Y\left(\frac{\partial E_e}{\partial z} + \frac{c_0^2}{2\imath k}\frac{\partial^2 E_e}{\partial x^2}\right) = -Y\frac{4\imath\pi\omega^2\mu}{c^2}\left[\chi_{2k}^0 E_k + \chi_{2k}^1 |\mathbf{E}|^2 E_k + \ldots\right]$$
$$+\varepsilon\frac{\partial}{\partial y}\frac{\partial E_k}{\partial x_k}. \tag{9}$$

Similarly,

$$Y\left(\frac{\partial E_m}{\partial z} + \frac{c_0^2}{2\imath k}\frac{\partial^2 E_m}{\partial x^2}\right) = -Y\frac{4\pi\imath\omega^2\mu}{c^2}\left[\chi_{3k}^0 E_k + \chi_{3k}^1 |\mathbf{E}|^2 E_k + \ldots\right]$$
$$+\varepsilon\frac{\partial}{\partial x}\frac{\partial E_k}{\partial x_k}. \tag{10}$$

The equation for the third ($z$) component in fact does not contribute the theory due to the input zero value of the TE mode electric field.

The functions $Y(y) = Y_0 \sin(\alpha y)$ satisfy the equation $Y_{yy} = -\alpha^2 Y$ and describe waveguide (transversal) orthonormal basis, $\alpha^2 = \omega^2/c_0^2 - k^2$. This connection is nothing but the dispersion relation arises from the linear version of the basic system (5) The spectrum of $\alpha$ is determined from boundary conditions that link Fourier components inside and outside the dielectric guide interval [-h,h]. We would comment it below. The one mode linear matching of $Y$ and $\exp[-py]$ yields

$$\alpha\tan(\alpha h) = p/\varepsilon, \quad \alpha^2 + p^2 = \omega^2(\varepsilon - 1)c^2, \tag{11}$$

where $p^2 = k^2 - \omega^2/c^2$.

Multiplying by the function Y and integrating both equations (9,10) across the slab, one meets the integrals to define nonlinear and dispersion constants in a way of [19]. Finally the system of nonlinear Schrödinger equation appear. We would write down the system in dimensionless form the rescaling the electric field components ($E_e \mapsto e_+, E_m \mapsto e_-$), coordinates and time ($x \mapsto \tau, z \mapsto \zeta$) that is made as below in the case of MB equations (see Sect.4.1, for fields 38).

The resulting evolution equation is the coupled nonlinear Schrödinger (CNS)

$$e_{-,\zeta} = e_{-,\tau\tau}/2 + (n_1|e_-|^2 + n_2|e_+|^2)e_-, \tag{12}$$

$$\imath e_{+,\zeta} = e_{-,\tau\tau}/2 + (n_3|e_-|^2 + n_4|e_+|^2)e_+. \tag{13}$$

There are more terms (linear and nonlinear) that may be neglected in some special conditions (see, e.g. [30,37]). The Manakov integrable case is here, when in both Eq.(12) and Eq.(13) the nonlinear terms enter with approximately equal weights ($n_1 \simeq n_2, n_3 \simeq n_4$).

A more exact matching condition seems more complicated: the relation Eq. (11) that connects the $\alpha$ with $\omega, k$ is valid only for the divided variables, otherwise some interval of the spectrum contributes [19]. There is a tradition to describe the electromagnetic field in waveguides by the Hertz vector. Let further the Hertz vector $\mathit{\Pi}$ is defined by electromagnetic potentials

$$A = \mathbf{\Pi}_t/c, \phi = -div\mathbf{\Pi} \tag{14}$$

plugging into the potential definitions one obtains the principal components of the field

$$\mathbf{B} = \text{rot}\mathbf{\Pi}_t/c, \mathbf{E} = \text{graddiv}\mathbf{\Pi} - \mathbf{\Pi}_{tt}/c^2 \tag{15}$$

Mention that the Lorentz condition holds automatically and the Eq.(15) allows to consider the arbitrary polarized field including processes of the field components interaction. The basic equation for the Hertz vector then introduces the weak cubic nonlinearity described by the parameter $\sigma$, equal to a dimensionless field amplitude.

To account this general matching, let us introduce the Green function for a half plane

$$G(x, y, \eta) = \frac{\imath\omega}{2c} H_1^{(1)}[\frac{\omega}{c}\sqrt{x^2 + (y - \eta)^2}] \tag{16}$$

Where the function $H_1^{(1)}$ is the Hankel function of the first kind. We shall account dependence on $x$, the corresponding dispersion neglecting the dispersion along $z$. Simplifying here the problem by the only one nonzero component ($\mathit{\Pi}_x = \mathit{\Pi}$) of the vector $\mathbf{\Pi}$ and one equation only for the amplitude A (with obvious generalization), we add the term $\hat{\mathit{\Pi}}$ to the Hertz vector component that describe this correction. The evolution equation(s) gain an additional term, that, after integration across the slab yields the extra dispersion term

$$\exp[-i(kz - \omega t)][Y(y)(\hat{\mathit{\Pi}}_y \pm p\frac{k^2c^2 + \varepsilon\omega^2}{k^2c^2 + \omega^2}\mathit{\Pi})]_{-h}^h. \tag{17}$$

The values of the derivative could be expressed by means of the Dirichlet problem solution for the stripe, or, for example, for upper boundary,

$$\hat{\mathit{\Pi}}_y|_h = \frac{\imath\omega}{2c}v.p. \int_{-\infty}^{\infty} H_1^{(1)}(\frac{\omega}{c}|z - \eta|)\frac{b_+ d\eta}{|z - \eta|}, \tag{18}$$

where $b_+ = Y(h)A\exp[\imath(k\eta - \omega t)]$; analogously for the lower boundary. The resulting equation may be written as in [19].

Finally the Hertz vector should be decomposed to the components for different polarizations to fit the boundary (in $z$) condition, that corresponds to the excited overlapped inputs likely, mentioning again the experiment of [30].

## 2.2    Novel Experiments in Dielectric Guides

In 1995 the paper [30] named *Observation of Manakov spatial soliton in AlGaAs Planar Waveguides* appeared. As it follows from the theory (see Sect.5), the specific conditions that lead to the possibility of such observation such as

i) unity ratio between self-phase modulation and cross-phase modulation (define the relative magnitudes of the tensor $\chi$ components from the Sect.2.1), and

ii) four wave mixing terms neglecting should be valid.

The first one is approximately good for the strongly birefringent material, used in the experiment; the second is averaged to zero due to fast phase changes between polarization components. The solitonic character of the phenomena relates directly to the spatial soliton dragging observed a bit earlier by the same group [7]. In this paper the link to Manakov paper [8] have not been mentioned, but there is the natural reference to the earlier pioneering work [31]. Namely this paper initiates investigation of self-focusing phenomena from the point of view of soliton theory. It was mentioned that the planar guide is applicable for the one-dimensional solitonic diffraction. Ref.[7] has already announced the possibility of such spatial soliton's use for the all-optical switch construction. Of course the temporal bright solitons dragging should be mentioned as well [6], but its high phase sensibility, perhaps, does not allow its practical applications for the switching devices. The experimental observation in 1999 of vector solitons dragging in a planar glass waveguide [32] promises the alternative approach to an optical NOT gate. The phenomena of the interaction between Gaussian input pulses with orthogonal polarizations was investigated numerically. The results of the simulation rise the question about integrability of the general case of the equation used in the isotropic (glass) medium.

The fiber guide propagation of the soliton of the two-component (vector) field had also been observed (see the mentioned paper [27]. The observation of bound multiple solitons generated again by the orthogonal polarized pulses in the (strongly birefringent) fibers and the good agreement with numerical simulation of the interaction supports the integrability hypothesis. Another type of polarization-locked vector soliton appears due to compensations produced while the soliton propagates due to randomly varying birefringence [33]. The collision property of the Manakov-like solitons [34] also exhibit the features usual for the integrable systems solutions.

Summarizing, the three different types of solitons of CNS could be specified [33]:

i)*unstable axis soliton*, [34,37] most close to the conventional one (NS) with two versions

ia) continuous wave [35],

ib) solitonic wave [36];

If the Kerr coefficient is positive (as in optical fibers) in both cases the fast axis becomes unstable;

ii) *Manakov soliton*, the classical integrable case [8]; the temporal version is the pulse in fibers [37], spatial one is observed in specially engineered planar anisotropic guides with the relation of nonlinear constants close to unit [30,32];

iii) *polarization locked vector soliton (PLVS)*, it has elliptic polarization, the stability "from self-phase modulation and cross-phase modulation to compensate for the differing phase velocities" of orthogonal components [33]. The CNS has the general form given by Eq.(67), but in the frame with the medium group velocity, the linear propagation velocities $c_i$ have opposite signs.

# 3   Nonlinear Resonance: N-wave Interactions

## 3.1   General Remarks and Standard DT Integration-Inclined Solitons as Illustration

In the case of fibers the classical N-wave interaction [38] is mainly one-dimensional: it may be either interaction between pulses inside one mode or intermode action [19]. There are lot of publications about the problem, the main results are reviewed in [40]. As a theoretical foundation that allow to account for absence of k-resonance (wave asynchronism) within a multisoliton construction we shall use DT technique [41]. In this review we follow mainly [5], where linear terms are included and pay the special attention to the illustrations of the wave asynchronism and attenuation influence [40].

SBS [43,44] and SRS systems [45], equivalent to the N-wave equations with linear terms, are also mentioned.

Let us consider a $n \times n$ matrix set $\{A\}$ and choose $n$ projectors $p_i^2 = p_i \in \{A\}$, $p_i p_k = 0$, i,k=1,2,...,n. The simplest example of such matrix is a diagonal one with the only (number i) nonzero element (that is equal to 1). A choice of numbers $a_i, b_k$ defines matrices

$$M = \sum_i a_i p_i;$$

$$N = \sum_i b_i p_i.$$

Such representation is convenient for generalizations (to operator case) [5]. The nonlinear equations for interacting waves appear as a compatibility condition if we start from the pair of Zakharov-Shabat (ZS) equations of the form

$$\Psi_t = MD\Psi + [H, M]\Psi,$$

$$\Psi_y = ND\Psi + [H, N]\Psi, \tag{19}$$

where the operator D may play the role of abstract differentiation, realized here as the commutator with some constant matrix $x$: $D\Psi = [x, \Psi]$. In analogy with [38] we put in potentials of the ZS equations the commutators [H,M] and [H,N]. Consider first the standard DT [41]. The existence of the standard DT restricts the choice of the matrix $x$,

$$M \times N = N \times M. \tag{20}$$

For elements $H_{ik} = p_i H p_k$ representation via projectors and the choice of $x = \sum_i x_i p_i$, that obviously satisfy Eq.(20), one gets the system

$$(a_k - a_i)H_{ik,y} - (b_k - b_i)H_{ik,t} = [(a_s - a_i)(b_k - b_s) - (b_s - b_i)(a_k - a_s)]H_{is}H_{sk}$$
$$-(H_{ik}x_i - x_k H_{ik})(b_i a_k - a_i b_k) \tag{21}$$

The solutions of the system may be constructed by means of the following proposition [5].

**Proposition 3.1** *The system 21 is integrable by means of standard matrix DT [41]*

$$\Psi[1] = D\Psi - (D\Phi)\Phi^{-1}\Psi,$$

*if $\Phi, \Psi$ are solutions of (19), when (20) holds. The DT of the matrix $H$ in a combination with some gauge transformation may be chosen as*

$$H[1] = H + (D\Phi)\Phi^{-1} + A, \tag{22}$$

*where $A$ is a matrix that commute with both matrices $M$ and $N$. This matrix is the gauge one guaranteeing $H_{ii}[1] = 0$. The obvious possible choice for a diagonal $M, N$ is*

$$A = -\mathrm{diag}(D\Phi)\Phi^{-1}.$$

**Proof.** The standard transforms [41] for the potentials $[H, M], [H, N]$ are $[H, M] + [(D\Phi)\Phi^{-1}, M], [H, N] + [(D\Phi)\Phi^{-1}, N]$, with the visible possibility to add $A$ to $H$.

If one treat the simplest three-wave case, the compatibility condition for equations (21) may be written with more details. For example the first one of them is

$$H_{12,t} - v_{12}H_{12,y} = n_{12}H_{13}H_{32} - k_{12}H_{12}.$$

and similar expressions for others with group velocities as $v_{12} = (b_2 - b_1)/(a_2 - a_1)$ and nonlinear constants as $n_{12} = [(a_3 - a_1)(b_2 - b_3)/(a_2 - a_1) - (b_3 - b_1)]$. The coefficient $k_{12} = (x_2 - x_1)(b_1a_2 - a_1b_2)/(a_2 - a_1)$ define either attenuation or may be equalized to the asynchronism parameter $\Delta k = k_1 - k_2 - k_3$ [3,19]. The last term seems to be interesting for N-wave systems even in a matrix case. Such a linear term for example may account for the mentioned SBS/SRS effects and a phase differences of waves that appears from wave asynchronism. Some damping may be accounted for as well. The mentioned physical systems appear if the reduction constraint is chosen as $H^+ = H$ which is hereditary while iterations are performed for some appropriate choice of $\Phi$.

The choice of the matrix $\Phi$ may be made by means of the idea of M. Salle [41] in the case of stationary solutions of the basic Lax equations with matrix spectral parameter $\Lambda = diag\{\lambda_1, ..., \lambda_n\}$. Namely, if

$$\Phi_t = \Phi\Lambda,$$

then

$$H[1] = H + \Phi\Lambda\Phi^{-1} + A, \tag{23}$$

The one-step transformation generate solutions that are illustrated by the plots below of the $H_{ik}(t)$ for different values of the parameters $x_i$.

## 3.2  Binary DT Application

It is useful to consider the other pair for N-wave system that is more convenient for the binary DT (bDT) use (cf. the Definition 2.2 of the elementary transforms and their binary combinations [5]).

$$D_t\Psi = -\lambda M\Psi + [H, M]\Psi,$$

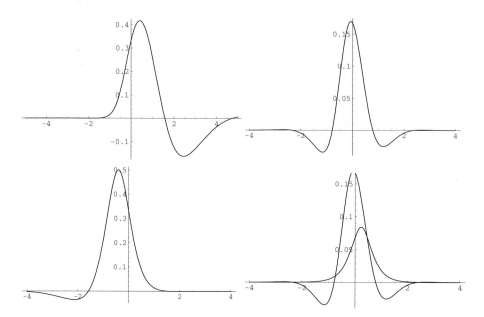

**Fig. 1.** The inclined solitons of a 3-wave system. Plots are given for the components $H_{12}, H_{23}, H_{31}$ as a function of $t$ at $y = 0$. The parameters of the equations are $a_1 = 1, a_2 = -1, b_1 = 1, b_2 = -1, b_3 = 0, a_3 = 0$. The last picture shows the component $H_{12}$ values with different (real) values of the parameter $k_{12}$. The symmetric one has $k_{12} = 0$. The asymmetric one corresponds to $x_2 = 0.5, x_3 = 0.6$.

$$D_y\Psi = -\lambda N\Psi + [H, N]\Psi \tag{24}$$

where differentiations are introduced by combinations of usual derivatives by parameters $t, y$ and the internal derivative

$$D_t\Psi = \Psi_t + R[x, \Psi],$$

$$D_y\Psi = \Psi_y + S[z, \Psi] \tag{25}$$

where $R, S, x, z$ are matrices. The general compatibility condition for this pair of linear equations is

$$[D_t, D_y] = \lambda(M[H, N] - N[H, N]) - [H_y, M]$$
$$+[[H, M], [H, N]] + R[x, [H, N] - S[z, [H, M]]$$

The commutator in the l.h.s of this equation is defined by

$$[D_y, D_t]\Psi = S[z, R][x, \Psi] - R[x, S][z, \Psi] + [SR][x, [z, \Psi]] + SR[[z, x], \Psi] \tag{26}$$

The Jacobi identity is used. The idempotents $p_i$ form the complete set in the $\{A\}$ in a sense that arbitrary matrix A may be represented as the $A = \sum_{ik} p_i A p_k$. The

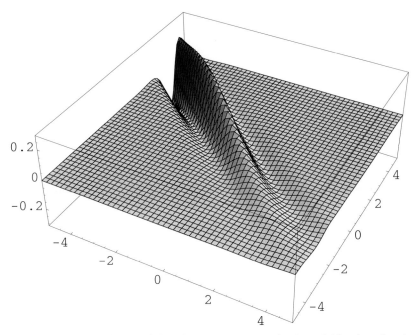

**Fig. 2.** The dependence of the $H_{12}$ component on both variables (ty plane)

simplest $t, y$-independent may be introduced by the equalities $x = x_i p_i, z = z_i p_i$, with a complex numbers $x_i, z_i \in C$.

Now we restrict ourselves to the case $R = \sum_i r_i p_i; S = \sum_i s_i p_i$. In this case the operators $D_{t,y}$ commute. The compatibility condition for the system (24) is then the nonlinear equation

$$[H_y, M] - [H_t, N] = [[H, M], [H, N]] + [x, [RH, N]] - [z, [SH, M] \qquad (27)$$

The form of identities given by Eq.(26) show the possibility to generalize our recent result [5] for this equation (27).

Let us study an example of reduction constraints (for general remarks see again [22,46]), [47]. In the matrix case we denote e.g. $\Psi^+_{pq} = \Psi^*_{qp}$ for any $\Psi \in \{A\}$, where $\Psi^*_{qp}$ is a linear combination of basis elements with complex conjugate coefficients. This reduction constraint corresponds to the classical N-wave system. In matrix notations given the conditions

$$H^+ = H \qquad (28)$$

the constraint Eq.(28) gives for the last terms of the equation (27) the following connection

$$[x^+, [HR^+, N]] - [z^+, [HS^+, M]] = -[x, [RH, N]] + [z, [SH, M] \qquad (29)$$

Or, in terms of "matrix elements", applying $p_i$ from left side and $p_k$ from the right, one has

$$(b_k - b_i)\{[x^+, H_{ik}R^+_k] + [x, R_i H_{ik}]\} + (a_k - a_i)\{[z^+ H_{ik}, S_k] + [z, S_i H_{ik}]\} = 0. \quad (30)$$

In the case of a diagonal $p_i$ for complex number coefficients the condition (30) in terms of new notations:

$$\xi_{ik} = R_i(x_i - x_k), \qquad \eta_{ik} = S_i(z_i - z_k)$$

and

$$v_{ik} = (a_i - a_k)/(b_i - b_k)$$

takes the form:

$$\xi_{ik} - \xi_{ki}^* = v_{ik}(\eta_{ik} - \eta_{ki}^*)$$

From the definitions of $M, N$ the term $M[H, N] - N[H, M]$ is zero. We should recall, that for the version of the procedure of the eDT construction we use here includes the conditions that restricts the choice of both operators $D_t$ and $D_y$.

$$[D_{t,y}, p_i] = 0$$

or

$$D_t p_i = \partial_t p_i + R[x, p_i] = D_y p_i = \partial_t p_i + S[z, p_i] = 0$$

As in the previous case originating from the pair given by (19), the last linear terms may account for the SRS effects, wave asynchronism and damping already in the simplest but physically significant Abelian three-wave case. This term in the Abelian case with $R, S \in Z$ may be generated by the gauge transformation (GT)

$$H_{ik} = h_{ik} \exp g_{ik};$$

with the substitution $g_{ik} = g_i - g_k$, where $g_i = x_i t - z_i y$. Existence of a Lax pair even in this case allows to construct multisolitons in more natural way by general formulas without a hard work of combining DT and GT.

Comparing the forms of binary DT [5] for both of ZS equations ( 24) one finds that the following proposition takes place:

**Proposition 3.2.** *Let* $\varphi_i$ *form a column solution of the system (24) and the functions* $\chi_k$ *belong to the row solutions of the conjugate system. The bDT is the symmetry for the compatibility condition (27) and the matrix $H$ is transformed as*

$$H[1] = H + (\mu - \nu)P \tag{31}$$

*where* $P_{ik} = \varphi_i(\chi^*, \varphi)^{-1}\chi_k$. *This transform, as usual, may be iterated literally or presented by means of a determinant-like representation (see Sect.5.2, [22,23]) that may be found similar to the case of the standard DT [41].*

Some solutions set of non-Abelian N-wave system with the reduction of Eq.(28) is also generated by the binary transformation (31) and the choice $\chi = \varphi^*$ and $\nu = -\mu^*$ so that

$$H[1] = H + (\mu + \mu^*)\Pi, \tag{32}$$

where $\Pi_{ik} = \varphi_i(\varphi^*, \varphi)^{-1}\varphi_k^*$ ([5]).

A soliton solution is obtained starting from the trivial seed solution of Eq.(27) i.e. H = 0. If one starts from the system given by Eq.(24) with zero H, the solution elements $\psi$ satisfy

$$\psi_{ik,t} + x_i\psi_{ik} - \psi_{ik}x_k + \lambda a_i\psi_{ik} = 0,$$

$$\psi_{ik,y} + z_i\psi_{ik} - \psi_{ik}z_k + \lambda a_i\psi_{ik} = 0, \qquad (33)$$

the function $\varphi$ satisfies the same system but with parameter $\mu$ and similar (conjugate) equation holds for $\chi$. The formal solution of Eq.(33) is given by the power series that defines the exponent

$$\psi_{ik} = \exp[-tad_x - yad_z - \lambda(a_it + b_iy)]c_{ik}.$$

The notation $ad_x$ means the internal derivative operator $ad_xc = [x,c]; x, c \in A$. The choice of a commutator algebra of $x, z, c_{ik}$ defines the resulting seed functions.

Going down to a matrix (with a commuting elements) N-wave system (27) with the reduction of Eq.(28) one may consider the second iteration formula for the two-soliton solution or infinitesimally perturbed soliton like it is outlined below (see also [5]).

The soliton-like solution is given by Eq.(31) $H_{ik} = 2(Re\mu)\Pi$, where

$$\Pi_{ik} = \varphi_k^*\varphi_i/(\varphi^*, \varphi).$$

Here $\varphi$ is the column solution of the ZS system (3.4) over the zero seed solution H = 0. The function $\varphi$ is linear combination of exponents with complex parameters that are composed of standard constants $a_i, b_i$ and new parameters $x_i, s_i, r_i$ introduced by the generalized differentiation in Eq.(25).

For a column solution $\psi$ of the same seed system (33) with the spectral parameter $\lambda$

$$\psi_i = \psi_{i1} = A_i \exp[-\lambda(a_it + b_iy) - (v_{ik}Re\eta_{i1} + iIm\xi_{i1})t + \eta_{i1}y)],$$

the solution $\varphi$ differs only by constants $A_i \rightarrow C_i$ and the parameter $\mu$ instead of $A_i$ and $\lambda$ respectively. Now we construct the bDT transform

$$\psi^{ec} = \psi + \varphi\frac{\mu + \mu^*}{\lambda + \mu}(\varphi^*, \psi)/(\varphi^*, \varphi).$$

In the case of commuting matrix elements

$$(\psi^{*ec}, \psi^{ec}) = (\psi^*, \psi) - f|(\psi^*, \varphi)|^2/(\varphi^*\varphi),$$

$$f = \frac{(\mu^* + \mu)(\lambda^* + \lambda)}{|\lambda + \mu^*|} \qquad (34)$$

The scalar products are similar

$$(\varphi*, \varphi) = |C_i|^2 \exp[-2Re\mu(a_1t + b_1y)] \exp[-2Re\mu(b_i - b_1)(v_{i1}t + y)]$$

$$(\psi *, \varphi) = A_i^* C_i \exp[-(\lambda + \mu)(a_i t + b_i y) - 2Re\xi_{i1}(v_{i1}t + y)]$$

By means of it the second iteration is immediately written

$$H[2] = H[1] - (\lambda + \lambda^*)P^{ec},$$

where

$$P_{ik}^{ec} = \frac{\psi_i^{ec}\psi_k^{*ec}(\varphi^*, \varphi)}{(\psi^*, \psi)(\varphi*, \varphi) - f|(\psi*, \varphi)|^2},$$

$$\psi_i^{ec} = \psi_i - \frac{\mu + \mu^*}{\lambda + \mu^*}\varphi_i \frac{(\varphi^*, \psi)}{(\varphi*, \varphi)},$$

with account of Eq.(34). This two-soliton solution has been analyzed numerically by a Mathematica programm. The results show the complicated behavior of wave components with asynchronism, energy exchange and losses parameters.

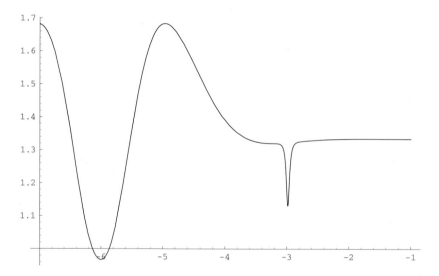

**Fig. 3.** The picture illustrates general two-soliton situation ($H[2]_{12}$ component is plotted), it shows a localized perturbations of a big soliton by a small one

The perturbation of the soliton solution H[1] in the last formula contains restricted functions for every wave multiplied by $2Re\lambda$ that may be chosen as the small parameter. The analysis shows stability of this solitons solution in the class of perturbations. Other solutions: bisolitons and periodic-seed generated ones are presented in the next section (plots are in the Sect.5). N-soliton solution construction in this matrix case will be describe in the section (5.2) in connection with the Manakov equations [22].

# 4   Linear Resonance: MB Equations

## 4.1   About Derivation and Rescaling

For real media the energy levels are usually degenerate that contribute to peculiarities of polarized pulses propagation. In [17] the integrability of MB equations for arbitrary polarization of light being resonant with quantum transitions $j_b = 0$ $- > j_a = 1$ have been established. It was shown that solitons with circular polarization are stable.

This problems of interest relates to the interaction of many-frequency laser pulses with a resonant many-level medium. For example such many-frequency action widen possibilities of isotope dividing or simulation of chemical reactions as well as spectroscopy investigations. The peculiarities of coherent pulses that propagate in many-level media may be used in the frequency transformation effectivisation [17].

Let's consider a medium that may be described in the two-level atoms approximation with energy levels $E_{a,b}$ that are degenerated by projections m and $\mu$ of total momenta $j_{a,b}$. Let a plane electromagnetic wave propagates along z and its electric field has the form

$$\mathbf{E} = \mathbf{E_0} \exp[i(kz - \omega t)] + c.c. \qquad (35)$$

where t is time. The carrier frequency $\omega = kc$ of the pulse is close to $\omega_0 = 2\pi(E_b - E_a)/h$ of the atom transition $j_a - > j_b$ between the energy levels with $E_b, E_a$.

The evolution of the pulse given by Eq.(35) in the semiclassical approach and coherent approximation is described by generalized MB equations [21]. The equations may be rewritten in the dimensionless variables $\tau = (t - z/c)/t_o$, $SS\zeta = z/ct_o$, $t_o = (3h/2\pi\omega d^2 N_o)^{1/2}$ is time dimension constant. Here d is the reduced dipole moment of the transition $j_b \rightarrow j_a$, $N_0$ - density of resonant particles, $\Delta = (\omega - \omega_o)t_o$ - resonance shift. Finally one have

$$\epsilon_{q;\zeta} = \sum_{\mu,m} R_{\mu,m} \left\langle J^q_{\mu,m} \right\rangle,$$

$$R_{\mu,m;\tau} = i \sum_{q,} \epsilon_q (\sum_{m'} J^q_{\mu,m'} R_{m',m} - \sum_{\mu'} R_{\mu,\mu'} J^q_{\mu',m})$$

$$R_{m,m';\tau} = i \sum_{q,m} (\epsilon_q^* J^q_{\mu,m} R_{\mu,m'} - \epsilon_q R_{m,\mu} J^q_{\mu,m'}),$$

$$R_{\mu,\mu';\tau} = i \sum_{q,m} (\epsilon_q J^q_{\mu,m} R_{m,\mu'} - \epsilon_q^* R_{\mu,m} J^q_{\mu',m}) \qquad (36)$$

The indices after the semicolon ; $\zeta$ and $\tau$ denote derivatives. The inhomogeneous broadening of the energy level is accounted: The brac and ket braces $<, >$ introduce the averaging procedure by a normalized partition function $f(\eta)$.

$$\langle A \rangle = \int_{-\infty}^{\infty} A(\eta) f(\eta) d\eta. \qquad (37)$$

In those equations the electric field components $E_{0q}$ of a light pulse, coefficients of the density matrix $\rho_{\mu,\mu}$, are proportional to the functions $\epsilon_q$, $R_{\mu,\mu'}$, $R_{m,m'}$, $R_{\mu,m}$:

$$E_q = h\epsilon_q/2\pi t_o d \tag{38}$$

$$\rho_{\mu,\mu'} = N_o f(\eta) R_{\mu,\mu'},$$

$$\rho_{m,m'} = N_o f(\eta) R_{m,m'},$$

$$\rho_{\mu,m} = N_o f(\eta) R_{\mu,m} \exp[i(kz - \omega t)]$$

The matrix $J^q_{\mu,m}$ is proportional to the Clebsch coefficients.

$$J^q_{\mu,m} = (-1)^{j_b - m} \begin{bmatrix} j_a & 1 & j_b \\ -m & q & \mu \end{bmatrix} 3^{1/2}$$

by the index q the components of the vector $\epsilon$ are labelled.

Let's consider the transition $j_b = 0 \to j_a = 1$ and introduce

$$e_+ = -i\epsilon_{+1}, e_- = -i\epsilon_{-1},$$

$$n_b = R_{\mu,\mu}|_{\mu=\mu'=0},$$

$$n_{a+} = R_{m,m'}|_{m=m'=1}, n_{a-} = R_{m,m'}|_{m=m'=-1},$$

$$\nu_a = R_{m,m'}|_{m=m'=-1}, \nu_+ = R_{\mu,m}|_{\mu=0,m=1}, \nu_- = R_{\mu,m}|_{\mu=0,m=-1}.$$

The quantization axis is chosen along the light propagation direction (axis $\zeta$).

As the interaction of a quantum system with a transverse electromagnetic waves is accompanied by transitions with the possible angular momentum changes (-1 or +1), the MB system may be simplified.

$$e_{-,\zeta} = -\langle \nu_- \rangle,$$

$$e_{+,\zeta} = -\langle \nu_+ \rangle,$$

$$n_{a-,\tau} = -\nu_- e_-^* - \nu_-^* e_-,$$

$$n_{a+,\tau} = -\nu_+ e_+^* - \nu_+^* e_+,$$

$$n_{b,\tau} = \nu_+ e_+^* - \nu_+^* e_+ + \nu_- e_-^* + \nu_-^* e_-,$$

$$\nu_{-,\tau} = -i(\eta - \Delta)\nu_- + (n_{a-} - n_b)e_- + \nu_a e_+,$$

$$\nu_{+,\tau} = -i(\eta - \Delta)\nu_+ + (n_{a+} - n_b)e_+ + \nu_a e_-,$$

$$\nu_{a,\tau} = -\nu_+ e_-^* - \nu_-^* e_+. \tag{39}$$

If the resonant atoms are in pure state, one may reduce the system (39) to only five equations

$$e_{-\zeta} = -\langle a_3 a_1^* \rangle,$$

$$e_{+\zeta} = -\langle a_3 a_2^* \rangle,$$

$$a_{1,\tau} = i(\eta - \Delta)a_1/2 - a_3 e_-^*,$$

$$a_{2,\tau} = i(\eta - \Delta)a_2/2 - a_3 e_+^*,$$

$$a_{3,\tau} = i(\eta - \Delta)a_3/2 + a_1 e_+ + a_2 e_+. \tag{40}$$

where $a_{1,2,3}$ are amplitudes of the probability density (population numbers) of the lower and two upper quantum states correspondingly.

Similar equations appear in a description of a propagation of two-frequency radiation in three-level medium where one of the transitions is forbidden and forces of oscillators are equal [21]. Moreover the shifts of frequencies from resonant states should be equal.

## 4.2   Integrability and Solutions of MB Equations

Below we generate soliton-like and quasi-periodic solutions of the MB system for arbitrary polarized light interacting with two-level atoms with a degenerate level. We follow results of [22] and [23] with some extra details and illustrations.

The general Maxwell - Bloch equations (36) represent a complicated system that is not solved analytically in the case of arbitrary angular momenta. However in the case of resonant transitions $j_b \rightarrow j_a = 1; j_b \rightarrow j_a = 0$ the equations (36) may be presented as the compatibility condition of some linear system (zero curvature representation). For the case of equations (39) it is the following system

$$\psi_\tau = V\psi$$

$$\psi_\zeta = \langle \alpha(\lambda)A \rangle \psi \tag{41}$$

where $V = U - \lambda J, J = diag\{1, 1, -1\}, \alpha(\lambda) = (2\lambda + i(\eta - \Delta))^{-1},$

$$U = \begin{pmatrix} 0 & 0 & -e_-^* \\ 0 & 0 & -e_+^* \\ e_- & e_+ & 0 \end{pmatrix}$$

$$A = \begin{pmatrix} n_{a-} & \nu_a & \nu_-^* \\ \nu_a^* & n_{a+} & \nu_+^* \\ \nu_- & \nu_+ & n_b \end{pmatrix}$$

Let's note that the first equation of the system (41) is the Zakharov - Shabat problem with the $3 \times 3$ matrix potential $U$ of a special structure (similar to one for the nonlinear Schrödinger equation). The independent variable here is the $\tau$ instead of x. The covariance of this equation (and the second one of the Lax pair) is the key observation for the use of the DT technique to construct solutions of MB system.

From the obvious equality $\psi_{\zeta\tau} = \psi_{\tau\zeta}$ we obtain the system

$$U_\zeta = [J, \langle A \rangle]/2,$$

$$A_\tau = [U, A] - i(\eta - \Delta)[A, J]/2 \tag{42}$$

from which it follows that the elements of U and A matrices should satisfy MB system (39). If matrix elements of A are representable in the form of $A_{kj} =$

$a_k a_j^*(k, j = 1, 2, 3)$ (the pure quantum states) then from Eq.(39) the system (40) follows.

The equations (39,40) appear also as the compatibility conditions of the conjugate Lax pair as well

$$\xi_\tau = -\xi W$$

$$\xi_\zeta = -\xi \langle \alpha(\kappa)A \rangle. \tag{43}$$

Here the potential and a new parameter $\kappa$ enter the conjugate ZS problem via $W = U - \kappa J$. The second key observation that allow to keep constraint conditions applying DT is the existence of automorphisms between direct and conjugate linear problems of a Lax representation. As the matrices U and A satisfy the following reduction conditions

$$A - A^+ = 0,$$

$$U + U^+ = 0; \tag{44}$$

one may check that there in the space of square matrix solutions of the ZS equations (wave functions - WFs) exists the automorphism in the sense of [46], see for a generalization [47].

$$\Psi(\lambda) \to [\Psi(-\lambda^*)]^{-1} \in \Psi(\lambda). \tag{45}$$

The reduction is provided by the coupling between the WF spaces of right and conjugate (left) zero curvature representation

$$\kappa = -\lambda^*, \, \xi = \psi^+. \tag{46}$$

Such functions are referred as coupled by an automorphism.

The following statement is valid:

**Statement 4.1** *The equations (39) are the corollary of Eq.(42) when the additional conditions (44) are posed.*

The DT technique that is developed in [22,48] may be applied for the construction of explicit solutions for the MB system (39). We specialize the scheme here. Let $\phi$ and $\chi$ are functions of the right and left (conjugate) problems Eq.(41). Spectral parameters (SP) are $\mu$ and $\nu$ correspondingly. We should suppose that those functions are also connected by the automorphism, or

$$\nu = -\mu^*, \chi = \phi^*.$$

The transformed WFs by the **binary** DT ( [5], look also for Sect.3.2)

$$\psi[1] = [1 - (\mu + \mu^*)P/(\lambda + \mu^*)]\psi$$

$$\xi[1] = \xi[1 - (\mu + mu^*)P/(\kappa - \mu)] \tag{47}$$

are also the solutions of right and conjugate ZS zero curvature representations with new potentials $V[1] = U[1] - \lambda J$, $W[1] = U[1] - \kappa J$ and

$$A[1] = A - 2(\mu + \mu^*)[\alpha(\mu)AP + \alpha(\mu)^*PA - (\alpha(\mu)A + \alpha(\mu)^*)PAP] \tag{48}$$

where
$$U[1] = U - (\mu + \mu^*)[J, P]. \tag{49}$$

The matrix P in the equations (47,48,49) is defined in analogy with [22] by

$$P_{kj} = \phi_k \phi_j^* / (\phi^+, \phi).$$

As $P^+ = P$ the reduction of Eq.(44) is conserved. It means that the elements of matrices U[1] and A[1] give new solutions of MB equations (39). For the transformed potential coefficients that have the sense of electromagnetic field components the following expressions are valid

$$e_-[1] = e_- + 2(\mu + \mu^*)\phi_3\phi_1^*/(\phi^+, \phi),$$

$$e_+[1] = e_+ + 2(\mu + \mu^*)\phi_3\phi_2^*/(\phi^+, \phi.) \tag{50}$$

If the initial state of the quantum system was pure the matrix $A_{kj} = a_k a_j^*$ and the result of the transformation is pure as well with

$$\mathbf{a}[\mathbf{1}] = \mathbf{a} - 2(\mu + \mu^*)\alpha(\mu)^* P \mathbf{a},$$

the normalization is conserved too due to the unitary character of the map given by Eq.(50), see, for the proof [13].

As by the map given by Eq.(45) the transforms of WFs that have been coupled by the automorphism should also be connected by $\xi[1] = \psi[1]^+$, the reduction (46) will be valid during iterations of BDT. Hence it may be realized via the use of $\phi_{(q)}$ and $\chi^{(q)}$ that are WFs of initial zero curvature representation to be coupled by automorphisms in pairs [22].

## 4.3   Solutions over the Nonzero Backgrounds

Starting from the following seed solutions of the equation (39) with zero functions $e_- = e_+ = 0, n_{a+,-} = \nu_a = \nu_{+,-} = 0$ but the nonzero $n_b = 1$, after first iteration we obtain the one-soliton solution that generalize $2\pi$ pulse of two-level system. This solutions have been built in the paper [17] within the "inverse scattering transform" (IST) method. A construction of two-soliton solution met difficulties and the analysis of soliton collisions was made by asymptotic methods [21]. By the way in [39,42] an algebraic approach that have features of Bäcklund and "Dressing" methods was developed and the breather two-soliton solution (with equal velocities) have been found. The validity of the DT (50) with reductions allows produce multisolitons by iterations in algorithmic way.

After a second iteration preserving the reduction (the WF have SP coupled by $\mu_{(2)} = \mu_{(1)*} = \mu$) we obtain the solution of the MB system (39) with the following elements of the potential matrix

$$e_-[2] = -2(a_1 c_1^* \Delta_2 \exp[-\imath\eta] - a_2 c_2^* \Delta_1 \exp[\imath\eta] - a_1 c_2^* \Delta^* - a_2 c_1^* \Delta)/\Delta[2]$$

$$e_+[2] = -2(b_1 c_1^* \Delta_2 \exp[-\imath\eta] - b_2 c_2^* \Delta_1 \exp[\imath\eta] - b_1 c_2^* \Delta^* - b_2 c_1^* \Delta)/\Delta[2]$$

$$\Delta[2] = \Delta_1 \Delta_2 - \mid \Delta \mid^2$$

$$\Delta_1 = [(\mid a_1 \mid^2 + \mid b_1 \mid^2) \exp[-\vartheta] + \mid c_1 \mid^2 \exp[\vartheta]]/(2\mu_R)$$

$$\Delta_2 = [(\mid a_2 \mid^2 + \mid b_2 \mid^2) \exp[-\vartheta] + \mid c_2 \mid^2 \exp[\vartheta]]/(2\mu_R)$$

where

$$\vartheta = 2\mu_R(\tau + \langle \mid \alpha(\mu) \mid^2 \rangle \xi),$$

$$\theta = 2\mu_I \tau - \langle (2\mu * I + \eta - \Delta) \mid \alpha(\mu) \mid^2 \rangle \xi),$$

where $\mu = \mu_R + \imath \mu_I, a_{1,2}, b_{1,2}, c_{1,2}$ are complex constants.

This solution generalizes the solution from [39]. It is characterized by eight real parameters: $\mu_R, \mu_I$, distance between pulse centers, electric field components ratio of the pulses and phase shifts of the same component of both pulses.

Let us consider the next example of arbitrary level populations ("nonzero dipole temperature" of the medium): $n_{a+}n_{a-}n_b \neq 0, e_- = e_+ = \nu_- = \nu_+ = \nu_a = 0$. WFs of the system (41) have the form

$$\psi_1 = C_1 \exp(-\lambda\tau + \langle \alpha(\lambda)n_{a-} \rangle \zeta),$$

$$\psi_2 = C_2 \exp(-\lambda\tau + \langle \alpha(\lambda)n_{a+} \rangle \zeta),$$

$$\psi_3 = C_3 \exp(\lambda\tau + \langle \alpha(\lambda)n_b \rangle \zeta).$$

Transforming the electric components by Eq.(50) with WF $\phi$ and spectral parameter $\mu$ one gets

$$e_-[1] = 4\mu_R \phi_3 \phi_1^* / (\phi^+, \phi),$$

$$e_+[1] = 4\mu_R \phi_3 \phi_2^* / (\phi^+, \phi),$$

where

$$\phi_1 = C_1 \exp(-\mu\tau + \langle \alpha(\mu)(n_{a-} - n_{a+}) \rangle \zeta/2)$$

$$\phi_2 = C_2 \exp(-\mu\tau + \langle \alpha(\mu)(n_{a-} - n_{a+}) \rangle \zeta/2)$$

$$\phi_3 = C_3 \exp(\mu\tau + \langle \alpha(\mu)(2n_b - n_{a-} - n_{a+}) \rangle \zeta/2)$$

Let $n_b > n_{a+} > n_{a-}$. While $\tau \to -\infty$, $e_+[1] \to 0$ at arbitrary $\zeta$ and if $\tau \to +\infty$, then $e_+[1] \to 0$. So the solution describes a transformation of a pulse resonant with the transition which has the minor population difference to one that corresponds to the larger population difference. that means that such pulses are not stable. The result seems to be physical: the transition from the basic state to more populated one would be more transparent (cf. [21]).

## 4.4    Periodic-Seed Solutions

Now we shall construct solutions of the MB system starting from a periodic seed background. Let along one transition the plane electromagnetic wave be propagated

$$e_+ = E \exp \imath(k\zeta + \omega\tau), \ e_- = 0,$$

where $E$ is the a complex constant. As in the previous example the shift from resonance is taken into account. Let also the second level be empty $n_{a-} = 0$, $n_b + n_{a+} = 1$. Then from the equations (39) one have

$$\nu_- = \nu_a = 0,$$

$$\nu_+ = \imath(2n_b - 1)(\omega + \eta - \Delta)^{-1} E \exp \imath(k\zeta + \omega\tau),$$

$$k = -\left\langle (2n_b - 1)(\omega + \eta - \Delta)^{-1} \right\rangle,$$

$$n_b = (1 \mp (\omega + \eta - \Delta)(4 \mid E \mid^2 + (\omega + \eta - \Delta)^{-1/2})/2.$$

During the derivation of the formula for the lower level population $n_b$ it was supposed that the quantum system had gone from one with free upper levels to the considered state. This condition fix the trace of the matrix $A^2$ that does not depend on $\tau$. It may be seen from the second equation of the system (42). The case of the arbitrary higher level population may be considered in similar way. Relaxation terms from spontaneous transitions may be neglected if the dimensionless electric field amplitude $\mid E \mid$ is big enough. The solution of the linear system is:

$$\psi_1 = C_1 \exp(-\lambda\tau),$$

$$\psi_2 = (C_+ \exp(\vartheta) + C_- \exp(-\vartheta)) \exp((\langle \alpha(\lambda) \rangle - ik)\zeta/2)$$

$$\psi_3 = -(C_+(\lambda + \sigma)) \exp(\vartheta) + C_-(\lambda - \sigma) \exp(-\vartheta)) \exp((\langle \alpha(\lambda) \rangle + ik)\zeta/2)/E^*,$$

where

$$\vartheta = \sigma(\tau + i \left\langle (2n_b - 1)\alpha(\lambda)(\eta - \Delta)^{-1} \right\rangle \zeta)$$

$$\sigma = (\lambda^2 - \mid E \mid^2)^{-1/2}.$$

Transforming by Eq.(50) and introducing a new parameter $\gamma = \gamma_R + i\gamma_I$ instead of the spectral parameter value $\mu$

$$\cosh(\gamma) = \mu/ \mid E \mid,$$

one obtains

$$e_-[1] = -4E \ \cosh(\gamma_R) \ \cos(\gamma_I)\bar{\phi}_3\bar{\phi}_1 \exp(ik\zeta/2)/(\bar{\phi}^+, \bar{\phi}),$$

$$e_+[1] = -E[1 - 4E \ \cosh(\gamma_R) \ \cos(\gamma_I)\bar{\phi}_3\bar{\phi}_2] \exp(ik\zeta)/(\bar{\phi}^+, \bar{\phi}),$$

where

$$\bar{\phi}_1 = C_1 \exp(- \mid E \mid \ \cosh(\gamma_R) \ \cos(\gamma_I)(\tau + \langle D \rangle \zeta) +$$

$$i(\mid E \mid \ \sinh(\gamma_R) \ \sinh(\gamma_I)\tau - \langle D(\eta - \Delta) \rangle \zeta/2)).$$

$$\bar{\phi}_2 = C_+ \exp(\vartheta) + C_- \exp(-\vartheta).$$

$$\bar{\phi}_3 = -E(C_+ \exp(\vartheta + \gamma) + C_- \exp(-\vartheta - \gamma)) / \mid E \mid .$$

is the solution of the same system with the SP $\mu$.

$$Re(\vartheta) = \mid E \mid \cos(\gamma_I)( \sinh(\gamma_R)\tau +$$

$$\langle (2n_b - 1)(\eta - \Delta)^{-1}((\eta - \Delta) \sinh(\gamma_R) - 2 \mid E \mid \sinh(\gamma_I))D \rangle \zeta),$$

$$Im(\vartheta) = \mid E \mid \cosh(\gamma_R) \left( \sinh(\gamma_I)\tau \right.$$

$$+ \left\langle \left\{ (2n_b - 1)(\eta - \Delta)^{-1}((\eta - \Delta) \sinh(\gamma_I) \right. \right.$$

$$\left. \left. +2 \mid E \mid \sinh(\gamma_R))D \right\} \right\rangle \zeta \right),$$

$$D = 2 \mid E \mid^2 [ \cosh(2\gamma_R) + \cos(2\gamma_I)] + (\eta - \Delta)^2$$

This solution is the function of the four complex parameters $E, \gamma, C_1, C^+/C_-$ and describe processes of a fission or a fusion of pulses resonant with different transitions. In order to verify this it is necessary to consider real parts of exponents arguments in the expressions for $\bar{\phi}_1, \bar{\phi}_2, \bar{\phi}_3$. The parameter $\gamma$ is introduced namely for a convenience of this operation. The expressions that determine the character of pulses behaviors are however complicated but allow to plot electric field components (see Fig.5). It is seen that the amplitude $e_+$ tends to $\mid E \mid$ at large $\zeta$ and the phase of the periodic wave undergo a shift.

So the obtained solutions describe processes of fission and fusion of radiation pulses that are resonant with different transitions. It shows the existence of pulses with different propagation velocities over the periodic wave background. The propagation of SIT pulses at arbitrary level population also demonstrate processes of frequency transformation. The technique developed here may be applied for many-level systems in the cases when the corresponding MB equations are integrable. The studying of more complicated reductions of the ZS problem potentials seems to be interesting looking for more weak constraints on quantum systems parameters. Another possibility for applications is investigation of the conditions of pulse propagation in thin films as well as reflection from them.

## 4.5   Nonreduced MB Equation Integrability

Here we develop approaches to a derivation [26] and solution [11] of a nonlinear evolution equation for density matrix $\rho$. We construct a class of integrable models and follow the restrictions that appear. The class of models formally is described by the Liouville - von Neumann (LvN) equation

$$\imath\rho_t = [H(\rho), \rho], \tag{51}$$

where $H(\rho)$ is a Hamiltonian operator. We are in a position when a Hamiltonian H of a subsystem depends on its density matrix. The dependence appear after application of some excluding procedure of variables of the rest part of some isolated system. A good example of such procedure is the Hartree-Fock method of quantum mechanics where we consider the Schrödinger equation for a one-particle wave function or density matrix with a Hamiltonian operator as a functional of other particles states. The others, in turn, are in a functional dependence of this "first" one. Similar theory of statistical physics is based on Vlasov equation.

Another important aspect of the construction relates to the so-called Landau - Ginzburg (or $\phi^4$) model: truncated Taylor expansion approximation for the Hamiltonian as a function of the field $\phi$ establishes a form for the effective Hamiltonian. The expansion we do is based on Frechet derivative (FD) notion and allows to extract linear (or multi-linear) in $\rho'$ terms of $H(\rho_0 + \rho')$. The self-conjugacy of H (it is Hermitian) is important: we evaluate left and right FD; the sum of linear in $\rho$ terms of the correspondent Taylor expansions should be Hermitian. The practical formula of Gâtaux may be used

$$DH(\rho_0, \rho) = \frac{\partial}{\partial t} H(\rho_0 + t\rho)|_{t=0} = H'_G(\rho_0)\rho, \tag{52}$$

if the result of differentiating is linear in $\rho$ and the limit exists in an operator norm.

Next we could fit the terms of equal powers in $\rho$ of expansion to integrable Hamiltonians reproduced below. Consider the following class of nonlinear LvN equations of the form given by Eq.(51)

$$i\dot{\rho} = [H(\rho), \rho] = \sum_{k=0}^{n}[A^{n-k}\rho A^k, \rho] = \sum_{k=0}^{n}[A^{n-k}, \rho A^k \rho] \tag{53}$$

where $A$ is a time-independent self-adjoint operator, the dot denotes the time derivative. The class of equations (53) is introduced in [11] and generalizes the set of Euler top model equations [49]. Let us stress that the problem we consider is generically infinite-dimensional and $\rho$ is positive, bounded, trace-class and Hermitian.

These constraints are Darboux-invariant in a class of solutions, if the binary transformation is used [11]. The covariance allows to construct and investigate the solutions [11,13]. Some plots are presented in [12,15]. Here we examine examples of such Hamiltonians and new approach that may give solutions satisfying the mentioned conditions.

The problem may be considered in abstract way but for the sake of clearance we take a simplest 2 by 2 matrix formulation in this first presentation. Let us consider a matrix element of $H^I(\rho_0 + \rho')$ as an analytical function of matrix elements of density matrix $\rho'_{ik}$. In the vicinity of $\rho = \rho_0$ we introduce a Taylor expansion considering $\rho'$ as a disturbance. (Further we omit the prime by $\rho$).

$$H^I_{ik} = H^I_{ik}(\rho_0) + \sum_{mn} \frac{\partial H^I_{ik}}{\partial \rho_{mn}}\rho_{mn} + ...,$$

where the derivatives are understood as the FD ones and could be calculated via Gâteau formula (52).

For the same reason of simplicity all the density elements are taken as a simple (scalar) functions of $x, t$. It is natural to relate $H^I_{ik}(\rho_0)$ to the Hamiltonian operator $H^0$ of equilibrium state and go to the interaction representation. Hence the approximate expression of the generic model Hamiltonian operator is taken as a linear combination of the density matrix elements, e.g.

$$H^I_{ik} \equiv h_{ik} = \sum_{mn} p^{ik}_{mn} \rho_{mn} + ...,$$

so that

$$p^{ik}_{mn} = \frac{\partial h_{ik}}{\partial \rho_{mn}}.$$

From the scope of the already mentioned simplest integrable model of [11] we have to link the model approximation with the Hermitian combination

$$H(\rho) = S\rho + \rho S \tag{54}$$

where $S^+ = S$. This case corresponds to $n = 1$ and $A = S$ from the Eq.(53), the nonlinear von Neumann equation becomes the Euler top type equation.

$$i\dot{\rho} = [S\rho + \rho S, \rho] = [S, \rho^2].$$

but with mentioned conditions for $\rho$ and $S$ being quantum operators.

The known integrable case however appear after the assumption about a waveform. Suppose that a wavetrain propagates in one direction and the simplified system corresponds to a reduced Maxwell-Bloch equation (rMB) we considered in previous Sects. 4.1,2. Within the formalism we cannot solve such problems as reflection from inhomogeneity or other one that include both directions of propagation. Below we start to develop an alternative approach to the problem that is based on ([11,12]) and Sect.4.6 technique that, perhaps, realize such a possibility.

Let us consider a medium that may be described in the two-level atoms approximation with energy levels $E_{a,b}$, wave functions $\psi_a$, $\psi_b$, and dipole approximation. Linear polarized electric field is directed along z axis

$$\mathbf{E} = E(x, t)\mathbf{k},$$

where t is the time. So the only component $d_z = d$ of polarization operator $\mathbf{d}$ is accounted, its matrix has the elements: $d_{11} = \int \psi^*_a d\psi_a dV = d_{22} = 0$,    $d^*_{21} = d_{12}$.

An evolution of a pulse in the semiclassical approach and coherent approximation is described by MB equations [9], (see also Sect.4.1) that may be represented in the dimensionless form, following the assumption that the total density matrix is $\rho_0 + \rho$, $\rho_0$ is an equilibrium one.

$$i\rho_t = [H(\rho_0 + \rho), \rho_0 + \rho] \tag{55}$$

where
$$H(\rho_0 + \rho) = H^0 + H^I(\rho),$$
the term $H_0$ does not depend on $\rho$. From the model expression for the interaction part $H^I(\rho) = (\mathbf{d}, \mathbf{E}(\rho))$ obviously depends on density matrix disturbance $\rho$. Let also hide $H_0$ going to interaction representation. The matrix $\rho$ is Hermitian and the complete $\rho_0 + \rho$ is normalized by the condition

$$Tr(\rho_0 + \rho) = 1 \tag{56}$$

or if $Tr(\rho_0) = 1$, the perturbation is traceless $Tr(\rho) = 0$. The Maxwell equations in this case of Eq.(55) and definition of $\mathbf{d}$ leads to the scalar D'Alambert one

$$c^2 E(x,t)_{xx} - E(x,t)_{tt} = 4\pi P_{tt} \tag{57}$$

where $P = N_0 Tr(\rho d_z) = N_0(d_{12}^* \rho_{12} + d_{12}\rho_{21})$, the $N_0$ is a number of resonant atoms. Or, taking into account the trace condition (56), and Hermicity, putting for real $q, r$ yields $\rho_{12} = q + \imath r$

$$P = 2N_0(d_1 q + d_2 r). \tag{58}$$

Here we put $d_{12} = d_1 + \imath d_2$. Finally, we substitute Eq.(58) into Eq.(57) and write a formal solution of the equation (57) with zero initial conditions:

$$E(x,t) = \int_0^t d\tau \int_{x-c(t-\tau)}^{x+c(t-\tau)} d\xi P_{\tau\tau}(\xi,\tau) = \hat{G}(d_1 q + d_2 r) \tag{59}$$

The integral operator $\hat{G}$ is hence defined by Eq.(59):

$$\hat{G} = 2N_0 \int_0^t d\tau \int_{x-c(t-\tau)}^{x+c(t-\tau)} d\xi \partial_\tau^2.$$

Substituting E(x,t) into the Hamiltonian of Eq.(55), one obtains

$$H^I = (d_1\sigma_1 + d_2\sigma_2)\hat{G}(d_1 q + d_2 r).$$

It is possible to proceed in the theory directly by means of the notion of Gâtaux derivative $H_G^I(0)'$ obtained by Eq.(52). We simply equalize the derivatives from both sides of Eq.(57), or

$$H^I = \{S, \rho\} = \{S, \sigma_1 q + \sigma_2 r\},$$

concluding in the system for the scalar (non-matrix) S:

$$(d_1^2\hat{G} - S)q - qS + d_1 d_2 \hat{G}r = 0,$$

$$(d_2^2\hat{G} - S)r - rS + d_1 d_2 \hat{G}q = 0$$

The DT technique that is developed in the [22,23] may be applied for a construction of solutions hierarchy for the RMB system with the interesting class of the self-scattered solutions [15]. The scheme for the case of the equation (55) is similar, with reductions as in [46].

## 4.6   The LvN Equation as Associated ZS Problem

The integration of the equation (53) in [11], [12] was made via U-V pair for some generalized equation with subsequent reduction restriction to stationary (by one of "times") solutions of the linear system. One more possibility may be recognized inside the following statement [26] that arise from operator version of dressing symmetry [50], that is generated by elementary Darboux transformations [22] and valid for operators S dependent on coordinates:

**Proposition 4.2** Let $\rho$ be a solution of the following operator equation;

$$i\rho_t = [U - \lambda J, \rho] \tag{60}$$

and let $\varphi$ and $\chi$ be solutions of a direct and conjugate Zakharov-Shabat equations.

$$(U - \mu J)\varphi = i\varphi_t, \tag{61}$$

$$\chi(U - \nu J) = -i\chi_t. \tag{62}$$

All entries belong to some operator (Banach) space B, and let an idempotent (projector) be $p \in B$.

If the scalar product analog $(\zeta, \varphi)_p = p\zeta\varphi p \in A_{pp}$ is invertible in the subspace determined by p and an idempotent P is introduced as before, but, perhaps, more general and convenient

$$P = \varphi(\chi, \varphi)_p^{-1}\chi.$$

Denoting $a = (\mu - \nu)(\nu - \lambda), b = -(\mu - \nu)(\mu - \lambda)$, one can state, that

$$\rho[1] = (1 + aP)\rho(1 + bP) \tag{63}$$

is the solution of the equation (60) but with the transformed operator $U[1] = U + (\mu - \nu)[P, J]$. Let us note [11] that the evolution of P is governed by the master equation

$$iP_t = (U - \mu J)P - P(U - \nu J) + (\mu - \nu)PJP. \tag{64}$$

The checking of the statements is straightforward by the use of Eq.(64).

Now go to the case $U = S\rho + \rho S$ and a constraint of covariance that appear from $U[1] = S\rho[1] + \rho[1]S$. The direct substitution yields

$$\{S, \nu\Delta P + \mu P\Delta - \lambda\Delta\} = (\lambda - \mu)(\nu - \lambda)[P, J]. \tag{65}$$

where $\Delta = [\rho, P]$, and $\{,\}$ denotes anticommutator. Either, in a more symmetrical form, introducing $\hat{S}$ such that $J = \hat{S}^2$ and taking into account the identity $[X, \hat{S}^2] = \{\hat{S}, [X, \hat{S}]\}$, the relation given by Eq.(65) goes to

$$\{S, (\mu - \nu)[P, \Delta]/2 + (\mu + \nu - 2\lambda)\Delta/2\} = (\lambda - \mu)(\nu - \lambda)\{\hat{S}, [P, \hat{S}]\}$$

Next natural idea is to put $S = \hat{S}$ that simply link the auxiliary operator $J$ and $S$ determined from equations of the previous section. We arrive at

$$q\{S, (\mu - \nu)[P, \Delta]/2 + (\mu + \nu - 2\lambda)\Delta/2 - (\lambda - \mu)(\nu - \lambda)[P, S]\} = 0$$

If, suppose, there exists some operator $G$ such that $\{S, G\} = 0$, then

$$[P, (\mu - \nu)\Delta/2 + (\mu + \nu - 2\lambda)\rho/2 - (\lambda - \mu)(\nu - \lambda)S] = G, \qquad (66)$$

that should be read as reduction constraint equation. The construction is the following. You have the expression for the $\rho[1]$ (63) via solutions of two ZS problems (61,62). The conditions of reduction (66) complicate the problem. The following possibility exist (66) may be transformed as

$$[P, P((\nu - \lambda)\rho - \kappa S) + ((\mu - \lambda)\rho - \kappa S)P] = G,$$

where the parameter $\kappa = (\lambda - \mu)(\nu - \lambda)$. The equation becomes an identity if

$$z_1\phi = (\rho + (\nu - \lambda)S)\phi)$$

$$z_2\chi = \chi(\rho + (\mu - \lambda)S)$$

so if the solutions of the ZS equations satisfy those new equations, the DT is hereditary. The reduction constraint is preserved.

## 5    Nonresonant Propagation (Kerr Effect): Manakov Solitons

### 5.1    Lax Pair and Periodic-Seed Solutions

A nonresonant polarized electromagnetic waves propagation in a medium with nonlinear dependence of refraction index on the electric field may be approximately described by the coupled nonlinear Schrödinger (CNS) equations as it was presented in the section 2.1. We would not consider the isolated NS equation separately but as the result of the natural limiting transition. The transition is almost obvious, one should put one of field component (e.g., $e_-$) equal to zero. So the integrable NS (Manakov) system is

$$\imath e_{-,\varsigma} = e_{-,\tau\tau}/2 + (|e_-|^2 + |e_+|^2)e_-$$

$$\imath e_{+,\varsigma} = e_{+,\tau\tau}/2 + (|e_-|^2 + |e_+|^2)e_+, \qquad (67)$$

where $e_-$ and $e_+$ are envelopes of left and right polarized waves. This system stands for processes of the two-dimensional stationary self-focusing and one-dimensional auto-modulation of electromagnetic waves with arbitrary polarization. In the following sections the analytical study of the systems (67) should be made. The observation of effects in connection with the NS system [27] gave rise to the growing interest for further experimental efforts and attempts to find new integrable cases of such systems [16,18]. The system (67) also may be written as the compatibility condition of the linear system

$$\psi_\varsigma + \lambda J\psi = U\psi \qquad (68)$$

$$\psi_\tau = (\lambda^2 A^{(2)} + \lambda A^{(1)} + A^{(0)})\psi \qquad (69)$$

with the diagonal $J = diag\{J_1, J_2, J_3\}$ or of the conjugate Lax pair as well

$$\xi_\zeta = -\xi(\lambda^2 A^{(2)} + \lambda A^{(1)} + A^{(0)})$$

$$\xi_\tau + \kappa\xi J = -\xi U$$

where

$$U = \begin{pmatrix} 0 & 0 & -e_-^* \\ 0 & 0 & -e_+^* \\ e_- & e_+ & 0 \end{pmatrix},$$

$$A^{(2)} = \imath J, \ A^{(1)} = -\imath U,$$

$$A^{(0)} = \imath \begin{pmatrix} \mid e_- \mid^2 & e_-^* e_+ & -e_{-,\zeta}^* \\ e_- e_+^* & \mid e_+ \mid^2 & -e_{+,\zeta}^* \\ -e_{-,\zeta} & -e_{+,\zeta} & -\mid e_- \mid^2 - \mid e_+ \mid^2 \end{pmatrix}$$

The potential matrix satisfy the reduction given above and matrices $A^{(i)}$ are chosen as follows

$$A^{(2)+} = -A^{(2)}, \quad A^{(1)+} = A^{(1)}, \quad A^{(0)+} = -A^{(0)}.$$

Therefore there exists the automorphism that is the direct analogue of the considered one (MB case).

$$\Psi(\lambda) \to [\Psi(-\lambda^*)^+]^{-1}$$

Hence there exist WFs that are coupled by the formula (46). The technique of DT with reductions described above may be applied to the case. The only difference is that in the formulas (47) and (48, 49, 50) the vector WF should be replaced by solutions of the problem (69).

By analogy one can built solutions of this NS (Manakov) system over the periodic background (the soliton case is considered at [21]). Let us choose the seed solution of the Eq.(67) as

$$e_- = 0, \qquad e_+ = E \exp(i(k\zeta + \omega\tau)) \tag{70}$$

Then from equations (67) it follows that

$$k = \omega^2 - \mid E \mid^2 .$$

Solving the system (69) one get expressions for WF

$$\begin{aligned} \psi_1 &= C_1 \exp(-\lambda\tau + i\lambda^2\zeta) \\ \psi_2 &= (C_+ \exp(\vartheta) + C_- \exp(-\vartheta)) \exp(-i(k\zeta + \omega\tau_0)/2) \\ \psi_3 &= -(C_+(\lambda + \sigma - ik/2)) \exp(\vartheta) + C_-(\lambda - \sigma - ik/2) \\ &\quad \times \exp(-\vartheta)) \exp(i(k\zeta + \omega\tau) \end{aligned} \tag{71}$$

where

$$\vartheta = \sigma(\tau + i(\lambda + ik/2)\zeta)$$

$$\sigma = ((\lambda - ik/2)^2 - \mid E \mid^2)^{-1/2}.$$

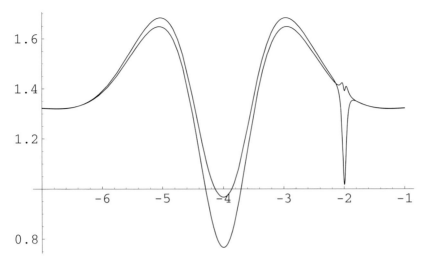

**Fig. 4.** The picture illustrates general two-soliton situation for both components $e_+, e_-$, the parameters are chosen in such way to demonstrate a possibility to describe interaction of TE and TM modes in the experiments of [25,7], when a big soliton in a component of electric field (laser beam) shifts (repulses) other beam. The small soliton is obviously suppressed in one of the components.

If you put here $\lambda = \mu$ and substitute the obtained formula for $\phi$ in the equation (49) you'll get the desired solution. Its difference from the solution of MB system is only in other dependence on the variable $\zeta$. The pictures and their description are placed below.

Let us consider the periodic-seed solution generated via Eq.(5). For a simplicity we choose the parameter $k = 0$, it means the reference frame that moves with the phase velocity of the background wave Eq.(70). We put also $s = \sqrt{2}$, $p = 1$, $C_1 = 1$, $C_+ = 1$, $C_- = 2$, $\mu = \imath$. The amplitude of the background wave $E$ is defined from the definition of $k = \omega^2 - |E|^2$.

## 5.2   Determinant Representation

The result of N-iterated DT for the ZS problem (68) we consider may be written in a determinant form by means of 2N seed eigen vector solutions $\phi^{(j)}$ and $\chi^i$ of the direct and conjugate ZS problems with the two sets of spectral values $\mu^{(i)}$ and $\mu_1^{(j)}$.

$$\psi_i[N] = \begin{vmatrix} \psi_i & \dfrac{(\chi^1,\psi)}{\lambda-\mu_1^{(1)}} & \cdots & \dfrac{(\chi^N,\psi)}{\lambda-\mu_1^{(N)}} \\ \phi_i^{(1)} & \dfrac{(\chi^1,\phi_i^{(1)})}{\mu^{(1)}-\mu_1^{(1)}} & \cdots & \dfrac{(\chi^N,\phi_i^{(1)})}{\mu^{(1)}-\mu_1^{(N)}} \\ \phi_i^{(N)} & \dfrac{(\chi^1,\phi_i^{(N)})}{\mu^{(N)}-\mu_1^{(N)}} & \cdots & \dfrac{(\chi^N,\phi_i^{(N)})}{\mu^{(N)}-\mu_1^{(N)}} \end{vmatrix} / \triangle[N],$$

$$\triangle[N] = \det[\frac{(\chi^i, \phi^{(j)})}{\mu^{(i)} - \mu_1^{(j)}}] \tag{72}$$

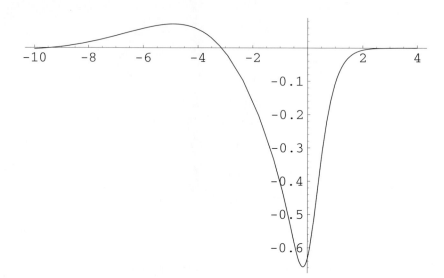

**Fig. 5.** The first component $e_-(0,t)$ of the solution over the periodic background

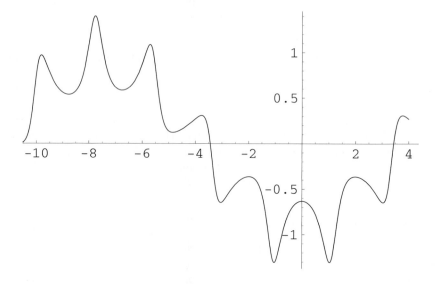

**Fig. 6.** The second component $e_+(y,t)$ of the solution over the periodic background

$$u_{ij}[N] = u_{ij} - (J_i - J_j)\sum_{p,s}(-1)^{p+s}\frac{\triangle_{ps}}{\triangle[N]}\chi_j^s\phi_i^{(p)} \tag{73}$$

where the determinants $\triangle_{ps}$ are obtained from the $\triangle[N]$ deleting of the p-th row and s-th column. The proof of the formulas may be delivered by the aid of

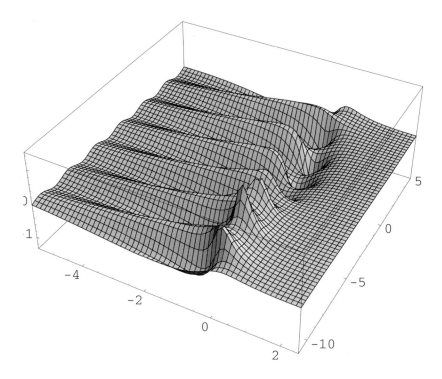

**Fig. 7.** On the last picture we show the surface of the $e_-(y,t)$ as the function of both variables.

the relation

$$\frac{(\phi[1],\psi[1])}{\lambda-\lambda_1}\frac{(\chi,\varphi)}{\mu-\mu_1} = \begin{vmatrix} \frac{(\phi,\psi)}{\lambda-\lambda_1} & \frac{(\chi,\psi)}{\lambda-\mu_1} \\ \frac{(\phi,\varphi)}{\mu-\lambda_1} & \frac{(\chi,\varphi)}{\mu-\mu_1} \end{vmatrix} \tag{74}$$

We have written the explicit formulas for iterated solutions of ZS equation (5.2) with additional information in Eq.(72) and the N-fold transforms for potentials Eq.(73) in general form for the binary DT. It is valid for Manakov system, MB system and N-wave system as well. Moreover as it had been mentioned the reduction for the first two systems (MB and Manakov) is the same [23]. Hence the difference is only in solutions of the seed equations of the corresponding Lax pair. The alternative (say - finite-gap) approach to the solution of the system, in particular, the Manakov system is developed in [53], where the solutions are expressed in terms of Kleinian hyper-elliptic functions.

## 5.3   Perturbations of Solutions

The general perturbation theory for soliton equation is under development. Rather promising way to understand the concept of the deformations inside integrable models leads to conservation laws via the Noether theorem [55]. Other important results account nonintegrable perturbations with and include such important effects as a dissipation and gain. For the NS equation it is made in [51], and for the Manakov equation - in [54]. The important results, including two-soliton solutions of Manakov equation is presented and illustrated in [54]. In the paper there are also numeric simulations of collisions of CLNS solitons (e.g. vector ones)including fission and fusion processes observed in [34]. A multi-soliton problem of a real sequence of pulses transmission with a weak interaction account was studied in [56]. The use of the soliton solutions may be sufficiently widen if some deformations of

    i) boundary (initial) conditions (see also Sect.3)

    ii) evolution equation perturbations may be accounted.

The interesting by itself direction of development of the theory unify MB and NS equations (NS-MB) [16] is generalized with account of inhomogeneous broadening of the resonance line [20]. In [18] some correction of the results is noted and multisolitons are studied in details.

## Acknowledgment

I would thank the organizers of OSTE-2002 conference for hospitality and the participants for discussions. I also acknowledge the priceless aid of M. Pietrzyk to find copies of recently published papers and R. Firla for the help in some plot production. The work was partially supported by the KBN Grant No. 2 P03B 163 15.

# References

1. H.A. Haus W.S. Wong: Rev. Mod. Phys. **68**, 423 (1996)
2. S. Turitsyn, N. Doran, J. Nijhof, V. Mezentsev T. Schäfer, W. Forysiak: Dispersion managed solitons. Les Houches Lectures, Lecture 7. p. 93, (Springer, Berlin 1999)
3. V. Cautaerts, Y. Kodama, A. Maruta, H. Sugavara: *Nonlinear Pulses in Ultra-Fast Communications*, (Les Houches Lectures, Lecture 9, p. 147, Springer, Berlin, 1999).
4. E. Seve, G. Millot S. Trillo: Phys. Rev. E **61**, 3139 (2000)
5. S. Leble: Computers Math. Applic., **35**, 73 (1998), S. Leble: Theor. J. Math. Phys, **22**, 239 (2000)
6. M. N. Islam: Opt. Lett. **15**, 417 (1990)
7. J.U. Kang, G.I. Stegeman, J.S. Atchison: Opt. Lett.**21**, 189 (1996)
8. S.V. Manakov: JETP, **65**, 505 (1973)
9. R.K.Boullough, P.M.Jack, P.W.Kitchenside and R.Saunders: Physica Scripta, **20**, 364 (1979)
10. H. Steudel, D.J. Kaup: J.Mod.Opt. **43**, 1851 (1996)
11. S.B. Leble, M. Czachor: *Darboux-integrable nonlinear Liouville-von Neumann equation* quant-ph/9804052, Phys. Rev. E **58** (1998)
12. N. Ustinov S. Leble M. Czachor M. Kuna *"Darboux-integration of $\imath\rho_t = [H, f(\rho)]$"* quant-ph/0005030, Phys. Lett. A. **279**, 333 (2001)
13. M Kuna, M Czachor S Leble: Phys. Lett. A, **255**, 42 (1999)
14. M. Czachor N. Ustinov *New Class of Integrable Nonlinear von Neumann-Type equations* arXiv:nlinSI/0011013 7Nov 2000.

15. M. Czachor, S.Leble M. Kuna and J.Naudts: 'Nonlinear von Neumann type equations'. In: *Trends in Quantum Mechanics" Proceedings of the International symposium* ed. H.-D. Doebner et al: (World Sci 2000) pp 209.
16. R.A. Vlasov, E.V. Doktorov. Covariant methods in theoretical physics, optics and acoustics. **94**, Minsk, (1981).
17. A.M. Basharov, A.I. Maimistov. JETP, **87**, 1595 (1984)
18. S. Kakei, J. Satsuma: J. Phys. Soc. Japan, **63**, 885 (1994)
19. S. Leble: *Nonlinear Waves in Waveguides*, (Springer-Verlag, Berlin, 1991)
20. A.I. Maimistov, E.A. Manykin: Sov. Phys. JETP, **58**, 685 (1983)
21. L.A. Bolshov, N.N. Elkin, V.V. Likhansky, M.I.Persiantsev. JETP, **94**, 101 (1988)
22. S.B. Leble and N.V. Ustinov: 'Solitons of Nonlinear Equations Associated with Degenerate Spectral Problem of the Third Order'. In *Nonlinear Theory and its Applications (NOLTA '93)*, eds. M. Tanaka and T. Saito (World Scientific, Singapore, 1993), pp 547-550.
23. S.B. Leble, N.V. Ustinov: On soliton and periodic solutions of Maxwell-Bloch system for two-level medium CSF **11**, 1763, (2000)
24. E.C. V. Malyshev J. Jarque: Opt. Soc. Am. B **12**, 1868 (1995); **14**, 1167 (1997).
25. R.K. Boullough, F.Ahmad: Phys. Rev. Lett. **27**, 330, (1971)
26. S. Leble 'Integrable models for Density Mathrix evolution'. In: *Proceedings of the workshop on Nonlinearity, Integrability and all that: Twenty Years after NEEDS'79* Ed. M.Boiti et al. World Sci 2000) p.311-317.
27. M.Chbat,C.Menyuk, I.Glesk, P.Prucnal: Opt. Lett. **20**, 258 (1995)
28. A Uthayakumar, K Porsezian, K Nakkeeran: Pure Appl. Opt. **7**, 1459 (1998)
29. Q-Han Park, H. J. Shin: Phys. Rev. E **59**, 2373 (1999).
30. J.U. Kang, G.I. Stegeman, J.S. Atchison, N. Akhmediev: PRL, **76**, 3699 (1996)
31. V.E. Zakharov, A.B Shabat: Sov. Phys. JETP, **34**, 62 (1972)
32. M. Bertolotti A. D'Andrea, E. Fazio et al: Opt. Comm. **168**, 399 (1999)
33. S.T. Cundiff,B.C. Collings, N.N. Achmediev, J.M. Soto-Crespo, K.Bergman, W.H. Knox: Phys. Rev. Lett. **2**, 3988 (1999)
34. C. Anastassiu, M. Segev, Steiglitz et al.: PRL, **83**, 2332 (1999)
35. H.G. Winful: Opt. Lett. **11**, 33 (1986)
36. K.J. Blow, N.J. Doran, D. Wood: Opt. Lett. **12**, 202 (1987)
37. C.Menyuk: IEEE J.Quantum Electron. **23**, 174 (1987)
38. S.P. Novikov S.V. Manakov L.P. Pitaevski V.E. Zakharov: *Theory of Solitons*, (Plenum, New York, 1984)
39. H. Steudel: In:'Proceedings of 3d International Workshop on Nonlinear Processes in Physics' Eds. V.G. Bar'yakhtar et al. (Kiev, Naukova Dumka, 1988) V.1 p.144.
40. ' Optical Solitons, Theoret. Challenges and Industr. Perspectives'; Les Houches Workshop, 1998, (Springer, Berlin, 1999.);Eds. V.E.Zakharov, S.Wabnitz,
41. V.B. Matveev, M. A. Salle:*Darboux transformations and solitons*, (Springer, Berlin, 1991)
42. H. Steudel: J. Mod. Opt. **35**, 693 (1988)
43. D.J. Kaup, J.Nonlinear Sci. **3**, 427 (1993)
44. I. Leonhardt, H. Steudel: Appl. Phys. B, **60**, S221 (1995)
45. J. Leon, A. Mikhailov: Phys. Lett. A, **253**, 33 (1999)
46. A.V. Mikhailov: Physica D **3**, 73 (1981)
47. V. Gerdjikov, A Yanovski J. Math. phys. **35**, 3687 (1994)
48. S. Leble, N.V. Ustinov: Inverse Problems, **N10**, 617 (1994)
49. S. Manakov: *Funktsional'nyi Analiz i Pril.* **10**, 93 (1976) M. Adler and P. van Moerbeke: Adv. Math. **38**, 267 (1980)

50. V. E. Zaharov, A. B. Šabat,, Funk. Anal. i Prilozh., **13**, 13 (1979)
51. D.J. Kaup: Phys. Rev. A, **42**, 5689 (1990)
52. T.L. Lakoba D.J. Kaup: Phys. Rev. E, **56**, 6147 (1997)
53. P.L. Christiansen, J.C. Eilbeck, V.Z. Enolski and N.A. Kostov: Proc. R. Soc. Lond. A, **456**, 2263 (2000)
54. J. Yang: Phys. Rev.E, **59**, 2393 (1999)
55. Steudel: H. Annalen der Physik (Leipzig) **32**, 205 (1975)
56. V. Gerdjikov et al: Phys. Lett. A **241**, 323 (1998)

# Solitons Around Us:
# Integrable, Hamiltonian and Dissipative Systems

N.N. Akhmediev and A. Ankiewicz

Optical Sciences Centre, Research School of Physical Sciences and Engineering, Institute of Advanced Studies, Australian National University, Canberra, ACT 0200, Australia.

**Abstract.** In recent years, the notion of solitons has been extended to various systems which are not necessarily integrable. We extend the notion of solitons and include a wider range of systems in our treatment. These include dissipative systems, Hamiltonian systems and a particular case of them, viz. integrable systems. We use the broad definition of solitons and give some examples to support this claim of ubiquity.

## 1   Introduction

A soliton is a concept which describes various physical phenomena ranging from solitary waves on a water surface to ultra-short optical pulses in an optical fiber. The main feature of solitons is that they can propagate long distances without visible changes. From a mathematical point of view, a soliton is a localized solution of a partial differential equation describing the evolution of a nonlinear system with an infinite number of degrees of freedom. Solitons are usually attributed to integrable systems. In this instance, solitons remain unchanged during interactions, apart from a phase shift. They can be viewed as 'modes' of the system, and, along with radiation modes, they can be used to solve initial-value problems using a nonlinear superposition of the modes [1]. However, in recent years, the notion of solitons has been extended to various systems which are not necessarily integrable. In fact, A. Boardman noted in [2] that it is "apparent that solitons are absolutely everywhere". Following the modern trend, we extend the notion of solitons and include a wider range of systems in our treatment. These include dissipative systems, Hamiltonian systems and a particular case of them, *viz.* integrable systems. We will use the broad definition of solitons and give some examples to support this claim of ubiquity.

Let us consider a trivial example from classical mechanics - a system with one degree of freedom, namely a pendulum [3]. A transition to a higher number of degrees of freedom could be made by taking several identical coupled pendula. Then we can make the following classification (see Fig.1). When the amplitude of the oscillations is small, the system can be approximated by a linear oscillator (Fig.1a). If we had several coupled oscillators, the general solution could be written as a linear superposition of normal modes. If the amplitude of the oscillations is not small, then the oscillations are nonlinear (Fig.1b). The exact analytic solution for this case does exist and it can be written in terms of elliptic Jacobi functions. However, for a coupled set of equations, the solution cannot be written as a linear superposition of modes. Finally, when losses are included (Fig.1c), the system becomes dissipative. The oscillations are undamped only if there is an external force pumping energy into the pendulum. For a system of coupled equations, the solution can only be found numerically in most cases.

A similar classification can be made in the case of systems with an infinite number of degrees of freedom. To be definite, we will mainly (but not exclusively) consider an equation which is widely-known as the complex cubic-quintic Ginzburg-Landau equation (CGLE):

$$i\psi_\xi + \frac{D}{2}\psi_{\tau\tau} + |\psi|^2\psi = i\delta\psi + i\epsilon|\psi|^2\psi + i\beta\psi_{\tau\tau} + i\mu|\psi|^4\psi - \nu|\psi|^4\psi, \quad (1)$$

The meaning of the various terms in this equation in photonics is as follows: $\xi$ is a normalized distance or number of trips around a laser cavity, $\tau$ is the retarded time, $\psi$ is the normalized envelope of the field, $D$ is the group velocity dispersion coefficient, $\delta$ is the linear gain coefficient (if negative, it represents loss), $i\beta\psi_{tt}$ accounts for spectral filtering, $\epsilon|\psi|^2\psi$ represents the nonlinear gain (which can arise from saturable absorption), the term with $\mu$ represents, if negative, the saturation of the nonlinear gain, and the one with $\nu$ corresponds to the nonlinear refractive index's deviation from the Kerr law.

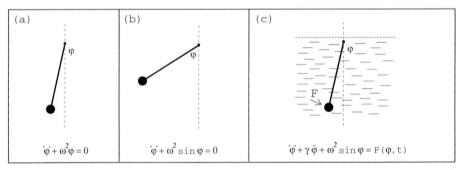

**Fig. 1.** A classification of a dynamical system with one degree of freedom (a pendulum) where $\varphi$ is the angle from the vertical and $\omega$ is a constant. (a) The amplitude is small (linear oscillations). (b) The amplitude is not small (nonlinear oscillations). The system is Hamiltonian. (c) Dissipative system. The pendulum is in a liquid and $\gamma$ is a constant. To maintain the oscillations, the pendulum needs to have an external periodic force (F).

By equating various terms on the right-hand-side of Eq.(1) to zero, we can have an equation which describes, as particular cases, integrable, Hamiltonian or dissipative systems. In particular, the system is Hamiltonian when $\delta = \epsilon = \beta = \mu = 0$. The system is integrable, when, in addition, $\nu = 0$. Then it follows that Hamiltonian systems can be considered as a subclass of dissipative ones, while integrable systems can be viewed as a subclass of Hamiltonian ones. This classification is illustrated in Fig.2.

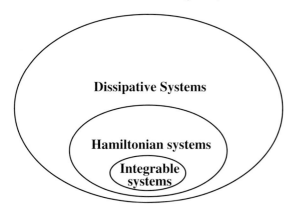

**Fig. 2.** A rough classification of nonlinear systems with an infinite number of degrees of freedom admitting soliton solutions. Hamiltonian systems can be considered as a subclass of dissipative ones, while integrable systems can be viewed as a subclass of Hamiltonian ones.

## 2   Integrable Systems

Let us start our analysis from the integrable case. The classical nonlinear equation found to be integrable [1] is the nonlinear Schrödinger equation (NLSE):

$$i\psi_\xi + \frac{1}{2}\psi_{\tau\tau} + |\psi|^2\psi = 0, \tag{2}$$

This is the standard way of writing this equation in the mathematical literature. The actual meaning of the independent variables $\xi$ and $\tau$ differs for different problems. For spatial solitons, both variables are spatial, so no confusion can arise. For problems related to dispersive waves, and, in particular, to optical fibers, $\xi$ is the distance along the fiber, and $\tau$ is the retarded time, i.e. a coordinate moving with the group velocity of the pulse. We will use this notation throughout this paper, bearing in mind that the solutions relate to various physically different situations. This form of the NLSE is for anomalous dispersion in the temporal (fiber) case and for self-focusing in the spatial case. In this section, we shall briefly describe methods for solving the NLSE and linear equations related to it. Clearly, we cannot cover the whole body of knowledge which exists today, and we refer the reader to existing books [4–8] for details.

The NLSE admits an infinite number of exact solutions. A complicated solution can be represented as a nonlinear superposition of simple ones. The structure of the superposition is defined by the spectrum of the inverse scattering technique (IST). One of the main points of interest is that Eq.(2) admits the soliton solution

$$\psi = \frac{e^{i\xi/2}}{\cosh\tau}. \tag{3}$$

This simple solution can be extended to include more parameters by using the symmetries of the NLSE. For example, if we know a solution of the NLSE, $\psi(\tau,\xi)$, then we can obtain a one-parameter family of solutions by the simple transformation

$$\psi'(\tau,\xi|q) = q\ \psi(q\tau, q^2\xi), \tag{4}$$

where $q$ is a real parameter. This transformation can be applied to an arbitrary solution of the NLSE. We can use it to extend the one-soliton solution, (3), from a fixed solution to a one-parameter family of solutions:

$$\psi'(\tau, \xi | q) = \frac{q}{\cosh\ q\tau}\ e^{i\omega\xi} \tag{5}$$

where $\omega = q^2/2$. Clearly, the parameter $q$ gives the amplitude of the soliton. It also determines the width of the soliton and its period in $\xi$.

This solution can be explained physically as follows. The dispersion (diffraction), described by the 2nd order derivative term in Eq.(2), causes the optical pulse (or beam) to spread out, while the self-focusing nonlinear term causes it to become more narrow. An arbitrary input pulse adjusts itself in such a way that these two effects balance. In fact, the sech shape achieves this. In the frequency domain, the self-phase modulation is proportional to the time derivative of the intensity. The group velocity dispersion produces a chirp of the same overall shape ( sech $\tau$ tanh $\tau$) but opposite sign and so, for the exact soliton, the two cancel and the soliton has no chirp across it. It can be viewed as an example of static equilibrium.

**Integrability**. Solutions of PDEs like Eq.(1) usually cannot be written in analytical form. Only in very special cases can the solution be written analytically for an arbitrary smooth initial condition. However, the NLSE is an example of such a rare PDE.

The inverse scattering technique is the main tool for solving initial-value problems related to integrable equations, including the NLSE. For the NLSE, the method has been developed by Zakharov and Shabat [1]. The inverse scattering technique is based upon the fact that the NLSE can be represented in the form of a compatibility condition between the linear equations of the following set:

$$\mathbf{R}_\tau = -\mathbf{JR\Lambda} + \mathbf{UR},$$

$$\mathbf{R}_\xi = -\mathbf{JR\Lambda}^2 + \mathbf{UR\Lambda} - \tfrac{1}{2}(\mathbf{JU}^2 - \mathbf{JU}_\tau)\mathbf{R}, \tag{6}$$

where $\mathbf{R}, \mathbf{J}$ and $\mathbf{U}$ are the following matrices:

$$\mathbf{R} = \begin{bmatrix} r_{11} & r_{12} \\ r_{21} & r_{22} \end{bmatrix}, \quad \mathbf{U} = \begin{bmatrix} 0 & \psi \\ \phi & 0 \end{bmatrix}, \quad \mathbf{J} = \begin{bmatrix} i & 0 \\ 0 & -i \end{bmatrix}, \tag{7}$$

and $\mathbf{\Lambda} = \begin{bmatrix} \lambda_1 & 0 \\ 0 & \lambda_2 \end{bmatrix}$ is an arbitrary diagonal complex matrix. The compatibility condition for these differential equations in $\mathbf{R}$, with non-constant coefficients depending on $\psi(\tau, \xi)$ and $\phi(\tau, \xi)$, is the equation

$$\mathbf{U}_\xi - \frac{1}{2}\mathbf{JU}_{\tau\tau} + \mathbf{JU}^3 = 0. \tag{8}$$

The point is that there is a one-to-one correspondence between the solutions of the linear system and the solutions of the NLSE. This means that if the solution of the linear system is a superposition of linear modes, then the solution of the NLSE is a nonlinear superposition of nonlinear modes (i.e. solitons and radiation waves). Any solution of the NLSE can be found (after some manipulations) using this correspondence. This is discussed further in [1,5,6,9]. Here we only mention that, for a dynamical system to be integrable, the system must have as many conserved quantities as degrees of freedom. The NLSE is a dynamical system with an infinite number of degrees of freedom, and so, to be integrable, it has to have an infinite number of conserved quantities (integrals). Their existence has indeed been demonstrated in [1].

Physically, integrability means that soliton solutions for such systems behave in a special way which is different from the behavior of solutions for other systems. Every localized solution is a nonlinear superposition of a finite number of solitons and a continuum of radiation waves. In special cases, radiation waves can be absent and then we have a purely solitonic solution. If there are several solitons, they have zero 'binding' energy. Solitons also have zero binding energy with radiation waves. Collisions of solitons are elastic. This means that, after a collision, they continue to propagate in the same direction as before the collision. The number of solitons is conserved. This does not happen in non-integrable systems.

**Coupled systems**. Another integrable system is described by a set of $N$ coupled nonlinear Schrödinger equations (NLSEs). In some special cases, these equations are found to be integrable [6,10–12]. Then, in analogy with the single (scalar) NLSE [1] (where the number of equations, $N$, is 1) and the Manakov case [13] ($N = 2$), the total solution consists of a finite number ($N$) of solitons plus small amplitude radiation waves. The former is defined by the discrete spectrum of linear ($L, A$) operators [1,13], while the latter are defined by the continuous spectrum. Most applications deal with the soliton part of the solution, since it contains the most important features of the problem.

The number of solitons can be arbitrary. Additionally, we assume that the components have independent phases. When the phases are independent, the soliton solution is a multi-parameter family. It can be called a 'multi-soliton complex' (MSC). The notion of the MSC can be applied to various physical problems [14]. These include an important recent development: incoherent solitons [15–18,20].

The set of equations describing the propagation of $N$ self-trapped, mutually-incoherent wave packets in a medium with a Kerr-like nonlinearity is

$$i\frac{\partial \psi_i}{\partial \xi} + \frac{1}{2}\frac{\partial^2 \psi_i}{\partial \tau^2} + \alpha \; \delta n(I)\psi_i = 0, \qquad (9)$$

where $\psi_i$ denotes the i-th component of the beam, $\alpha$ is a coefficient representing the strength of nonlinearity, $\tau$ is the transverse co-ordinate (a moving time frame in fibers), $\xi$ is the co-ordinate along the direction of propagation, and $\delta n(I) = f\left(\sum_{i=1}^{N} |\psi_i|^2\right)$ is the change in refractive index profile created by all the incoherent components of the light beam. The response time of the nonlinearity is assumed to be long compared to temporal variations of the mutual phases of all the components, so the medium responds to the average light intensity, and this is just a simple sum of modal intensities.

Interestingly enough, the set of equations (9) has $N$ quantities $Q_i = \int_{-\infty}^{\infty} |\psi_i|^2 \, d\tau$ which are conserved separately from the conservation of the total energy $Q = \sum_{i=1}^{N} Q_i$. This occurs because there is no energy transfer mechanism between the components. In fact, this is the main difference from the phase-dependent components case, where only the total energy is conserved.

If the function $\delta n(y) = y$, then the set of equations (9) is a generalized Manakov set, which has been shown to be integrable [12]. This means that all solutions, in principle, can be written in analytical form [17,19].

## 3    Hamiltonian Systems

Reductions to integrable systems are extreme simplifications of the complex systems existing in nature. They can be considered as a subclass of the more general Hamiltonian systems (see Fig.2). Indeed, such a simplification allows us to analyze the systems quantitatively and to completely understand the behavior of the solitons. Solitons in Hamiltonian (but nonintegrable) systems can also be regarded as nonlinear modes, but in the sense that they allow us to describe the behavior of systems with an infinite number of degrees of freedom in terms of a few variables, thus allowing us effectively to reduce the number of degrees of freedom.

When the nonlinear term in the modified NLSE differs from that of the Kerr law, the resulting equation is no longer integrable. Instead of an infinite number of conservation laws, the equation now has only a few. Although stationary solitons exist, and some solutions can be written in analytic form, their behavior differs from that of solutions of the NLSE. As a result, the behavior of the whole system with arbitrary initial conditions becomes different. In particular:

1. In contrast to solitons of the NLSE, soliton-like solutions can be stable or unstable, depending on the parameter of the family.

2. A collision between two stable solitons is inelastic, in contrast to the collision between solitons of the NLSE. The amplitudes of the output pulses can differ from the amplitudes of the input solitons. There is a loss of energy from the soliton component of the solution.

3. The number of solitons after the collision can be different from the number of solitons before the collision.

4. Part of the energy of the colliding solitons can be transformed into radiation.

5. Superposition states of two solitons (e.g. pulsating solutions like breathers) do not exist.

Non-integrability is not necessarily related to the nonlinear term. Higher - order dispersion or birefringence, for example, also make the system non -

integrable, while it remains Hamiltonian. However, terms associated with higher-order dispersion or birefringence introduce other physical effects to the behavior of one-soliton solutions. If the part of the equation which corresponds to the deviation can be considered small, then some features of integrable systems can still be observed. However, the general approach should be an extension of the one used in the theory of dynamical systems with a finite number of degrees of freedom. Thus, as a first step, we should find stationary solutions (solitons); secondly, we should study their stability, and, as a third step, we can study the dynamics of solutions in the vicinity of stationary solutions. The third step usually requires numerical simulations. Examples of applications of this approach have been presented in several works [21,22]. Soliton solutions can be found relatively easily, either analytically or numerically. Their stability has also been analyzed, and the fact that the system is Hamiltonian allows us to apply some theorems.

The Hamiltonian ($H$) is one of the fundamental ideas in mechanics [3], and, more specifically, in the theory of conservative dynamical systems with a finite (or even infinite) number of degrees of freedom. The Hamiltonian formalism has become one of the most universal in the theory of integrable systems [24] and nonlinear waves in general [23]. If the Hamiltonian $H$ exists, we can present evolution equations in a canonical form:

$$i\psi_\xi = \frac{\delta H}{\delta \psi^*}, \qquad i\psi_\xi^* = -\frac{\delta H}{\delta \psi}. \qquad (10)$$

where the right hand sides are written as Frechet derivatives of the Hamiltonian relative to function (or functions) defining the evolution of the system. In the case of non-integrable systems, the Hamiltonian exists whenever the system is conservative, and it is useful for stability analysis [25,26]. It turns out that the most useful approach in the soliton theory of conservative non-integrable Hamiltonian systems is a representation on the plane of conserved quantities, *viz.* Hamiltonian-versus-energy [9]. A three-dimensional plot (Hamiltonian - energy - momentum) is useful when dealing with two-parameter families of solutions [27].

As in the integrable case, the solitons still form a one- (or few-) parameter family of solutions. Recently, Hamiltonian-versus-energy curves have been used effectively to study families of solitons and their properties, *viz.* range of existence, stability and general dynamics. Specific problems considered up to now include scalar solitons in non-Kerr media [9], vector solitons in birefringent waveguides [28], radiation phenomena from unstable soliton branches [29], optical couplers [30], general principles of coupled nonlinear Schrödinger equations [31,32], parametric solitons in quadratic media [33] and the theory of Bose-Einstein condensates [34]. Moreover, Hamiltonian-versus-energy curves are useful not only for studying single soliton solutions, but also for analyzing the stability of bound states (when they exist) [35]. Other examples could be mentioned as well.

In most publications, soliton families have been studied using plots of energy versus propagation constant. These curves allow the soliton families to be presented graphically and, moreover, allow predictions of their stability properties. We believe that the first example of their application was presented in [36]. Kusmartsev [37] was the first person to understand the importance of projecting curves on the plane of conserved quantities. He applied catastrophe theory and a mapping technique to represent soliton families with diagrams and to show that the critical points on these diagrams define the bifurcations where the soliton stability changes. In [38], a direct approach to analyze the $H(Q)$ soliton curves has been presented and, additionally, the concept has been enhanced with a stability theorem. This theorem turns the employment of $H(Q)$ curves into a powerful tool for analyzing soliton solutions, their stability and their dynamics. In particular, a theorem which relates the concavity of the $H$ - $Q$ curves to the stability of the solitons has been proved. The main advantages of this approach are its simplicity, clarity and the fact that it provides the possibility of predicting simple dynamics of evolution for solitons on unstable branches.

For simplicity, let us consider scalar wave fields $\psi(t, \xi)$. The nonlinear Schrödinger equation (NLSE) for a general nonlinearity law is [9,25,26]:

$$i\psi_\xi + \frac{1}{2}\psi_{\tau\tau} + N(|\psi|^2)\psi = 0, \tag{11}$$

where $N$ is the nonlinearity law. It indicates that the change in refractive index depends on the local intensity. Localized solutions satisfy the ansatz

$$\psi(\tau, \xi) = f(\tau) \; exp \, (iq\xi) \tag{12}$$

where $f(\tau)$ is a real field profile, and $q$ is the propagation constant.

The total energy associated with an arbitrary solution, $\psi(t, \xi)$, is $Q = \int\limits_{-\infty}^{\infty} I \, d\tau$, where the intensity is $I = |\psi|^2 = f^2$. In spatial problems, $Q$ is the power or power flow. In problems related to pulse propagation in optical fibers, where $t$ is regarded as a retarded time, $Q$ is the total pulse energy. For localized solutions (Eq.(12)), $Q$ is finite and it is one of the conserved quantities of Eq.(11).

Similarly, the Hamiltonian is another conserved quantity:

$$H = \int\limits_{-\infty}^{\infty} \left[ \frac{1}{2}f_\tau^2 - F(I) \right] d\tau, \tag{13}$$

with $F$ given by $F(I) = \int\limits_{0}^{I} N(I') \, dI'$. The Hamiltonian plays a major role in the dynamics of the infinite-dimensional system. In fact, stationary solutions of equation (9) can be derived from the Hamiltonian using the variational principle $\delta H = 0$.

Now, substituting Eq.(12) into Eq.(11) and integrating once, we have $f_\tau^2 = 2(qI - F)$ or

$$H = qQ - 2K, \tag{14}$$

where $K = \int\limits_{-\infty}^{\infty} F(I)\,d\tau$. This expression can be used to calculate $H$ versus $Q$ curves explicitly. It is easy to show that in the case of a Kerr medium $H(Q) = -\frac{Q^3}{24}$.

One of the advantages of using $H - Q$ curves is that they can predict the stability of solitons. It is apparent that, if there is more than one branch at a given $Q$, then the lowest branch (i.e. the one with the minimum Hamiltonian) is stable. This conclusion follows directly from the nature of the Hamiltonian and does not need a separate proof. However, the stability condition can take a more direct form. There is a useful theorem in this regard. For solitons in media with local nonlinearities, we have,

$$\frac{dH}{dq} = -q\frac{dQ}{dq}. \tag{15}$$

Then it follows that

$$\frac{dH}{dQ} = -q. \tag{16}$$

If we start at $q = 0$ and traverse the curve so that $q$ is increasing, then the magnitude of the slope always increases. Furthermore,

$$\frac{d^2H}{dQ^2} = \frac{-1}{dQ/dq}. \tag{17}$$

The denominator on the right-hand-side defines the stability of the lowest-order modes (fundamental solitons) [25,38-42]. Hence, the stability is directly related to the concavity of the $H$ versus $Q$ curve. Specifically, the solitons with $H''(Q) < 0$ are stable while those with $H''(Q) > 0$ are unstable.

Another consequence of Eq.(15) is that

$$\frac{dH}{dq} = 0 \;\Rightarrow\; \frac{dQ}{dq} = 0 \text{ or } q = 0. \tag{18}$$

Thus if $Q$ has a stationary point, then so does $H$. For $q > 0$, this produces a cusp on the $H - Q$ diagram. However, we can have $dH/dq = 0$ with $q = 0$ and $dQ/dq \neq 0$. This produces a rounded maximum on the $H$ vs. $Q$ plot and not a cusp.

Clearly, from Eq.(15), if we have $q > 0$, then $H$ decreases as $Q$ increases, meaning that $\frac{dH}{dQ} < 0$. On the other hand, if $q < 0$ is allowable, then $H$ and $Q$ have the same slope, so that $\frac{dH}{dQ} > 0$.

Thus, we can conclude that, for the lowest order modes:

1. **Solitons with $H''(Q) < 0$ are stable while those with $H''(Q) > 0$ are unstable.**

2. **Stability changes only at cusps.**

This criterion for stability can be more general than $dQ/dq > 0$, because it involves only conserved quantities which always exist in conservative systems; this is in contrast to $q$, which may not be defined uniquely.

*Example: Dual power law nonlinearity.* This nonlinearity is given by $N = I^b + \nu I^{2b}$. When $\nu$ is positive, the refractive index increases monotonically with $I$. We consider $\nu < 0$, so that the dependence of $N$ on $I$ is not monotonic and we can expect qualitatively new effects.

For $b > 2$, $Q$ has a minimum and $H$ has a maximum at $q > 0$, thus producing a cusp in the $H$ versus $Q$ plot (see Figure 3). Solitons exist only above some threshold energy in this case. The important conclusion from this case is that the upper branch should be unstable, because the Hamiltonian is concave upwards while the lower branch should be stable as it is concave downwards. Numerical simulations show that this is indeed the case.

The exact soliton profile is given by

$$f(\tau) = \left[ \frac{1}{2q(1+b)} \left[ 1 + \sqrt{1 + \frac{4q\nu}{(1+2b)}(1+b)^2} \ \cosh\left(2b\sqrt{2q}\,\tau\right) \right] \right]^{-\frac{1}{2b}}. \tag{19}$$

To illustrate further the usefulness of the $H(Q)$ diagrams in predicting dynamics, let us consider a simple example. In Fig.3, the upper unstable branch of solitons corresponds to the range $0 < q < q_c$. The lower stable branch corresponds to the interval $q_c < q < q_{max} = 6/49$. The cusp appears at $q = q_c = 0.0492$. An example of propagation is shown in Fig.4(a). It shows the instability of the upper branch.

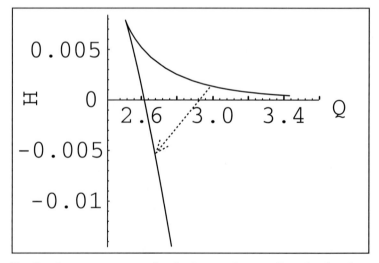

**Fig. 3.** Hamiltonian-versus-energy for dual-power-law nonlinearity for the values of the parameters $b = 5/2$ and $\nu = -1$. The dotted arrow shows a transformation which occurs from the unstable branch to the stable one, due to the soliton's interaction with radiation. The cusp occurs at $q = q_c = 0.0492$ and corresponds to the soliton's minimum energy of $Q = 2.51$ and maximum Hamiltonian, viz. $H = 0.00783$.

Numerical simulations start with the exact solution, Eq.(19), as the initial condition, and take $q = 0.005$, which corresponds to $Q = 2.936$. This soliton is unstable, and due to interaction with radiation, it evolves into a soliton of the stable branch. The initial and the final soliton profiles are shown in Fig.4. The final state, after the radiation waves have dispersed, is a soliton with parameters $q = 0.094$ and $Q = 2.69$. The course of the above transformation is clearly seen in Fig.4. It is represented by the dotted arrow in Fig.3. A physically similar process has been considered in [29] for solitons in birefringent fibers. As a general rule, this analysis shows that the transformation always takes place from an upper right point on the $H(Q)$ diagram to a lower left point on the diagram. Hence the direction of the arrow in Fig.3 must be down and to the left. Other examples are presented in [38].

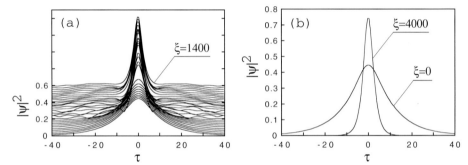

**Fig. 4.** (a) Evolution of an unstable soliton. The result of this evolution is shown schematically by the arrow in Fig.3. (b) Initial ($\xi = 0$) and final ($\xi > 1400$) soliton profiles. Initially the stationary soliton solution ($q = 0.005$) is unstable, but it evolves into a soliton on the stable branch while emitting small amplitude radiation waves (note ripples in (a)).

The instability eigenvalues of the linearized equations for the upper soliton branch must be complex, as they have real parts which correspond to the deviation from the unstable soliton and imaginary parts which correspond to interactions with radiation. Complex eigenvalues have been proved to exist for Hamiltonian systems in [43–46].

This approach can be generalized to include more complicated Hamiltonian nonlinear systems, including cases with two [28,30] or more coupled NLSEs [47–49], parametric solitons [50,51] and examples of higher-order dimensionality [52]. For example, the curves $H(Q)$ calculated numerically in [32] show clearly that our stability criterion can be applied to a system of coupled NLSEs. The results obtained in [34] also show that this principle can be generalized to the case of (1+3)-D solitons. It is quite obvious, then, that (1+2)-D cases and spatio-temporal solitons ((1+3)-D) [53–55] could also be handled with our approach. This means that, independent of their physical nature, single-soliton solutions of Hamiltonian systems can be well-understood and analyzed using the concavity of the $H(Q)$ curves.

## 4    Dissipative Systems

The next level of generalization is to consider solitons in dissipative systems (see Fig.2). The main feature of these systems is that they include energy exchange with external sources. These are no longer Hamiltonian and the solitons in these systems are also qualitatively different from those in Hamiltonian systems. In Hamiltonian systems, soliton solutions appear as a result of a balance between diffraction (dispersion) and nonlinearity. Diffraction spreads a beam while nonlinearity will focus it and make it narrower. The balance between the two results in a stationary solution, which is usually a one-parameter family. In systems with gain and loss, in order to have stationary solutions, gain and loss must be also balanced. This additional balance results in solutions which are fixed. Then the shape, amplitude and the width are all fixed and depend on parameters of the equation. This situation is presented schematically in Fig.5. However, the common feature is that solitons, when they exist, can again be considered as 'modes' of dissipative systems.

We will concentrate here on equation (1), the complex Ginzburg-Landau equation (CGLE). Many non-equilibrium phenomena, such as convection instabilities [66], binary fluid convection [67] and phase transitions [68], can be described by this equation. In optics, this equation (or a generalization of it) describes the essential features of processes in lasers [69–71], optical parametric oscillators [73], spatial soliton lasers [59,74], Fabry-Perot cavities filled with nonlinear material and driven by an external field [60–62] and all-optical transmission lines [63].

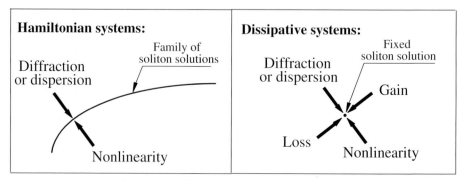

**Fig. 5.** Qualitative difference between the soliton solutions in Hamiltonian and dissipative systems. In Hamiltonian systems, soliton solutions are the result of a single balance, and comprise one- or few-parameter families, whereas, in dissipative systems, the soliton solutions are the result of a double balance and, in general, are isolated. There can be exceptions to this rule [56–58], but, usually, the solutions are fixed (i.e. isolated from each other). On the other hand, it is quite possible for several isolated soliton solutions to exist for the same equation parameters. This is valid for (1+1)-dimensional as well as for (2+1)-dimensional cases. In the latter case, the terms "localized structures" [59], "bullets" [60,61] or "patterns" [72] are also used along with the term "solitons" [62].

This equation is essentially the nonlinear Schrödinger equation (NLSE) with gain and loss, where both gain and loss are frequency- and intensity-dependent, and we allow for a non-Kerr nonlinearity. Equation (1) has been written in such a way that if the right-hand side of it is set to zero, we obtain the standard NLSE. The equation can also have additional terms related to a finite aperture [74] and other forms of local [59] or nonlocal [76,77] nonlinearities. Here we concentrate on the (1+1) dimensional case, as it is the fundamental one and it allows us to understand some of the features of solitons in (2+1) dimensional cases.

Another simple qualitative picture is presented in Fig.6a. In order to be stationary, solitons in dissipative systems need to have regions where they extract energy from an external source, as well as regions where energy is dissipated to the environment. A stationary soliton is the result of a dynamical process of continuous energy exchange with the environment and its redistribution between various parts of the pulse. Hence the soliton by itself is an object which is far from static equilibrium. We can say that it is in a state of dynamical equilibrium.

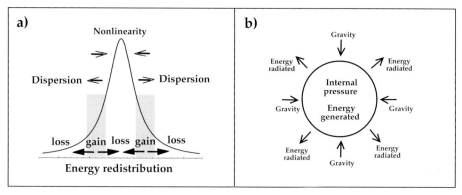

**Fig. 6.** (a) Qualitative description of solitons in dissipative systems. The soliton has areas of consumption as well as dissipation of energy which can be both frequency (spatial or temporal) and intensity dependent. Arrows show the energy flow across the soliton. The soliton is a result of a complicated dynamical process of energy exchange with the environment and between its own parts. (b) The sun as an example of an object in dynamic (quasi-) equilibrium. Gravitational forces balance the internal pressure and the generated energy is balanced by the emitted radiation.

There are many systems in physics and in nature which share the feature of being in a state of dynamic equilibrium, so that externally they look stationary. Consider, for example, the sun (or any other star). Energy in the sun is generated internally, redistributed through the layers and radiated (see Fig.6b). There is continuous energy flow from inner to outer layers, but the density and temperature profiles do not change. Indeed, on our time scale, the sun is in dynamic equilibrium, as its size and output remain the same. Of course, more complicated processes produce more complicated structures.

The sun is still an object of the inanimate world. Living things can also, to some extent, be analogous to solitons, as they remain in a state of stationary dynamic equilibrium. An animal has material and energy inputs and outputs and complicated internal dynamics. There are forces which try to 'disperse' an animal in nature and this happens if it dies. We can also view an animal as a highly nonlinear dynamical system. Hence all the physical features which are present in the CGLE are applicable. As a result, we have a living object which stays in dynamic equilibrium for a long time.

We can further this analogy if we take into account the fact that the CGLE admits a multiplicity of solutions for the same set of external parameters. This resembles various species existing in the same environment. Just as solitons are fixed (or isolated) in their properties, so are animal species in biology.

Recently, we have found pulsating soliton solutions of the CGLE which constitute a novel and unexpected phenomenon. An example is shown in Fig.7a. It is remarkable and surprising that pulsating solutions are fixed in the same way as stationary pulses. This means that a solution $\psi(\xi, \tau)$ is a unique function of $\xi$ and $\tau$ for each set of equation parameters. We recall that Hamiltonian systems do not have pulsating soliton solutions. Even if excited initially, pulsating solitons are subject to restructuring and evolve into stationary solitons [33]. Some exceptions to this rule are the integrable models where pulsating structures are nonlinear superpositions of fundamental solitons [75]. Dissipative systems, in contrast to Hamiltonian ones, admit pulsating solitons. Interestingly enough, they do not appear from the integrable limit and hence do not have anything in common with the nonlinear superposition of fundamental solitons of the NLSE [75]. The parameters of the CGLE have to be far enough from the NLSE limit in order to obtain pulsating solitons. The variety of these solutions and their region of existence is huge. In either case, pulsating solutions cannot be found by extrapolating from those in the integrable limit.

To illustrate the fact that the CGLE allows us to describe behavior even more complicated than just a state of dynamical equilibrium, we give here an example of an exploding soliton (see details in [81]). Fig.7b shows that, for a certain set of parameters, the 'stationary' soliton of the CGLE remains stationary only for a limited period of time, and then explodes (or, in other words, erupts like a volcano). After a 'cooling' period, the solution becomes 'stationary' again. This happens periodically, as with many other phenomena in nature. An interesting fact is that this behavior occurs not just in some very specific case, but for a wide range of parameters of the CGLE. We can consider the CGLE as the model of minimum complexity which still allows us to describe quite complicated dynamics in the world of systems far from equilibrium.

Equation (1) is nonintegrable, and only particular exact solutions can be obtained. In general, initial value problems with arbitrary initial conditions can only be solved numerically. The cubic CGLE, obtained by setting $\mu = \nu = 0$ in Eq.(1), has been studied extensively [78-80]. Exact solutions to this equation can be obtained using a special ansatz [78], Hirota bilinear method [79] or reduction

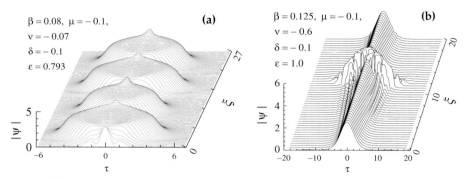

**Fig. 7.** (a) An example of pulsating soliton of the CGLE. (b) Erupting soliton of the (cubic-quintic) CGLE. The parameters are shown in each case.

to systems of linear PDEs [82]. However, it was realized many years ago that the soliton-like solutions of this equation are unstable to perturbations.

The case of the quintic CGLE has been considered in a number of publications using numerical simulations, perturbative analysis and analytic solutions. Originally, this equation was used mainly as a model for binary fluid convection [83–85]. The existence of soliton-like solutions of the quintic CGLE in the case $\epsilon > 0$ has been demonstrated numerically [84,85]. A qualitative analysis of the transformation of the regions of existence of the soliton-like solutions, when the coefficients on the right-hand-side change from zero to infinity, has been made in [86]. An analytic approach, based on the reduction of Eq.(1) to a three-variable dynamical system, which allows us to obtain exact solutions for the quintic equation, has been developed in [57,87]. The most comprehensive mathematical treatment of the exact solutions of the quintic CGLE, using Painlevé analysis and symbolic computations, is given in [88]. The general approach used in that work is the reduction of the differential equation to a purely algebraic problem. The solutions include solitons, sinks, fronts and sources. The great diversity of possible types of solutions requires a careful analysis of each class of solutions separately. In the brief review which follows, we will concentrate solely on soliton-like solutions.

The CGLE has no known conserved quantities. This contrasts with the infinite number for the NLSE and the small number for Hamiltonian systems. Here, for an arbitrary initial condition, even the energy is not conserved, as there is energy exchange with the external world. Instead, the energy associated with solution $\psi$ is $Q = \int_{-\infty}^{\infty} |\psi|^2 \, d\tau$, and its rate of change with respect to $\xi$ is [9]:

$$\frac{d}{d\xi}Q = F[\psi], \tag{20}$$

where the real functional $F[\psi]$ is given by

$$F[\psi] = 2 \int\limits_{-\infty}^{\infty} \left[ \delta|\psi|^2 + \epsilon|\psi|^4 + \mu|\psi|^6 - \beta|\psi_\tau|^2 \right] \, d\tau.$$

Similarly [9], the momentum is $M = Im \left( \int\limits_{-\infty}^{\infty} \psi_\tau^* \psi \, d\tau \right)$, and its rate of change is defined by

$$\frac{d}{d\xi} M = J[\psi], \tag{21}$$

where

$$J[\psi] = 2 \, Im \int\limits_{-\infty}^{\infty} \left[ (\delta + \epsilon|\psi|^2 + \mu|\psi|^4)\psi + \beta\psi_{\tau\tau} \right] \, \psi_\tau^* \, d\tau.$$

By definition, this functional is the force acting on a soliton along the $\tau$-axis. There are only two rate equations, *viz.* Eq.(20) and Eq.(21), which can be derived for the CGLE. They can be used for solving various problems related to CGLE solitons.

If the coefficients $\delta$, $\beta$, $\epsilon$, $\mu$ and $\nu$ on the right-hand side are all small, then soliton-like solutions of Eq.(1) can be studied by applying perturbative theory to the soliton solutions of the NLSE. Let us consider the right-hand side of equation (1), with $D = +1$, as a small perturbation, and write the solution as a soliton of the NLSE, *viz.*

$$\psi(\tau, \xi) = \frac{\eta}{\cosh[\eta(\tau + \Omega\xi)]} \exp[-i\Omega\tau + i(\eta^2 - \Omega^2)\xi/2]. \tag{22}$$

In the presence of the perturbation, the parameters of the soliton, *viz.* the amplitude $\eta$ and frequency (or velocity) $\Omega$, change adiabatically. The equations for them can be obtained from the balance equations for the energy and momentum. Using Eq.(22) and Eq.(20), we have the equations for the evolution of $\eta(\xi)$ and $\Omega(z)$:

$$\frac{d\eta}{d\xi} = 2\eta \left[ \delta - \beta\Omega^2 + \frac{1}{3}(2\epsilon - \beta)\eta^2 + \frac{8}{15}\mu\eta^4 \right], \quad \frac{d\Omega}{d\xi} = -\frac{4}{3} \beta\Omega \, \eta^2. \tag{23}$$

The dynamical system of equations, (23), has two real dependent variables and the solutions can be presented on the plane. An example is given in Fig.8. It has a line of singular points at $\eta = 0$, and, depending on the equation parameters, may have one or two singular points on the semi-axis $\Omega = 0$, $\eta > 0$. The values of $\eta^2$ for singular points are defined by finding the roots of the biquadratic polynomial in the square brackets in Eq.(23). When the roots are negative (hence, $\eta$ is imaginary), there are no singular points and hence no soliton solution. If both roots of the quadratic polynomial (in $\eta^2$) are positive (so that both $\eta$ are real), then there are two fixed points and two corresponding soliton solutions. Both

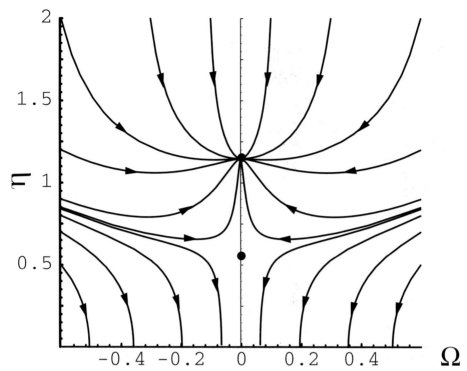

**Fig. 8.** The phase portrait of the dynamical system (23) for $\delta = -0.03$, $\beta = 0.1$, $\epsilon = 0.2$ and $\mu = -0.11$. The upper fixed point is a sink which defines the parameters of a stable approximate soliton-like solution of the quintic CGLE. Any soliton-like initial condition in close proximity to a fixed point will converge to a stable stationary solution. The points on the line $\eta = 0$ are stable when $\delta < 0$ and $\beta > 0$. This condition is needed for the background state $\psi = 0$ to be stable.

roots are positive when either $\beta < 2\epsilon$, $\mu < 0$ and $\delta < 0$ or $\beta > 2\epsilon$, $\mu > 0$ and $\delta > 0$.

The stability of at least one these fixed points requires $\beta > 0$. Moreover, the stability of the background requires $\delta < 0$. In the latter case, we necessarily have $\beta < 2\epsilon$, $\mu < 0$ and the upper fixed point is a sink (as shown in Fig.8) which defines the parameters of a stable approximate soliton solution of the quintic CGLE. The background $\psi = 0$ is also stable, so that the whole solution (soliton plus background) is stable. Finally, when only one of the roots is positive, there is a singular point in the upper half-plane and there is a corresponding soliton solution. However either the background or the soliton itself is unstable, so that the total solution is unstable. The term with $\nu$ in the CGLE does not influence the location of the sink. It only introduces an additional phase term, $\exp(8i\nu\eta^4\xi/15)$, into the solution of Eq.(22).

In the case of cubic CGLE, $\mu = 0$ and $\nu = 0$. The stationary point is then

$$\eta = \sqrt{3\delta/(\beta - 2\epsilon)}, \qquad \Omega = 0. \tag{24}$$

It is stable provided that $\delta > 0$, $\beta > 0$ and $\epsilon < \beta/2$. Clearly, in this perturbative case the soliton and the background cannot be stable simultaneously. Hence, this approach shows that to have both the soliton and the background stable, we need to have quintic terms in the CGLE (see also [64]).

This simple approach shows that, in general, the CGLE has stationary soliton-like solutions, and that, for the same set of equation parameters, there may be two of them simultaneously (one stable and one unstable). Moreover, this approach shows that soliton parameters are **fixed**, as depicted in Fig.5. This occurs because the dissipative terms in Eq.(1) break the scale invariance associated with the conservative system.

Despite its simplicity and advantages in giving stability and other properties of solitons, the perturbative analysis has some serious limitations. Firstly, it can be applied only if the coefficients on the right-hand-side of Eq.(1) are small, and this is not always the case in practice. Correspondingly, it describes the convergence correctly only for initial conditions which are close to the stationary solution. Secondly, the standard perturbative analysis cannot be applied to the case $D < 1$, when the NLSE itself does not have bright soliton solutions. However, Eq.(1) has stable soliton solutions for this case as well.

Exact analytical solutions can be found only for certain combinations of the values of the parameters [9,65]. In general, we need to use some numerical technique to find stationary solutions. One way to do this is to reduce Eqs. (9) to a set of ODEs. We achieve this by seeking solutions in the form:

$$\psi(t, \xi) = \psi_o(\tau) exp(-i\omega\xi) = a(\tau) \, exp[i\phi(\tau) - i\omega\xi], \qquad (25)$$

where $a$ and $\phi$ are real functions of $\tau = t - v\xi$, $v$ is the pulse velocity and $\omega$ is the nonlinear shift of the propagation constant. Substituting Eq.(25) into Eq.(9), we obtain an equation for two coupled functions, $a$ and $\phi$. We separate real and imaginary parts, get a set of two ODEs and transform them.

The resultant set contains all stationary and uniformly translating solutions. The parameters $v$ and $\omega$ are the eigenvalues of this problem. In the $(M, a)$ plane, where $M = a^2\phi'(\tau)$, the solutions corresponding to pulses are closed loops starting and ending at the origin. The latter happens only at certain values of $v$ and $\omega$. If $v$ and $\omega$ differ from these fixed values, the trajectory cannot comprise a closed loop. By properly adjusting the eigenvalue $\omega$, it is possible to find the soliton solution with a "shooting" method.

Numerical simulations show that a multitude of soliton solutions of the CGLE exist. They have a variety of shapes and stability properties. They can even be partly stable and partly unstable. Trajectories on the phase portrait are deflected from their smooth motion near the singular points. As a result, the trajectory can have additional loops and the soliton shape can become multi-peaked. Even when the soliton shape is smooth (bell-shaped), the singular points can influence the soliton evolution in $\xi$.

Singular points have a certain physical meaning. They define a continuous wave solution with the same propagation constant as the soliton. Thus, there is a close relation between solitons and CW solutions. When the CW solution is stable itself, it may lead to the existence of "front" solutions. Even if it is not stable, it still has a role is in influencing the soliton shape.

The vast majority of soliton solutions lack stability. In fact, in some regions of the parameter space, all of them can be unstable. However, even in this case, unstable solutions can emerge from an arbitrary pulse and exist in this form for some distance. Surprisingly, some solutions are lost due to explosions [81], but reappear repeatedly after this visibly chaotic stage of evolution, as discussed earlier and shown in Fig.7b. The explosion phenomenon never occurs if solitons are stable for the given values of the parameters.

Clearly, dissipative solitons are qualitatively different from Hamiltonian solitons. We now summarize some of the main features of solitons in dissipative systems.

1. *Solitons in dissipative systems are fixed solutions*, in contrast to Hamiltonian solitons which are one- or two-parameter families.

2. *There is a multiplicity of soliton solutions in dissipative systems* for a single equation (CGLE). In the case of Hamiltonian systems, one equation usually has one family of solitons. The number of soliton families increases with the number of coupled equations [9].

3. *In addition to stationary solutions, there are pulsating solitons.* These solutions are also fixed.

4. In dissipative systems, *several solitons which belong to different 'branches' can be stable simultaneously.* For the cubic-quintic nonlinearity, the stable solitons belong to the subclass of high amplitude solitons. In contrast, in the case of Hamiltonian systems, only the lowest branch of solitons (the fundamental mode) is usually stable [9]. In other words, the branch of solitons with the lowest Hamiltonian is stable.

5. *In dissipative systems, the singular points of the equations for nonlinear modes partition the soliton into pieces with different stability properties.* This can happen even to a "ground state" soliton with the perfect "bell-shape" profile. Such partitions can lead to quite complicated behavior for some dissipative solitons. The most striking example of complicated dynamics is the "soliton explosion" [81]. In contrast, the "ground-state" solitons of Hamiltonian systems must be viewed as single particles, and are stable (or unstable) as a whole.

# References

1. V. E. Zakharov and A. B. Shabat: Sov. Phys. JETP **34**, 62 (1971).
2. A. D. Boardman: Editorial, Photonics Science News **5**, 2 (1999).
3. H. Goldstein: *Classical Mechanics*, (2-nd ed., Addison-Wesley, New York, 1980).
4. R. K. Dodd, J. C. Eilbeck, J. D. Gibbon and H. C. Morris: *Solitons and Nonlinear Wave Equations*, (Academic Press, London, 1984).
5. A. C. Newell: *Solitons in Mathematics and Physics*, (Society of Industrial and Applied Mathematics, Arizona, 1985).
6. M. J. Ablowitz and P. A. Clarkson: *Solitons, Nonlinear Evolution Equations and Inverse Scattering*, (London Mathematical Society Lecture Notes Series **149**, Cambridge University Press, Cambridge, 1991).

7. R. K. Bullough and P. J. Caudrey: *Solitons*, (Springer–Verlag, Berlin, 1980).
8. S. P. Novikov, S. V. Manakov, L. P. Pitaevskii and V. E. Zakharov: *Theory of Solitons–The Inverse Scattering Method*, (Plenum, New York, 1984).
9. N. N. Akhmediev and A. Ankiewicz: *Solitons: Nonlinear Pulses and beams*, (Chapman & Hall, London, 1997).
10. V. S. Gerdzhikov and M. I. Ivanov: Teor. Mat. Fiz. **52**, 89 (1982) Theor. Math. Phys. **52**, 676 (1982).
11. K.Nakkeeran, K. Porsezian, P. Shanmugha Sundaram, and A. Mahalingam: Phys. Rev. Lett. **80** 1425 (1998).
12. V. G. Makhan'kov and O. K. Pashaev: Teor. Mat. Fiz. **53**, 55 (1982) Theor. Math. Phys. **53**, 979 (1982).
13. C. B. Manakov: Sov. Phys. JETP **38**, 248 (1974).
14. N. Akhmediev and A. Ankiewicz: Chaos **10**, 600 (2000).
15. M. Mitchell, Z. Chen, M. Shih and M. Segev: Phys. Rev. Lett. **77**, 490 (1996).
16. M. Segev and G. Stegeman: Physics Today No 8, 42 (1998).
17. N. N. Akhmediev and A. Ankiewicz: Photonics Science News **5**, 13 (1999).
18. D. N. Christodoulides, T. H. Coskun and R. I. Joseph: Opt. Lett. **22**, 1080 (1997).
19. A. Ankiewicz, W. Królikowski, and N. N. Akhmediev: Phys. Rev. E **59**, 6079 (1999).
20. A. W. Snyder, D. J. Mitchell: Phys. Rev. Lett. **80**, 1422 (1998).
21. N. N. Akhmediev and J. M. Soto-Crespo: Phys. Rev. E **49**, 4519 (1994).
22. N. N. Akhmediev and J. M. Soto-Crespo: Phys. Rev. E **49**, 5742 (1994)
23. V. E. Zakharov: Izv. Vyssh. Uchebn. Zaved., Radiofiz. **17**, 431 (1974).
24. L. D. Faddeev and L. A. Takhtadjan: *Hamiltonian Methods in the Theory of Solitons*, (Springer–Verlag, Berlin, 1987).
25. V. E. Zakharov and A. M. Rubenchik: Sov. Phys. JETP **38**, 494 (1973).
26. J. J. Rasmussen and K. Ripdal: Physica Scripta **33**, 481 (1986)
27. N. N. Akhmediev: Optical and Quantum Electronics **30**, 535 (1998).
28. E. A. Ostrovskaya, N. N. Akhmediev G. I. Stegeman, J. U. Kang, J. S. Aitchison: J.Opt.Soc.Am. B **14**, 880 (1997).
29. A. V. Buryak and N. N. Akhmediev: Opt. Comm. **110**, 287 (1994).
30. J. M. Soto-Crespo, N. N. Akhmediev and A. Ankiewicz: J.Opt.Soc.Am. B **12**, 1100 (1995).
31. N. N. Akhmediev, A. V. Buryak, J. M. Soto-Crespo and D. R. Andersen: J.Opt.Soc.Am. B **12**, 434 (1995).
32. Y. Chen: Phys. Rev. E **57**, 3542 (1998).
33. D. Artigas, L. Torner and N. N. Akhmediev: Opt. comm. **143**, 322 (1997).
34. N. N. Akhmediev, M. P. Das and A. Vagov: Condensed Matter Theories **12**, 17 (1997). Ed. J. W. Clark and P. V. Panat
35. A. V. Buryak and N. N. Akhmediev: Phys. Rev. E **51**, 3572 (1995).
36. N. N. Akhmediev: Sov. Phys. JETP **56**, 299 (1982).
37. F. V. Kusmartsev: Phys. Rep. **183**, 1 (1989).
38. N. Akhmediev, A. Ankiewicz and R. Grimshaw: Phys. Rev. E **59**, 6088 (1999).
39. A. A. Kolokolov: Zh. Prikl. Mekh. Tekh. Fiz. **3**, 152 (1973). English translation pp.426
40. M. Grillakis, J. Shatah and W. Strauss: J. Functional Analysis **74**, 160 (1987).
41. C. K. R. T. Jones and J. V. Moloney: Phys. Lett. A **117**, 175 (1986).
42. D. J. Mitchell and A. W. Snyder: J.Opt.Soc.Am. B **10**, 1572 (1993) .
43. I. V. Barashenkov, D. E. Pelinovsky and E. V. Zemlyanaya: Phys. Rev. Lett, **80**, 5117 (1998).

44. H. T. Tran, J. D. Mitchell, N. N. Akhmediev and A. Ankiewicz: Opt. Comm. **93**, 227 (1992).

45. I. V. Barashenkov, M. M. Bogdan and T. Zhanlav: in "*Nonlinear World*", Proceedings of the Fourth International Workshop on Nonlinear and Turbulent Processes in Physics, Kiev 1989. Edited by V. G. Bar'yakhtar et al. World Scientific, Singapore, 1990, p.3; I. V. Barashenkov, M. M. Bogdan and V. E. Korobov: Europhysics Letters **15**, 113 (1991).

46. N. N. Akhmediev, A. Ankiewicz and H. T. Tran: J.Opt.Soc.Am. B **10** 230 (1993).

47. A. D. Boardman and K. Xie: Phys. Rev. E **55**, 1899 (1997).

48. A. D. Boardman, P. Bontemps and K. Xie: J. Opt. Soc. Am. B **14**, 1 (1997).

49. N. N. Akhmediev and A. V. Buryak: J.Opt.Soc.Am. B **11**, 804 (1994).

50. L. Torner and G. I. Stegeman: J. Opt. Soc. Am. B (special issue), **14**, 3127 (1997).

51. L. Torner, D. Mihalache, M. C. Santos and N. N. Akhmediev: J.Opt.Soc.Am. B **15**, 1476 (1998).

52. P. Agin and G.I. Stegeman: J. Opt. Soc. Am. B (special issue), **14**, 3162 (1997).

53. N. N. Akhmediev and J. M. Soto-Crespo: Phys. Rev. A **47**, 1358 (1993).

54. D. E. Edmundson and R. H. Enns: Opt. Lett. **17**, 586 (1992).

55. S. Blair and K. Wagner: Optical and Quantum Electronics **30**, 697 (1998).

56. N. Bekki and K. Nozaki: Phys. Lett. A **110**, 133 (1985).

57. W. Van Saarloos and P. C. Hohenberg: Physica D **56**, 303 (1992).

58. N. N. Akhmediev and V. V. Afanasjev: Phys. Rev. Lett. **75**, 2320 (1995).

59. V. B. Taranenko, K. Staliunas and C. O. Weiss: Phys. Rev. A **56**, 1582 (1997).

60. W. J. Firth and A. J. Scroggie: Phys. Rev. Lett. **76**, 1623 (1996).

61. N. A. Kaliteevstii, N. N. Rozanov and S. V. Fedorov: Optics and Spectroscopy **85**, 533 (1998).

62. D. Michaelis, U. Peschel and F. Lederer: Opt. Lett. **23**, 1814 (1998).

63. A. Hasegawa, and Y. Kodama: *Solitons in optical communications*, (Oxford University Press, New York, 1995).

64. J. D. Moores: Opt. Comm. **96**, 65 (1993).

65. N. N. Akhmediev, V. V. Afanasjev and J. M. Soto - Crespo: Phys. Rev. E **53**, 1190 (1996).

66. C. Normand, Y. Pomeau: Rev. Mod. Phys. **49**, 581 (1977).

67. P. Kolodner: Phys. Rev. A **44**, 6466 (1991).

68. R. Graham: *Fluctuations, Instabilities and Phase Transitions*, (ed. T. Riste, Springer-Verlag, Berlin, 1975).

69. C. O. Weiss: Phys. Rep. **219**, 311 (1992).

70. P. K. Jakobsen, J. V. Moloney, A. C. Newell and R. Indik: Phys. Rev. A **45**, 8129 (1992).

71. G. K. Harkness, W. J. Firth, J. B. Geddes, J. V. Moloney and E. M. Wright: Phys. Rev. A, **50**, 4310 (1994).

72. G.-L. Oppo, G. D'Alessandro and W. J. Firth: Phys. Rev. A **44**, 4712 (1991).

73. P.-S. Jian, W. E. Torruellas, M. Haelterman, S. Trillo, U. Peschel and F. Lederer: Opt. Lett. **24**, 400 (1999).

74. A. M. Dunlop, E. M. Wright and W. J. Firth: Opt. Comm. **147**, 393 (1998).

75. J. Satsuma and N. Yajima: Progr. Theor. Phys. Suppl. **55**, 284 (1974).

76. V. S. Grigoryan and T. C. Muradyan: J.Opt.Soc.Am. B **8**, 1757 (1991).

77. N. N. Akhmediev, M. J. Lederer and B. Luther-Davies: Phys. Rev. E **57**, 3664 (1998).

78. N. R. Pereira and L. Stenflo: Phys. Fluids **20**, 1733 (1977).

79. K. Nozaki and N. Bekki: Phys. Soc. Japan. **53**, 1581 (1984).

80. P.-A. Bélanger, L. Gagnon and C. Paré: Opt. Lett. **14**, 943 (1989).
81. J. M. Soto-Crespo, N. Akhmediev and A. Ankiewicz: Phys. Rev. Lett. **85**, 2937 (2000).
82. R. Conte and M. Musette: Physica D **69,** 1 (1993).
83. B. A. Malomed: Physica D **29**, 155 (1987).
84. O. Thual and S. Fauve: J. Phys. **49**, 1829 (1988).
85. H. R. Brand and R. J. Deissler: Phys. Rev. Lett. **63**, 2801 (1989).
86. V. Hakim, P. Jakobsen and Y. Pomeau: Europhys. Lett. **11**, 19 (1990).
87. W. Van Saarloos and P. C. Hohenberg: Phys. Rev. Lett. **64**, 749 (1990).
88. P. Marcq, H. Chaté and R. Conte: Physica D **73**, 305 (1994).

# System Analysis
# Using the Split Operator Method

K.J. Blow

Aston University, Aston Triangle, Birmingham, B4 7ET, UK

**Abstract.** The split step operator technique underlies both numerical and analytical approaches to optical communication system studies. In this paper I will show how these are related and how they give rise to real systems applications of the principles. The technique can also be related to other analytical methods such as the Lie transform. The predicted special points of systems are also equivalent to pre-chirping techniques.

## 1   Introduction

The use of optical solitons in communication systems is now a well established field of research. Many techniques have been developed to study the complex propagation phenomena including numerical simulation  [3]  [5], perturbation theory, variational methods [2], transforms and inverse scattering theory. Here I will describe the split operator method that was originally used to derive a higher order numerical integration scheme but has since been used to derive analytic results in the average soliton regime  [4]. The split operator method can be used to achieve two things. First, we can demonstrate that under certain conditions pulse propagation can be described by the lossless nonlinear Schrodinger equation [NLS]. Second, we can derive corrections to the NLS that give additional information about system performance. The method is based on a formal operator representation of the nonlinear differential equations describing the propagation of light in a monomode optical fiber.

### 1.1   Operator Representations

We begin with a brief review of operator representations of differential equations. The NLS can be written in the following form where $u$ is the optical field and the two terms on the RHS represent the dispersion and nonlinearity respectively.

$$i\frac{\partial u}{\partial z} = \frac{1}{2}\frac{\partial^2 u}{\partial t^2} + |u|^2 u \tag{1}$$

In order to simplify much of the following discussion we will represent this in the following way

$$i\frac{\partial u}{\partial z} = \hat{D}u + \hat{N}u. \tag{2}$$

The dispersion term is now represented by the operator $\hat{D}$ and the nonlinearity by the operator $\hat{N}$. With this representation we can now formally write the solution of the NLS as

$$u(z) = e^{-i(\hat{D}+\hat{N})z}u(0). \tag{3}$$

So far we have not really achieved anything as we cannot actually evaluate the exponentiated operators. What we have done is to write the formal solution in a way that can be manipulated using other techniques. The most important result we will use is the Baker-Campbell-Hausdorf theorem which can be expressed in the following way

$$e^{-i(\hat{D}+\hat{N})z} = e^{-i\hat{D}z}e^{-i\hat{N}z}e^{[\hat{D},\hat{N}]\frac{z^2}{2}+\cdots\cdots}. \tag{4}$$

This theorem tells us how to separate out the exponents but involves the introduction of commutators of the two operators. The remaining terms involve higher order commutators associated with high powers of $z$. The important point here is that for sufficiently small $z$ we can separate the effects of the two operators.

## 1.2   Simple Numerical Integration

Let us now look at the issue of accuracy in a very simple example, numerical integration. Suppose we want to evaluate the following integral numerically

$$y = \int_a^b f(x)dx. \tag{5}$$

The simplest scheme is to use Simpson's rule as illustrated graphically in Fig. 1a. The area under the curve is approximated by a rectangle and the formula is given by

$$y_S = \frac{(b-a)}{n}\sum_{i=0}^{n} f(a + i\frac{(b-a)}{n}) + O(\frac{(b-a)}{n}). \tag{6}$$

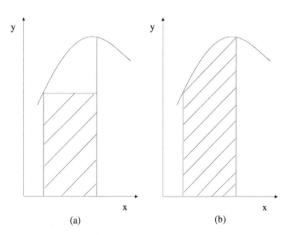

(a)                           (b)

**Fig. 1.** a) Graphical illustration of Simpson's rule, b) Graphical illustration of the Trapezium rule

Giving an estimate of the integral that is accurate to first order in the step size $(b-a)/n$. If we want a higher order scheme then we can replace the rectangle by the trapezium shown in Fig. 1b and obtain the following equation for the estimate of the integral

$$y_T = \frac{(b-a)}{n}\left(\frac{f(a)}{2} + \sum_{i=1}^{n-1} f(a+i\frac{(b-a)}{n}) + \frac{f(b)}{2}\right) + O(\frac{(b-a)^2}{n^2}) \quad (7)$$

which can now be written in terms of the Simpson's rule estimate as

$$y_T = -\frac{(b-a)f(a)}{2n} + y_S - \frac{(b-a)f(b)}{2n}. \quad (8)$$

Note that the difference between $y_S$ and $y_T$ occurs only at the end points of the integration range. The remarkable fact is that we can change from a first order to a second order scheme simply by adjusting the end points and this is true no matter how large the intervening range. The importance of the end points is the theme of this chapter and it will reappear in both the section on numerical integration of the NLS and the section on systems.

## 2    Numerical Modelling

Numerical modelling has played an important role in the study of optical solitons and we will now outline the most commonly used method: the split step Fourier method. The pure NLS (1) contains the two terms describing dispersion and nonlinearity. If we ignore the dispersion then we can solve analytically for the nonlinearity to obtain

$$u(z,t) = e^{-i|u_0|^2 z}u_0 \quad (9)$$

where $u_0$ is the field $u(t)$ at $z = 0$. Equation (9) is the explicit representation of the operator form $e^{-i\hat{N}z}u_0$.

Similarly if we ignore the nonlinearity we can solve analytically for the effect of dispersion in frequency space to obtain

$$\tilde{u}(z,\omega) = e^{i\omega^2 z}\tilde{u}_0 \quad (10)$$

where the solution has now been expressed in terms of the Fourier transform of the initial field $\tilde{u}(z = 0,\omega)$. Equation (10) is the explicit representation of the operator form $e^{-i\hat{D}z}u_0$.

So we now need some approximate way of expressing the solution of the NLS in terms of the solutions to the individual terms describing dispersion and nonlinearity. We have already seen how to do this by using the Baker-Campbell-Hausdorf theorem (4). However, the local error, arising from the commutators between the dispersion term and the nonlinear term, is proportional to the square of the propagation length. Therefore, in order to simulate arbitrarily long systems we need to break the system up into smaller pieces of length $\delta$ and to use the approximate solution of Eq.(4) without the commutator term as follows

$$e^{-i(\hat{D}+\hat{N})\delta} = e^{-i\hat{D}\delta}e^{-i\hat{N}\delta} \tag{11}$$

which corresponds physically to treating the combined effects of dispersion and nonlinearity alternately so that the small length $\delta$ of fiber is replaced by a length $\delta$ of fiber with no nonlinearity followed (or preceded) by a length $\delta$ of fiber with no dispersion. This is shown schematically in Fig. 2.

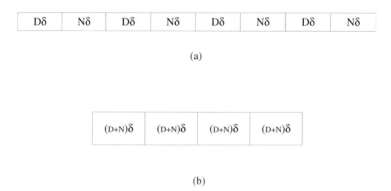

(a)

(b)

**Fig. 2.** a) Schematic representation of the split step method, b) Equivalent representation of the actual fiber simulated

So far, we have described the simplest implementation of the spit step method which is equivalent to the Simpson's rule for numerical integration in that the local error is $O(\delta^2)$ and the global error is therefore $O(\delta)$. We can do better than this by using a modified form of the split as we will now show.

## 2.1    Second Order Split Step Fourier Method

In order to construct a numerical scheme that is globally second order we need a local approximation that is accurate to third order. First we note that the error is proportional to the commutator $[\hat{D}, \hat{N}]$ and that this commutator changes sign if we interchange $\hat{D}$ and $\hat{N}$. Thus the sign of the local error [1] depends on the order in which the individual operators are applied. So if we take two sections of fiber and apply the dispersion and nonlinearity operators in opposite order in those sections then for sufficiently small lengths the error term will cancel making the local error accurate to third order. In terms of the operator representation this becomes

$$e^{-i(\hat{D}+\hat{N})\delta} = e^{-i(\hat{D}+\hat{N})\frac{\delta}{2}}e^{-i(\hat{D}+\hat{N})\frac{\delta}{2}} \approx e^{-i\hat{D}\frac{\delta}{2}}e^{-i\hat{N}\frac{\delta}{2}}e^{-i\hat{N}\frac{\delta}{2}}e^{-i\hat{D}\frac{\delta}{2}} \tag{12}$$

---

[1] In order to avoid any confusion, let me emphasise that the errors we are discussing are systematic as they arise from mathematical approximations and are not errors arising from finite precision in numerical computation.

where the fiber of length $\delta$ has been split into two sections of length $\delta/2$ and the operators applied in a different order in each section. We now note that the middle pair are both the same operator and so, since every operator commutes with itself, we can combine these two exactly into one to obtain

$$e^{-i(\hat{D}+\hat{N})\delta} \approx e^{-i\hat{D}\frac{\delta}{2}}e^{-i\hat{N}\delta}e^{-i\hat{D}\frac{\delta}{2}} \tag{13}$$

which is similar to the Trapezium rule we have already seen. If we apply this form again then the final operator of one length can also be combined (exactly) with the first operator of the subsequent length. The pictorial representation of the numerical scheme is now shown schematically in Fig. 3

| D$\delta$/2 | N$\delta$ | D$\delta$ | N$\delta$ | .............. | D$\delta$ | N$\delta$ | D$\delta$/2 |
|---|---|---|---|---|---|---|---|

**Fig. 3.** Schematic representation of the second order split step method

Note that the only difference with the first order method is that the first and last steps are of half the normal length. Simply by changing the end points we have made the first order method into a more accurate second order method, just as with the trapezium rule. The end points matter!

Finally, if you consider the errors involved into this symmetric version of the operator splitting you will see that all the even order errors have now been eliminated.

## 2.2   Fourth Order Split Step Fourier Method

If we want a more accurate scheme then we have to turn to another approach. The local error using the symmetric split step is of order $\delta^3$. Wood and Gibbons, see discussion in ([5]), pointed out that changing the sign of the step length would reverse the sign of the local error and allow the possibility of arranging cancellation of that term. In particular, consider the following sequence of applications of the symmetric operator split over a length x, denoted $S_2(x)$

$$S_2(\delta)S_2(\delta)S_2(\delta)S_2(\delta)S_2(-2\delta)S_2(\delta)S_2(\delta)S_2(\delta)S_2(\delta). \tag{14}$$

The total length of fiber that has been simulated is $8\delta - 2\delta = 6\delta$ whereas if we write the local error of a single term as $E\delta^3$ then the total error is $8E\delta^3 + E(-2\delta)^3 = 0$. In practice, the fourth order method has an overhead of performing more Fourier transforms but the possible reduction in step length can make this viable. One note of caution [11] is that while the global error in the second order scheme reduces uniformly with step length, in some cases there can be oscillations in the error when using the fourth order scheme.

## 3   Analytic Results

We will now turn our attention to the application of the split step operator method to analytic studied of systems. We begin with the study of a simple system with loss, periodic amplification and a constant dispersion [4]. We will show that the renormalized soliton obeys the NLS and also calculate the size of the field renormalization. This derivation is an alternative approach to the Lie transformation method [9].

## 3.1  Average Solitons

We will now consider a periodically amplified system in which all the links are the same length and all the amplifiers are perfect with a constant gain. The evolution of the field in the fiber is described by the NLS with the addition of a loss term as follows

$$i\frac{\partial E}{\partial z} = \frac{1}{2}\frac{\partial^2 E}{\partial t^2} + |E|^2 E - i\Gamma E \tag{15}$$

where $E$ is the normalized optical electric field, $t$ is the time normalized to the pulse width in a frame moving with the group velocity and $z$ is the distance normalized to the soliton period [10] [6] [1]. One unit cell of the periodic system consists of a fiber of length $L$ followed by an amplifier of gain $g$. If we denote the input field to the fiber by $E_0$, the output field by $E_1$ and the field after the amplifier by $E_2$ then we have the following operator forms relating the fields

$$E_1 = e^{-i(\hat{D}+\hat{N}+\hat{\Gamma})L}E_0 \, , \tag{16a}$$

$$E_2 = e^g E_1 \, . \tag{16b}$$

In order to proceed we need to ensure that the exponents in Eq.(16a) are all small so that we can apply the Campbell-Baker-Hausdorff theorem. This condition is satisfied, for single solitons, if the system length is shorter than the soliton period (small $z$). Since the soliton period is proportional to the square of the pulse width we can always satisfy this condition by working with sufficiently long pulses and we now assume this condition to be satisfied. For our present purposes it is also sufficient to use the first order splitting of the operators.

The exponent of Eq.(16a) contains three terms which need to be separated as two terms. It is convenient to treat the nonlinearity together with the loss as we will now see. Taking those two terms together they correspond to the following differential equation

$$i\frac{\partial E}{\partial E} = -i\Gamma E + |E|^2 E \tag{17}$$

which can be solved analytically to give the following solution

$$E(z) = e^{-\Gamma z}e^{-i\frac{1-e^{-2\Gamma z}}{2\Gamma}|E(0)|^2}E(0). \tag{18}$$

Note that the form of the exponent in this solution is similar to that of the self phase modulation solution 9 characteristic of the pure nonlinearity and therefore we can rewrite this solution as a rescaled nonlinear operator as follows

$$E(z) = e^{-\Gamma z}e^{-i\Lambda^2 \hat{N}z}E(0). \tag{19}$$

We can now combine this new form of the operator solution for loss with nonlinearity with the dispersion and gain to obtain the full solution for a single span of the system as

$$E_2 = e^g e^{-i\hat{D}z} e^{-\Gamma z} e^{-i\Lambda^2 \hat{N}z} E_1. \tag{20}$$

The first three terms, gain, dispersion and loss, all commute with each other and can be rearranged. At the same time we make the choice $g = -\Gamma z$ to cancel the loss and gain terms to obtain

$$E_2 = e^{-i\hat{D}z} e^{-i\Lambda^2 \hat{N}z} E_1. \tag{21}$$

Since we used the first order split to separate the operators we can recombine them to the same accuracy to obtain a renormalized NLS. We can make this more explicit by scaling the field as follows

$$E = \Lambda u \tag{22}$$

to obtain the final form of the renormalized, lossless NLS

$$u_2 = e^{-i(\hat{D}+\hat{N})L} u_0. \tag{23}$$

Thus we have now shown that the rescaled field $u$ obeys the lossless NLS provided we observe the field periodically (sometimes referred to as stroboscopically) and therefore that if we want the rescaled field to be a single soliton then the real field needs to be enhanced by the factor $\Lambda$.

## 3.2   Special Points in Constant Dispersion Systems

In the previous section we showed that in the limit that the soliton period is very large compared to the amplifier spacing then there is a rescaled field that obeys the NLS. This limit is achieved by making the pulse width large but in order to achieve high bit rates we need to make the pulse width small. We are therefore left with the question of how the system behaves as the soliton period becomes comparable with the amplifier spacing. In order to answer this question we need more information about the correction terms to the NLS in this limit. These corrections come from the commutators that we have so far ignored. However, it is important to distinguish between the errors that are a result of the operator splitting and errors that are due to the system architecture. Clearly, if we use an operator splitting accurate to order $n$ then we can only calculate corrections attributable to the system which are of lower order. Since the NLS operator is of order $z$ then we need to calculate system errors of order $z^2$ and therefore we need to use an operator splitting which is locally of order $z^3$, the symmetric second order split.

We begin by calculating the local error at an arbitrary point in the periodic unit cell. This is shown schematically in Fig. 4.

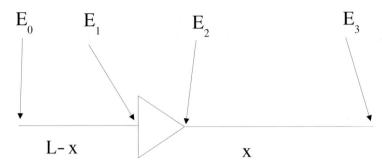

**Fig. 4.** Schematic representation of the unit cell illustrating the notation for the fields

We can now write down the relationships between the input and output fields as follows

$$E_3 = e^{-i(\hat{D}+\hat{N}+\hat{\Gamma})x} E_2, ; \tag{24a}$$

$$E_2 = e^{\Gamma L} E_1, ; \tag{24b}$$

$$E_1 = e^{-i(\hat{D}+\hat{N}+\hat{\Gamma})(L-x)} E_0. \tag{24c}$$

where we have already assumed that the gain will exactly compensate the loss over the full segment path.

As we have pointed out, we need to use the symmetric split and we also keep the nonlinearity and loss terms together as before so that the equation for $E_3$ in terms of $E_2$ now becomes

$$E_3 = e^{-\Gamma x} e^{-i\hat{D}\frac{x}{2}} e^{-i\Lambda^2(\Gamma x)\hat{N}x} e^{-i\hat{D}\frac{x}{2}} E_2 \tag{25}$$

where the field renormalization factor is given in terms of the function $\Lambda(p)$ defined by

$$\Lambda^2(p) = \frac{1-e^{-2p}}{2p}. \tag{26}$$

Similarly, for $E_1$ we obtain

$$E_1 = e^{-\Gamma(L-x)} e^{-i\hat{D}\frac{L-x}{2}} e^{-i\Lambda^2(\Gamma(L-x))\hat{N}(L-x)} e^{-i\hat{D}\frac{L-x}{2}} E_0. \tag{27}$$

We can now express $E_3$ in terms of $E_0$

$$E_3 = e^{-\Gamma x} e^{-i\hat{D}\frac{x}{2}} e^{-i\Lambda^2(\Gamma x)\hat{N}x} e^{-i\hat{D}\frac{x}{2}} e^{\Gamma L} e^{-\Gamma(L-x)}$$

$$e^{-i\hat{D}\frac{L-x}{2}} e^{-i\Lambda^2(\Gamma(L-x))\hat{N}(L-x)} e^{-i\hat{D}\frac{L-x}{2}} E_0. \tag{28}$$

Note that Eq.(28) contains a mixture of terms representing dispersion, non-linearity and loss. We need to move all the gain or loss terms to the left, where they will then cancel out, but to do this we need to commute a nonlinear operator with a loss term. In our case this is simple to do as loss followed by nonlinearity is equivalent to a rescaled nonlinearity followed by loss as follows

$$e^{-i\Lambda^2 \hat{N}x} e^{-\Gamma z} = e^{-\Gamma z} e^{-ie^{-2\Gamma z}\Lambda^2 \hat{N}x}. \tag{29}$$

Using this relation we can now simplify Eq.(28) to obtain

$$E_3 = e^{-i\hat{D}\frac{x}{2}} e^{-ie^{2\Gamma x}\Lambda^2(\Gamma x)\hat{N}x} e^{-i\hat{D}\frac{x}{2}} e^{-i\hat{D}\frac{L-x}{2}} e^{-i\Lambda^2(\Gamma(L-x))\hat{N}(L-x)} e^{-i\hat{D}\frac{L-x}{2}} E_0 \tag{30}$$

In order to calculate the deviation from the NLS we need to reorder the terms in the equation so that the leading order terms become a symmetric split operator representation of the NLS while retaining any second order commutators that result from the reordering. Remember that because we used the symmetric split in the first place there are no second order commutators that have been missed in Eq.(28). Thus the ones that result from the reordering are truly a property of the system. Once this is done we obtain the following

$$E_3 = e^{-i\hat{D}\frac{L}{2}} e^{-i\Pi^2 \hat{N}L} e^{-i\hat{D}\frac{L}{2}} e^{f(x)[\hat{D},\hat{N}]} E_0 \tag{31}$$

where

$$\Pi^2 = e^{2\Gamma x}\Lambda^2(\Gamma L) \tag{32}$$

and

$$f(x) = \frac{Lx}{2}e^{2\Gamma x}[\Lambda^2(\Gamma x) - \Lambda^2(\Gamma L)]. \tag{33}$$

Finally we re-scale the fields by the factor $\Pi$, $E = u/\Pi$, and recombine the symmetric operators to recover the renormalized NLS together with the correction factor as follows

$$u_3 = e^{-i(\hat{D}+\hat{N})L} e^{\frac{f(x)}{\Pi^2}[\hat{D},\hat{N}]} u_0. \tag{34}$$

The prefactor to the commutator in Eq.(34) represents the amplitude of the local error at various points in the unit cell. This function is zero at $z = 0$ and $z = L$ which at first sight would suggest that the system should be operated by starting either immediately before or after an amplifier. However, since the error is always positive, this choice would lead to a gradual accumulation of the error term as the solitons propagate in the system. It is therefore better to launch the soliton at points in the system where

$$\frac{f(z)}{\Pi^2} = \bar{F} \tag{35}$$

where $\bar{F}$ is the average error over a unit cell

$$\bar{F} = \frac{1}{L} \int_0^L \frac{f(z)}{\Pi^2} dz. \tag{36}$$

In Fig. 5 we plot the function $f(z)/\Pi^2 - \bar{F}$ for the case $L = 2$ and $\Gamma = .5$ when the function takes the simple form

$$2\left(1 - s - \frac{e^{-2s}}{1 - e^{-2}}\right) \tag{37}$$

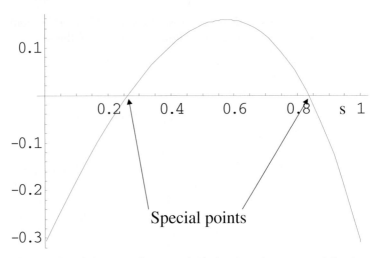

**Fig. 5.** Plot of the error function (37) showing the two special points

where $s = z/L$. The figure shows that there are two points at which the function is zero. These points are known as special points in the system and are the positions where it is optimum to launch a soliton pulse.

Further details of numerical simulations illustrating the effect of launching at the special points can be found in [7].

We can understand the nature of the special points in relation to another concept, pre-chirping [8]. Since the system is no longer an ideal lossless fiber, it is not necessary that the pulse should retain the simple soliton shape throughout propagation. We can therefore ask what is the best way to operate the system and this question can be posed in two complementary ways. First, we can ask at what point in the system is the actual pulse closest to the normal soliton pulse? Second, we can ask what is the optimum pulse to launch at the start of the system for best performance? The answer to the first question is provided by the positions of the special points whereas the answer to the second question is usually given in terms of the optimum chirp, or prechirp, that must be applied to the pulse. In Fig 6 we can see how these are simply different views, in (a) the pulse is launched at a special point whereas in (b) the extra fiber is considered to be part of the transmitter.

### 3.3   Special Points in Dispersion Managed Systems

dispersion-managed soliton is a simple technique used to keep the local dispersion in a system high, to avoid four wave mixing, while keeping the average system dispersion low, to avoid effects such as Gordon-Haus jitter. An extensive list of references to this topic can be found in [12]. Here we will look at the special points of these dispersion managed systems as shown schematically in Fig. 7.

The system average dispersion is $D$ and the local dispersion alternates between the values of $D + \delta$ and $D - \delta$. Consider first the simple lossless case when

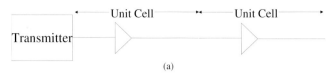

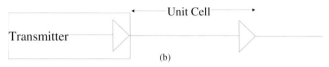

**Fig. 6.** Schematic illustration of the difference between (a) special points and (b) prechirping

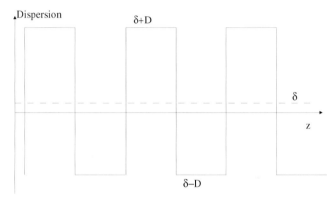

**Fig. 7.** Dispersion Management showing the local dispersion, full lines, and the system average dispersion, dashed line

the two fiber sections have equal lengths. A calculation similar to the one performed in the previous section can now be performed except that the local error needs to be calculated separately for the point $z$ located in both fibers with the average being taken over the full unit cell. The error function corresponding to Eq.(37) is now

$$1 - 2s \qquad (38)$$

so that the special points are now located at half the length of the fiber sections. This is exactly analogous to the trapezium rule (7) and the symmetric split operator (13) so here is a real system implication of the fact that the end points matter! For perfect lossless dispersion managed systems we should add a half length to the beginning of the system and a half length at the end for optimum operation and this is independent of the total length of the system.

The general case of dispersion management in the presence of loss can also be solved in the same way. For a system in which the amplifiers are placed after both the anomalous and normal fibers we obtain

$$\frac{1}{4(-1+e^{4G})G} \times$$
$$(-(-1-e^{2G}+2e^{4G}+(-1+e^{2G}+2e^{4G}-8e^{2G(1+s)})G)(-D+\delta) \qquad (39)$$

$$+ ((-2+3e^{2G}+e^{4G}+4e^{2G(1+s)})G - (-2+e^{2G}+e^{4G})(1+4Gs))(D+\delta))$$

where $G = \Gamma L$. In Fig 8 the error function is shown for three values of the map depth $D/\delta$.

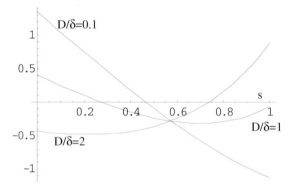

**Fig. 8.** Error function in dispersion managed systems for various values of the map strength as a function of s

There is always at least one special point but it can move essentially anywhere within the fiber length. This degree of flexibility leads on to the next topic of magic points.

## 3.4    Magic Points

We have seen various error functions depending on the type of system under consideration. In dispersion managed systems the error will always be of the following form when first order commutators are allowed for

$$\delta F_\delta(\Gamma, z, L) + D F_D(\Gamma, z, L) \qquad (40)$$

Turitsyn et al [12] pointed out that for such a form there might be values of z for which the two functions $F_\delta$ and $F_D$ are simultaneously zero. These 'magic points' would have the property of being special points that are independent of both the local dispersion and the system average dispersion and hence independent of wavelength making them ideal for use in WDM systems.

If we take the result of the previous section and look for such magic points then we find that they only exist for zero loss. In order to obtain the magic points we need a more general map in which the two fiber segments can have different lengths. The calculations can become very lengthy and so we will now introduce a combining theorem. Consider the case where the unit cell consists of a fiber section followed by a single amplifier followed by another fiber section with both fibers having the same loss coefficient $\Gamma$. We want to replace this by an effective fiber followed by an amplifier. If we let $P(N, L, D)$ be the propagator for a fiber of length $L$, dispersion $D$ and nonlinear coefficient $N$ then we can obtain the following result for the propagator of the effective fiber

$$P(N_1, L_1, D_1)gP(N_2, L_2, D_2) =$$
$$gP(N_{eff}(g, L_1, N_1, L_2, N_2), L_1 + L_2, \frac{D_1L_1 + D_2L_2}{L_1, L_2}). \tag{41}$$

Note that the effective fiber dispersion is equal to the path average dispersion and that the effective fiber length is equal to the total length as we would expect. The effective nonlinearity is now given by

$$N_{eff}(g, L_1, N_1, L_2, N_2) = \frac{e^{2g}(e^{2\Gamma L_1} - 1)N_1 + e^{2\Gamma L_1}(e^{2\Gamma L_2} - 1)N_2}{e^{2\Gamma(L_1+L_2)} - 1}. \tag{42}$$

This result can now be used to calculate the propagator for arbitrary combinations of fibers and amplifiers by applying it recursively to the map until a single effective propagator is found. At each step in the recursion we generate some commutators and the contribution at each step is given by

$$C(g, L_1, D_1, N_1, L_2, D_2, N_2) =$$
$$\frac{1}{4\Gamma}(D_2L_2N_1e^{2g-2\Gamma L_2}(1 - e^{-2\Gamma L_1}) - D_1L_1N_2(e^{-2\Gamma L_2} - 1)). \tag{43}$$

This is now all we need to calculate the correction terms for an arbitrary map.

## 4    Conclusions

We have shown that a central theme, the end points matter, is common to the areas of simple numerical integration, numerical integration and real systems applications. By making simple changes to the way the ends of a system or calculation are treated we can affect the global properties such as accuracy of a scheme or reduce the transients experienced during pulse propagation. The split step operator method is a general technique that can be used to obtain a variety of results in the studies of soliton systems.

## References

1. G.P.Agrawal: *Nonlinear fiber optics*, (Academic Press, New York, 1989)
2. D.Anderson: Phys. Rev. A **27**, 3185 (1983)
3. K.J.Blow, N.J.Doran: Opt. Comm. **42**, 403 (1982)

4. K.J.Blow, N.J.Doran: IEEE Photon Tech Letts **3**, 369 (1991)
5. K.J.Blow, D.Wood: IEEE J Quant Elect **25**, 2665 (1989)
6. N.J.Doran and K.J.Blow: IEEE J Quant Elect **19**, 1883 (1983)
7. W.Forysiak, N.J.Doran, F.M.Knox, K.J.Blow: Opt. Comm. **117**, 65 (1995)
8. T.Georges and B.Charbonnier: IEEE Photon Tech Letts **9**, 127 (1997)
9. A.Hasegawa, Y.Kodama: Phys Rev Letts **66**, 161 (1991)
10. A.Hasegawa, F.P.Tappert: Appl. Phys. Lett. **23**, 142 (1973)
11. R.Hawkins: private communication
12. S.K.Turitsyn, M.Fedoruk, T-S.Yang and W.L.Kath: IEEE J Quant Elect **36**, 290 (2000)

# Multicomponent Higher Order Bright Soliton Solutions and Shape Changing Collisions in Coupled Nonlinear Schrödinger Equations

M. Lakshmanan and T. Kanna

Centre for Nonlinear Dynamics, Department of Physics, Bharathidasan University, Tiruchirapalli 620 024, India

**Abstract.** Optical soliton propagation through multimode fibers, photorefractive materials and so on is governed by a set of coupled nonlinear Schrödinger (CNLS) equations. Here we present the exact bright one-, two- and three- soliton solutions of the integrable (Manakov) 2-CNLS equations and generalize them to arbitrary integrable N-CNLS equations. We also point out that these soliton solutions of CNLS equations undergo a fascinating shape changing collision, which is an uncommon feature of $(1+1)$ dimension systems. From an application point of view, we briefly point out the role of this shape changing collision in the context of optical computing.

## 1 Introduction

The theoretical and experimental studies on optical solitons have shown the possibility to revolutionize the long distance optical communication due to the capability of solitons propagating over considerable distances without losing their identities[1,2]. It is found that such temporal optical soliton propagation in single mode fibers is governed by the ubiquitous nonlinear Schrödinger equation [2]. In the case of birefringent fibers and two mode fibers, the propagation equation is a set of 2-CNLS equations which is nonintegrable [1,2] in general. One can also verify that the optical soliton propagation in a multimode fiber is governed by a set of N-CNLS equations [2] which is also generally not integrable. However, it becomes integrable for specific choices of parameters [3]. The CNLS equations also appear in the theory of soliton wavelength division multiplexing [4], multi-channel bit parallel-wavelength optical fiber network [5], bio-physics [6], and so on.

Very recently, it has been noted that in the context of beam propagation in a Kerr-like photorefractive medium, which generally exhibits very strong nonlinear effects with extremely low optical powers, the governing equations are again a set of N-CNLS equations [7]. In recent years, there is an increasing interest in studying soliton propagation through such photorefractive media after the observation of so called partially incoherent solitons through excitation by partially coherent light [8] and also with an ordinary incandescent light bulb [9]. As a consequence, it has been observed both experimentally and theoretically that the N-CNLS equations support a class of stationary partially coherent solitons (PCS), which are of variable shape and have been interpreted as multisoliton complexes [7,10].

Though a large number of investigations exist in the literature on the study of soliton propagation in CNLS equations, exact results are scarce, except for a special parametric choice of the two coupled nonlinear Schrödinger equations, namely the Manakov model [11]. In a recent work Radhakrishnan, Lakshmanan and Hietarinta [12] have obtained the explicit 2-soliton solution of the integrable Manakov system and identified novel shape changing collision property in it. In a very recent study the present authors [13] have extended the analysis to the case of integrable 3-CNLS and arbitrary N-CNLS systems, where a similar property has been identified, but with much more exciting possibilities. We have also shown in our work [13] that the PCS solutions available in the literature are special cases of these shape changing solitons. Further, Jakubowski, Steiglitz and Squier [14] have exploited the above shape changing collision property of the 2-CNLS (Manakov) system, by identifying the collision as corresponding to a linear fractional transformation, in performing logical operations. Later Steiglitz [15] has extended the idea even further and successfully shown that universal logic gates and an all optical computer in a homogeneous bulk medium equivalent to a Turing machine can be constructed at least in a mathematical sense.

Our aim in this paper is to briefly review the various aspects of the fascinating shape changing collision dynamics exhibited by the CNLS equations and to point out its role in developing optical computing. In sec.2, as a background information, we illustrate how the CNLS equations result as governing equations for intense light propagation in multimode fibers and photorefractive materials. The bright one-, two- and three- soliton solutions of the CNLS equations are obtained in sec.3. Section 4 deals with a detailed analysis of the two-soliton solution of 2-, 3- and N-CNLS equations. The conclusion is presented in sec.5.

## 2    CNLS Equations as Governing Equations for Intense Light Propagation in Multimode Fibers and Photorefractive Materials

As a prelude to the CNLS equations, let us consider the intense electromagnetic wave propagation in a birefringent fiber (see for example, [2]). Due to birefringence a single mode fiber can support two distinct modes of polarization which are orthogonal to each other. These two modes can be viewed as one the ordinary ray (O-ray) for which the refractive index of the medium is constant along every direction of the incident ray and the other as the extraordinary ray (E-ray) whose refractive index for the medium varies with the direction of the incident ray. Then the nonlinear phase variation of a particular mode not only depends on its own intensity but also on that of the co-propagating mode.

The study of such a system starts straight away from the Maxwell's equations for electromagnetic wave propagation in a dielectric medium,

$$\nabla^2 \vec{E} - \frac{1}{c^2}\frac{\partial^2 \vec{E}}{\partial t^2} = -\mu_0 \frac{\partial^2 \vec{P}}{\partial t^2} \ , \tag{1}$$

where $\vec{E}(\vec{r}, t)$ is the electric field, $\mu_0$ represents the free space permeability, $c$ is the velocity of light and $\vec{P}$ is the induced polarization which can be separated into two parts : $\vec{P} = \vec{P}_L + \vec{P}_{NL}$, where $\vec{P}_L(\vec{r}, t)$ and $\vec{P}_{NL}(\vec{r}, t)$ represent the linear and nonlinear parts of the induced polarization. It is well known that they can be expressed in terms of the electric field as [2]

$$\vec{P}_L(\vec{r}, t) = \epsilon_0 \int_{-\infty}^{\infty} \chi^{(1)}(t - t') \vec{E}(\vec{r}, t') dt' , \tag{2a}$$

$$\vec{P}_{NL}(\vec{r}, t) = \epsilon_0 \int \int \int_{-\infty}^{\infty} \chi^{(3)}(t - t_1, t - t_2, t - t_3) :$$
$$\vec{E}(\vec{r}, t_1) \vec{E}(\vec{r}, t_2) \vec{E}(\vec{r}, t_3) dt_1 dt_2 dt_3 , \tag{2b}$$

where $\epsilon_0$ is the free space permittivity and $\chi^{(j)}$ is the $j$th order susceptibility tensor of rank $(j + 1)$.

Considering wave propagation along elliptically birefringent optical fibers, the electric field $\vec{E}(\vec{r}, t)$ can be written in the quasi-monochromatic approximation as

$$\vec{E}(\vec{r}, t) = \frac{1}{2} [\hat{e}_1 E_1(z, t) + \hat{e}_2 E_2(z, t)] e^{-i\omega_0 t} + c.c , \tag{3}$$

where $z$ is the direction of propagation, $t$ represents the retarded time and c.c stands for the complex conjugate and the orthonormal polarization vectors $\hat{e}_1$ and $\hat{e}_2$ can be expressed in terms of the unit polarization vectors $\hat{x}$ and $\hat{y}$ along the $x$ and $y$ directions respectively as

$$\hat{e}_1 = \frac{\hat{x} + ir\hat{y}}{(1 + r^2)^{1/2}} , \tag{4a}$$

$$\hat{e}_2 = \frac{r\hat{x} - i\hat{y}}{(1 + r^2)^{1/2}} , \tag{4b}$$

in which the parameter $r$ is a measure of the extent of ellipticity. In equation (3) $E_1$ and $E_2$ are the complex amplitudes of the two polarization components at frequency $\omega_0$. Considering the medium to be isotropic, the nonlinear polarization $\vec{P}_{NL}(\vec{r}, t)$ can be obtained by substituting (3) and (4) in (2b).

Under a slowly varying approximation, $E_1$ and $E_2$ can be written as

$$E_j(z, t) = F_j(x, y) Q_j(z, t) e^{iK_{0j}z}, \quad j = 1, 2 , \tag{5}$$

where $F_j(x, y)$ and $K_{0j}$ , $j = 1, 2$, are the fiber mode distributions in the transverse directions and the propagation constants for the two modes, respectively. Then the resulting evolution equations for $Q_j(z, t)$ can be deduced as

$$iQ_{1z} + \frac{i}{v_{g1}} Q_{1t} - \frac{k''}{2} Q_{1tt} + \mu(|Q_1|^2 + B|Q_2|^2) Q_1 = 0 ,$$

$$iQ_{2z} + \frac{i}{v_{g2}} Q_{2t} - \frac{k''}{2} Q_{2tt} + \mu(|Q_2|^2 + B|Q_1|^2) Q_2 = 0 , \tag{6}$$

where $v_{g1}$ and $v_{g2}$ are the group velocities of the two co-propagating waves respectively, $k'' = \left[\frac{\partial^2 K}{\partial \omega^2}\right]_{\omega=\omega_0}$ accounts for the group velocity dispersion, $\mu$ is the nonlinearity coefficient and $B = \frac{2+2\sin^2\vartheta}{2+\cos^2\vartheta}$ is the XPM coupling parameter ($\vartheta$ - birefringence ellipticity angle which varies between 0 and $\frac{\pi}{2}$ ). Here we have considered the fiber to be lossless and neglected the third order dispersion term. Further, we have also treated the fiber as strongly birefringent and hence the four-wave mixing terms also can be neglected.

Introducing now the transformation $T = t - \frac{z}{2}(\frac{1}{v_{g1}} + \frac{1}{v_{g2}})$, $Z = z$, equation (6) becomes

$$iQ_{1Z} + \frac{i}{2}\hat{\rho}Q_{1T} - \frac{k''}{2}Q_{1TT} + \mu(|Q_1|^2 + B|Q_2|^2)Q_1 = 0\,,$$

$$iQ_{2Z} - \frac{i}{2}\hat{\rho}Q_{2T} - \frac{k''}{2}Q_{2TT} + \mu(|Q_2|^2 + B|Q_1|^2)Q_2 = 0\,, \tag{7}$$

where $\hat{\rho} = (\frac{1}{v_{g1}} - \frac{1}{v_{g2}})$. Then, by using the transformation $\tilde{q}_j = \left(\frac{T_0^2}{k''}\right)^{\frac{1}{2}} Q_j$, $j = 1, 2$, $z' = \frac{|k''|Z}{T_0^2}$, $t' = \frac{T}{T_0}$ and redefining $z'$ and $t'$ as $z$ and $t$ respectively, in the anomalous dispersion regime ($k''$ is negative), we end up with the following set of equations,

$$i\tilde{q}_{1z} + i\rho\tilde{q}_{1t} + \frac{1}{2}\tilde{q}_{1tt} + \mu(|\tilde{q}_1|^2 + B|\tilde{q}_2|^2)\tilde{q}_1 = 0\,,$$

$$i\tilde{q}_{2z} - i\rho\tilde{q}_{2t} + \frac{1}{2}\tilde{q}_{2tt} + \mu(|\tilde{q}_2|^2 + B|\tilde{q}_1|^2)\tilde{q}_2 = 0\,, \tag{8}$$

where $\rho = \frac{T_0\hat{\rho}}{2|k''|}$ and $T_0$ is a measure of the pulse width.

Equations (8) can be further simplified with the transformation

$$\tilde{q}_1 = q_1 e^{i\left(\frac{\rho^2 z}{2} - \rho t\right)}\,, \tilde{q}_2 = q_2 e^{i\left(\frac{\rho^2 z}{2} + \rho t\right)} \tag{9}$$

and then redefining $z$ as $2z$, we obtain

$$iq_{1z} + q_{1tt} + 2\mu(|q_1|^2 + B|q_2|^2)q_1 = 0\,,$$

$$iq_{2z} + q_{2tt} + 2\mu(|q_2|^2 + B|q_1|^2)q_2 = 0\,. \tag{10}$$

Equation (10) is the 2-CNLS equation in the standard form. It is in general nonintegrable and fails to satisfy the Painleve' property unless $B = 1$ [3,16]. In the later case, we have the celebrated Manakov system of equations [11]

$$iq_{1z} + q_{1tt} + 2\mu(|q_1|^2 + |q_2|^2)q_1 = 0\,,$$

$$iq_{2z} + q_{2tt} + 2\mu(|q_1|^2 + |q_2|^2)q_2 = 0\,, \tag{11}$$

which is a completely integrable soliton system .

Besides the above, there exists other situations also where the above type of CNLS equations arise. For example, in the case of propagation of two optical fields which are having different frequencies, the governing equations have a similar form [2] as Eq.(10). Similarly let us consider the simultaneous propagation of N-optical fields (beams) with different wavelengths in a single mode fiber. This kind of simultaneous propagation of multiple beams in a single mode fiber is known as wavelength division multiplexed (WDM) transmission. Here also, by extending the above analysis, one can find that the governing equations are related to a set of N-CNLS equations [1,2],

$$iq_{jz} + q_{jtt} + 2\mu(|q_j|^2 + B \sum_{p=1;\ p \neq j}^{N} |q_p|^2)q_j = 0, \quad j = 1, 2, ...N , \qquad (12)$$

where $q_j(z,t)$ is the slowly varying amplitude of the jth wave. In addition, it is also found that a similar set of N-CNLS equations represent the soliton propagation through optical fiber array, with $q_j$ representing the envelope in the jth core. System (12) is also found to be nonintegrable, except for the special case $B = 1$, when the Painleve' property is satisfied [3,16]. Also Lax pair exists for this case [3] and the corresponding integrable equations can be written as

$$iq_{jz} + q_{jtt} + 2\mu \sum_{p=1}^{N} |q_p|^2 q_j = 0, \quad j = 1, 2, ..., N . \qquad (13)$$

Further, it has been found that the governing equation of N-self-trapped mutually incoherent wave packets in such a media is given by the N-CNLS equations (13) [7]. In this context, $q_j$ represents the jth component of the beam, $z$ and $t$ represent the normalized co-ordinates along the direction of propagation and the transverse co-ordinate respectively, and $\sum_{p=1}^{N} |q_p|^2$ represents the change in the refractive index profile created by all the incoherent components of the light beam and $2\mu$ represents the strength of nonlinearity.

## 3    Soliton Solutions of the CNLS Equations

Now in order to facilitate the understanding of the underlying dynamics of the above integrable CNLS equations (11) as well as (13) it is essential to obtain the soliton solutions associated with these integrable systems. In this regard, by applying Hirota's technique [17], we point out that the most general bright one-soliton and two-soliton solutions for the Manakov system (11) can be easily obtained and novel properties deduced [12, 18, 20]. The procedure can be straightforwardly extended to obtain N-soliton solutions as well [13, 19]; however, we will confine ourselves to the study of one-, two- and three-soliton solutions alone here.

Considering Eq.(13) and by making the bilinear transformation $q_j = \frac{g^{(j)}}{f}$, $j = 1, 2, ..., N$, where $g^{(j)}(z,t)$'s are complex functions while f(z,t) is a real function, the following bilinear equations can be obtained,

$$(iD_z + D_t^2)(g^{(j)}.f) = 0 \ , j = 1, 2, ..., N \ , \tag{14a}$$

$$D_t^2(f.f) = 2\mu \sum_{n=1}^{N} g^{(n)} g^{(n)*} \ , \tag{14b}$$

where the Hirota's bilinear operators $D_z$ and $D_t$ are defined by

$$D_z^n D_t^m(a.b) = \left(\frac{\partial}{\partial z} - \frac{\partial}{\partial z'}\right)^n \left(\frac{\partial}{\partial t} - \frac{\partial}{\partial t'}\right)^m a(z,t)b(z',t')\Big|_{(z=z',t=t')} . \tag{15}$$

The above set of equations can be solved by introducing the following power series expansions to $g^{(j)}$'s and f:

$$g^{(j)} = \lambda g_1^{(j)} + \lambda^3 g_3^{(j)} + ..., \quad j = 1, 2, ..., N \ , \tag{16a}$$
$$f = 1 + \lambda^2 f_2 + \lambda^4 f_4 + ... \ , \tag{16b}$$

where $\lambda$ is the formal expansion parameter. The resulting set of equations, after collecting the terms with the same power in $\lambda$, can be solved to obtain the forms of $g^{(j)}$ and $f$. For illustrative purpose we explain below the procedure to obtain the one-, two- and three-soliton solutions.

## 3.1   N=2 Case

### (a) One-Soliton Solution of the 2-CNLS Equation

In order to get the one-soliton solution of the 2-CNLS equation, the power series expansions for $g^{(1)}$, $g^{(2)}$ and f are terminated as follows:

$$g^{(1)} = \lambda g_1^{(1)} \ , \tag{17a}$$

$$g^{(2)} = \lambda g_1^{(2)} \ , \tag{17b}$$

$$f = 1 + \lambda^2 f_2. \tag{17c}$$

Substituting Eq.(17) in the bilinear equations (14) we obtain the following linear partial differential equations at various powers of $\lambda$:

$$\lambda: \quad \hat{D}_1(g_1^{(j)}.1) = 0 \ , \tag{18a}$$

$$\lambda^2: \quad \hat{D}_2(f_2.1 + 1.f_2) = 2\mu(g_1^{(1)}.g_1^{(1)*} + g_1^{(2)}.g_1^{(2)*}) \ , \tag{18b}$$

$$\lambda^3: \quad \hat{D}_1(g_1^{(j)}.f_2 + g_3^{(j)}.1) = 0 \ , \tag{18c}$$

$$\lambda^4: \quad \hat{D}_2(f_2.f_2) = 0, \quad j = 1, 2 \ , \tag{18d}$$

where $\hat{D}_1 = (iD_z + D_t^2)$ and $\hat{D}_2 = D_t^2$. The solution which is consistent with the above system is

$$g^{(1)} = \alpha_1^{(1)} e^{\eta_1} \ , \tag{19a}$$

$$g^{(2)} = \alpha_1^{(2)} e^{\eta_1} \ , \tag{19b}$$

$$f = 1 + e^{\eta_1 + \eta_1^* + R}, \quad e^R = \frac{\mu(|\alpha_1^{(1)}|^2 + |\alpha_1^{(2)}|^2)}{(k_1 + k_1^*)^2} \ , \tag{19c}$$

where $\eta_1 = k_1(t + ik_1z)$, $\alpha_1^{(1)}$, $\alpha_1^{(2)}$ and $k_1$ are complex parameters. Then the resulting bright one-soliton solution is obtained as

$$\begin{pmatrix} q_1 \\ q_2 \end{pmatrix} = \begin{pmatrix} \alpha_1^{(1)} \\ \alpha_1^{(2)} \end{pmatrix} \frac{e^{\eta_1}}{1 + e^{\eta_1 + \eta_1^* + R}} = \begin{pmatrix} A_1 \\ A_2 \end{pmatrix} \frac{k_{1R} e^{i\eta_{1I}}}{\cosh\left(\eta_{1R} + \frac{R}{2}\right)}. \tag{20}$$

Here $\sqrt{\mu}(A_1, A_2) = \sqrt{\mu}(\alpha_1^{(1)}, \alpha_1^{(2)})/(\mu(|\alpha_1^{(1)}|^2 + |\alpha_1^{(2)}|^2))^{1/2}$ represents the unit polarization vector, $k_{1R}A_j$, $j = 1, 2$, gives the amplitude of the $j$th mode and $2k_{1I}$ is the soliton velocity.

## (b) Two-Soliton Solution

Here the series expansions given by (16) are terminated as

$$g^{(1)} = \lambda g_1^{(1)} + \lambda^3 g_3^{(1)}, \tag{21a}$$

$$g^{(2)} = \lambda g_1^{(2)} + \lambda^3 g_3^{(2)}, \tag{21b}$$

$$f = 1 + \lambda^2 f_2 + \lambda^4 f_4, \tag{21c}$$

to obtain the bright two-soliton solution of equation (11). Then the resulting partial differential equations at various powers of $\lambda$ are as follows:

$$\lambda: \quad \hat{D}_1(g_1^{(j)}.1) = 0, \tag{22a}$$

$$\lambda^2: \quad \hat{D}_2(1.f_2 + f_2.1) = 2\mu(g_1^{(1)}g_1^{(1)*} + g_1^{(2)}g_1^{(2)*}), \tag{22b}$$

$$\lambda^3: \quad \hat{D}_1(g_1^{(j)}.f_2 + g_3^{(j)}.1) = 0, \tag{22c}$$

$$\lambda^4: \quad \hat{D}_2(1.f_4 + f_2.f_2 + f_4.1) = 2\mu \sum_{j=1}^{2} g_1^{(j)}g_3^{(j)*} + g_3^{(j)}g_1^{(j)*}, \tag{22d}$$

$$\lambda^5: \quad \hat{D}_1(g_1^{(j)}.f_4 + g_3^{(j)}.f_2) = 0, \tag{22e}$$

$$\lambda^6: \quad \hat{D}_2(f_2.f_4 + f_4.f_2) = 2\mu(g_3^{(1)}g_3^{(1)*} + g_3^{(2)}g_3^{(2)*}), \tag{22f}$$

$$\lambda^7: \quad \hat{D}_1(g_3^{(j)}.f_4) = 0, \tag{22g}$$

$$\lambda^8: \quad \hat{D}_2(f_4.f_4) = 0, \quad j = 1, 2. \tag{22h}$$

The solutions compatible with (22) are

$$g_1^{(1)} = \alpha_1^{(1)}e^{\eta_1} + \alpha_2^{(1)}e^{\eta_2}, \tag{23a}$$

$$g_1^{(2)} = \alpha_1^{(2)}e^{\eta_1} + \alpha_2^{(2)}e^{\eta_2}, \tag{23b}$$

$$g_3^{(1)} = e^{\eta_1 + \eta_1^* + \eta_2 + \delta_{11}} + e^{\eta_1 + \eta_2 + \eta_2^* + \delta_{21}}, \tag{23c}$$

$$g_3^{(2)} = e^{\eta_1 + \eta_1^* + \eta_2 + \delta_{12}} + e^{\eta_1 + \eta_2 + \eta_2^* + \delta_{22}}, \tag{23d}$$

$$f_2 = e^{\eta_1 + \eta_1^* + R_1} + e^{\eta_1 + \eta_2^* + \delta_0} + e^{\eta_1^* + \eta_2 + \delta_0^*} + e^{\eta_2 + \eta_2^* + R_2}, \tag{23e}$$

$$f_4 = e^{\eta_1 + \eta_1^* + \eta_2 + \eta_2^* + R_3}, \tag{23f}$$

where

$$\eta_i = k_i(t + ik_i z) , \quad e^{\delta_0} = \frac{\kappa_{12}}{k_1 + k_2^*}, \quad e^{R_1} = \frac{\kappa_{11}}{k_1 + k_1^*}, \quad e^{R_2} = \frac{\kappa_{22}}{k_2 + k_2^*} ,$$

$$e^{\delta_{11}} = \frac{(k_1 - k_2)(\alpha_1^{(1)}\kappa_{21} - \alpha_2^{(1)}\kappa_{11})}{(k_1 + k_1^*)(k_1^* + k_2)}, \quad e^{\delta_{21}} = \frac{(k_2 - k_1)(\alpha_2^{(1)}\kappa_{12} - \alpha_1^{(1)}\kappa_{22})}{(k_2 + k_2^*)(k_1 + k_2^*)},$$

$$e^{\delta_{12}} = \frac{(k_1 - k_2)(\alpha_1^{(2)}\kappa_{21} - \alpha_2^{(2)}\kappa_{11})}{(k_1 + k_1^*)(k_1^* + k_2)}, \quad e^{\delta_{22}} = \frac{(k_2 - k_1)(\alpha_2^{(2)}\kappa_{12} - \alpha_1^{(2)}\kappa_{22})}{(k_2 + k_2^*)(k_1 + k_2^*)},$$

$$e^{R_3} = \frac{|k_1 - k_2|^2}{(k_1 + k_1^*)(k_2 + k_2^*)|k_1 + k_2^*|^2}(\kappa_{11}\kappa_{22} - \kappa_{12}\kappa_{21}) \tag{23g}$$

and

$$\kappa_{ij} = \frac{\mu(\alpha_i^{(1)}\alpha_j^{(1)*} + \alpha_i^{(2)}\alpha_j^{(2)*})}{k_i + k_j^*}, \quad i,j = 1,2 . \tag{23h}$$

Then we can write the final form of the bright two-soliton solution of equation (11) as

$$q_1 = \frac{\alpha_1^{(1)}e^{\eta_1} + \alpha_2^{(1)}e^{\eta_2} + e^{\eta_1+\eta_1^*+\eta_2+\delta_1} + e^{\eta_1+\eta_2+\eta_2^*+\delta_2}}{D} , \tag{24a}$$

$$q_2 = \frac{\alpha_1^{(2)}e^{\eta_1} + \alpha_2^{(2)}e^{\eta_2} + e^{\eta_1+\eta_1^*+\eta_2+\delta_1'} + e^{\eta_1+\eta_2+\eta_2^*+\delta_2'}}{D} , \tag{24b}$$

$$D = 1 + e^{\eta_1+\eta_1^*+R_1} + e^{\eta_1+\eta_2^*+\delta_0} + e^{\eta_1^*+\eta_2+\delta_0^*} + e^{\eta_2+\eta_2^*+R_2}$$
$$+ e^{\eta_1+\eta_1^*+\eta_2+\eta_2^*+R_3} . \tag{24c}$$

The above most general bright two-soliton solution with six arbitrary complex parameters $k_1$, $k_2$, $\alpha_1^{(j)}$ and $\alpha_2^{(j)}$, $j = 1,2$ corresponds to a shape changing (inelastic) collision of two bright solitons which will be explained in the following section. The above solution (24) can also be written in a more compact form as

$$q_j = \frac{\alpha_1^{(j)}e^{\eta_1} + \alpha_2^{(j)}e^{\eta_2} + e^{\eta_1+\eta_1^*+\eta_2+\delta_{1j}} + e^{\eta_1+\eta_2+\eta_2^*+\delta_{2j}}}{D} , \quad j = 1,2 . \tag{25}$$

Note that in [12], $\delta_{11}$, $\delta_{12}$, $\delta_{21}$ and $\delta_{22}$ are called as $\delta_1$, $\delta_1'$, $\delta_2$ and $\delta_2'$, respectively. The redefined quantities $\delta_{ij}$'s, $i,j = 1,2$ are now used for notational simplicity.

## (c) Three-Soliton Solution

The two-soliton solution itself is very difficult to derive and complicated to analyze [12]. So obtaining the three-soliton solution is a more laborious and tedious task. However we have successfully obtained the explicit form of the bright three-soliton solution also [19]. In order to obtain the three-soliton solution of Eq.(13) for the $N = 2$ case we terminate the power series (16a) and (16b) as

$$g^{(j)} = \lambda g_1^{(j)} + \lambda^3 g_3^{(j)} + \lambda^5 g_5^{(j)} , \tag{26a}$$

$$f = 1 + \lambda^2 f_2 + \lambda^4 f_4 + \lambda^6 f_6, \quad j = 1,2 . \tag{26b}$$

Substitution of (26) into the bilinear equations (14a) and (14b) yields the following set of linear partial differential equations at various powers of $\lambda$.

$$\lambda: \quad \hat{D}_1(g_1^{(j)}.1) = 0 \,, \tag{27a}$$

$$\lambda^2: \quad \hat{D}_2(1.f_2 + f_2.1) = 2\mu \sum_{j=1}^{2}(g_1^{(j)}g_1^{(j)*}) \,, \tag{27b}$$

$$\lambda^3: \quad \hat{D}_1(g_1^{(j)}.f_2 + g_3^{(j)}.1) = 0 \,, \tag{27c}$$

$$\lambda^4: \quad \hat{D}_2(1.f_4 + f_2.f_2 + f_4.1) = 2\mu \sum_{j=1}^{2}(g_1^{(j)}g_3^{(j)*} + g_3^{(j)}g_1^{(j)*}) \,, \tag{27d}$$

$$\lambda^5: \quad \hat{D}_1(g_1^{(j)}.f_4 + g_3^{(j)}.f_2 + g_5^{(j)}.1) = 0 \,, \tag{27e}$$

$$\lambda^6: \quad \hat{D}_2(1.f_6 + f_2.f_4 + f_4.f_2 + f_6.1)$$
$$= 2\mu \sum_{j=1}^{2}(g_1^{(j)}g_5^{(j)*} + g_3^{(j)}g_3^{(j)*} + g_5^{(j)}g_1^{(j)*}) \,, \tag{27f}$$

$$\lambda^7: \quad \hat{D}_1(g_1^{(j)}.f_6 + g_3^{(j)}.f_4 + g_5^{(j)}.f_2) = 0 \,, \tag{27g}$$

$$\lambda^8: \quad \hat{D}_2(f_2.f_6 + f_4.f_4 + f_6.f_2) = 2\mu \sum_{j=1}^{2}(g_3^{(j)}g_5^{(j)*} + g_5^{(j)}g_3^{(j)*}) \,, \tag{27h}$$

$$\lambda^9: \quad \hat{D}_1(g_3^{(j)}.f_6 + g_5^{(j)}.f_4) = 0 \,, \tag{27i}$$

$$\lambda^{10}: \quad \hat{D}_2(f_4.f_6 + f_6.f_4) = 2\mu \sum_{j=1}^{2}(g_5^{(j)}g_5^{(j)*}) \,, \tag{27j}$$

$$\lambda^{11}: \quad \hat{D}_1(g_5^{(j)}.f_6) = 0 \,, \tag{27k}$$

$$\lambda^{12}: \quad \hat{D}_2(f_6.f_6) = 0 \,. \tag{27l}$$

As in the cases of one- and two- soliton solutions, here also by solving the above set of coupled linear partial differential equations recursively one can obtain the three-soliton solution as given below.

$$q_j = \frac{\alpha_1^{(j)}e^{\eta_1} + \alpha_2^{(j)}e^{\eta_2} + \alpha_3^{(j)}e^{\eta_3} + e^{\eta_1+\eta_1^*+\eta_2+\delta_{1j}} + e^{\eta_1+\eta_1^*+\eta_3+\delta_{2j}}}{D}$$
$$+ \frac{e^{\eta_2+\eta_2^*+\eta_1+\delta_{3j}} + e^{\eta_2+\eta_2^*+\eta_3+\delta_{4j}} + e^{\eta_3+\eta_3^*+\eta_1+\delta_{5j}} + e^{\eta_3+\eta_3^*+\eta_2+\delta_{6j}}}{D}$$
$$+ \frac{e^{\eta_1^*+\eta_2+\eta_3+\delta_{7j}} + e^{\eta_1+\eta_2^*+\eta_3+\delta_{8j}} + e^{\eta_1+\eta_2+\eta_3^*+\delta_{9j}} + e^{\eta_1+\eta_1^*+\eta_2+\eta_2^*+\eta_3+\tau_{1j}}}{D}$$
$$+ \frac{e^{\eta_1+\eta_1^*+\eta_3+\eta_3^*+\eta_2+\tau_{2j}} + e^{\eta_2+\eta_2^*+\eta_3+\eta_3^*+\eta_1+\tau_{3j}}}{D} \,, \quad j = 1,2 \,, \tag{28a}$$

where
$$D = 1 + e^{\eta_1+\eta_1^*+R_1} + e^{\eta_2+\eta_2^*+R_2} + e^{\eta_3+\eta_3^*+R_3} + e^{\eta_1+\eta_2^*+\delta_{10}} + e^{\eta_1^*+\eta_2+\delta_{10}^*}$$
$$e^{\eta_1+\eta_3^*+\delta_{20}} + e^{\eta_1^*+\eta_3+\delta_{20}^*} + e^{\eta_2+\eta_3^*+\delta_{30}} + e^{\eta_2^*+\eta_3+\delta_{30}^*} + e^{\eta_1+\eta_1^*+\eta_2+\eta_2^*+R_4}$$
$$+e^{\eta_1+\eta_1^*+\eta_3+\eta_3^*+R_5} + e^{\eta_2+\eta_2^*+\eta_3+\eta_3^*+R_6} + e^{\eta_1+\eta_1^*+\eta_2+\eta_3^*+\tau_{10}}$$
$$+e^{\eta_1+\eta_1^*+\eta_3+\eta_2^*+\tau_{10}^*} + e^{\eta_2+\eta_2^*+\eta_1+\eta_3^*+\tau_{20}} + e^{\eta_2+\eta_2^*+\eta_1^*+\eta_3+\tau_{20}^*}$$
$$+e^{\eta_3+\eta_3^*+\eta_1+\eta_2^*+\tau_{30}} + e^{\eta_3+\eta_3^*+\eta_1^*+\eta_2+\tau_{30}^*} + e^{\eta_1+\eta_1^*+\eta_2+\eta_2^*+\eta_3+\eta_3^*+R_7} \ . \quad (28b)$$

Here
$$\eta_i = k_i(t + ik_i z) \, , i = 1, 2, 3 \, , \qquad\qquad\qquad (28c)$$

$$e^{\delta_{1j}} = \frac{(k_1 - k_2)(\alpha_1^{(j)}\kappa_{21} - \alpha_2^{(j)}\kappa_{11})}{(k_1 + k_1^*)(k_1^* + k_2)}, \quad e^{\delta_{2j}} = \frac{(k_1 - k_3)(\alpha_1^{(j)}\kappa_{31} - \alpha_3^{(j)}\kappa_{11})}{(k_1 + k_1^*)(k_1^* + k_3)} \, ,$$

$$e^{\delta_{3j}} = \frac{(k_1 - k_2)(\alpha_1^{(j)}\kappa_{22} - \alpha_2^{(j)}\kappa_{12})}{(k_1 + k_2^*)(k_2 + k_2^*)}, \quad e^{\delta_{4j}} = \frac{(k_2 - k_3)(\alpha_2^{(j)}\kappa_{32} - \alpha_3^{(j)}\kappa_{22})}{(k_2 + k_2^*)(k_2^* + k_3)} \, ,$$

$$e^{\delta_{5j}} = \frac{(k_1 - k_3)(\alpha_1^{(j)}\kappa_{33} - \alpha_3^{(j)}\kappa_{13})}{(k_3 + k_3^*)(k_3^* + k_1)}, \quad e^{\delta_{6j}} = \frac{(k_2 - k_3)(\alpha_2^{(j)}\kappa_{33} - \alpha_3^{(j)}\kappa_{23})}{(k_3^* + k_2)(k_3^* + k_3)} \, ,$$

$$e^{\delta_{7j}} = \frac{(k_2 - k_3)(\alpha_2^{(j)}\kappa_{31} - \alpha_3^{(j)}\kappa_{21})}{(k_1^* + k_2)(k_1^* + k_3)}, \quad e^{\delta_{8j}} = \frac{(k_1 - k_3)(\alpha_1^{(j)}\kappa_{32} - \alpha_3^{(j)}\kappa_{12})}{(k_1 + k_2^*)(k_2^* + k_3)} \, ,$$

$$e^{\delta_{9j}} = \frac{(k_1 - k_2)(\alpha_1^{(j)}\kappa_{23} - \alpha_2^{(j)}\kappa_{13})}{(k_1 + k_3^*)(k_2 + k_3^*)} \, ,$$

$$e^{\tau_{1j}} = \frac{(k_2 - k_1)(k_3 - k_1)(k_3 - k_2)(k_2^* - k_1^*)}{(k_1^* + k_1)(k_1^* + k_2)(k_1^* + k_3)(k_2^* + k_1)(k_2^* + k_2)(k_2^* + k_3)}$$
$$\times \left[ \alpha_1^{(j)}(\kappa_{21}\kappa_{32} - \kappa_{22}\kappa_{31}) + \alpha_2^{(j)}(\kappa_{12}\kappa_{31} - \kappa_{32}\kappa_{11}) \right.$$
$$\left. +\alpha_3^{(j)}(\kappa_{11}\kappa_{22} - \kappa_{12}\kappa_{21}) \right] \, ,$$

$$e^{\tau_{2j}} = \frac{(k_2 - k_1)(k_3 - k_1)(k_3 - k_2)(k_3^* - k_1^*)}{(k_1^* + k_1)(k_1^* + k_2)(k_1^* + k_3)(k_3^* + k_1)(k_3^* + k_2)(k_3^* + k_3)}$$
$$\times \left[ \alpha_1^{(j)}(\kappa_{33}\kappa_{21} - \kappa_{31}\kappa_{23}) + \alpha_2^{(j)}(\kappa_{31}\kappa_{13} - \kappa_{11}\kappa_{33}) \right.$$
$$\left. +\alpha_3^{(j)}(\kappa_{23}\kappa_{11} - \kappa_{13}\kappa_{21}) \right] \, ,$$

$$e^{\tau_{3j}} = \frac{(k_2 - k_1)(k_3 - k_1)(k_3 - k_2)(k_3^* - k_2^*)}{(k_2^* + k_1)(k_2^* + k_2)(k_2^* + k_3)(k_3^* + k_1)(k_3^* + k_2)(k_3^* + k_3)}$$
$$\times \left[ \alpha_1^{(j)}(\kappa_{22}\kappa_{33} - \kappa_{23}\kappa_{32}) + \alpha_2^{(j)}(\kappa_{13}\kappa_{32} - \kappa_{33}\kappa_{12}) \right.$$
$$\left. +\alpha_3^{(j)}(\kappa_{12}\kappa_{23} - \kappa_{22}\kappa_{13}) \right] \, ,$$

$$e^{R_m} = \frac{\kappa_{mm}}{k_m + k_m^*} \, , \quad m = 1, 2, 3 \, ,$$

$$e^{\delta_{10}} = \frac{\kappa_{12}}{k_1 + k_2^*} \, , \quad e^{\delta_{20}} = \frac{\kappa_{13}}{k_1 + k_3^*} \, , \quad e^{\delta_{30}} = \frac{\kappa_{23}}{k_2 + k_3^*} \, ,$$

$$\qquad\qquad\qquad\qquad\qquad\qquad\qquad\qquad\qquad (28d)$$

$$e^{R_4} = \frac{(k_2 - k_1)(k_2^* - k_1^*)}{(k_1^* + k_1)(k_1^* + k_2)(k_1 + k_2^*)(k_2^* + k_2)} \left[\kappa_{11}\kappa_{22} - \kappa_{12}\kappa_{21}\right],$$

$$e^{R_5} = \frac{(k_3 - k_1)(k_3^* - k_1^*)}{(k_1^* + k_1)(k_1^* + k_3)(k_3^* + k_1)(k_3^* + k_3)} \left[\kappa_{33}\kappa_{11} - \kappa_{13}\kappa_{31}\right],$$

$$e^{R_6} = \frac{(k_3 - k_2)(k_3^* - k_2^*)}{(k_2^* + k_2)(k_2^* + k_3)(k_3^* + k_2)(k_3 + k_3^*)} \left[\kappa_{22}\kappa_{33} - \kappa_{23}\kappa_{32}\right],$$

$$e^{\tau_{10}} = \frac{(k_2 - k_1)(k_3^* - k_1^*)}{(k_1^* + k_1)(k_1^* + k_2)(k_3^* + k_1)(k_3^* + k_2)} \left[\kappa_{11}\kappa_{23} - \kappa_{21}\kappa_{13}\right],$$

$$e^{\tau_{20}} = \frac{(k_1 - k_2)(k_3^* - k_2^*)}{(k_2^* + k_1)(k_2^* + k_2)(k_3^* + k_1)(k_3^* + k_2)} \left[\kappa_{22}\kappa_{13} - \kappa_{12}\kappa_{23}\right],$$

$$e^{\tau_{30}} = \frac{(k_3 - k_1)(k_3^* - k_2^*)}{(k_2^* + k_1)(k_2^* + k_3)(k_3^* + k_1)(k_3^* + k_3)} \left[\kappa_{33}\kappa_{12} - \kappa_{13}\kappa_{32}\right],$$

$$e^{R_7} = \frac{|k_1 - k_2|^2 |k_2 - k_3|^2 |k_3 - k_1|^2}{(k_1 + k_1^*)(k_2 + k_2^*)(k_3 + k_3^*)|k_1 + k_2^*|^2 |k_2 + k_3^*|^2 |k_3 + k_1^*|^2}$$
$$\times \left[(\kappa_{11}\kappa_{22}\kappa_{33} - \kappa_{11}\kappa_{23}\kappa_{32}) + (\kappa_{12}\kappa_{23}\kappa_{31} - \kappa_{12}\kappa_{21}\kappa_{33})\right.$$
$$\left. + (\kappa_{21}\kappa_{13}\kappa_{32} - \kappa_{22}\kappa_{13}\kappa_{31})\right]$$

and

$$\kappa_{il} = \frac{\mu\left(\alpha_i^{(1)}\alpha_l^{(1)*} + \alpha_i^{(2)}\alpha_l^{(2)*}\right)}{(k_i + k_l^*)}, \quad i, l = 1, 2, 3.$$

The above three-soliton solution represents three soliton interaction in the 2-CNLS equations and is characterized by nine arbitrary complex parameters $\alpha_i^{(j)}$'s and $k_i$'s, $i = 1, 2, 3$, $j = 1, 2$. The form in which we have presented the solution eases the complexity in generalizing the solution to multicomponent case as well as to higher order soliton solutions. For details see [19].

## 3.2 N=3 Case

The results are scarce for Eq.(13) with $N > 2$ and there exists a large class of physical systems in which the N-CNLS equations occur naturally. In order to study the solution properties of such systems first we consider the integrable 3-CNLS equations,

$$iq_{jz} + q_{jtt} + 2\mu(|q_1|^2 + |q_2|^2 + |q_3|^2)q_j = 0, \quad j = 1, 2, 3. \tag{29}$$

### (a) One-Soliton Solution

The one-soliton solution of equation (29) can be obtained by terminating the power series (16) as

$$g^{(j)} = \lambda g_1^{(j)}, j = 1, 2, 3, \quad f = 1 + \lambda^2 f_2. \tag{30}$$

Substituting this into the bilinear equations(14), and solving the resulting differential equations resulting at various powers of $\lambda$ we obtain $g^{(j)}$ and $f$ as

$$g^{(j)} = \alpha_1^{(j)} e^{\eta_1} \ , j = 1, 2, 3 \ , \quad f = 1 + e^{\eta_1 + \eta_1^* + R} \ , \tag{31}$$

where now $e^R = \mu \sum_{j=1}^3 |\alpha_1^{(j)}|^2 / (k_1 + k_1^*)^2, \ j = 1, 2, 3$. Then the bright one-soliton solution can be written as

$$(q_1, q_2, q_3)^T = \frac{e^{\eta_1}}{1 + e^{\eta_1 + \eta_1^* + R}} \left( \alpha_1^{(1)}, \alpha_1^{(2)}, \alpha_1^{(3)} \right)^T \ ,$$

$$= k_{1R} e^{i\eta_{1I}} \operatorname{sech} \left( \eta_{1R} + \frac{R}{2} \right) (A_1, A_2, A_3)^T \ , \tag{32}$$

where $\eta_1 = k_1(t + ik_1 z)$, $A_j = \alpha_1^{(j)} / \Delta$ and $\Delta = (\mu \sum_{j=1}^3 |\alpha_1^{(j)}|^2)^{1/2}$. Note that $|A_1|^2 + |A_2|^2 + |A_3|^2 = \frac{1}{\mu}$. Here $\alpha_1^{(j)}$, $k_1$, $j = 1, 2, 3$, are four arbitrary complex parameters. Further $k_{1R} A_j$ gives the amplitude of the $j$th mode ($j = 1, 2, 3$) and $2k_{1I}$ is the soliton velocity in all the three components.

## (b) Two-Soliton Solution

The general bright two-soliton solution of equation (29) can be generated by terminating the series (16) as

$$g^{(j)} = \lambda g_1^{(j)} + \lambda^3 g_3^{(j)} \ , \quad f = 1 + \lambda^2 f_2 + \lambda^4 f_4 \tag{33a}$$

and then solving the resulting set of linear partial differential equations at various powers of $\lambda$ recursively. It is given by

$$q_j = \frac{(\alpha_1^{(j)} e^{\eta_1} + \alpha_2^{(j)} e^{\eta_2} + e^{\eta_1 + \eta_1^* + \eta_2 + \delta_{1j}} + e^{\eta_1 + \eta_2 + \eta_2^* + \delta_{2j}})}{D} \ , \tag{34a}$$

where

$$D = 1 + e^{\eta_1 + \eta_1^* + R_1} + e^{\eta_1 + \eta_2^* + \delta_0} + e^{\eta_1^* + \eta_2 + \delta_0^*} + e^{\eta_2 + \eta_2^* + R_2}$$

$$+ e^{\eta_1 + \eta_1^* + \eta_2 + \eta_2^* + R_3} \ , j = 1, 2, 3 \ ,$$

$$\eta_i = k_i(t + ik_i z) \ , \ e^{R_1} = \frac{\kappa_{11}}{(k_1 + k_1^*)} \ , \ e^{R_2} = \frac{\kappa_{22}}{(k_2 + k_2^*)} \ , \ e^{\delta_0} = \frac{\kappa_{12}}{(k_1 + k_2^*)} \ ,$$

$$e^{\delta_{1j}} = \frac{(k_1 - k_2)(\alpha_1^{(j)} \kappa_{21} - \alpha_2^{(j)} \kappa_{11})}{((k_1 + k_1^*)(k_1^* + k_2))} \ , \ e^{\delta_{2j}} = \frac{(k_2 - k_1)(\alpha_2^{(j)} \kappa_{12} - \alpha_1^{(j)} \kappa_{22})}{((k_2 + k_2^*)(k_1 + k_2^*))}$$

$$e^{R_3} = \frac{|k_1 - k_2|^2 (\kappa_{11} \kappa_{22} - \kappa_{12} \kappa_{21})}{(k_1 + k_1^*)(k_2 + k_2^*)|k_1 + k_2^*|^2}$$

and

$$\kappa_{il} = \frac{\mu \sum_{n=1}^3 \alpha_i^{(n)} \alpha_l^{(n)*}}{(k_i + k_l^*)} \ , \ i, l = 1, 2, j = 1, 2, 3 \ . \tag{34b}$$

This solution contains eight arbitrary complex parameters $\alpha_1^{(1)}$, $\alpha_1^{(2)}$, $\alpha_1^{(3)}$, $\alpha_2^{(1)}$, $\alpha_2^{(2)}$, $\alpha_2^{(3)}$, $k_1$ and $k_2$ and represents two soliton interaction in 3-CNLS equations.

## (c) Three-Soliton Solution

Following the procedure for obtaining the 1- and 2-soliton solutions of the 3-CNLS equations and the procedure to obtain the 3-soliton solution of the 2-CNLS equations in sec.3.1c, we find the form of the 3-soliton solution of 3-CNLS equations is similar to that of (28). It reads as

$$
q_j = \frac{\alpha_1^{(j)}e^{\eta_1} + \alpha_2^{(j)}e^{\eta_2} + \alpha_3^{(j)}e^{\eta_3} + e^{\eta_1+\eta_1^*+\eta_2+\delta_{1j}} + e^{\eta_1+\eta_1^*+\eta_3+\delta_{2j}} + e^{\eta_2+\eta_2^*+\eta_1+\delta_{3j}}}{D}
$$

$$
+ \frac{e^{\eta_2+\eta_2^*+\eta_3+\delta_{4j}} + e^{\eta_3+\eta_3^*+\eta_1+\delta_{5j}} + e^{\eta_3+\eta_3^*+\eta_2+\delta_{6j}} + e^{\eta_1^*+\eta_2+\eta_3+\delta_{7j}}}{D}
$$

$$
+ \frac{e^{\eta_1+\eta_2^*+\eta_3+\delta_{8j}} + e^{\eta_1+\eta_2+\eta_3^*+\delta_{9j}} + e^{\eta_1+\eta_1^*+\eta_2+\eta_2^*+\eta_3+\tau_{1j}}}{D}
$$

$$
+ \frac{e^{\eta_1+\eta_1^*+\eta_3+\eta_3^*+\eta_2+\tau_{2j}} + e^{\eta_2+\eta_2^*+\eta_3+\eta_3^*+\eta_1+\tau_{3j}}}{D} , \quad j = 1,2,3 . \tag{35}
$$

Here $\delta_{mj}$, $m = 1,2,...,9$, $j = 1,2,3$ and the quantity D are as defined in (28), except that $j$ now varies from 1 to 3 instead of 1 and 2. Further here the $\kappa_{il}$'s are redefined as

$$
\kappa_{il} = \frac{\mu\left(\alpha_i^{(1)}\alpha_l^{(1)*} + \alpha_i^{(2)}\alpha_l^{(2)*} + \alpha_i^{(3)}\alpha_l^{(3)*}\right)}{k_i + k_l^*} , \quad i,l = 1,2,3 . \tag{36}
$$

It can be observed from the above expression that as the number of solitons increases the complexity also increases and the present three-soliton solution is characterized by twelve arbitrary complex parameters $\alpha_1^{(1)}, \alpha_1^{(2)}, \alpha_1^{(3)}, \alpha_2^{(1)}, \alpha_2^{(2)}, \alpha_2^{(3)}, \alpha_3^{(1)}, \alpha_3^{(2)}, \alpha_3^{(3)}, k_1, k_2$ and $k_3$.

## 3.3   N-CNLS Equations

Now it is straightforward to extend the above analysis for the 2 and 3-CNLS equations to N-CNLS equations. In the following we present the one-, two- and three- soliton solutions of the N-CNLS equations (13).

## (a) One-Soliton Solution

$$
(q_1, q_2, \ldots, q_N)^T = k_{1R}e^{i\eta_{1I}}\text{sech}\left(\eta_{1R} + \frac{R}{2}\right)(A_1, A_2, \ldots, A_N)^T , \tag{37}
$$

where $\eta_1 = k_1(t+ik_1z)$, $A_j = \alpha_1^{(j)}/\Delta$, $\Delta = (\mu(\sum_{j=1}^N |\alpha_1^{(j)}|^2))^{1/2}$, $e^R = \Delta^2/(k_1 + k_1^*)^2$, $\alpha_1^{(j)}$ and $k_1$, $j = 1,2\ldots,N$, are $(N+1)$ arbitrary complex parameters.

## (b) Two-Soliton Solution

$$q_j = \frac{(\alpha_1^{(j)} e^{\eta_1} + \alpha_2^{(j)} e^{\eta_2} + e^{\eta_1+\eta_1^*+\eta_2+\delta_{1j}} + e^{\eta_1+\eta_2+\eta_2^*+\delta_{2j}})}{D} \; , \; j=1,2,3,...,N \; ,$$

(38a)

where

$$D = 1 + e^{\eta_1+\eta_1^*+R_1} + e^{\eta_1+\eta_2^*+\delta_0} + e^{\eta_1^*+\eta_2+\delta_0^*} + e^{\eta_2+\eta_2^*+R_2}$$
$$+ e^{\eta_1+\eta_1^*+\eta_2+\eta_2^*+R_3} \; ,$$

$$\eta_i = k_i(t + ik_i z) \; , \; e^{R_1} = \frac{\kappa_{11}}{(k_1+k_1^*)} \; , \; e^{R_2} = \frac{\kappa_{22}}{(k_2+k_2^*)} \; ,$$

$$e^{\delta_0} = \frac{\kappa_{12}}{(k_1+k_2^*)} \; , e^{\delta_{1m}} = \frac{((k_1-k_2)(\alpha_1^{(m)}\kappa_{21} - \alpha_2^{(m)}\kappa_{11}))}{((k_1+k_1^*)(k_1^*+k_2))} \; ,$$

$$e^{\delta_{2m}} = \frac{((k_2-k_1)(\alpha_2^{(m)}\kappa_{12} - \alpha_1^{(m)}\kappa_{22}))}{((k_2+k_2^*)(k_1+k_2^*))} \; ,$$

$$e^{R_3} = \frac{(|k_1-k_2|^2(\kappa_{11}\kappa_{22} - \kappa_{12}\kappa_{21}))}{((k_1+k_1^*)(k_2+k_2^*)|k_1+k_2^*|^2)} \; , m = 1,2,...,N$$

and

$$\kappa_{il} = \mu \sum_{n=1}^{N} \alpha_i^{(n)} \alpha_l^{(n)*} / (k_i + k_l^*) \; , \; i,l = 1,2 \; .$$

(38b)

One may also note that the above two-soliton solution depends on $2(N+1)$ arbitrary complex parameters $\alpha_1^{(m)}$, $\alpha_2^{(m)}$, $k_1$ and $k_2$, $m = 1,2,...,N$.

## (c) Three-Soliton Solution

Following the procedure given in the previous sections we obtain the 3-soliton solution to the N-CNLS equations as

$$q_j = \frac{\alpha_1^{(j)} e^{\eta_1} + \alpha_2^{(j)} e^{\eta_2} + \alpha_3^{(j)} e^{\eta_3} + e^{\eta_1+\eta_1^*+\eta_2+\delta_{1j}} + e^{\eta_1+\eta_1^*+\eta_3+\delta_{2j}} + e^{\eta_2+\eta_2^*+\eta_1+\delta_{3j}}}{D}$$

$$+ \frac{e^{\eta_2+\eta_2^*+\eta_3+\delta_{4j}} + e^{\eta_3+\eta_3^*+\eta_1+\delta_{5j}} + e^{\eta_3+\eta_3^*+\eta_2+\delta_{6j}} + e^{\eta_1^*+\eta_2+\eta_3+\delta_{7j}}}{D}$$

$$+ \frac{e^{\eta_1+\eta_2^*+\eta_3+\delta_{8j}} + e^{\eta_1+\eta_2+\eta_3^*+\delta_{9j}} + e^{\eta_1+\eta_1^*+\eta_2+\eta_2^*+\eta_3+\tau_{1j}}}{D}$$

$$+ \frac{e^{\eta_1+\eta_1^*+\eta_3+\eta_3^*+\eta_2+\tau_{2j}} + e^{\eta_2+\eta_2^*+\eta_3+\eta_3^*+\eta_1+\tau_{3j}}}{D}$$

(39)

Here also the denominator $D$ and all the other quantities are the same as those given under (28) with the redefinition of $\kappa_{il}$'s as

$$\kappa_{il} = \frac{\mu \sum_{n=1}^{N} \alpha_i^{(n)} \alpha_l^{(n)*}}{k_i + k_l^*} \; , \; i,l = 1,2,3 \; .$$

(40)

One can check that this three-soliton solution of N-CNLS equation depends on $3(N+1)$ arbitrary complex parameters $\alpha_1^{(m)}$, $\alpha_2^{(m)}$, $\alpha_3^{(m)}$, $k_1$, $k_2$ and $k_3$, $m = 1, 2, ..., N$. Generalizing the above analysis to N-soliton solution of N-CNLS equations it can be verified that it depends on $N(N+1)$ arbitrary complex parameters.

## 4 Shape Changing Collisions in Coupled Nonlinear Schrödinger Equations

### 4.1 N=2 Case

The novel collision properties associated with the CNLS equations can be revealed by analyzing the asymptotic forms of the two-soliton solutions [12,18]. In this connection, let us first consider the two-soliton solution(24) of the Manakov system, which is an integrable 2-CNLS system. Without loss of generality we assume that $k_{jR} > 0$ and $k_{1I} > k_{2I}$, $k_j = k_{jR} + ik_{jI}$, $j = 1, 2$, which corresponds to a head-on collision of the solitons. One can easily check that asymptotically the two-soliton solution becomes two well separated solitons $S_1$ and $S_2$. For the above parametric choice, the variables $\eta_{jR}$'s for the two solitons ($\eta_j = \eta_{jR} + i\eta_{jI}$) behave asymptotically as (i)$\eta_{1R} \sim 0$, $\eta_{2R} \to \pm\infty$ as $z \to \pm\infty$ and (ii)$\eta_{2R} \sim 0$, $\eta_{1R} \to \mp\infty$ as $z \to \pm\infty$. This leads to the following asymptotic forms for the two-soliton solution.

(i) Before Collision (Limit $z \to -\infty$)

In the limit $z \to -\infty$, the solution(24) can be easily seen to take the following forms.

(a) $S_1$ ($\eta_{1R} \sim 0$, $\eta_{2R} \to -\infty$) :

$$\begin{pmatrix} q_1 \\ q_2 \end{pmatrix} \to \begin{pmatrix} \alpha_1^{(1)} \\ \alpha_1^{(2)} \end{pmatrix} \frac{e^{\eta_1}}{1 + e^{\eta_1 + \eta_1^* + R_1}} = \begin{pmatrix} A_1^{1-} \\ A_2^{1-} \end{pmatrix} k_{1R} e^{i\eta_{1I}} \operatorname{sech}\left(\eta_{1R} + \frac{R_1}{2}\right), (41)$$

where $(A_1^{1-}, A_2^{1-}) = [\mu(\alpha_1^{(1)}\alpha_1^{(1)*} + \alpha_1^{(2)}\alpha_1^{(2)*})]^{-\frac{1}{2}}(\alpha_1^{(1)}, \alpha_1^{(2)})$. In $A_i^{1-}$, $i = 1, 2$, superscript 1- denotes $S_1$ at the limit $z \to -\infty$ and subscripts 1 and 2 refer to the modes $q_1$ and $q_2$, respectively.

(b) $S_2$ ($\eta_{2R} \sim 0$, $\eta_{1R} \to \infty$):

$$\begin{pmatrix} q_1 \\ q_2 \end{pmatrix} \to \begin{pmatrix} e^{\delta_{11} - R_1} \\ e^{\delta_{12} - R_1} \end{pmatrix} \frac{e^{\eta_2}}{1 + e^{\eta_2 + \eta_2^* + R_3 - R_1}}$$

$$= \begin{pmatrix} A_1^{2-} \\ A_2^{2-} \end{pmatrix} k_{2R} e^{i\eta_{2I}} \operatorname{sech}\left(\eta_{2R} + \frac{(R_3 - R_1)}{2}\right), \qquad (42)$$

where

$$(A_1^{2-}, A_2^{2-}) = \left(\frac{a_1}{a_1^*}\right) c \, [\mu(\alpha_2^{(1)}\alpha_2^{(1)*} + \alpha_2^{(2)}\alpha_2^{(2)*})]^{-\frac{1}{2}}$$

$$\left[(\alpha_1^{(1)}, \alpha_1^{(2)})\kappa_{11}^{-1} - (\alpha_2^{(1)}, \alpha_2^{(2)})\kappa_{21}^{-1}\right],$$

in which

$$a_1 = (k_1 + k_2^*) \left[ (k_1 - k_2)(\alpha_1^{(1)*}\alpha_2^{(1)} + \alpha_1^{(2)*}\alpha_2^{(2)}) \right]^{\frac{1}{2}}$$

and

$$c = \left[ \frac{1}{|\kappa_{12}|^2} - \frac{1}{\kappa_{11}\kappa_{22}} \right]^{-\frac{1}{2}} .$$

Other quantities have been defined earlier.

(ii) After Collision (Limit $z \to \infty$)

Similarly, for $z \to \infty$, we have the following forms.

(a) $S_1$ ($\eta_{1R} \sim 0, \quad \eta_{2R} \to \infty$):

$$\begin{pmatrix} q_1 \\ q_2 \end{pmatrix} \to \begin{pmatrix} e^{\delta_{21} - R_2} \\ e^{\delta_{22} - R_2} \end{pmatrix} \frac{e^{\eta_1}}{1 + e^{\eta_1 + \eta_1^* + R_3 - R_2}}$$

$$= \begin{pmatrix} A_1^{1+} \\ A_2^{1+} \end{pmatrix} k_{1R} e^{i\eta_{1I}} \operatorname{sech}\left( \eta_{1R} + \frac{(R_3 - R_2)}{2} \right), \qquad (43)$$

where

$$(A_1^{1+}, A_2^{1+}) = (\frac{a_2}{a_2^*}) \, c \, [\mu(\alpha_1^{(1)}\alpha_1^{(1)*} + \alpha_1^{(2)}\alpha_1^{(2)*})]^{-\frac{1}{2}}$$

$$\left[ (\alpha_1^{(1)}, \alpha_1^{(2)})\kappa_{12}^{-1} - (\alpha_2^{(1)}, \alpha_2^{(2)})\kappa_{22}^{-1} \right]$$

in which

$$a_2 = (k_2 + k_1^*) \left[ (k_1 - k_2)(\alpha_1^{(1)}\alpha_2^{(1)*} + \alpha_1^{(2)}\alpha_2^{(2)*}) \right]^{\frac{1}{2}} .$$

(b) $S_2$ ($\eta_{2R} \sim 0, \quad \eta_{1R} \to -\infty$):

$$\begin{pmatrix} q_1 \\ q_2 \end{pmatrix} \to \begin{pmatrix} \alpha_2^{(1)} \\ \alpha_2^{(2)} \end{pmatrix} \frac{e^{\eta_2}}{1 + e^{\eta_2 + \eta_2^* + R_2}} = \begin{pmatrix} A_1^{2+} \\ A_2^{2+} \end{pmatrix} k_{2R} e^{i\eta_{2I}} \operatorname{sech}\left( \eta_{2R} + \frac{R_2}{2} \right), (44)$$

where $(A_1^{2+}, A_2^{2+}) = [\mu(\alpha_2^{(1)}\alpha_2^{(1)*} + \alpha_2^{(2)}\alpha_2^{(2)*})]^{-\frac{1}{2}}(\alpha_2^{(1)}, \alpha_2^{(2)})$.

The asymptotic forms of the solitons $S_1$ and $S_2$ after interaction, given respectively by (43) and (44), can be related to the forms before interaction, given by (41) and (42) respectively, by introducing a transition matrix $T_j^l$ such that

$$A_j^{l+} = A_j^{l-} T_j^l, \quad j, l = 1, 2, \qquad (45)$$

where the superscripts $l\pm$ represent the solitons designated as $S_1$ and $S_2$ at $z \to \pm\infty$. Here

$$|T_j^1|^2 = |1 - \lambda_2(\alpha_2^{(j)}/\alpha_1^{(j)})|^2/|1 - \lambda_1\lambda_2|, \qquad (46a)$$

$$|T_j^2|^2 = |1 - \lambda_1\lambda_2|/|1 - \lambda_1(\alpha_1^{(j)}/\alpha_2^{(j)})|^2, \quad j = 1, 2, \qquad (46b)$$

$$\lambda_1 = \kappa_{21}/\kappa_{11} \quad \text{and} \quad \lambda_2 = \kappa_{12}/\kappa_{22}. \qquad (46c)$$

Besides the above changes in the amplitudes, there is a phase shift of the soliton positions as indicated below.

The expressions (45-46) clearly show that there is an intensity redistribution among the two modes of the solitons $S_1$ and $S_2$ and that the transition matrix acts as a measure of this redistributed intensity. However, we can also check easily that in the special case in which the parameters $\alpha_i^{(j)}, i, j = 1, 2$, satisfy the relation $\alpha_1^{(1)}/\alpha_2^{(1)} = \alpha_1^{(2)}/\alpha_2^{(2)}$, the absolute of the transition matrix takes the value $|T_i^j| = 1$. In this case, we have the standard pure shape preserving, elastic collision of solitons that is familiar in the literature. However for all other values of $\alpha_i^{(j)}, i, j = 1, 2$, such that relation $\alpha_1^{(1)}/\alpha_2^{(1)} \neq \alpha_1^{(2)}/\alpha_2^{(2)}$, the amplitudes of the solitons do undergo changes and we have shape changing (inelastic) collisions.

## (a)Intensity (Amplitude) Redistribution during the Shape Changing Collision Process

The amplitude change of individual modes of $S_1$ and $S_2$ during the collision process shows some interesting features. Actually, from (41) to (44) we observe that the initial amplitudes of the two modes of the two solitons $(A_1^{1-}k_{1R}, A_2^{1-}k_{1R})$ and $(A_1^{2-}k_{2R}, A_2^{2-}k_{2R})$ undergo a redistribution among them and the solitons emerge with amplitudes $(A_1^{1+}k_{1R}, A_2^{1+}k_{1R})$ and $(A_1^{2+}k_{2R}, A_2^{2+}k_{2R})$, respectively, where $A_i^{l\pm}, i, l = 1, 2$, are given above. The changes in the amplitudes due to collision are essentially given by the transition elements $T_i^j$, see (45-46). It can be easily checked that for suitable choice of parameters it is even possible to make one of the $T_i^j$'s vanish so that one of the modes after collision (or before collision) has zero intensity (see figure 1 below). Another noticeable observation of this interaction process is that even though there is a redistribution of intensity among the modes the total intensity of each mode is also conserved separately, that is, $|A_1^{n\pm}|^2 + |A_2^{n\pm}|^2 = \frac{1}{\mu}$, $n = 1, 2$, see (41)-(44). More specifically, from the equation of motion(11) itself, one can observe that the total intensity of each mode is also conserved, that is,

$$\int_{-\infty}^{\infty} |q_j|^2 dz = \text{constant} , \quad j = 1, 2 . \tag{47}$$

This can also be inferred from Fig. 1.

## (b)Phase Shift of Solitons during the Collision Process

Further, from the asymptotic forms of the solitons $S_1$ and $S_2$, it can be observed that the phases of solitons $S_1$ and $S_2$ also change and that the phase shifts are now not only functions of the parameters $k_1$ and $k_2$ but also dependent on $\alpha_i^{(j)}$'s, $i, j = 1, 2$. The phase shift suffered by soliton $S_1$ during collision is [20]

$$\Phi^1 = \frac{(R_3 - R_1 - R_2)}{2}$$
$$= \left(\frac{1}{2}\right) \log \left[ \frac{|k_1 - k_2|^2 (\kappa_{11}\kappa_{22} - \kappa_{12}\kappa_{21})}{|k_1 + k_2^*|^2 \kappa_{11}\kappa_{22}} \right] . \tag{48}$$

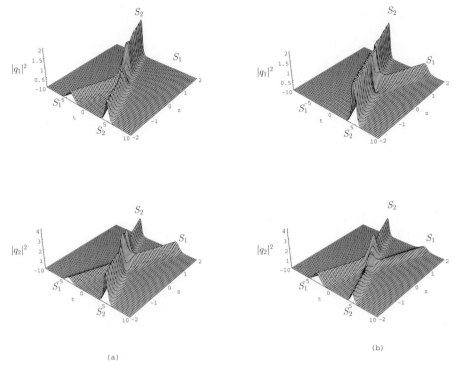

**Fig. 1.** Two distinct possibilities of the dramatic intensity redistribution in the two soliton collision in 2-CNLS system

Similarly soliton $S_2$ suffers a phase shift

$$\Phi^2 = -\frac{(R_3 - R_2 - R_1)}{2} = -\Phi^1$$
$$= -\left(\frac{1}{2}\right) \log\left[\frac{|k_1 - k_2|^2(\kappa_{11}\kappa_{22} - \kappa_{12}\kappa_{21})}{|k_1 + k_2^*|^2\kappa_{11}\kappa_{22}}\right] . \tag{49}$$

Then the absolute value of phase shift suffered by the two solitons is

$$|\Phi| = |\Phi^1| = |\Phi^2| . \tag{50}$$

From the above expression the absolute phase shift for the pure elastic collision case ($\alpha_1^{(1)} : \alpha_2^{(1)} = \alpha_1^{(2)} : \alpha_2^{(2)}$) corresponding to parallel modes can be obtained as

$$|\Phi| = \left|\log\left[\frac{|k_1 - k_2|^2}{|k_1 + k_2^*|^2}\right]\right| = 2\left|\log\left[\frac{|k_1 - k_2|}{|k_1 + k_2^*|}\right]\right| . \tag{51}$$

Similarly for the case corresponding to orthogonal modes ($\alpha_1^{(1)} : \alpha_2^{(1)} = \infty$, $\alpha_1^{(2)} : \alpha_2^{(2)} = 0$) the absolute phase shift is found to be

$$|\Phi| = \left|\log\left[\frac{|k_1 - k_2|}{|k_1 + k_2^*|}\right]\right| . \tag{52}$$

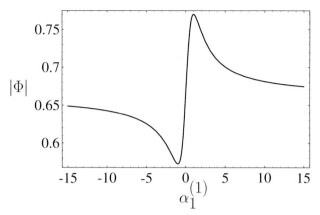

**Fig. 2.** Plot of the magnitude of phase shift as a function of the parameter $\alpha_1^{(1)}$, when it is real (for illustrative purpose), see (48-50). The other parameters are chosen as $k_1 = 1 + i$, $k_2 = 2 - i$, $\alpha_1^{(2)} = \alpha_2^{(2)} = 1$ and $\alpha_2^{(1)} = \frac{39+80i}{89}$.

Thus phase shifts do vary depending on $\alpha_i^{(j)}$'s (amplitudes) for fixed $k_i$'s. In Fig. 2, we plot the change of $|\Phi|$ as a function of $\alpha_1^{(1)}$, when it is real, at $\alpha_1^{(2)} = \alpha_2^{(2)} = 1$, $\alpha_2^{(1)} = \frac{39+80i}{89}$, $k_1 = 1 + i$ and $k_2 = 2 - i$.

### (c) Change in the Relative Separation Distance of the Solitons

As a consequence of the above phase shift, the relative separation distances $t_{12}^{\pm}$ (position of $S_2$ (at $z \to \pm\infty$)minus position of $S_1$ (at $z \to \pm\infty$)) also do vary during the collision process. In such a pair-wise collision, the change in the relative separation distance is found to be

$$\Delta t_{12} = t_{12}^- - t_{12}^+ = \frac{(k_{1R} + k_{2R})}{k_{1R}k_{2R}}\Phi^1 \ . \tag{53}$$

Again $\Delta t_{12}$ depends upon the parameters $\alpha_i^{(j)}$'s and so on the amplitudes of the modes.

### (d) Role of $\alpha_i^{(j)}$'s in the Collision Process

It can be straightforwardly seen that the above mentioned three properties, intensity (amplitude) redistribution, phase shift and relative separation distances, characterizing the CNLS soliton collisions, are nontrivially dependent on the complex parameters $\alpha_i^{(j)}$'s, which in turn determine the quantities $A_i^{j\pm}$ defining the amplitudes of the modes. These parameters play a pivotal role in the shape changing collision process. In particular, the change in the amplitudes of the two modes of $S_1$ and $S_2$ can be varied dramatically by changing $\alpha_i^{(j)}$'s and even the amplitudes before and after interaction can be made equal, a case corresponding to elastic collision, for the particular choice $\frac{\alpha_1^{(1)}}{\alpha_2^{(1)}} = \frac{\alpha_1^{(2)}}{\alpha_2^{(2)}}$ as the absolute of the transition elements $|T_i^j|$, $i, j = 1, 2$, are equal to one in this special case. For all other choices, the amplitudes undergo changes due to collision and under suitable circumstances the amplitude of one of the modes (either before or after collision) can even vanish, showing in a dramatic way the shape changing nature of the collisions.

It can also be observed from ( 48-53) that not only the amplitudes of the solitons but also the phases and hence the relative separation distances between them depend on the complex parameters $\alpha_i^{(j)}$'s. As a result, their variation during collision is also determined predominantly by $\alpha_i^{(j)}$'s.

Now let us look at the possible ways by which such shape changing collision can occur in the Manakov system. We can identify two distinct types of interactions for each of the soliton as a consequence of conservation of total intensity of individual solitons and individual modes. The first possibility is an enhancement of intensity in any one of the modes of either one of the solitons (say $S_1$) and suppression in the remaining mode of the corresponding soliton with commensurate changes in the other soliton. The other possibility is an interaction which allows one of the modes of either one of the solitons (say $S_1$) to get suppressed while the other mode of the corresponding soliton to get enhanced (with corresponding changes in $S_2$). In either of the cases the intensity may be completely or partially suppressed (enhanced), as determined by the entries of the transition matrix $T_i^j$.

For illustrative purpose, we have shown a dramatic case of the head-on collision [12] of two solitons for the parametric values, $k_1 = 1 + i$, $k_2 = 2 - i$, $\alpha_1^{(1)} = \alpha_1^{(2)} = \alpha_2^{(2)} = 1$, and $\alpha_2^{(1)} = \frac{(39+i80)}{89}$ in Fig. 1a. Here initially the time profiles of the two solitons are evenly split between the two components $q_1$ and $q_2$. At the large positive $z$ end the profile of the $S_1$ soliton is almost completely suppressed in the $q_1$ component while it is suitably enhanced in the $q_2$ component. Also there is a rearrangement of the amplitudes in the second soliton $S_2$ in both the modes also. In Fig. 1b, we have shown the reverse possibility for $k_1 = 1 + i$, $k_2 = 2 - i$, $\alpha_1^{(1)} = 0.02 + 0.1i$,r $\alpha_1^{(2)} = \alpha_2^{(1)} = \alpha_2^{(2)} = 1$. Finally, Fig. 3 shows the possibility of elastic collision for the same $k_1$ ,$k_2$ values as in Fig. 1 but with $\alpha_i^{(j)} = 1$, $i, j = 1, 2$. For other choices of parameters there will be in general partial suppression or enhancement of solitons after collision.

## 4.2   N=3 and Arbitrary Cases

The above analysis can be analogously extended to study the collision properties of the 3-CNLS equations and N-CNLS equations [13]. For example, for the 3-CNLS system, we find that the above kind of shape changing collisions of the Manakov system occur in this case also but with many possibilities of intensity exchange among the three modes. During the inelastic (shape changing) interaction among the two one solitons $S_1$ and $S_2$ of the 3- CNLS, the soliton $S_1(S_2)$ has the following six possible combinations to exchange the intensity among its

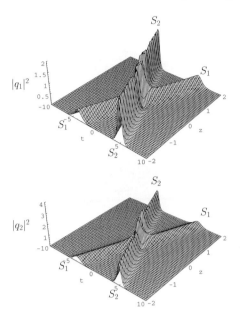

**Fig. 3.** Elastic collision of two solitons in the Manakov system for a specific choice of the parameters.

modes : $(q_1, q_2, q_3) \rightarrow (q_1^a, q_2^b, q_3^c)_i$, [$a, b, c = S$ (suppression),$E$, (enhancement)], with $i = 1$, $a = E$, $b = S$, $c = S$; $i = 2$, $a = S$, $b = E$, $c = S$; $i = 3$, $a = S$, $b = S$, $c = E$; $i = 4$, $a = S$, $b = E$, $c = E$; $i = 5$, $a = E$, $b = S$, $c = E$ and $i = 6$, $a = E$, $b = E$, $c = S$. The conservation of total intensity of the soliton and mode restricts the collision scenario to take place with these six possibilities. All these six possibilities are shown in Fig. 4, for suitable parametric choices. For example, Fig. 4a is plotted for the parametric values $k_1 = 1 + i$, $k_2 = 2 - i$, $\alpha_2^{(1)} = \alpha_2^{(2)} = (39 + i80)/89$, $\alpha_1^{(1)} = \alpha_1^{(2)} = \alpha_1^{(3)} = \alpha_2^{(3)} = 1$ and $\mu = 1$. Generalizing the above analysis for the 2-CNLS and 3-CNLS equations, to N-CNLS equations, one can verify that the shape changing interaction can lead to an intensity redistribution among the modes of each of the soliton of the N-CNLS system in $2^N - 2$ ways.

## 4.3   Shape Changing Collisions and Higher Order Solitons

The collision dynamics of multicomponent higher order solitons is more intricate. However one can proceed in the same manner as in the case of the two-soliton solution of $N = 2$ case as well as that of the arbitrary $N$ case, given in the previous sections. For more details, one can look into [19] in which such higher order soliton interactions are discussed in detail. One important property which has to be noticed in the three (and higher order solitons as well) soliton collision processes is the state restoration property of one of the solitons among the three

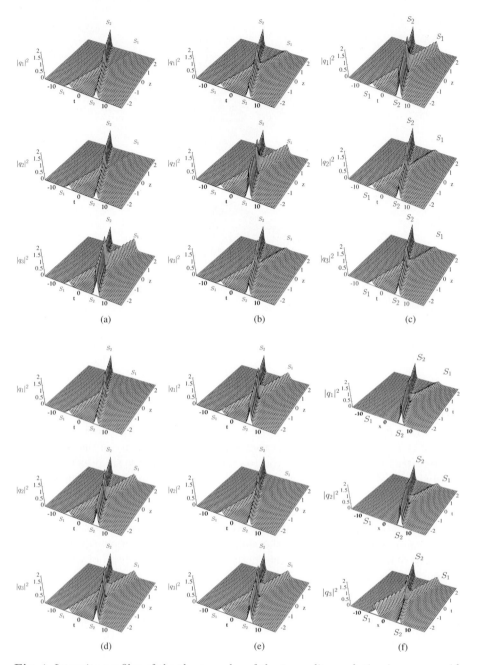

**Fig. 4.** Intensity profiles of the three modes of the two-soliton solution in a waveguide described by the CNLS equations (13) with $n = 3$ showing six different dramatic scenarios of the shape changing collision, for the parametric choice indicated in the text.

after collision with the rest of the two. That is, it is possible to restore the state of a particular soliton during its collision with other two solitons whose states change after collision [19]. This property has further important consequences leading to the construction of logic gates and universal Turing equivalent computing machines (For details, see [15,21]) .

# 5   Conclusion

In the above, we have presented the bright soliton solutions of the integrable CNLS equations and pointed out that this system exhibits a novel kind of shape changing collision characterized by the properties such as intensity redistributions among the solitons in different modes, amplitude dependent phase shifts and relative separation distances. In fact this type of collision property makes the CNLS system a very attractive one. In particular, as mentioned in the introduction, in [14,15, see also 21] this property has been used advantageously for performing logical operations and constructing various logic gates including the universal NAND gate. The key property which helps in the construction of such logic operations and gates is the state restoration of certain of the solitons in the multisoliton collision process characterized by three and higher order soliton solutions. These developments in turn show the possibility of constructing a Turing machine at least in a mathematical sense which seems to be feasible in a practical sense with the observation of spatial solitons in photorefractive materials [8,9].

The idea behind the aforementioned construction of logic gates directly follows from the asymptotic analysis of two soliton solution of 2-CNLS system presented in sec.3. It can be shown that the intensity redistribution in a collision process can be represented equivalently through a linear fractional transformation [14,20] which possesses many interesting properties including the existence of inverse transformations and group property. This is the starting point of the development of constructing logic gates in the context of shape changing soliton collision in CNLS equations [14, 15, 21]. We believe that such studies will shed new light in the areas such as optical computing, long distant optical communication, optical switching, etc. and we can hope to achieve considerable further progress in the near future.

**Acknowledgment**

The work reported here has been supported by the Department of Science and Technology, Govt. of India and the Council of Scientific and Industrial Research (CSIR), Government of India in the form of research projects to M.L. T.K. wishes to acknowledge CSIR for the award of a Senior Research Fellowship.

# References

1. E. Iannone, F. Matera, A. Mecozzi and M. Settembre: *Nonlinear Optical Communication Networks*,(Wiley-Interscience, New York, 1998).
2. G. P. Agrawal: *Nonlinear Fiber Optics* (Academic Press, New York, 1995).
3. R. Sahadevan, K. M. Tamizhmani and M. Lakshmanan: J. Phys. A **19**, 1783 (1986);V. G. Makhan'kov and O. K. Pashaev: Theor. Math. Phys. **53**, 979 (1982); K. Nakkeeran: Phys. Rev. E **62**, 1313 (2000).
4. S. Chakravarty, M. J. Ablowitz, J. R. Sauer, and R. B. Jenkins: Opt. Lett. **20**, 136 (1995).
5. C. Yeh and L. Bergman: Phys. Rev. E **57**, 2398 (1998).
6. A.C. Scott: Phys. Scr. **29**, 279 (1984).
7. A. Ankiewicz, W. Krolikowski, and N. N. Akhmediev: Phys. Rev. E **59**, 6079 (1999); N. Akhmediev, W. Krolikowski, and A. W. Snyder: Phys. Rev. Lett. **81**, 4632 (1998).
8. M. Mitchell, Z. Chen, M. F. Shih and M. Segev: Phys. Rev. Lett. **77**, 490 (1996).
9. M. Mitchell and M. Segev: Nature, **387**, 880 (1997).
10. N. Akhmediev and A. Ankiewicz, Chaos **10**, 600 (2000).
11. S. V. Manakov: Zh. Eksp. Teor. Fiz. **65**, 505 (1973) [Sov. Phys. JETP **38**, 248 (1974)].
12. R. Radhakrishnan, M. Lakshmanan, and J. Hietarinta: Phys. Rev. E **56**, 2213 (1997).
13. T. Kanna and M. Lakshmanan: Phys. Rev. Lett, **86**, 5043 (2001).
14. M. H. Jakubowski, K. Steiglitz and R. Squier: Phys. Rev. E **58**, 6752 (1998).
15. K. Steiglitz: Phys. Rev. E. **63**,016608 (2000).
16. R. Radhakrishnan, R.Sahadevan, and M. Lakshmanan: Chaos, Solitons and Fractals **5**, 2315 (1995).
17. R. Hirota: J. Math. Phys. (N.Y.) **14**, 805 (1973).
18. M.Lakshmanan, T.Kanna and R.Radhakrishnan: Rep. Math. Phys. **46**, 143 (2000).
19. T. Kanna and M. Lakshmanan: (Submitted to Phys. Rev. E).
20. T. Kanna and M. Lakshmanan: Eur. Phys. J. B (2002) to appear.
21. M. Lakshmanan and T. Kanna: Pramana-J. Phys. **57**, 885 (2001).

# Mathematical Modelling
# in Fiber and Waveguide Grating Structures

A.B. Aceves

Department of Mathematics and Statistics,
University of New Mexico, Albuquerque, NM 87131

**Abstract.** In this chapter, I give an overview of mathematical modelling in fiber and waveguide grating structures. Studies on fiber and waveguide gratings and fiber arrays are presented. The possible formation of soliton-like and bullet-like pulses in such media are discussed.

## 1   Introduction

Theory and experiments of optical solitons in fiber Bragg gratings have a similar history to that of optical phenomena described by the integrable nonlinear Schrödinger equation (except for the fact that Gap solitons are not true solitons, since the governing equations are nonintegrable). The theoretical discovery of the soliton and its integrability properties lead to several applications. Since the soliton propagation in an optical fiber was first demonstrated 20 years ago, we have seen an explosion of ideas, experiments and designs of novel all optical communications lines based on the robustness of the soliton (see papers on this volume). We remind the readers that the soliton property is based on the interplay between temporal dispersion and the nonlinear refractive index which causes focusing. Typical (dispersion, nonlinear) lengths at which this balance is achieved varies from a few hundred meters to several kilometers, which are ideal for several applications. In similar fashion, early theoretical work on periodic structures [1] and later specifically to optical gratings [2,3] was later experimentally demonstrated [4]. These experiments show that what Bragg gratings and the gap soliton brings at sufficiently high powers, is realizing this balance in a few centimeters. We can clearly see how this would facilitate future all optical switches as it was suggested in a recent paper [5]. Other possible applications include novel soliton lasers and pulse compressors. Furthermore, the unique dispersion relation of the fiber grating, and the corresponding solitons, allows in theory all velocities from zero, to the speed of light in the bare fiber. An ongoing challenge of trapping light on a defect is currently a topic of great interest [6], One would hope to achieve zero velocity by a clever tailoring of the Bragg grating. This research goes beyond its intellectual value; all optical buffers and storing devices can be based on such fibers. Here, we briefly discuss model approaches to study pulse dynamics in nonuniform structures. Other interesting gap soliton studies, just to name some, include work on gap solitons in quadratic media [7]; there is also work on gap solitons in active media [8], and Raman gap solitons [9], where the authors suggest the use of the Raman effect as a way to build a Stationary gap soliton. There is also work outside the classical optics regime; two examples are Quantum gap solitons [10] as well as gap solitons in Bose-Einstein condensates [11].

Planar nonlinear waveguides with an additional periodic structure in the direction of propagation have drawn less attention, perhaps due to a lack of experimental realizations. It is nevertheless an interesting and potentially useful geometry where optical bullets can propagate. In this article we describe an asymptotic model describing optical bullets in Bragg grating waveguides.

## 2   Mathematical Formulation for Pulses in Fiber Bragg Gratings

Fiber gratings have had wide applications in particular as sensors, filters and in laser systems where they are used as mirrors or pulse compressors. In all these cases the grating acts as a linear optical element. In the mid 90's, experiments on fiber gratings demonstrated the existence of Bragg grating. At high intensities, these solitons (or solitary waves), are the result of a balance between the grating induced dispersion and the nonlinearity. The extraordinary feature is that this balance is achieved in distances of the order of a few centimeters, which opens the possibility of applications in ultrafast phenomena or as a distributed feedback pulse generator.

A fiber grating refers to a standard fiber with linear refractive index $\bar{n}$, where a periodic perturbation $n_p = \Delta n \cos(\frac{2\pi}{d} z)$ is super-imposed on it. The period of the index of refraction $d$ is of the order of the light inverse wavelength; a consequence of this is that in such dielectric, strong back reflection of light happens at frequencies $\omega$ inside the photonic bandgap,

$$\omega_B (1 - \frac{\Delta n}{2\bar{n}} \eta) < \omega < \omega_B (1 + \frac{\Delta n}{2\bar{n}} \eta)$$

where $\omega_B = \frac{\pi c}{\bar{n} d}$ is the Bragg frequency and $\eta \approx 0.8$, is the fraction of energy in the core. The addition of a Kerr-type nonlinearity $n_2 I$ in the index of refraction, compensates at high intensities the linear behavior generating the grating solitons. These solitons are a superposition of pulsed like forward and backward fields whose envelopes satisfy the nonlinear coupled mode equations (NLCME) [12]

$$\frac{\bar{n}}{c} \frac{\partial E_1}{\partial T} + \frac{\partial E_1}{\partial Z} = i\kappa e^{2i\nu Z} E_2 + i(2\Gamma_X |E_2|^2 + \Gamma_S |E_1|^2) E_1, \tag{1}$$

$$\frac{\bar{n}}{c} \frac{\partial E_2}{\partial T} - \frac{\partial E_2}{\partial Z} = i\kappa e^{-2i\nu Z} E_1 + i(2\Gamma_X |E_1|^2 + \Gamma_S |E_2|^2) E_2 \tag{2}$$

where $E_{1(2)}$ is the backward (forward) field envelope, $\nu$ is the detuning between the inverse wavelength of light and $2\pi/d$; $\Gamma_S = \Gamma_X = \frac{4\pi \bar{n} n_2}{\lambda Z_0}$ are the self and cross phase modulation coefficients, with $\lambda$ been the free space wavelength and $Z_0$ the vacuum impedance.

We will now study these coupled mode equations, but in dimensionless units. coupled equations for the complex amplitudes of the field

$$\partial_t e_1 + \partial_z e_1 - i e_2 - i(|e_2|^2 + \sigma|e_1|^2)e_1 = 0$$
$$\partial_t e_2 - \partial_z e_2 - i e_1 - i(|e_1|^2 + \sigma|e_2|^2)e_2 = 0 \qquad (3)$$

In the absence of the self-phase modulation term ($\sigma = 0$), the above equations are integrable by means of the inverse scattering transform (IST). Since, in the physically relevant values for $\sigma$ in nonlinear dielectrics is far from zero, this is not a near integrable system, or better said, it cannot be viewed as a perturbed integrable system. Nevertheless, we will show how the symmetry properties (ie conserved quantities) that this system has in common with the integrable one, allows us to find large classes of soliton-like solutions. In the end, we can summarize the behavior of the nonintegrable system, by saying that what we find is that two properties of the integrable system; the existence of a Lax pair [13,14] and of a Bäcklund transformation (BT) [15,16] remain for the optics model with a constraint.

From the computational point of view, multi-soliton solutions can be obtained by both the IST and the BT. The IST method is more useful in determining the soliton and radiation components of the initial conditions, whereas the BT gives a simpler approach to construct the hierarchy of multi-soliton solutions. What we show is that multi-soliton solutions also exist in the optics model, we do so by studying the solution form in the Thirring model as it is obtained using the BT [16]. Finally, we also discuss how the given constraint is applied in the derivation of quasi-periodic solutions. For the integrable system, the explicit finite derivation of the finite-gap solutions can be found in [17].

But before doing this, simple, but interesting solutions are obtained by direct substitution. We consider first the linear problem. If the intensities of both modes are small, they will solve to good approximation the linear problem

$$\partial_t e_1 + \partial_z e_1 - i e_2 = 0$$
$$\partial_t e_2 - \partial_z e_2 - i e_1 = 0$$

The solutions are of the form $e_1 = A e^{i(kz-\Omega t)}$, $e_2 = \frac{A}{k-\Omega} e^{i(kz-\Omega t)}$, where $k$ satisfies the dispersion relation $k^2 = \Omega^2 - 1$. From the dispersion relation, one can see that for frequencies in the gap $-1 < \Omega < 1$, there will be a decay in the field amplitude as it propagates, proportional to $e^{-\sqrt{1-\Omega^2}z}$. This decay is due to strong backscattering and was first studied by Brillouin [18].
At high intensities however, the continuous wave (cw) solutions will include the nonlinear effects. We obtain these type of solutions by substituting in equations ( 3) for $e_1, e_2$ the following form,

$$e_1 = A e^{i(kz-\Omega t)+i\phi_0}$$
$$e_2 = B e^{i(kz-\Omega t)+i\phi_0}$$

If we assume the amplitudes $A$ and $B$ to be the free parameters, then

$$\Omega = -\frac{A^2 + B^2}{2AB} - \frac{(1+\sigma)}{2}(A^2 + B^2) \tag{4}$$

$$k = \frac{B^2 - A^2}{2AB} + \frac{(1-\sigma)}{2}(B^2 - A^2) \tag{5}$$

We then have a three parameter family of solutions. The particular choice of parameters, $k = 0$ which implies $B = \pm A$ and $\Omega = \mp 1 - (1 + \sigma)A^2$, represent temporal solutions whose stability characteristics [19,20] indicate that modulational instabilities appear. These instabilities lead to the generation of pulses which are most likely to be one soliton like solutions as given in the next section and in [2,3].

The derivation of other solutions is given by establishing a connection between the model studied here and the massive Thirring model in classical field theory. The first observation is that three invariances of the integrable model,

(a) *phase invariance*    $e_i \to e_i e^{i\phi_0}$

(b) *translational invariance in $t$ and $z$*    $(z \to z + z_0, \quad t \to t + t_0)$

are preserved in the general case $\sigma \neq 0$, thus one expects at least three conserved quantities. The first conserved quantity of ( 3) is,

$$\int (|e_1|^2 + |e_2|^2)dz$$

It comes from the conservation equation

$$\partial_t(|e_1|^2 + |e_2|^2) = \partial_z(|e_2|^2 - |e_1|^2) \tag{6}$$

which gives a relation between the total energy and its flux. This conservation law, which is independent of $\sigma$, will be an important tool in the construction of soliton-like solutions. Other two conserved quantities are, the Hamiltonian

$$H = \int ([\frac{i}{2}(e_1\partial_z e_1^* + e_2\partial_z e_2^*) + e_2 e_1^*] + c.c + |e_1|^2|e_2|^2 + \frac{\sigma}{2}(|e_1|^4 + |e_2|^4)dz$$

and the total momentum flux

$$M = i \int (e_1\partial_z e_1^* + e_2\partial_z e_2^* - c.c)dz$$

Whether there exists more conserved quantities, is not known, but one only needs equation ( 6) for this work.

The first step in our construction is to assume a solution of the form $e_j = \alpha\psi_j e^{i\theta}$, where $\psi_j$ is a solution of the Thirring model and $\theta$ depends on $z$ and $t$. Substitution of this ansatz in equations ( 3) gives two equations for $\theta$,

$$\partial_\tau\theta = \sigma\alpha^2|\psi_1|^2 + (\alpha^2 - 1)|\psi_2|^2 \tag{7}$$

$$\partial_\eta\theta = \sigma\alpha^2|\psi_2|^2 + (\alpha^2 - 1)|\psi_1|^2 \tag{8}$$

where $\tau = \frac{1}{2}(t + z)$ and $\eta = \frac{1}{2}(t - z)$ are the cone coordinates.

Therefore, for the ansatz to work, a consistency condition on ( 7, 8) must be imposed. Let us obtain this condition by first integrating ( 7) and then differentiating the result with respect to $\eta$,

$$\partial_\eta \theta = \sigma\alpha^2 \int^\tau \partial_\eta |\psi_1|^2 d\tau' + (\alpha^2 - 1) \int^\tau \partial_\eta |\psi_2|^2 d\tau'$$

the right hand side can now be simplified to

$$\sigma\alpha^2 \int^\tau \partial_\eta |\psi_1|^2 d\tau' - (\alpha^2 - 1)|\psi_1|^2 \tag{9}$$

if we use equation ( 6), which in the cone variables read $\partial_\tau |\psi_1|^2 = -\partial_\eta |\psi_2|^2$. Then from equations ( 8) and ( 9) the consistency condition now reads

$$\int^\tau \partial_\eta |\psi_1|^2 d\tau' = \frac{2(\alpha^2 - 1)}{\sigma\alpha^2}|\psi_1|^2 + |\psi_2|^2 \tag{10}$$

and the conditions for which it is satisfied are given in the following proposition

**Proposition**

Equation ( 10) is satisfied if $|\psi_1|^2 = K|\psi_2|^2$ and $\alpha = \sqrt{\frac{2}{2+\sigma(K+\frac{1}{K})}}$.

*Proof*

If the above conditions are satisfied, then the right hand side of ( 10)

$$\frac{2(\alpha^2 - 1)}{\sigma\alpha^2}|\psi_1|^2 + |\psi_2|^2 = [\frac{2(\alpha^2 - 1)}{\sigma\alpha^2} + \frac{1}{K}]|\psi_1|^2$$

which after using the condition on $\alpha$ becomes $-K|\psi_1|^2$. Substitution of the above conditions on the left hand side of ( 10) also gives $-K|\psi_1|^2$ since there

$$\int^\tau \partial_\eta |\psi_1|^2 d\tau' = K \int^\tau \partial_\eta |\psi_2|^2 d\tau'$$

which is equal to $-K \int^\tau \partial_{\tau'} |\psi_1|^2 d\tau' = -K|\psi_1|^2$ after we use equation ( 6). QED. To summarize, we have found that those solutions of the Thirring model that satisfy the condition $|\psi_1|^2 = K|\psi_2|^2$, give solutions to equations ( 3) which describe waves in nonlinear periodic media of the form $e_{1(2)} = \alpha\psi_{1(2)}e^{i\theta}$, where $\alpha = \sqrt{\frac{2}{2+\sigma(K+\frac{1}{K})}}$ and $\theta$ is obtained by quadrature from any of the consistent equations ( 7, 8). In the next two subsections, we will discuss when the above condition for $\psi_i$ is satisfied for the multisoliton and the quasiperiodic solutions.

## 2.1    Multi-soliton Solutions

Let us discuss now which of the soliton solutions of the integrable model satisfy the condition of proportionality of the intensities. As it has already been shown

[3], all of the one soliton solutions satisfy the condition so that one obtains soliton-like solutions (now known as Bragg solitons) of ( 3) and they read

$$
e_1 = \alpha \left( \frac{1+v}{1-v} \right)^{\frac{1}{4}} \sin Q \exp \left( -i \left( \cos Q \frac{t - vz}{\sqrt{1 - v^2}} - \theta(\zeta) - \phi \right) \right)
$$
$$
\times \operatorname{sech} \left( \sin Q \zeta - \frac{i}{2} Q \right)
$$
(11)

$$
e_2 = \alpha \left( \frac{1-v}{1+v} \right)^{\frac{1}{4}} \sin Q \exp \left( -i \left( \cos Q \frac{t - vz}{\sqrt{1 - v^2}} - \theta(\zeta) - \phi \right) \right)
$$
$$
\times \operatorname{sech} \left( \sin Q \zeta + \frac{i}{2} Q \right)
$$
(12)

where $\theta(\zeta) = -\dfrac{4 \sigma v \alpha^2}{1 - v^2} \arctan \left[ \left| \cot g \dfrac{1}{2} Q \right| \coth(\sin Q \zeta) \right]$, $\zeta = \dfrac{z - vt - z_0}{\sqrt{1 - v^2}}$ and $\alpha = \sqrt{\dfrac{1 - v^2}{(1 - v^2) + \sigma(1 + v^2)}}.$

Equations ( 11, 12), represent a two parameter family of solutions of ( 3), where the parameters $-1 < v < 1$ and $0 < Q < \pi$, are respectively given in terms of the amplitude and phase of the eigenvalue of the scattering problem associated to the Thirring model [13,14]. Physically, these solutions describe pulses propagating in periodic nonlinear fibers. Through the nonlinearity, the grating created by the two modes $e_1, e_2$ is such that it "erases" that of the medium, allowing the field to propagate in what would be a regime of strong Bragg scattering for linear fields. The case $v = 0$, gives the profile of a stationary localized wave occurring in the forbidden frequency zones as given by the linear dispersion relation. In the study of nonlinear superlattices, this particular case is also called gap soliton [1,21].

A detailed description of the characteristics of these solutions can be found in [3]. Here we will simply indicate that numerical simulations have shown the same stability characteristics than for the integrable model. Figure 1 shows that after some radiation losses, initial conditions close to those corresponding to (10,11) eventually go into a member of this family. In [3], it was shown that collisions between these pulses was inelastic, with a permanent long term interaction evidenced by a persistent breathing of the pulses after the interaction. This is consistent with the fact that (1,2) is not integrable in the IST sense.

We now turn to the possible existence of multi-soliton-like solutions of ( 3). To determine the conditions under which the consistency condition is satisfied, we follow the construction procedure of multi-solitons by use of multi-Bäcklund transformations given in [16]. One has that the n-soliton solution of the integrable model is

$$
\psi_1 = -\nu e^{-i\beta}, \quad \psi_2 = \rho e^{-i\kappa}
$$

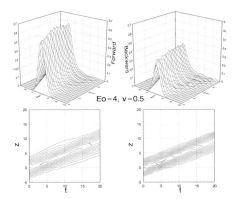

**Fig. 1.** Stable pulse dynamics leading to a soliton like solution [3]

where $e^{i\kappa} = \sum_{i=1}^{n} [\frac{\lambda_i}{\lambda_i}]$, $e^{i\beta} = (1 - 2\sum_{i,j=1}^{n} \frac{f_i \gamma_{ij} \bar{f}_j}{\lambda_j})$

$$\nu = 4i \sum_{i,j=1}^{n} f_i \gamma_{ij} \bar{g}_j \tag{13}$$

$$\rho = 4i \sum_{i,j=1}^{n} \frac{f_i \gamma_{ij} \bar{g}_j}{\lambda_j^2} \tag{14}$$

In the above expressions $f_i = a_i e^{i(\bar{\lambda}^2 \tau + \bar{\lambda}^{-2} \eta)}$, $g_i = b_1 e^{-i(\bar{\lambda}^2 \tau + \bar{\lambda}^{-2} \eta)}$ where $a_i, b_i$ are arbitrary complex constants as well as $\lambda_i$. Finally the matrix $\gamma = \mathbf{\Gamma}^{-1}$, where $\Gamma_{ij} = \frac{2\lambda_i \bar{f}_i f_j - 2\bar{\lambda}_j \bar{g}_i g_j}{\lambda_i^2 - \bar{\lambda}_j^2}$. The parameters $\lambda_i$ are the eigenvalues of the associated scattering problem and their amplitudes give the velocity of each soliton component, so if all $\lambda's$ are of different amplitudes, the initial state will decompose into n separated one soliton solutions. It is clear that equations ( 3) also have an asymptotic solution that has the form of n separated soliton-like solutions as given by ( 11, 12), therefore we now consider the case of the bound n-soliton solution where $|\lambda_i| = |\lambda|, i = 1, 2, ...N$.

Although the actual computation of the solution is rather cumbersome, we can easily see the conditions under which $|\psi_1|^2 = K|\psi_2|^2$. From ( 13) and ( 14), one can see that if we impose the $n - 1$ constraints, $g_i = 0, i = 2, 3, ...n$ then

$$|\psi_1|^2 = |\nu|^2 = 16|g_1|^2| \sum_{i=1}^{n} \gamma_{i1} f_i|^2$$

$$|\psi_2|^2 = |\rho|^2 = \frac{16}{|\lambda|^4} |g_1|^2| \sum_{i=1}^{n} \gamma_{i1} f_i|^2$$

In other words $|\psi_1|^2 = \frac{1}{|\lambda|^4}|\psi_2|^2$, so in the solution form for $e_i = \alpha \psi_i e^{i\theta}$, $\alpha$ takes the value $\sqrt{\frac{2}{2+\sigma(|\lambda|^4 + \frac{1}{|\lambda|^4})}}$.

These constraints on the coefficients $g_i$, do not give a trivial solution as it can be seen in the Appendix A, where we carry out the calculations when $n = 2$. For the case where $|\lambda| = 1$, the intensities, are proportional to $F(z)[1 - \cos(ct)]$, where $F$ and $c$ are given in $A.2$. Following on what was said for the one soliton-like solutions, we find that the intensity fields not only take a stationary form in the forbidden frequency zones, but they can also be breathers.

## 2.2   Quasi-periodic Solutions

We now proceed to look for conditions on the parameters of the quasiperiodic solutions of the Thirring model which would satisfy the constraint of proportionality of intensities. The derivation of the quasi-periodic, or finite gap solutions of the integrable model can be found in [17]. Here we simply present the intensities of the two components. They are,

$$|\psi_1|^2 = C_1 \frac{|\theta(r - \Omega - D + n)|^2}{|\theta(r - \Omega - D)|^2} \tag{15}$$

$$|\psi_2|^2 = C_2 \frac{|\theta(2n - r - \Omega - D)|^2}{|\theta(n - r - \Omega - D)|^2} \tag{16}$$

where $\theta$ is the usual Riemann theta function. The arguments in $\theta$ are vectors in $\mathbf{C}^g$ ($g$ is the genus of the corresponding Riemann surface), with $n, r, D$ being constant vectors (or parameters in the solution), $\Omega = \Omega_1 t + \Omega_2 z$ are the $g$ complex phases in the solutions. Finally, $C_1, C_2$ are proportionality constants.

The conditions under which the constraint $|\psi_1|^2| = K|\psi_2|^2$ is satisfied, depend on the proper choice of the vectors $r$ and $n$, given the general properties of the theta functions. To find the most general constraints on $r, n$ that lead to solutions of the nonintegrable model is not an easy task. The only cases we can show at this time are first, when $n = 2r$, since then, the argument of both numerators is $3r - \Omega - D$ and the argument of both denominators is $r - \Omega - D$, therefore $|\psi_1|^2 = \frac{C_2}{C_1}|\psi_2|^2$. This proves that a very general finite-gap type of solutions of ( 3) exist, although it is not known if they arise from general type of initial conditions, except possibly for the simpler genus one case.

When the genus $g = 1$, the solution takes the form of a cnoidal wave in a travelling wave coordinate. Here, $\theta$ is the elliptic Jacobi theta function $\theta_3$. From known properties of this function, we have that

$$\frac{|\theta(\Omega + S_1)|^2}{|\theta(\Omega + S_2)|^2} = \frac{K_{S_1}}{K_{S_2}} \tag{17}$$

where $S_1, S_2 \epsilon \mathbf{C}$ are constant vectors and $K_{S_1}, K_{S_2}$ and constants. Then the condition ( 10) is always satisfied for the $g = 1$ gap solutions in a similar way than the case of the one soliton solutions. This is not so surprising as one can see from the reduction of ( 3) to ordinary differential equations, by going into travelling coordinates (see Appendix B for details). One finds that all the solutions in this reduced system, correspond to either the one soliton-like solution or the genus one type of solution.

## 2.3    Study of Pulse Dynamics in Nonuniform Gratings

One of the most exciting possible applications of fiber gratings is that of an all optical buffer. Indeed, as the bit rate in communications increase, there is a need to improve other network components; these include elements to re-route, insert and extract data. Current networks have electronic buffers, with a known limited capacity.

To date, the best experiments performed on fiber gratings, have pulses travelling at 52% of the speed of light [22]. The possibility of further slowing and even trapping pulses in nonuniform gratings is both challenging and exciting.

By nonuniform gratings, one means gratings where the period and/or the grating strength varies along the direction of propagation. An example is when the overall grating consists of a link of several fibers each with a uniform grating, but different that the others. Then, assuming at all time that the pulse central frequency lies outside the frequency gap of every fiber, the governing equation is the NLSE where the coefficients are piecewise constant. A different grating structure of interest is if the chirping consists of adiabatic changes in the grating, either by a shift of the center frequency gap and/or a variation of the size of the gap. These properties are reflected in the model by having adiabatic variations on the coefficients of the NLSE. For example if the chirping is such that the frequency of the field slowly approaches the edge of the bandgap, this reduces the value of $v_g$ and increases the effective dispersion. Adiabatic pulse compression in such fibers has been experimentally demonstrated and explained by studying the perturbed NLSE,

$$i\frac{\partial U}{\partial T} + iv_g(Z)\frac{\partial U}{\partial Z} - \frac{1}{2}\Omega''(Z)\frac{\partial^2 U}{\partial Z^2} + c|U|^2 U = 0 \qquad (18)$$

and by using soliton perturbation methods.

A more interesting scenario recently proposed in [6], where they add a defect to the periodic perturbation in the refractive index $n_p = \Delta n \cos(\frac{2\pi}{d}z) + \epsilon W(z)$, where the last term represents a defect. In this case, the governing equations are like (1,2) but with the coefficients being $Z$ dependent as well as having an additional detuning term. In [6], the authors show efficient light trapping in the defect by means of energy transfer from a moving solitary wave to a linear defect mode.

# 3    Gap Soliton Bullets

While much research has dealt with the so called 1-1 problem, very little is known on existence of stable two-dimensional or three-dimensional gap solitons [24–28]. In [25,26] research has focused on two-dimensional periodic structures in the directions transverse to propagation. In contrast, in this study we consider propagation in a planar waveguide Kerr medium with a periodic refractive index profile in the direction of propagation.

Under the envelope approximation, the equations for fully localized optical pulses (optical bullets) propagating in Kerr-type planar waveguides or in bulk media are the two-dimensional and the three-dimensional cubic NLSE, respectively. In both cases the NLSE model predicts at low intensities a dispersion/diffraction dominated dynamics, whereas at high intensities the field amplitude reaches infinite values in finite propagation distances (collapse). In particular, in both cases optical bullets are unstable [29].

We recall that the two-dimensional cubic NLSE is the *critical* case for collapse, in the sense that it is the "boundary" between the *subcritical* one-dimensional case where stable soliton dynamics occurs and the *supercritical* three-dimensional case for which collapse is typically not arrested by small perturbations. In contrast, it has been shown that collapse in the two-dimensional critical NLS can be arrested by various mechanisms (e.g., nonlinear saturation, vectorial effect and nonparaxiality) even when they are small [30]. Therefore, it is more reasonable to try to realize stable bullet propagation (i.e., no collapse) in the planar waveguide two-dimensional case rather than in the bulk media three-dimensional case.

Here we review the first asymptotic study of localized pulse dynamics in a grating waveguide, following [24]. By use of careful multiple-scales analysis, one derives a perturbed 2-dimensional NLS for the amplitude of the gap bullets. Asymptotic analysis and simulations show that solutions of this perturbed NLS do not collapse. Thus, a unique feature of this model is that collapse arrest is solely due to the specific dispersion relation associated with the grating. While this study is just a first step in the overall study of gap soliton dynamics, we believe that it indicates the potential for realizing gap solitons in planar waveguides.

## 3.1   Slowly Varying Envelope Approximation

We consider electromagnetic waves propagating in a Kerr medium with a planar waveguide geometry, i.e., the field is confined in one ($y$) transverse direction and diffracts in the other ($x$) transverse direction. We assume the usual envelope approximation, which in this case consists of two components, ($E_+, E_-$) each describing the envelope of plane waves propagating in the $\pm z$-directions, respectively.

Nonlinear wave phenomena results from the balance between dispersion and nonlinearity and in many instances the description is given by the NLSE. In this section we show this is the case for gap bullets, at least within a certain range of frequencies.

Our starting point are the nonlinear coupled mode equations in a waveguide configuration [27]:

$$i(\partial_T + c_g \partial_z)E_+(T,x,z) + \kappa E_- + \partial_{x^2}^2 E_+ + \Gamma(|E_+|^2 + 2|E_-|^2)E_+ = 0 \, ,$$

$$i(\partial_T - c_g \partial_z)E_-(T,x,z) + \kappa E_+ + \partial_{x^2}^2 E_- + \Gamma(|E_-|^2 + 2|E_+|^2)E_- = 0 \, .$$

In the linear case (i.e., $\Gamma = 0$) the solution of these equations is given by

$$\begin{pmatrix} E_+ \\ E_- \end{pmatrix} = \mathbf{U} e^{i(k_z z + k_x x - \Omega T)} + c.c \, , \qquad \mathbf{U} = \begin{pmatrix} U_+ \\ U_- \end{pmatrix} ,$$

where $\Omega, k_z, k_x$ satisfy the dispersion relation $(\Omega - k_x^2)^2 = \kappa^2 + c_g k_z^2$. In particular, when $k_x = k_z = 0$ then $\Omega = \pm\kappa$, and the linear problem has the solution:

$$\begin{pmatrix} E_+ \\ E_- \end{pmatrix} = c \begin{pmatrix} 1 \\ -1 \end{pmatrix} e^{-i\kappa T} + c.c. .$$

We note that this solution satisfies

$$L \begin{pmatrix} E_+ \\ E_- \end{pmatrix} = 0 ,$$

where $L$ is the operator

$$L = \begin{bmatrix} i\partial_T & \kappa \\ \kappa & i\partial_T \end{bmatrix} .$$

## 3.2   Weakly Nonlinear Theory

Envelope equations are derived for frequencies in the vicinity of $\Omega = \kappa$, $k_x = k_z = 0$, when dispersion and nonlinearity are of the same order. Let $\epsilon$ be the distance between $\Omega$ and $\kappa$, i.e., $\Omega = \kappa + O(\epsilon)$. From the dispersion relation it follows that $k_x, k_z = O(\epsilon^{1/2})$. Therefore, using the method of multiple scales (see, e.g., [31]) we look for solutions of the form

$$\begin{pmatrix} E_+ \\ E_- \end{pmatrix} = \epsilon^{1/2} A(\tau_1, \tau_2, X, Z) \begin{pmatrix} 1 \\ -1 \end{pmatrix} e^{-i\kappa T} + \epsilon \mathbf{U}_1 + \epsilon^{3/2} \mathbf{U}_2 + \epsilon^2 \mathbf{U}_3 + \dots ,$$

where $\tau_1 = \epsilon T$, $\tau_2 = \epsilon^2 T$, $X = \epsilon^{1/2} x$ and $Z = \epsilon^{1/2} z$.

We now proceed to solve for $(E_+, E_-)$ for successive orders in $\epsilon$. Balancing the $\mathbf{O}(\epsilon)$ terms gives:

$$L\mathbf{U}_1 = -ic_g \partial_Z A \begin{pmatrix} 1 \\ 1 \end{pmatrix} e^{-i\kappa T} .$$

This linear problem has the solution $\mathbf{U}_1 = -i\frac{c_g}{2\kappa}\partial_Z A \begin{pmatrix} 1 \\ 1 \end{pmatrix} e^{-i\kappa T}$. In order to go to higher orders we first need a careful computation of the nonlinear terms. We have that

$$(|E_+|^2 + 2|E_-|^2)E_+ = \left( \left| \epsilon^{1/2} A - i\epsilon\frac{c_g}{2\kappa}\partial_Z A \right|^2 + 2\left( \left| -\epsilon^{1/2} A - i\epsilon\frac{c_g}{2\kappa}\partial_Z A \right|^2 \right) \right)$$
$$\times \left( \left| \epsilon^{1/2} A - i\epsilon\frac{c_g}{2\kappa}\partial_Z A \right|^2 \right) e^{-i\kappa T} .$$

Expanding the square modulus terms gives

$$\left| \epsilon^{1/2} A - i\epsilon\frac{c_g}{2\kappa}\partial_Z A \right|^2 = \left( \epsilon^{1/2} A - i\epsilon\frac{c_g}{2\kappa}\partial_Z A \right) \left( \epsilon^{1/2} A^* + i\epsilon\frac{c_g}{2\kappa}\partial_Z A^* \right)$$
$$= \epsilon|A|^2 + i\epsilon^{3/2}\frac{c_g}{2\kappa}(A\partial_Z A^* - A^*\partial_Z A) + \epsilon^2\frac{c_g^2}{4\kappa^2}\partial_Z A\partial_Z A^* .$$

Similarly,

$$2\left|\epsilon^{1/2}A + i\epsilon\frac{c_g}{2\kappa}\partial_Z A\right|^2 = 2\left(\epsilon^{1/2}A + i\epsilon\frac{c_g}{2\kappa}\partial_Z A\right)\left(\epsilon^{1/2}A^* - i\epsilon\frac{c_g}{2\kappa}\partial_Z A^*\right)$$

$$= 2\left(\epsilon|A|^2 + i\epsilon^{3/2}\frac{c_g}{2\kappa}(A^*\partial_Z A - A\partial_Z A^*)\right.$$

$$\left. + \epsilon^2\frac{c_g^2}{4\kappa^2}\partial_Z A\partial_Z A^*\right).$$

Thus the nonlinear terms read:

$$(|E_+|^2 + 2|E_-|^2)\, E_+ = \left(3\epsilon|A|^2 + i\epsilon^{3/2}\frac{c_g}{2\kappa}(A^*\partial_Z A - A\partial_Z A^*) + 3\epsilon^2\frac{c_g^2}{4\kappa^2}\partial_Z A\partial_Z A^*\right)$$

$$\left(\epsilon^{1/2}A - i\epsilon\frac{c_g}{2\kappa}\partial_Z A\right)e^{-i\kappa T} + \text{c.c}$$

$$= \left(3\epsilon^{3/2}|A|^2 A - i\epsilon^2\frac{c_g}{2\kappa}\left(2|A|^2\partial_Z A + A^2\partial_Z A^*\right)\right.$$

$$\left. + \epsilon^{5/2}\left(\frac{c_g^2}{2\kappa^2}A\partial_Z A\partial_Z A^* + \frac{c_g^2}{4\kappa^2}A^*(\partial_Z A)^2\right)\right.$$

$$\left. - i\epsilon^3\frac{3c_g^3}{8\kappa^3}(\partial_Z A)^2(\partial_Z A^*)\right)e^{-i\kappa T} + \text{c.c.}$$

Similarly,

$$(|E_-|^2 + 2|E_+|^2)\, E_- = -\left(3\epsilon^{3/2}|A|^2 A + i\epsilon^2\frac{c_g}{2\kappa}(2|A|^2\partial_Z A + A^2\partial_Z A^*)\right.$$

$$\left. + \epsilon^{5/2}\left(\frac{c_g^2}{2\kappa^2 A}\partial_Z A\partial_Z A^* + \frac{c_g^2}{4\kappa^2}A^*(\partial_Z A)^2\right)\right.$$

$$\left. + i\epsilon^3\frac{3c_g^3}{8\kappa^3}(\partial_Z A)^2(\partial_Z A^*)\right)e^{-i\kappa T} + \text{c.c.}$$

We now continue computing higher order corrections to $(E_+, E_-)$. Balancing the $\mathbf{O}(\epsilon^{3/2})$ terms gives

$$L\mathbf{U_2} = \left(-i\partial_{\tau_1}A - \partial_{X^2}^2 A - \frac{c_g^2}{2\kappa}\partial_{Z^2}^2 A - 3\Gamma|A|^2 A\right)\begin{pmatrix}1\\-1\end{pmatrix}e^{-i\kappa T} + \text{c.c}.$$

Note that the right hand side has slowly varying terms arising from $\mathbf{U_1}$. Since $\begin{pmatrix}1\\-1\end{pmatrix}e^{-i\kappa T}$ is in the null space of $L$, the physical requirement that

$$i\partial_{\tau_1}A + \partial_{X^2}^2 A + \frac{c_g^2}{2\kappa}\partial_{Z^2}^2 A + 3\Gamma|A|^2 A = 0 . \tag{19}$$

accounts for the removal of secular terms, or solvability condition. Once this condition is imposed, one finds $\mathbf{U_2} = 0$.

The derivation of the two-dimensional NLSE for the bullets amplitude is not surprising, given that the one-dimensional NLSE has been previously derived for fiber gratings with no diffraction due to the transverse waveguide structure at frequencies close to but outside the gap. We recall that for the focusing critical NLSE

$$i\psi_t(t,x,y) + \psi_{xx} + \psi_{yy} + |\psi|^2\psi = 0 \ , \qquad \psi(0,x,y) = \psi_0(x,y) \ ,$$

collapse can occur when the input power $\int |\psi_0|^2 \, dxdy$ is above the critical power $N_c$ [32], whereas no collapse occurs for the mixed-signs NLSE

$$i\psi_t(t,x,y) + \psi_{xx} - \psi_{yy} + |\psi|^2\psi = 0 \ ,$$

which is similar to the hydrodynamics problem. Therefore, Eq.(19) has blowup solutions for pulses whose center frequency is at $\Omega \sim \kappa$, but not when $\Omega = -\kappa$, in which case the NLSE is given by

$$i\partial_{\tau_1} A + \partial_{X^2}^2 A - \frac{c_g^2}{2\kappa}\partial_{Z^2}^2 A + 3\Gamma|A|^2 A = 0 \ .$$

Returning to the case of more interest ($\Omega \sim \kappa$), the possibility of collapse indicates a breakdown of the asymptotic expansion. From physical considerations, however, we do not expect the bullet to collapse. Indeed, since pulse compression leads to spectral broadening, at some point the down-frequency side of the spectral pulse will "see" the edge of the gap in the dispersion relation, thus preventing further broadening and most likely arresting the collapse. Of interest is then to see if this could be modelled by considering higher order corrections in the asymptotic expansion. To do so we introduce a slower time variable $\tau_2 = \epsilon^2 T$. Continuing the expansion we have to $O(\epsilon^2)$,

$$L\mathbf{U_3} = \left( i\frac{\Gamma c_g}{2\kappa}(2|A|^2\partial_Z A + A^2\partial_Z A^*) - i\frac{c_g}{2\kappa}\partial_Z(-i\partial_{\tau_1} - \partial_{X^2}^2)A \right)\begin{pmatrix} 1 \\ 1 \end{pmatrix} e^{-i\kappa T} + cc \ .$$

Using Eq.(19) we have that

$$\partial_Z(-i\partial_{\tau_1} - \partial_{X^2}^2)A = \frac{c_g^2}{2\kappa}\partial_{Z^3}^3 A + \partial_Z(3\Gamma|A|^2 A) \ .$$

Therefore, the equation for $\mathbf{U_3}$ can be rewritten as

$$L\mathbf{U_3} = -\left[ i\frac{\Gamma c_g}{2\kappa}(4|A|^2\partial_Z A + 2A^2\partial_Z A^*) + \frac{c_g^3}{4\kappa^2}\partial_{Z^3}^3 A \right]\begin{pmatrix} 1 \\ 1 \end{pmatrix} e^{-i\kappa T} + cc \ ,$$

whose solution is

$$\mathbf{U_3} = -i\frac{c_g}{4\kappa^2}\left[ \Gamma(4|A|^2\partial_Z A + 2A^2\partial_Z A^*) + \frac{c_g^2}{2\kappa}\partial_{Z^3}^3 A \right]\begin{pmatrix} 1 \\ 1 \end{pmatrix} e^{-i\kappa T} + cc \ .$$

The $\mathbf{O}(\epsilon^{5/2})$ terms gives

$$
\mathbf{U_4} = -\left( i\partial_{\tau_2} A + \frac{c_g^2}{4\kappa^2}(\Gamma\partial_Z(4|A|^2\partial_Z A + 2A^2\partial_Z A^*) + \frac{c_g^4}{8\kappa^3}\partial_{Z^4}^4 A \right.
$$
$$
\left. + \frac{\Gamma c_g^2}{4\kappa^2}(2A\partial_Z A\partial_Z A^* + A^*(\partial_Z A)^2) \right) \begin{pmatrix} 1 \\ -1 \end{pmatrix} e^{-i\kappa T} + cc .
$$

As in the $O(\epsilon^{3/2})$ case, in order to prevent secular terms we impose the condition

$$
i\partial_{\tau_2} A + \frac{\Gamma c_g^2}{4\kappa^2}\left( 5A^*(\partial_Z A)^2 + 10A\partial_Z A\partial_Z A^* + 4|A|^2\partial_{Z^2}^2 A + 2A^2\partial_{Z^2}^2 A^* \right)
$$
$$
+ \frac{c_g^4}{8\kappa^3}\partial_{Z^4}^4 A = 0 . \tag{20}
$$

We finish by adding Eqs.(19) and (20) and defining the slow time $\tau = \tau_1 + \epsilon\tau_2$, leading the perturbed two-dimensional NLSE

$$
i\partial_\tau A + \partial_{X^2}^2 A + \frac{c_g^2}{2\kappa}\partial_{Z^2}^2 A + 3\Gamma|A|^2 A = -\epsilon F(A) , \tag{21}
$$

$$
F(A) = \frac{\Gamma c_g^2}{4\kappa^2}(5A^*(\partial_Z A)^2 + 10A\partial_Z A\partial_Z A^* + 4|A|^2\partial_{Z^2}^2 A + 2A^2\partial_{Z^2}^2 A^*) + \frac{c_g^4}{8\kappa^3}\partial_{Z^4}^4 A .
$$

## 3.3    Analysis of the 2D Perturbed NLSE

In the previous section we derived the perturbed NLSE (21) as the slowly-varying envelope approximation of the coupled mode equations. While we do not expect Eq.(21) to remain valid over very long distances, a natural question is whether solutions of Eq.(21) can collapse. We now show that collapse is arrested in Eq.(21), resulting instead in stable focusing-defocusing oscillations.

In order to bring Eq.(21) to the standard form we make the change of variables

$$
y = \frac{\sqrt{2\kappa}}{c_g} z , \psi = \sqrt{3\Gamma} A .
$$

This leads to the nondimensional equation

$$
i\psi_\tau(\tau, x, y) + \Delta\psi + |\psi|^2\psi = -\frac{\epsilon}{12\kappa}\left( 10\psi|\psi_y|^2 + 5\psi^*\psi_y^2 \right.
$$
$$
\left. + 4|\psi|^2\psi_{yy} + 2\psi^2\psi_{yy}^* + \frac{1}{2}\psi_{yyyy} \right) \tag{22}
$$

where $\Delta = \partial_{xx} + \partial_{yy}$. When the input power is close to the critical power we can analyze the effects of the small terms on the right-hand-side of Eq.(22) on collapse and on the propagation dynamics using *Modulation theory* [30]. Modulation theory is based on the observation that a self-focusing pulse rearranges itself as a modulated Townesian, i.e., $|\psi| \sim L^{-1}(z)R(r/L(z))$, where $R(r)$, the so-called *Townes soliton*, is the ground-state positive solution of

$$\Delta R(r) - R + R^3 = 0 , \qquad R'(0) = 0 , \quad \lim_{r \to \infty} R(r) = 0 . \qquad (23)$$

Application of modulation theory to Eq. (22) leads to the following result

**Proposition 1.** *When $\epsilon/\kappa \ll 1$, self-focusing dynamics in Eq.(22) is given, to leading order, by the reduced system of ODEs*

$$\begin{cases} L_{\tau\tau}(\tau) = -\frac{\beta}{L^3}, \\ \beta_\tau(\tau) = -\frac{\epsilon}{\kappa}C_{\mathrm{GS}}\frac{N_c}{2M}\left(\frac{1}{L^2}\right)_\tau, \end{cases} \qquad (24)$$

*where $N_c = \int_0^\infty R^2\, rdr \approx 1.86$, $M = \dfrac{1}{4}\int_0^\infty r^2 R^2\, rdr \approx 0.55$, and $C_{\mathrm{GS}} \approx 55/48$ .*

In [30] it is shown that Eq.(24) is the generic reduced equation in critical self-focusing. Analysis of this reduced system shows that its solutions do not collapse. Rather, they undergo stable focusing-defocusing oscillations. Indeed, we observe this behavior when we solve Eq.(22) numerically (see Fig. 2).

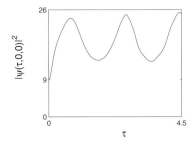

**Fig. 2.** On-axis intensity of the solution of Eq.(22) with $\epsilon/\kappa = 0.075$. The initial condition is $\psi(\tau = 0, x, y) = 2\sqrt{N(0)}\, e^{-(x^2+y^2)}$, and $N(0) = 1.2N_c$, [24]

Direct integration of the coupled mode equations on sec.3.1 clearly indicate an instability leading to an initial collapse type behavior, but as it has been explained here, this collapse is finally arrested (see Fig. 3)

## 4   Conclusions

In this article we presented an overview of the theory of pulse propagation phenomena in fiber and waveguide Bragg gratings. While one can only say that for fibers with uniform gratings, most of the relevant theory and experiments have been developed to a great extent, it is still intriguing the fact that even though there is every indication that the governing coupled mode equations are

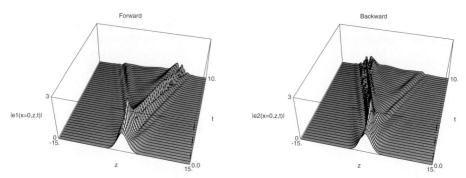

**Fig. 3.** Evolution of the forward (left) and backward (right) field intensities as described by the coupled mode equations [27]

nonintegrable, there is a richer class of solutions compared to what one typically encounters in Hamiltonian nonintegrable systems. One would hope that some of the experimentally observed periodic pulse trains are well represented by the multi-soliton and/or quasiperiodic solutions presented here.

In terms of experiments, the ability to custom engineer grating structures in fibers, opens many possibilities for applications. To model such structures, one needs to include specific propagation ($z$) dependence in the coefficients of either the coupled mode equations, or when appropriate on the approximate nonlinear Schrödinger-like equation. It is of great interest to find suitable functional forms for these coefficients to model novel applications. The recent work of [6] in which defects are added to the grating to stop light is both intriguing and potentially very useful. It has yet to be demonstrated experimentally. Another model not discussed here is to couple two or more fiber gratings by cross-talk of the transverse waveguide modes. If one can balance the three relevant length scales (coupling, dispersion and nonlinear lengths), ultrafast switching could be realized in this coupler. More analysis is needed for this model.

With respect to waveguide gratings, we have shown that by use of a multiple-scales analysis, one derives the extended NLSE approximation for optical bullet dynamics in a waveguide with a grating in the direction of propagation. We then showed that in this extended NLSE, collapse is arrested. An intuitive explanation for this is as follows. When the field frequency lies outside of the gap in the dispersion relation, the dynamics is initially well approximated by the 2-dimensional NLSE and for sufficiently high input power a collapse-type dynamics occurs (see Fig. 1). Since critical collapse occurs in a radially symmetric way [33,34], the resulting spectral broadening occurs in all wavenumber directions. Our analysis thus shows that the fact that broadening of the $z$-component of the wavenumber is prevented by the presence of the band-gap is enough to prevent the collapse,[1] leading instead to nontrivial dynamics which is yet to be fully understood. This suggests, nevertheless, that the periodic waveguide

---

[1] It remains to see whether such a small effect would be able to prevent the collapse in bulk media, where collapse is supercritical.

structure is a suitable candidate for first observing gap-soliton bullets in short propagation, if only as a metastable object. Better understanding of the overall dynamics, however, will require a direct analysis of the coupled mode equations.

### Acknowledgements

I have been fortunate enough that during the last 10 years, I have had research collaborations on this field with the main players. Special thanks go to S. Wabnitz, C. De Angelis, B. Costantini, S. Trillo, M. de Sterke, N. Broderick, J. Sipe, B. Eggleton, R. Slusher, G. Lenz and M. Weinstein and G. Fibich.

## Appendix A

For the bound two-soliton solution with the constraint $g_2 = 0$ defining $\lambda_i = |\lambda|e^{iQ_i}, i = 1, 2$ and following the notation of section 2, one has

$$\Gamma_{11} = \frac{|a_1|^2 e^{2\sin 2Q_1(|\lambda|^2\tau - \frac{1}{|\lambda|^2}\eta)+iQ_1} - |b_1|^2 e^{-2\sin Q_1(|\lambda|^2\tau - \frac{1}{|\lambda|^2}\eta)-iQ_1}}{i|\lambda|\sin 2Q_1}$$

$$\Gamma_{22} = \frac{|a_2|^2 e^{2\sin 2Q_2(|\lambda|^2\tau - \frac{1}{|\lambda|^2}\eta)+iQ_2}}{i|\lambda|\sin 2Q_2}$$

$$\Gamma_{12} = \frac{2\bar{a}_1 a_2 e^{i(|\lambda|^2\tau + \frac{1}{|\lambda|^2}\eta)(\cos 2Q_1+\cos 2Q_2)} e^{(\sin 2Q_1+\sin 2Q_2)(|\lambda|^2\tau - \frac{1}{|\lambda|^2}\eta)} e^{iQ_1}}{|\lambda|e^{2i(Q_1-Q_2)}}$$

and

$$\Gamma_{21} = \bar{\Gamma}_{12}e^{i(Q_1-Q_2)}$$

Substitution of $\gamma = \mathbf{\Gamma}^{-1}$ in the expressions for $|\psi_j|^2, j = 1, 2$ gives

$$|\psi_2|^2 = |\frac{1}{\lambda}|^4|\psi_1|^2 = \frac{16|b_1a_1a_2|^2|\lambda|^2}{|\Delta|^2}[\frac{1}{\sin^2 Q_2} + 4(1 - \cos(\Phi + Q_1 - Q_2))] \quad (25)$$

where

$$\Delta = -|a_1|^2(\frac{1}{\sin Q_1 \sin Q_2} + 4)e^{2\sin 2Q_1(|\lambda|^2\tau - \frac{1}{|\lambda|^2}\eta)+i(Q-1+Q_2)}$$

$$+|b_1|^2 \exp\left(-2\sin 2Q_1(|\lambda|^2\tau - \frac{1}{|\lambda|^2}\eta) + i(Q_2 - Q_1)\right) \quad (26)$$

and $\Phi = 2\cos Q_2(|\lambda|^2\tau + \frac{1}{|\lambda|^2}\eta) + 2arg(a_1 + a_2)$.

Looking back at equation ( 25), for the particular case $|\lambda| = 1$ and replacing $\tau = \frac{1}{2}(t + z)$ and $\eta = \frac{1}{2}(t - z)$, the intensities become proportional to

$$\frac{e^{-4\sin Q_1 z}}{1 + D_1 e^{-4\sin Q_1 z} + D_2 e^{-8\sin Q_1 z}}[1 + 4\sin^2 Q_2 - 4\sin^2 Q_2 \cos(2\cos(Q_2 t) + D_3)] \quad (27)$$

where $D_1, D_2$ and the constant of proportionality depend on all the parameters. This, as one can see, is a breather type of solution. Figure 3, shows the sum of the their intensities. The actual computation of the phase $\theta$ was not performed, but the most important part of the problem is the behavior of the intensities, which are plotted for this solution in figure 3.

## Appendix B

Here we present a reduction of the governing equations to a system of ordinary differential equations (ODE). Following the form of the soliton solutions ( 11), let us consider the ODE reduction of equations ( 3) by assuming the following travelling wave form for the modes $e_1, e_2$ [3],

$$e_j = K_j \psi_j(y) e^{-i \cos Q(\frac{t-vz}{\sqrt{1-v^2}})+i\phi_0} \quad j = 1,2$$

where,

$$K_1 = 1/K_2 = \left(\frac{1+v}{1-v}\right)^{\frac{1}{4}} \quad y = \frac{z - vt}{\sqrt{1 - v^2}}$$

with $0 < Q < \pi$ and $-1 < v < 1$.
The pair of ordinary differential equations that one obtains is

$$\frac{d\psi_1}{dy} = i \cos Q \psi_1 + i\psi_2 + i|\psi_2|^2\psi_1 + i\sigma\left(\frac{1+v}{1-v}\right)|\psi_1|^2\psi_1$$

$$\frac{d\psi_2}{dy} = -i \cos Q \psi_2 - i\psi_1 - i|\psi_1|^2\psi_2 - i\sigma\left(\frac{1-v}{1+v}\right)|\psi_2|^2\psi_2$$

which can be written in Hamiltonian form $\frac{du_i}{dy} = i\frac{\delta H}{\delta u_i^*}$, where $u_1 = \psi_1$, $u_2 = \psi_2^*$ and

$$H = \cos Q(|u_1|^2 + |u_2|^2) + (u_1u_2 + c.c) + |u_1|^2|u_2|^2$$
$$+ \frac{\sigma}{2}\left[\left(\frac{1+v}{1-v}\right)|u_1|^4 + \left(\frac{1-v}{1+v}\right)|u_2|^4\right]$$

A second constant of the motion $C = |u_1|^2 - |u_2|^2$, makes the system completely integrable in $\mathbf{R}^2$. It is interesting to observe that for the original equations ( 3), integrability in the IST sense only happens when $\sigma = 0$, whereas in this ODE reduction, complete integrability always happens.

The reduction to the phase plane is given in terms of $p = |\psi_2|^2$ and $q = arg(\psi_1) - arg(\psi_2)$ and reads,

$$\dot{q} = \frac{\partial h}{\partial p}, \quad \dot{p} = -\frac{\partial h}{\partial q}$$

with $h = 2p \cos Q + 2\sqrt{p(C+p)} \cos q + p(C + p) + \frac{\sigma}{2}[(\frac{1+v}{1-v})(C+p)^2 + (\frac{1-v}{1+v})p^2]$.

For the $C = 0$ case, whose phase portrait is shown in figure, two hyperbolic points exit at $(\pi + Q, 0)$ and $(\pi - Q, 0)$ and an elliptic point at $(\pi, \alpha^2(1 - \cos Q))$, where $\alpha$ is given in (1.). The two hyperbolic points are connected by a heteroclinic trajectory which corresponds, in the original variables $e_{1,2}$ to the one soliton-like solutions (2.1a,b). This orbit for the case $v = 0$, was first obtained by Chang, Ellis and Lee [23] in their study of fermion confinement in a chiral symmetry theory in 1-1 dimensions. Their confined and time independent solutions of the Thirring model were obtained before the integrability of the model was determined. The same equations were also derived by Mills and Trullinger in their description of stationary localized waves inside nonlinear superlattices. However, neither of these studies discuss the full phase plane or the periodic orbits. A further analysis of the $C = 0$ case shows that the system decouples into

$$\dot{q} = \sqrt{4\cos^2 q + 8\cos Q \cos q - d} \tag{28}$$

where $d$ is a constant of integration. After integrating $q$, $p$ is obtained by quadrature,

$$p(y) = p_0 e^{2\int^y \sin q \, dy'} \tag{29}$$

The general solutions of equations (28,29) are given in terms of elliptic functions and they correspond in the original variables to cnoidal waves. An explicit form of the solutions can be obtained for the case $Q = \frac{\pi}{2}$; they are,

$$q = \sin^{-1}(sn(u|m))$$

and

$$p = p_0(\, dn(u|m) - m^{\frac{1}{2}} cn(u|m)\, ) \tag{30}$$

where $cn, dn$ are the usual cnoidal functions and $u = (4 - d)^{\frac{1}{2}} y$, $m$ a constant of integration.

# References

1. W. Chen and D. L. Mills: Phys. Rev. Lett. **58**, 160 (1987).
2. D. N. Christodoulides and R. I. Joseph: Phys. Rev. Lett. **62**, 1746 (1989).
3. A. B. Aceves and S. Wabnitz: Phys. Lett. A **141**, 37 (1989).
4. B.J. Eggleton, R.E. Slusher, C.M. de Sterke, P.A. Krug, and J.E. Sipe:Phys. Rev. Lett. **76**, 1627 (1996); B.J. Eggleton:J. Opt. Soc. Am. B **14**, (1997).
5. N. G. R. Broderick, D. J. Richardson and M. Ibsen:Opt. Lett. **25**, 536 (2000).
6. R. H. Goodman, R. E. Slusher and M. I. Weinstein, "Stopping Light on a Defect:J. Opt. Soc. Am. B (to appear)
7. See for example the special Optics Express issue, **3**, 11 (1998).
8. N. Aközbek and S. John: Phys. Rev. E **58**, 3876(1998).
9. H. G. Winful and V. Perlin: Phys. Rev. Lett. **84**, 3586 (2000).
10. V. Rupasov and M. Singh:Phys. Rev. A **54**, 3614 (1996).
11. O. Zobay, S. Potting, P. Meystre and E. M. Wright: Phys. Rev. A **59**, 643 (1999).

12. C.M. de Sterke and J.E. Sipe,: In:*Progress in Optics*, Vol XXXIII (North-Holland, Amsterdam), 203 (1994).

13. E. A. Kuznetsov, and A. V. Mikhailov: Teor. Mat. Fiz. **30**, 193 (1977).

14. D. J. Kaup, and A. C. Newell: Lett. Nuovo Cimento **20**, 325 (1977).

15. A.V. Mikhailov: Pis'ma Zh. Eksp. Teor. Fiz. **23**, 356 (1976)

16. D. David, J. Harnad and S. Shnider: Lett. Math. Phys. **8**, 27 (1984).

17. R.F. Bikbaev:Teor. Mat. Fiz. **63**, 377 (1985).

18. L. Brillouin, *Wave propagation in periodic structures*, (McGraw-Hill, New York, 1946).

19. C. M. de Sterke and J. Sipe: Phys. Rev. A **42**, 2858 (1990).

20. A. B. Aceves, C. De Angelis, and S. Wabnitz:Opt. Lett. **17**, 1566 (1992).

21. D. L. Mills and S. E. Trullinger:Phys. Rev. B **36**, 947 (1987).

22. B. J. Eggleton, C. M. de Sterke and R. E. Slusher:J. Opt. Soc. Am. B **16**, 587 (1999).

23. S. J. Chang, S. D. Ellis and B. W. Lee: Phys. Rev D **11**, 3572 (1975).

24. A. B. Aceves, G. Fibich and B. Ilan: Physica D (to appear).

25. N. Aközbek and S. John: Phys. Rev. E **57**, 2287 (1998).

26. S. F. Mingaleev and Y. S. Kivshar:Phys. Rev. Lett. **86**, 5474 (2001).

27. A. B. Aceves, B. Costantini and C. De Angelis: J. Opt. Soc. Am. B **12**, 1475 (1995).

28. N. M. Litchinitser, C. J. McKinstrie, C. M. de Sterke:J. Opt. Soc. Am. B **18**, 45 (2001).

29. C. Sulem and P.L. Sulem, *The nonlinear Schrödinger equation*, (Springer, New-York, 1999).

30. G. Fibich and G. C. Papanicolaou:SIAM J. of Appl. Math. **60**, 183 (1999).

31. C.M. Bender and S. Orszag, *Advanced Mathematical Methods for Scientists and Engineers.* (McGraw-Hill New York, 1978).

32. G. Fibich and A. Gaeta: Opt. Lett. **25**, 335 (2000)

33. M.J. Landman, G.C. Papanicolaou, C. Sulem, P.L. Sulem and X.P. Wang: Physica D **47**, 393 (1991)

34. G. Fibich and B. Ilan:J. Opt. Soc. Am. B **17**, 1749 (2000)

# Theory of Gap Solitons in Short Period Gratings

S. Trillo[1,2] and C. Conti[2]

[1] University of Ferrara, Via Saragat 1, 44100 Ferrara, Italy
[2] INFM-RM3, Via della Vasca Navale 84, 00146 Roma, Italy

**Abstract.** We review the basic properties of gap solitons propagating along one-dimensional short-period gratings. The main emphasis is on Bragg gratings with Kerr nonlinearity, though we briefly discuss also other soliton-bearing structures.

## 1   Introduction

It is well known that media which possess periodic properties over the wavelength scale, exhibit a response which is strongly sensitive to the frequency of the incoming radiation [1]. In particular, so-called photonic bandgaps (gaps for briefness) open up around characteristic frequencies which yield the strongest reflection of light propagating along a chosen direction. The most simple of these structures is the one-dimensional (1D) Bragg grating or distributed feedback (DFB) device, where a $\Lambda$-periodic modulation of the linear properties, i.e., the linear refractive index $n = n(\lambda)$ (or effective index in waveguide geometries) is built in to induce overall reflection around so-called Bragg frequencies $\omega_{Bm}$ such that waves reflected from successive elementary corrugations add in phase, i.e. $2\beta(\omega_{Bm})\Lambda = 2\pi m$, $\beta$ and $m = 1, 2, \ldots$ denoting the wavevector (propagation constant in waveguides) and the reflection order, respectively. In terms of vacuum wavelength $\lambda_0$ of the optical field, such linear resonances arise at

$$\lambda_0 = \lambda_{0m} \equiv \frac{2\Lambda}{m} n(\lambda_{0m}). \tag{1}$$

Here $\lambda_{0m}$ mark the center of the gaps within which the low-power forward propagation is inhibited (damped). A good insight into the physics of the nonlinear regime of propagation in a Bragg grating can be gained with reference to the Kerr effect due to cubic nonlinearities and responsible for the refractive index change $\Delta n = n_{tot} - n = n_{2I}I$ proportional to the optical intensity $I$. Its effect on the Bragg grating response was investigated [2–4], independently from Gap soliton concept introduced later by Chen and Mills [5]. Assuming $n_{2I} > 0$ i.e. focusing materials, the intensity-dependent index change induces a red-shift of the bandgap [see Eq. (1)]. As a result, at sufficiently high intensity, the resonance (gap) can be detuned from the operating wavelength, thus resulting into self-induced transparency. Gap solitons (GSs) can be understood as bell-shaped envelopes which can travel along the grating thanks to the high intensity core which bleaches the reflectivity, while the pulse is taken together by the low-intensity reflectivity which prevents photons to escape from the tails. This simple picture is supported by more thorough analysis of GSs in Kerr media [6–14], and their experimental realization [15–22] (for a recent review see also Ref. [23]). More recent investigations have shown that also quadratic nonlinearities can give rise to GSs [24–41].

As far as localization phenomena in periodic media are concerned, however, the linear (Bragg) distributed feedback is not the only structure of interest. While Bragg coupling entails a coupling of linear origin between the forward and backward wave, one could conceive structures where such coupling arises from the nonlinear response of condensed matter, as for instance in stratified Kerr materials [42–45], or in quadratic $[\chi^{(2)}]$ media via backward second-harmonic generation (B-SHG) [46–58] . In the latter case, a forward-propagating beam at $\omega$ generates a beam at $2\omega$ in the backward direction. Following what is known as quasi-phase-matched (QPM) technique [59], SHG in the backward direction can be phase-matched by introducing a periodic nonlinear susceptibility, that is a grating of nonlinearity. While the QPM method is routinely and successfully employed to phase-match SHG in the forward direction by means of relatively long-period gratings [60], also B-SHG induced by short-period QPM gratings has been experimentally observed [61–66]. This type of structures are *linearly transparent* due to the absence of any linear Bragg effect, while they become strongly reflective at high intensities. Therefore they behave as *nonlinear reflectors*, that is they can be viewed 1D implementation of photonic crystals of pure nonlinear origin [67,68]. Though less intuitive than the Bragg case, propagation of GSs can occur also in this type of structure when an additional nonlinear effect intervene to bleach the reflectivity in the high-intensity portion of the pulses [58].

In the following we analyze in detail the physics of GSs in Bragg gratings drawing the reader's attention to the main similarities and differences which can be expected when the localized modes are due to a gap of nonlinear origin. Although this covers the basic physics of trapping mechanisms taking place in 1D optical gratings, we emphasize that GSs arise also in a variety of physical contexts involving different material responses and geometries such as chains of discrete oscillators [69–71] or Heisenberg ferromagnetic chains [72], photon-atom coupling [73], Josephson junctions [74], electrical circuits [75], propagation in planar structures [76] or propagation of multicomponent (vector) waves [77,78], media exhibiting absorptive (two-level) response [79] or scattering (Raman) processes [80], quantum systems [81,82], and Bose-Einstein condensation [83], whose common basic ingredients are the presence of nonlinearity and periodicity.

The chapter is organized as follows. In sec.2 we discuss the existence of stopbands in the frequency response of gratings and their underlying physical mechanisms. In sec.3 we discuss GS solutions of Bragg gratings, while sec.4 and sec.5 are devoted to discuss briefly GS existence in quadratic materials and GS stability problems, respectively. In sec.6 we overview the properties of GSs in media with periodic nonlinearities. Our conclusions are given in sec.7. Finally, we warn the reader that the term soliton is used to mean any localized solution of the grating models, rather than in the spirit of a more strict mathematical definition related to integrability.

# 2   Reflection Versus Transparency in Short Period Gratings

In general two modes with different propagation constants (or wavevectors for plane-waves) $\beta_{1,2}$ are efficiently coupled by a periodic modulation of the linear optical properties with period $\Lambda$ (pitch $\beta_0 = 2\pi/\Lambda$) when the grating fulfils the $m$-th resonance condition $\Delta\beta \equiv \beta_2 - \beta_1 \approx m\beta_0$. Therefore forward propagating modes can be coupled by gratings with relatively long wavelength $\Lambda = 2\pi m/\Delta\beta$ (we point out that also this type of coupling gives rise to solitons [25,84], though we will not address them further in this chapter). Conversely, in the particular case $\beta_2 = -\beta_1 \equiv \beta$ such that coupling occurs between two modes which are identical except for their forward and backward propagating directions, the grating is of the Bragg type and has short pitch $\Lambda = m\pi/\beta$. The latter condition is equivalent to Eq. (1), once we recall the general expression for the propagation constant $\beta = 2\pi n(\lambda_0)/\lambda_0$.

Counterpropagating modes, however, can be also coupled in structures which exhibit short periodicity of the nonlinear properties. In Kerr media this can be obtained by making a dielectric stack by alternating media with the same linear index and different nonlinear index $n_{2I}$, for instance of focusing ($n_{2I} > 0$) and defocusing ($n_{2I} < 0$) type, respectively [42–45]. As mentioned above, in quadratic media nonlinear coupling can occur between modes with different carrier frequency through short wavelength QPM gratings, made by alternating $\chi^{(2)}$ nonlinearity. Unlike the Bragg grating, all these structures are transparent at low power, whereas the grating becomes effective only at high power and lead to a stopband of nonlinear origin.

In the next two subsections, we discuss in more detail the origin of the linear and nonlinear stopbands in the framework of the stationary (or continuous-wave, CW) coupled-mode theory (CMT).

## 2.1   Stopband of Linear Origin

In the presence of a modulation of the linear index along $Z$, which can be Fourier expanded as

$$\Delta n(Z) = \sum_{m>0} c_m \exp(\pm im\beta_0 Z), \qquad (2)$$

the CMT can be worked out by separating the forward ($E_m^+$) and backward ($E_m^-$) contributions to the total field

$$E(Z,T) = [E_m^+(Z)\exp(i\beta Z) + E_m^-(Z)\exp(-i\beta Z)]\exp[-i(\omega_{Bm} + \Delta\omega)T], \quad (3)$$

where $\beta = \beta(\omega_{Bm} + \Delta\omega)$ is the light wavevector (propagation constant) evaluated at frequency detuning $\Delta\omega$ from $m$−th Bragg resonance $\omega_{Bm}$ (implicitly defined by the condition $\beta(\omega_{Bm}) = m\beta_0/2$). The CMT problem reads as

$$-i\frac{dE_m^+}{dZ} = \Gamma_m E_m^- \, e^{-i\Delta\beta_m Z} \quad ; \quad i\frac{dE_m^-}{dZ} = \Gamma_m^* E_m^+ \, e^{i\Delta\beta_m Z}, \qquad (4)$$

where $\Delta\beta_m = 2\beta - m\beta_0$ stands for the detuning from Bragg condition, and $\Gamma_m \propto c_m$ is the $m$-th order coupling coefficient. In terms of the vector $U = (U_+, U_-)^T$ of phase-shifted amplitudes $U_\pm = E_m^\pm \exp(\pm i\Delta\beta Z/2)$, Eqs. (4) can be recast in the form

$$\frac{dU}{dz} = AU \quad ; \quad A = \begin{pmatrix} i\delta_m & i\kappa_m \\ -i\kappa_m & -i\delta_m \end{pmatrix}, \tag{5}$$

where we define $z = Z/Z_0$, $\delta_m = \Delta\beta_m Z_0/2$, $\kappa_m = \Gamma_m Z_0$ (without loss of generality taken real and positive), $Z_0$ being a suitable length scale (either the device length $Z_0 = L$, or $Z_0 = \Gamma_1^{-1}$ which yields $\kappa_1 = 1$).

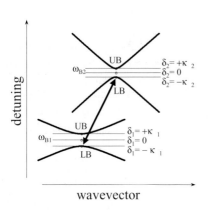

**Fig. 1.** First two Bragg resonances in an anharmonic Bragg grating. Two forbidden gaps of normalized width $\kappa_1$ and $\kappa_2$ are centered around Bragg frequencies $\omega_{B1}$ and $\omega_{B2}$. LB and UB indicate the lower- and upper-band edge, respectively. In the presence of a single optical carrier frequency only one resonance is effective. However, the nonlinear coupling between fields at two (or more) carriers can induce coupling between different resonances as indicated by the arrow.

The solution of the linear problem (5), yields a characteristic equation $\lambda_m^2 = \kappa_m^2 - \delta_m^2$ for the eigenvalues $\lambda_m$ [i.e., $U \propto \exp(\lambda_m z)$] of $A$. The two roots $\lambda_m^\pm = \pm i\sqrt{\delta_m^2 - \kappa_m^2}$ are imaginary as long as $|\delta_m| > \kappa_m$, representing a wavevector shift of the linear modes of the structure. Conversely, $\lambda_m^\pm = \pm\sqrt{\kappa_m^2 - \delta_m^2}$ are real for $|\delta_m| < \kappa_m$, entailing a forbidden gap of frequencies. Inside this stopband the solutions $U^\pm(Z)$ are of hyperbolic type and describe the optical power dropping off along the DFB structure due to *linear* reflection into the backward mode. One can easily recover the usual dispersion relation $\delta\omega_m = \delta\omega_m(\lambda_m)$ explicitly in wavevector-frequency $(\lambda, \delta\omega_m)$ plane, by expanding at first-order the wavevector around the $m$-th order Bragg frequency $\omega_{Bm}$ as

$$k(\omega) = k(\omega_{Bm}) + \frac{dk}{d\omega}\Delta\omega_m = \frac{m\beta_0}{2} + \frac{1}{V_m}\Delta\omega_m, \tag{6}$$

where $V_m = (dk/d\omega|_{\omega_{Bm}})^{-1}$ is the group-velocity of the host material at frequency $\omega_{Bm}$, and $\Delta\omega_m \equiv \omega - \omega_{Bm}$. Substitution in the expression (characteristic equation) $\delta_m^2 = \kappa_m^2 - \lambda_m^2$ yields, in terms of normalized frequency deviation $\delta\omega_m \equiv \Delta\omega_m Z_0/V_m$, the lower-bound (LB, $\delta\omega_m^-$) and upper-bound (UB, $\delta\omega_m^+$) of the dispersion relation

$$\delta\omega_m(\lambda_m) = \delta\omega_m^{\pm} = \pm\sqrt{\kappa_m^2 - \lambda_m^2}, \tag{7}$$

shown in Fig. 1 for the two lowest Bragg resonances $m = 1, 2$. Getting back to real-world units, the dispersion relation is symmetric around the wavevectors $\pi/\Lambda$ ($m = 1$) and $2\pi/\Lambda$ ($m = 2$), at which two frequency gaps $|\Delta\omega_1| < \Gamma_1 V_1$ and $|\Delta\omega_2| < \Gamma_2 V_2$ open up around Bragg frequencies $\omega_{B1}$ and $\omega_{B2}$, respectively. At frequencies falling inside one of such gaps the grating behaves as a *linear reflector*. In Fig. 2 we show the reflectivity of the Bragg grating around one generic resonance, as obtained from the analytical solution of the linear problem (5). As shown, the reflectivity grows larger when the normalized length $z_L = \Gamma_m L$ increases, and eventually saturates to unity within the whole gap when $z_L$ is large enough ($z_L > 4$).

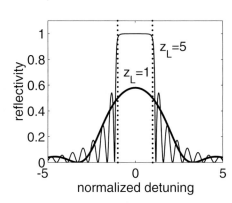

**Fig. 2.** Spectral response of a Bragg grating around the $m$-th resonance: reflectivity against the normalized detuning $\delta\omega$ (dropping the $m$ subscript for simplicity) from Bragg resonance, for two different values of Bragg coupling strength. Normalizing with $Z_0 = \Gamma^{-1}$, the coupling strength is measured by the normalized grating length $z_L \equiv \Gamma L = 2$ (thin solid line) and $z_L = 5$ (thick solid line). Recall also that $\delta\omega$ is equivalent to the first-order Taylor expansion of the Bragg detuning $\delta = \Delta\beta/(2\Gamma)$. The dotted lines delimit the stopband $|\delta| = |\delta\omega| \leq 1$

## 2.2 Stopbands due to Periodic Nonlinearities

In a quadratic medium where the $\chi^{(2)}$ susceptibility is periodic, the forward propagating field $E_1$ at frequency $\omega = \omega_0 + \Delta\omega$ and backward propagating $E_2$ at frequency $2\omega$ obey the following nonlinear CMT problem

$$-i\frac{dE_1}{dZ} = \chi_2(Z)E_2E_1^* \, e^{i\Delta kZ} \quad ; \quad i\frac{dE_2}{dZ} = \chi_2(Z)E_1^2 \, e^{-i\Delta kZ}, \tag{8}$$

where $\Delta k = \Delta k(\omega) = \beta_2(2\omega) + 2\beta(\omega)$ is the nonlinear wavevector (propagation constant in waveguides) mismatch, and $\chi_2(Z)$ is $\Lambda$-periodic susceptibility normalized in such a way that $|E_{1,2}|^2$ give the intensity (power in waveguides). By referring to the Fourier expansion of the nonlinear grating

$$\chi_2(Z) = \hat{\chi}_2 \sum_{m>0} c_m \exp(\pm im\beta_0 Z), \tag{9}$$

where $\beta_0 = 2\pi/\Lambda$ and $\hat{\chi}_2$ characterizes the nonlinearity of the host medium, perfect (QPM) phase-matching at frequency $\omega \equiv \omega_0$ can be achieved when, at some order $m = \overline{m}$ (usually high, due to technological limits), the grating wavevector $\overline{m}\beta_0$ compensates for the mismatch $\Delta k(\omega_0)$, i.e. when the period of the grating is

$$\Lambda = \frac{2\overline{m}\pi}{\Delta k(\omega_0)}. \tag{10}$$

Retaining only the effective term of the grating and introducing the dimensionless variables $u_2 = [E_2/\sqrt{I_r}]\exp(i\Delta k Z)$, $u_1 = \sqrt{2}E_1/\sqrt{I_r}$, Eqs. (8) can be cast in the following form

$$\frac{du_1}{dz} = i\frac{\partial H_{bshg}}{\partial u_1^*} = i\, u_2 u_1^* \;\; ; \;\; \frac{du_2}{dz} = -i\frac{\partial H_{bshg}}{\partial u_2^*} = -i\,\delta k\, u_2 - i\,\frac{u_1^2}{2}, \tag{11}$$

where $H_{bshg} = H_{bshg}(u_{1,2}, u_{1,2}^*) = \delta k|u_2|^2 + (u_1^2 u_2^* + u_1^{*2}u_2)/2$ is the conserved Hamiltonian, $z = Z/Z_{nl}$ ($Z_{nl} = (\hat{\chi}_2 c_{\overline{m}}\sqrt{I_r})^{-1}$ being the nonlinear length scale associated with the reference intensity (or power) $I_r$), and $\delta k = (\Delta k - 2\overline{m}\pi/\Lambda)Z_{nl}$ is the residual mismatch due to the nonzero detuning $\Delta\omega$ from phase-matching frequency $\omega_0$. Following the approach of Ref. [85] and exploiting the conservation of photon flux (Poynting vector) $P = |u_1|^2/2 - |u_2|^2$, Eqs. (11) can be solved by reducing them to the 1D oscillator (set $\dot{y} = dy/dz$)

$$\dot{\eta} = -2(\eta + P)\sqrt{\eta}\sin\phi = \frac{\partial H_r}{\partial\phi},$$

$$\dot{\phi} = -\delta k + \frac{3\eta + P}{\sqrt{\eta}}\cos\phi = -\frac{\partial H_r}{\partial\eta}, \tag{12}$$

$$H_r = \delta k\,\eta + 2(\eta + P)\sqrt{\eta}\cos\phi,$$

for the effective (action-angle like) variables $\eta = |u_2|^2$ and $\phi = \phi_2 - 2\phi_1 = \mathrm{Arg}(u_2) - 2\mathrm{Arg}(u_1)$. By expressing $\phi$ through $H_r$, one can easily decouple Eqs. (11) and obtain a single equation for $\eta(z)$ (assuming in this context the equivalent meaning of coordinate of an ideal particle moving in a potential $U$ with energy $E$), which can be solved by a quadrature formula as follows

$$\dot{\eta} = \pm\sqrt{2[E - U(\eta)]} \;\; \Rightarrow \;\; z - z_L = \pm\int_0^{\eta(z)}\frac{d\eta}{\sqrt{2[E - U(\eta)]}}, \tag{13}$$

where $z_L \equiv L/Z_{nl}$ is the normalized grating length, $2[U(\eta) - E] = -4\eta^3 + (\delta k^2 - 8P)\eta^2 - 4\left(P^2 + H_0\delta k\right)\eta + 4H_0^2$, $H_0 = H_r(\eta(0), \phi(0)) = H_r(\eta(z_L), \phi(z_L))$, and $\pm$ sign must be chosen accordingly with the initial value of $\sin\phi$ in Eq. (12). By inverting the elliptic integral (13) [86], we can calculate the reflectivity of the structure $\mathcal{R} = |E_2(0)|^2/|E_1(0)|^2$) for increasing values of $z_L$ (i.e., larger intensities, or longer gratings). As shown in Fig. 3 the device behaves as a *nonlinear reflector*, i.e. in the low intensity limit the grating is transparent ($\mathcal{R} \simeq 0$), while $\mathcal{R}$ increases dramatically with the intensity in a stopband (gap) centered around phase-matching. A clear signature of the nonlinear origin of this gap

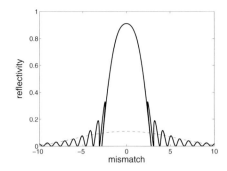

**Fig. 3.** Reflectivity $\mathcal{R}$ vs. residual wavevector mismatch $\delta k$ in a BSHG nonlinear reflector, in the quasi-linear ($z_L = L/Z_{nl} = 0.5$, dashed curve), intermediate ($z_L = 1$, thick solid curve) and high-intensity regimes ($z_L = 6$, thin solid curve)

is the bistability displayed by the sidelobes in Fig. 3. Transparency and reflection are associated to oscillating and decaying fields, respectively [41]. At fixed mismatch, the transition between the two regimes occur abruptly at a critical value of output power $P_c = \delta k^2/16$, and the BSHG reflector acts as limiter with clamped output power $P \simeq P_c$ [49,56,58]. It is worth-noting that also a stratified Kerr structure can behave as a hard-limiter [42–45].

## 2.3 Nonlinearity-Induced Transparency in Short-Period Gratings

At high intensities, when the Kerr effect is no longer negligible, the CMT (4) must be generalized to include the cubic terms, which yield [2,4]

$$
\begin{aligned}
-i\frac{dU_+}{dZ} &= \frac{\partial H_{bragg}}{\partial U_+^*} = \Gamma U_- + \frac{\Delta\beta}{2}U_+ + \chi\left(|U_+|^2 + 2|U_-|^2\right)U_+, \\
+i\frac{dU_-}{dZ} &= \frac{\partial H_{bragg}}{\partial U_-^*} = \Gamma U_+ - \frac{\Delta\beta}{2}U_- + \chi\left(|U_-|^2 + 2|U_+|^2\right)U_-,
\end{aligned}
\tag{14}
$$

where we drop the subscript $m$ for simplicity, and $\chi \propto n_{2I}$ is a nonlinear coefficient. Equations (14) have a Hamiltonian structure with $H_{bragg}(U_\pm, U_\pm^*) = \Gamma(U_+U_-^* + U_+U_-^*) + (\Delta\beta/2)(|U_+|^2 - |U_-|^2) + \chi\left(|U_+|^4 + |U_-|^4 + 2|U_+U_-|^2\right)$, which permits to obtain explicit solutions (see also Refs. [2,4]) through a procedure formally identical to that followed to arrive at Eqs. (12-13). Such solutions show that Bragg gratings become bistable devices, and exhibit self-transparency induced by the nonlinear response. These features are clearly shown in Fig. 4, where we display a typical curve of reflectivity $\mathcal{R} = |U_-(0)|^2/|U_+(0)|^2$ as a function of the input illumination (made in the forward direction). The strong linear reflectivity of the grating is bleached by the intensity-dependent index change, until eventually a first transparent state ($\mathcal{R} = 0$) is reached at the point $T$ in Fig. 4(a). Note that, due to the multivalued response, the point $T$ can be reached through a hysteresis cycle only by decreasing the input intensity. In the state $T$ the field is strongly localized inside the grating, as shown in Fig. 4(b) where we display the intensity of the total field. Roughly speaking such field is a stationary (i.e., zero-velocity) GS. Strictly speaking, however, the field in Fig. 4(b) appears to be the truncation, due to the finite grating, of a true periodic solution (indeed sequences of localized bumps appear when the grating length is a multiple of such period and correspond to other transparency points). Conversely GSs are localized solutions which exist ideally in an infinite grating, or in practice in a grating much longer than their physical width. To gather the possibility that such localized structures might move along the grating we need, however, to consider the nonstationary problem, which will be addressed in the next section.

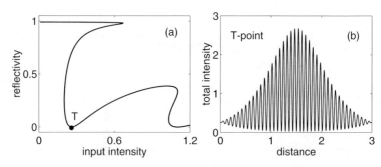

**Fig. 4.** Nonlinear response of a perfectly resonant ($\delta_m = 0$) Bragg grating of length $z_L = \Gamma L = 3$. (a) Reflectivity versus normalized input intensity $|u_+(Z = 0)|^2 = \chi\Gamma|U_+(Z = 0)|^2$; (b) Total intensity versus distance (inside a grating with 40 periods) at transparency point [T in (a)]

As far as BSHG is concerned, let us point out that the net effect of the grating is not limited to make the phase-matching achievable. By using a multiscale approach, where the distance and the fields are expanded in power series of a smallness parameter $\varepsilon \sim \beta_0^{-1} \sim \Delta k^{-1}$, we find that, at the lowest relevant order, BSHG is described by the following CMT system

$$
\begin{aligned}
-i\frac{dE_1}{dZ} &= \hat{\chi}_2 c_{\overline{m}} E_2 E_1^* \;+\; |\hat{\chi}_2|^2 \gamma_3 (|E_1|^2 + |E_2|^2) E_1, \\
+i\frac{dE_2}{dZ} &= \hat{\chi}_2 c_{\overline{m}} E_1^2 \;+\; 2|\hat{\chi}_2|^2 \gamma_3 |E_1|^2 E_2,
\end{aligned}
\tag{15}
$$

which shows that the dynamics is affected, besides the quadratic phase-matched terms which are present also in Eqs. (11), by effective Kerr terms of self- and cross-induced nature, weighted by the coefficient

$$
\gamma_3 = \sum_{m \neq \overline{m}} \frac{\overline{m} c_m^2}{\Delta k (m - \overline{m})}.
\tag{16}
$$

The form of such effective Kerr terms is quite universal for any (mismatched) SHG mixing interaction, as shown for standard QPM [87], or mismatched SHG in homogeneous bulk media [88]. In this context it is important to point out that the Kerr terms in Eqs. (15), together with the intrinsic Kerr material response, might frustrate the reflection. In other words they can be responsible for inducing self-transparency similarly (except for the low-intensity limit where nonlinear reflectors are transparent) to the Bragg case analyzed above.

# 3    Gap Solitons in Bragg Gratings with Kerr Nonlinearity

In order to analyze the main properties of gap solitons, it is necessary to extend
the CW coupled-mode models to deal with the temporal case. Since the material
dispersion turns out to be negligible in comparison with the dispersion associated
with the periodic structure (related to the curvature of the dispersion relation,
see Fig. 1), the temporal evolution is accounted for by restricting the attention to
those terms that account for the group-velocity of the envelopes, i.e. first-order
derivatives, dropping higher-order derivatives. In a medium which exhibits a
(genuine or effective) Kerr response, the simplest model that generalize the CW-
CMT (14) and governs the propagation of the forward ($u_+$) and backward ($u_+$)
propagating envelopes at Bragg frequency $\omega_{Bm}$, takes the following well-known
form (for a simple derivation see, e.g. Ref. [23])

$$
\begin{aligned}
i\partial_t u_+ &= \frac{\delta H}{\delta u_+^*} = -i\partial_z u_+ - u_- - \left( X\,|u_-|^2 + S\,|u_+|^2 \right) u_+, \\
i\partial_t u_- &= \frac{\delta H}{\delta u_-^*} = +i\partial_z u_- - u_+ - \left( X\,|u_+|^2 + S\,|u_-|^2 \right) u_-,
\end{aligned}
\tag{17}
$$

where the dimensionless space and time variables $z, t$ are linked to the real-world
ones $Z, T$ as $z = \Gamma Z$ and $t = \Gamma_1 V_1 T$, $\Gamma_1$ and $V_1$ being the $m = 1$ Bragg coupling
coefficient, (a straightforward extension is possible for $m \neq 1$) and the group-
velocity at Bragg frequency $\omega_{B1}$, respectively. Moreover $u_\pm = (\chi/\Gamma)^{1/2} E_1^\pm$ are
normalized envelopes, and $\chi \propto n_{2I}$ is a common nonlinear coefficient for self-
and cross-phase modulation, whose relative weight is given by the normalized
*external* coefficients $(S, X)$. For instance, *scalar* mode coupling in fiber gratings is
described by $S = 1$ and $X = 2$ [7–10,12]. However Eqs. (17) are more general and
describe also situations where the two fields might have different polarizations,
or the cubic nonlinearity originates from cascading in quadratic media and either
$S$ or $X$ can be vanishing [25,29]. Therefore we analyze Eqs. (17) with generic
coefficients $S, X$. The model (17) has the following properties

- it is not integrable by the inverse scattering method, except in the $S = 0$
  case for which it reduces to the integrable Massive Thirring model (MTM),
  developed in relativistic field theory [90,91]
- it has neither Galileian nor Lorentz gauge invariance, which would permit
  to generate moving solutions from the rest ones. Again an exception arises
  in the $S = 0$ case where the system becomes Lorentz-invariant
- it is a conservative model which possesses a Hamiltonian structure $\partial_t u =
  J\nabla_u H$. Equations (17) can be cast in this form with $u = [u_+\ u_-]^T$, $J =
  \mathrm{diag}(i\ i)$, the gradient is given in terms of variational derivatives as $\nabla =
  [\delta/\delta u_+^*\ \delta/\delta u_-^*]^T$, while the conserved (because of $t$-translational symmetry)
  Hamiltonian $H = H(u_\pm, u_\pm^*, \partial_z u_\pm, \partial_z u_\pm^*)$ reads as

$$
H = \int_{-\infty}^{+\infty} dz \; \frac{i}{2} \left[ u_-(\partial_z u_-)^* - u_+(\partial_z u_+)^* - u_-^*(\partial_z u_-) + u_+^*(\partial_z u_+) \right]
$$
$$
- (u_- u_+^* + u_+ u_-^*) - S(|u_+|^4 + |u_-|^4) - X|u_+ u_-|^2 \tag{18}
$$

- as can be shown (equivalently by means of Noether theorem, Lie group analysis, or invariance of $H$ under the action of translational and rotational groups), the model possess also invariance of photon flux $Q$ (mass in the context of field theory) and momentum $M$, associated with rotational and $z$-translational symmetry, respectively. Their expressions are

$$Q = \int_{-\infty}^{+\infty} dz \left( |u_+|^2 + |u_-|^2 \right),$$

$$M = \frac{i}{2} \int_{-\infty}^{+\infty} dz \left[ u_+ (\partial_z u_+)^* + u_- (\partial_z u_-)^* - c.c. \right].$$

A remarkable property of Eqs. (17) is that the entire family of localized solutions can be constructed analytically. In other words, in spite of the fact that the PDE model (17) is not integrable, a fully integrable system of ODEs can be obtained for the complex amplitudes of the solitary waves. Although GS solutions of Eqs. (17) have been obtained earlier [7–9], it is only recently that the problem of their existence has been fully addressed [13]. The analysis reveals the occurrence of bifurcations between qualitatively different types of GSs which exist in domains of the parameter space directly linked to the gap discussed in sec.2. Below we outline the main steps of such analysis.

Generally speaking, the rotational and translational symmetries of Eqs. (17) imply (see Ref. [89] for a general theory) that GSs constitute a family characterized by two *internal* parameters (at variance with the *externally* controllable ones $S, X$ in Eqs. (17), or $\Gamma, V_1, \chi$ in their dimensional counterpart), namely the normalized detuning $\Delta$ from Bragg frequency ($\Gamma V_1 \Delta$ in real-world units), and the normalized soliton velocity $v$ ($v V_1$ in real-world units). In terms of these parameters, GS solutions of Eqs. (17) can be sought for in the following form (on the basis, for instance, of the expression of solitons in the integrable case $S = 0$)

$$u_+(z,t) = U_+ \sqrt{\eta(\zeta)} \, \exp\left[ -i\Delta t + i\beta\zeta + i\phi_+(\zeta) \right],$$

$$u_-(z,t) = U_- \sqrt{\eta(\zeta)} \, \exp\left[ -i\Delta t + i\beta\zeta + i\phi_-(\zeta) \right],$$

(19)

where both the intensity $\eta$ and phase (chirp) $\phi_\pm$ profiles depend on the soliton-frame variable $\zeta \equiv \gamma(z - vt)$, with $\gamma = (1 - v^2)^{-1/2}$ being the (subluminal, $|v| < 1$) Lorentz factor. Furthermore, for the ansatz (19) to be self-consistent, one finds upon substitution into Eqs. (17) that the peak amplitudes $U_+ = A\sqrt[4]{\frac{1+v}{1-v}}$ and $U_- = -sA\sqrt[4]{\frac{1-v}{1+v}}$ are $v$-dependent, and accounts for the absolute value of the nonlinearity and its sign through the overall coefficients $A = \gamma^{-1} \left| 2X \left(1 - v^2\right) + 2S \left(1 + v^2\right) \right|^{-1/2}$, and $s \equiv \mathrm{sign} \left[ X(1 - v^2) + S(1 + v^2) \right]$, respectively. Finally $\beta \equiv \gamma v \Delta$ represents the ($\Delta$ and $v$ dependent) GS propagation constant.

With the ansatz (19), it is easy to show that Eqs. (17) reduce to the following one-dimensional fully-integrable ODE model (set $\dot{y} = dy/d\zeta$ and $\phi \equiv \phi_+ - \phi_-$)

$$\dot{\eta} = 2\eta \sin \phi = -\frac{\partial H}{\partial \phi},$$

$$\dot{\phi} = 2\delta + 2\cos \phi - \eta = \frac{\partial H_r}{\partial \eta}, \qquad (20)$$

$$H_r = H_r(\eta, \phi) = 2\eta \cos \phi + 2\delta\eta - \eta^2/2,$$

which is ruled by the single parameter $\delta \equiv \gamma\Delta$. The GS solutions of Eqs. (17) correspond to the separatrix trajectories of Eqs. (20) which emanates from the unstable fixed points $\eta_s, \phi_s$. The bifurcation analysis of Eqs. (20) shows that two possible classes of GSs exist: (i) those of the bright type, which requires $\eta(\zeta = \pm\infty) = \eta_s = 0$ and can be shown to exist for $\delta^2 \leq 1$ where the point $\eta_s = 0$ is unstable (ii) those with nonvanishing pedestal which emanate from an unstable point $\eta_s \neq 0$ for $\delta < 1$ (focusing case, $s = 1$) or $\delta > 1$ (defocusing case, $s = -1$).

Analytical expressions for the intensity and phases can be obtained from Eqs. (20) by inverting a quadrature integral of the form (13) calculated with $H_r = H_r(\eta_s, \phi_s)$ (see Ref. [13] for details). The situation is summarized in Fig. 5 for the focusing case (for the defocusing case just apply the transformation $\Delta \to -\Delta$). The domain $\delta^2 < 1$ of existence of bright GSs corresponds in the parameter plane $(\Delta, v)$ to the inner domain bounded by the unit circle,

$$\Delta^2 + v^2 < 1 \quad \Rightarrow \quad \delta\omega^2 < \gamma^2 = \frac{1}{1 - v^2}. \qquad (21)$$

In Eq. (21) we have also expressed this existence condition in terms of frequency detuning $\delta\omega$ (already introduced in sec.2) of waves $u_\pm \sim \exp(-i\delta\omega t)$, as measured in the rest (laboratory) frame. The condition fulfilled by $\delta\omega$ in Eq. (21) follows from the relation $\delta\omega = \gamma^2\Delta$, which can be easily obtained from Eq. (19) by grouping all the exponentials that contain $t$ in a single term $\exp(-i\delta\omega t)$. Note that $\delta\omega$ and $\Delta$ coincide only at zero velocity ($v = 0$).

We refer to the unit circle (19) as the *dynamical gap* since it is easy to show that frequency-detuned *travelling*-waves which lie in this domain are exponentially damped. Note that the dynamical gap (19) is wider than the gap $|\delta\omega| < 1$, henceforth referred to as *rest gap* and measured in the lab frame through the reflectivity (see Fig. 2).

The most important consequence of Eq. (21) is that bright GSs are *slow* localized waves which can travel with any velocity $|v| \leq 1$, or in real-world units with velocity $V$ such that $-V_1 \leq V \leq V_1$, i.e. they can be still in the lab frame ($V = 0$) or move with any velocity smaller than the group-velocity of light in the host medium. In the former case GSs are symmetric ($U_- = U_+$), while moving GSs are asymmetric (AS in Fig. 5) with dominant component in the direction of motion. Moreover, since a nonlinearity of given sign tends always to shift the gap in a given direction (e.g. red-shift for focusing nonlinearities), GSs exhibit a strongly asymmetric behavior against frequency (i.e., reversal of detuning $\Delta$).

With focusing nonlinearities, bright GSs change gradually from low-amplitude ones (LA in Fig. 5) to high-amplitude (HA) ones as the frequency $\Delta$ is spanned across the gap from the high-frequency (UB in Fig. 1) to the low-frequency (LB in Fig. 1) edge. LA-GSs tend, for $\delta \sim 1, \Delta > 0$, to hyperbolic secant soliton solutions of the nonlinear Schrödinger (NLS) model, which represents a good description (alternative to Eqs. (17), and derivable from this) close to the upper band edge [6,20]. Right on the circle ($\delta = 1$) LA-GSs vanish. Viceversa HA-GSs have a finite-amplitude limit for $\delta \to 1$ with Lorentzian (LZ in Fig. 5) intensity profile. Optical Lorentzian solitons are characterized by slower than exponential decay and are known also in other mixing interactions [92–95]. In this case LZ-GSs mark the bifurcation of HA-GSs into a pair of dark-antidark (DK-AK) GSs which coexist at any point on the left semiplane outside the circle (see Fig. 5).

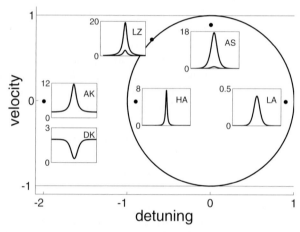

**Fig. 5.** Gap solitons in the detuning-velocity ($\Delta, v$) plane in the focusing case $S, X > 0$ (superluminal solutions, i.e. $|v| > 1$, are not considered). The insets show GS intensity profiles $|u_{\pm}(\zeta)|^2$ sampled at points marked by a filled circle (the horizontal scale for $\zeta$ is the same for all insets). Bright GSs exist within the unitary circle, and are gradually changing from low-amplitude (LA) to high-amplitude (HA) ones close to the high- and low-frequency edge of the gap, respectively. Bright GSs with non-zero velocity are asymmetric (AS). HA solitons become finite-amplitude Lorentzian (LZ) over the left semi-circle, and then bifurcate into dark (DK) and antidark (AK) pairs which coexist for frequencies below the LB edge of the dynamical gap

Although also GSs with nonzero background might affect the dynamics of Bragg gratings [39], GSs of the bright type are more relevant because they can be easily excited by standard pulsed illumination. To gain a better insight into their properties it is useful to map their existence domain (dynamical gap) onto the plane $v, \delta\omega$, since $\delta\omega$ determines the actual laser frequency $\omega = \omega_{Bm} + \delta\omega \Gamma_m V_m$ that it is needed to excite a GS with velocity $v$ and frequency detuning $\Delta = \delta\omega/\gamma^2$. In this plane the dynamical gap $|\delta\omega| < \gamma$ (see Eq. (21)) corresponds to the whole (lighter and darker) shaded area in Fig. 6(a). In Fig. 6(a) the lighter area corresponds to the rest gap $|\delta\omega| < 1$, where the reflectivity is high [Fig. 6(b)]. The dynamical gap reduces to the rest gap only for stationary ($v = 0$) GSs. A moving GS "sees" a bandgap which is wider (the faster the soliton the wider the gap) than the reflectivity bandwidth, and become infinite as the velocity of light ($|v| = 1$) is approached. Talking about the gap seen by a moving soliton is justified by the fact that the gap calculated in Lorentz transformed variables gives indeed the dynamical gap [13]. It is also worth pointing out that GSs excited by means of a laser operating at a frequency detuning $\delta\omega$ lying outside the reflection bandwidth or rest gap are usually referred to in the literature as Bragg solitons [15]. However, in light of the results discussed above, in our opinion, there is no reason to distinguish between Bragg and gap solitons. All the existing bright solitons are GSs when referred to the bandgap seen in the moving frame.

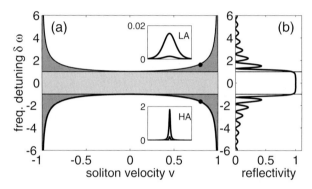

**Fig. 6.** (a) The dynamical photonic bandgap (whole shaded domain) $|\delta\omega| < \gamma$ in the parameter plane of velocity and rest detuning ($v, \delta\omega$). Such domain is mapped back onto the inner domain bounded by the circle $\Delta^2 + v^2 = 1$ in the plane ($\Delta, v$), see Fig. 5. The rest bandgap $|\delta\omega| < 1$ is the smaller region between the two dashed lines $\delta\omega = \pm 1$ (light shaded area), and corresponds to the bandwidth of the linear reflectivity curve shown with the same vertical scale in (b) for a grating of normalized length $z_L = \Gamma L = 4$. The insets show intensity profiles $|u_{\pm}(\zeta)|^2$ of two moving ($v = 0.8$) LA and HA solitons

From Fig. 6(a) it is also immediately clear that, when $|\delta\omega| > 1$ so that the laser operates outside the rest gap (as usual in fiber-grating experiments), one has a lower bound $v_{low}^2 = 1 - \delta\omega^{-2}$ on the square velocity of the excitable GSs. In other words only GSs travelling with velocity $|v| > v_{low} = (1 - \delta\omega^{-2})^{1/2}$ can be excited. Far from Bragg resonance ($|\delta\omega| \to \infty$), $v_{low} \to 1$, and GSs tend to loose their property to be slow waves. Conversely, inside the rest gap ($|\delta\omega| < 1$), such limitation does not hold, and GSs with any velocity $|v| < 1$ can be excited. Ideally, a particular value of velocity is selected by matching the input beam profile (intensity and phase) to the solution which corresponds to that particular value of velocity. However, this can be hardly done in practice because the actual field profile inside the grating is affected by strong reflection at the grating boundary, and the velocity, in general, cannot be predicted with simple arguments. Moreover, a GS requires to be excited to be also stable, a subject that is addressed in sec.5.

# 4   Gap Solitons Supported by Frequency Conversion Processes

We briefly mention the fact that GSs exist and can propagate also when a nonlinear mechanism different from the self-action due to the Kerr effect is present. In particular GS are supported also by nonlinear mixing processes which entail exchange of energy between envelopes at different carrier frequencies. The $m = 1$ and $m = 2$ resonances can be coupled via SHG, while $m = 1$ and $m = 3$ resonances can be coupled via third-harmonic generation in a cubic medium. The former mixing process can be envisaged to give the most efficient coupling, which is ruled by the following set of four equations [26–28]

$$\pm i\partial_z u_1^\pm + i\partial_t u_1^\pm + u_1^\mp + u_2^\pm (u_1^\pm)^* = 0,$$

$$\pm i\partial_z u_2^\pm + i\frac{1}{v_2}\partial_t u_2^\pm + \delta k\ u_2^\pm + \kappa_2 u_2^\mp + \frac{(u_1^\pm)^2}{2} = 0,$$

$$(22)$$

for the normalized counterpropagating envelopes $u_m^\pm$ at carrier frequency $m\omega_{B1}$, $m = 1, 2$. In Eqs. (22) we make use of the normalization $Z_0 \equiv \Gamma_1^{-1}$ so that $z = \Gamma_1 Z$ and $t = \Gamma_1 V_1 T$. The dynamics depend on three external parameters which depends on the choice of material, phase-matching geometry, and grating characteristics: the velocity ratio $v_2 = V_2/V_1$, the coupling ratio $\kappa_2 = \Gamma_2/\Gamma_1$, and the SHG mismatch $\delta k = [k(2\omega_{B1}) - 2k(\omega_{B1})]Z_0$. Note that the choice of the fundamental Bragg frequency as carrier frequency leads to have $\delta\beta_1 = 0$ and $\delta\beta_2 = -2\delta k$, i.e. perfect Bragg resonance at second harmonic requires nonlinear phase-matching.

Similarly to the Kerr case, GSs constitute a two-parameter family of localized waves, which depend on two internal parameters $\delta\omega$ and $v$ playing the role of normalized frequency detuning ($u_m^\pm \sim \exp(-im\delta\omega t)$) and soliton velocity, respectively. The main similarities and differences of parametric GS solutions of Eqs. (22) with GSs of the Kerr case are summarized below

- Eqs. (22) have the same symmetries and invariants ($Q, M, H$) of the Kerr case, and this explain why the number of internal parameters does not change in spite of the fact that GSs are now four-component waves
- At fixed soliton velocity the GS profiles depend on two effective detunings $\delta_1 = \delta\omega$ and $\delta_2 = 2\delta\omega/v_2 + \delta k$ which represent the Taylor expanded Bragg detunings from $m = 1$ and $m = 2$ linear resonances
- The entire family of GSs and their bifurcations can no longer be characterized analytically. Several different approaches (exact, numerical, perturbative, Kerr-effective limits etc. [26–33,36,39–41]) have been applied to characterize the GS solutions of Eqs. (22)

- Remarkably GSs do not require a linear damping mechanism at both carrier frequencies or, in other words, they exist also outside the intersection region between the two dynamical gaps $\frac{\delta_m^2}{\kappa_m^2} + \frac{v^2}{v_m^2} < 1$ ($m = 1, 2$ and we set $v_1 = \kappa_1 = 1$), which generalize Eq. (21) This case, where the damping mechanism for the second-harmonic originates from the nonlinearity, shows no similarity to the Kerr case.

# 5    Stability and Excitation

Since the localized solutions discussed above are indeed solitary waves it is crucial to assess their stability, by studying the evolution linearized around the soliton. The most important result in the stability theory of solitary waves is the so-called Vakhitov-Kolokolov criterion for one-parameter families of solitary waves of the generalized (nonintegrable) NLS equation [96]. It relates the stability property to the sign of the derivative $\frac{dQ}{d\beta}$, where $Q$ is the soliton power (mass) invariant and $\beta$ is the soliton internal parameter (propagation constant). In this case, the unstable eigenvalue of the linearized problem turns out to be real and crosses through the origin in the complex plane (zero eigenvalue) at threshold $\frac{dQ}{d\beta} = 0$. Since the origin is always an eigenvalue of the linearized problem associated with the translational symmetry of the original equation, this mechanism is often refered to as *translational* instability. Other vector models of the NLS-type possessing two-parameter families of solitons exhibit similar translational instabilities with a threshold condition given in terms of a determinant criterion which involves derivatives of the invariants with respect to the soliton internal parameters (see, e.g. Refs. [97,98]).

The GS model (17) contains a different linear operator with respect to the second-order one present in Schrödinger-type models, thus bearing more similarities to spinorial (Dirac-type) models developed in the context of the relativistic field theory, for which soliton stability was addressed to some extent [99]. Nevertheless also GS models exhibit translational instabilities whose threshold is given by the following condition [100,101]

$$\begin{vmatrix} \partial_\Delta Q & \partial_\Delta M \\ \partial_v Q & \partial_v M \end{vmatrix} = \frac{\partial Q}{\partial \Delta} \frac{\partial M}{\partial v} - \left( \frac{\partial Q}{\partial v} \right)^2 = 0. \tag{23}$$

However, GS models exhibit also a different, namely *oscillatory*, instability mechanism associated with complex eigenvalues bifurcating into the right semiplane after collisions of imaginary eigenvalues [100–102]. It is worth pointing out that oscillatory instabilities are common in dissipative models [103], but are rather unusual in conservative systems, where they have been found in other systems after their discovery for GSs [104,105]. Oscillatory instabilities of GSs dominate over translational ones, the latter affecting only a small domain of the parameter space. Although, the threshold condition for oscillatory instabilities must be determined numerically, as a rule of thumb high-amplitude (HA) are unstable while low-amplitude (LA) are stable. Figure 7 shows the oscillatory decay of a high amplitude GS in a fiber grating.

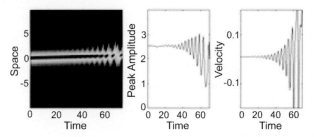

**Fig. 7.** Oscillatory decay of an unstable high-amplitude ($\Delta = -0.9$) stationary GS in a fiber Bragg grating ($S = 1$, $X = 2$). From left to right: contour of $|u_+(z,t)|^2$, temporal evolution of its peak amplitude, temporal evolution of the instantaneous velocity $v(t)$

Note that, as far as the stability problem is concerned, other mechanisms such as modulational instabilities and self-pulsing [106–110] might affect the propagation of GSs with nonvanishing background. Also the stability of GSs against the effect of material dispersion was addressed [111].

Finally when GSs are stable one can show that they can be excited by unidirectional pulsed illumination of the grating. An example of GS generation at the interface between a homogeneous medium ($z < 0$) and a Bragg grating ($z > 0$) is shown in Fig. 8. This particular situation refer to SHG in a singly-resonant Bragg grating (described by Eqs. (22) with $\kappa_2 = 0$) in a regime for which the nonlinearity yields an effective defocusing Kerr nonlinearity, and the grating operate in the region where GSs are of the LA type. As shown, while in the high intensity case a slow GS is clearly formed in spite of the strong grating reflection at the interface $z = 0$, at low intensity the field does not penetrate inside the structure.

## 6    Localization in Gap of Nonlinear Origin

Similarly to the Bragg grating case we might consider the time-dependent propagation in the BSHG reflector, described by the dimensionless model

$$i(\partial_t + \partial_z)u_1 + u_2u_1^* + (X_1|u_2|^2 + S_1|u_1|^2)u_1 = 0,$$

$$i(\frac{1}{v_2}\partial_t - \partial_z)u_2 + \frac{u_1^2}{2} + (X_2|u_1|^2 + S_2|u_2|^2)u_2 = 0,$$

(24)

where $u_m$, $m = 1, 2$, are envelopes at grating-assisted phase-matching frequency $m\omega_0$, scaled as in Eqs. (11), and additionally $t = V_1T/Z_{nl}$ is a normalized time, $v_2 = V_2/V_1$ is the velocity ratio, and $S_{1,2}, X_{1,2}$ are cubic coefficients which account for the material Kerr effect and the grating correction [see Eqs. (15-16)].

As shown in Ref. [58], Eqs. (24) support localized envelopes (GSs of pure nonlinear origin) which span the bandwidth of the $\chi^{(2)}$ mixing process (see Fig. 3). These constitute a two-parameter family of GSs of pure nonlinear origin which stems from the balance of two nonlinear effect: the reflectivity induced by

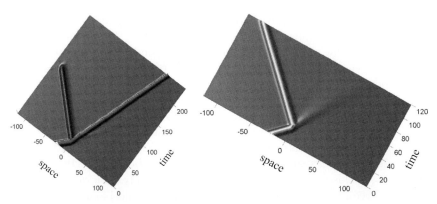

**Fig. 8.** Left: simulation of a typical process of GS generation showing $|u_1^+(z,t)|^2 + |u_1^-(z,t)|^2$ vs. spatial $z$ and temporal $t$ coordinates, as obtained in SHG with parameters $\delta_1 = -0.9$, $\delta_2 = 5$, $v_2 = 0.5$, $\kappa_2 = 0$. $z = 0$ is the interface between a linear homogeneous medium ($z < 0$) and a quadratically nonlinear Bragg grating ($z > 0$). An incident gaussian pulse $u_1^+(t = 0) = 10\exp(-\frac{z+25}{5})^2)$, when reaching the grating is partially reflected, while about 50 % of the energy is trapped in a slowly propagating GS. A strongly dispersing linear wave is also present for $z > 0$ due to the part of the input pulse spectrum lying outside the photonic bandgap. The reflected pulse vanishes in proximity of $z = -100$ due to absorbing boundary conditions.
Right: As above except for a low-power excitation $u_1^+(t = 0) = 1\exp(-\frac{z+25}{5})^2)$. The input energy is mostly reflected and a small strongly dispersing linear waves is transmitted. Note the different time scale respect to the left frame

the quadratic mixing, and the shift of the phase-matching bandwidth due to the cubic nonlinearities. Such GS solutions can be sought as

$$u_1 = \sqrt{(1-v)(1+v/v_2)}u(\zeta)\exp\left[i\delta\omega(z-t)\right],$$

$$u_2 = (1-v)w(\zeta)\exp\left[i2\delta\omega(z-t)\right], \tag{25}$$

where $\zeta = z - vt$, $v$ is the normalized soliton velocity, and $\delta\omega$ ($\Delta\omega = \delta\omega V_1/Z_{nl}$ in real-world units) is the frequency detuning from phase-matching frequency. Substitution of the ansatz (25) into Eqs. (24) leads after lengthy but straightforward algebra to an oscillator of the form (12) with the Hamiltonian

$$H_{bshg} = \Delta\eta + \sigma\frac{\eta^2}{2} + 2(P+\eta)\sqrt{\eta}\cos\phi, \tag{26}$$

where $\eta = |w|^2$, $\phi = \phi_2 - 2\phi_1 = \text{Arg}(w) - 2\text{Arg}(u)$, and the dot stands for $d/d\zeta$. In Eq. (26) $\sigma = \sigma(S_{1,2}, X_{1,2}, v, v_2)$ is an effective nonlinear coefficient, and $\Delta$ depend on the mismatch $\delta k(\delta\omega)$ and the (conserved) imbalance $P = |u|^2/2 - |w|^2$ [58]. GS solutions can be found, similarly to the Bragg grating case, by analyzing the unstable fixed points of the Hamiltonian (26) and by constructing the separatrix trajectories which emanate from them. Their features can be summarized as follows

- They present many similarities with GSs of the Bragg grating. First, they are slow waves which propagate with velocity $-v_2 < v < 1$, and have dominant component in the direction of propagation. Second, the existence of GSs is strongly asymmetric with respect to reversal of frequency detuning.
- An important difference with Bragg GSs is that in general GSs of the BSHG reflector possess a pedestal which is necessary to induce the reflection and take together the pulse. However, as for out-gap Bragg GSs, dark and anti-dark GSs [shown in Fig. 9(a1-a2-b1-b2)] coexist.
- Also in this case Lorentzian solitons exist [see Fig. 9(c1-c2)] and mark the transition between two different families of GSs (both with DK-AK pairs). However, in this case Lorentzian solitons are the only achievable bright solitons and require perfect phase-matching ($\delta\omega = 0$) and balance ($P = 0$).

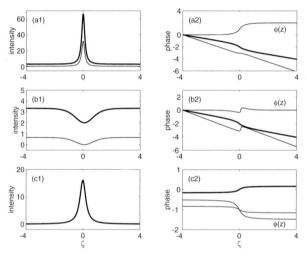

**Fig. 9.** Intensity ($|u(\zeta)|^2$, $|w(\zeta)|^2$) and phase profiles of GSs of a BSHG reflector in the presence of the Kerr effect. Thick and thin lines refer to the fundamental and second-harmonic frequency, respectively. We show also the overall phase $\phi = \phi_2 - 2\phi_1$ (thin line labeled $\phi(z)$). (a1-a2) antidark GS corresponding to $P = 1$ and $\delta k = -5$; (b1-b2) coexisting dark GS for the same values of parameters; (c1-c2) Lorentzian GS (intensity $|w(\zeta)|^2 = |u(\zeta)|^2/2$) corresponding to $P = \delta k = 0$

# 7   Summary and Further Developments

To summarize, we have discussed the physics of optical gap solitons in short period gratings. The main features of GSs in Kerr Bragg gratings such as their existence, bifurcations, stability, and excitation, have been discussed in details, while we have briefly mentioned other situations of experimental interest. The importance of such studies will become clear if gap solitons will ultimately find their way through several envisaged applications which encompass delay lines, memories, all-optical signal reshaping, enhancement of frequency conversion, etc.

While the theory is pretty well established for gratings with Kerr-like response, which have been experimentally tested in fibers and semiconductors, many possibilities are left open for the full understanding of gap solitons in other structures for which experiments have not been conducted so far.

## Acknowledgment

We are indebted to our collaborators Gaetano Assanto and Alfredo De Rossi who have contributed to part of the work discussed in this chapter.

# References

1. J.D. Joannopoulos, R.D. Meade, J.N. Winn: *Photonic crystals: molding the flow of light* (Princeton University Press, Princeton, 1995)
2. H.G. Winful, J. H. Marburger, E. Garmire: Appl. Phys. Lett. **35**, 379 (1979)
3. H.G. Winful, G.D. Cooperman: Appl. Phys. Lett. **40**, 298 (1982)
4. A. Mecozzi, S. Trillo, S. Wabnitz: Opt. Lett. **12**, 1008 (1987)
5. W. Chen, D. L. Mills: Phys. Rev. Lett. **58**, 160 (1987)
6. C. M. de Sterke, J. E. Sipe: Phys. Rev. A **38**, 5149 (1988); **39**, 5163 (1989); **42**, 550 (1990)
7. A. B. Aceves, S. Wabnitz: Phys. Lett. A **141**, 37 (1989)
8. D. N. Christodoulides, R. I. Joseph: Phys. Rev. Lett. **62**, 1746 (1989)
9. J. Feng, F. K. Kneubül: IEEE J. Quantum Electron. QE-29, 590 (1993)
10. C. M. De Sterke, J. E. Sipe: in *Progress in Optics XXXIII*, edited by E. Wolf, (Elsevier, Amsterdam, 1994), Chap. III
11. Q. Li, C. T. Chen, K.M. Ho, C.M. Soukoulis: Phys. Rev. B **53**, 15577 (1996)
12. D. L. Mills: *Nonlinear optics* (Springer, New York, 1998)
13. C. Conti, S. Trillo: Phys. Rev. E **64**, 036617 (2001)
14. R.H. Goodman, M.I. Weinstein, P.J. Holmes: arXiv:nlin/0012020 v3 (2001)
15. B. J. Eggleton, R. E. Slusher, C. M. de Sterke, P. A. Krug, J. E. Sipe: Phys. Rev. Lett. **76**, 1627 (1996)
16. B. J. Eggleton, C. M. de Sterke, R. E. Slusher: Opt. Lett. **21**, 1223 (1996)
17. B. J. Eggleton, C. M. de Sterke, R. E. Slusher: J. Opt. Soc. Am. B **14**, 2980 (1997)
18. N. G. R. Broderick, D. Taverner, D. J. Richardson, M. Ibsen, R.I. Laming: Opt. Lett. **22**, 1837 (1997); Phys. Rev. Lett. **79**, 4566 (1997)
19. P. Millar, R. M. De La Rue, T. F. Krauss, J. S. Aitchison, N. G. R. Broderick, D.J. Richardson: Opt. Lett. **24**, 685 (1999)
20. B. J. Eggleton, C. M. de Sterke, R.E. Slusher: J. Opt. Soc. Am. B **16**, 587 (1999)
21. N. G. R. Broderick, D. Taverner, D. J. Richardson, M. Ibsen: J. Opt. Soc. B **17**, 345 (2000)
22. S. Pitois, M. Haelterman, G. Millot: Opt. Lett. **26**, 780 (2001)
23. C. M. de Sterke, B. J. Eggleton, J.E. Sipe: "Bragg Solitons: Theory and Experiments". In *spatial solitons*, ed. by S. Trillo, W.E. Torruellas (Springer, Heidelberg 2001)
24. Y. S. Kivshar: Phys. Rev. E **51**, 1613 (1995)

25. S. Trillo: Opt. Lett. **21**, 1732 (1996)
26. C. Conti, S. Trillo, G. Assanto: Phys. Rev. Lett. **78**, 2341 (1997); Opt. Lett. **22**, 445 (1997)
27. H. He, P. D. Drummond: Phys. Rev. Lett. **78**, 4311 (1997); Phys. Rev. E **58**, 5025 (1998)
28. T. Peschel, U. Peschel, F. Lederer, B. A. Malomed: Phys. Rev. E **55**, 4730 (1997)
29. C. Conti, G. Assanto, S. Trillo: Opt. Lett. **22**, 1350 (1997)
30. C. Conti, S. Trillo, G. Assanto: Phys. Rev. E **57**, R1251 (1998)
31. C. Conti, S. Trillo, G. Assanto: Opt. Lett. **23**, 334 (1998); C. Conti, G. Assanto, S. Trillo: Electron. Lett. **34**, 689 (1998)
32. C. Conti, A. De Rossi, S. Trillo: Opt. Lett. **23**, 1265 (1998)
33. C. Conti, G. Assanto, S. Trillo: Opt. Exp. **3**, 389 (1998)
34. W.C.K. Mak, B. A. Malomed, P.L. Chu: Phys. Rev. E **58**, 6708 (1998)
35. A. Arraf, C. M. de Sterke: Phys. Rev. E **58**, 7951 (1998)
36. C. Conti, G. Assanto, S. Trillo: Phys. Rev. E **59**, 2467 (1999)
37. T. Iizuka, Y. S. Kivshar: Phys. Rev. E **59**, 7148 (1999)
38. A. V. Buryak, I. Towers, S. Trillo: Phys. Lett. A **267**, 319 (2000)
39. S. Trillo, C. Conti, G. Assanto, A. V. Buryak: Chaos, **10** 590 (2000)
40. T. Iizuka, C. M. de Sterke: Phys. Rev. E **62**, 4246 (2000)
41. C. Conti, S. Trillo: "Self-transparency and localization in gratings with quadratic nonlinearity". In *Nonlinear photonic crystal*, ed. by B. Eggleton, R.E. Slusher (Springer, New York, 2002), to be published
42. L. Brzozowski, E.H. Sargent, IEEE J. Quantum Electron. **36**, 550 (2000)
43. L. Brzozowski, E. H. Sargent, D.E. Pelinovsky: Phys. Rev. E **62**, R4536-R4539 (2000)
44. D.E. Pelinovsky, L. Brzozowski, E. H. Sargent: IEEE J. Lightwave Tech. **19**, 114 (2001)
45. D.E. Pelinovsky, J. Sears, L. Brzozowski, E. H. Sargent: J. Opt. Soc. Am. B **19**, 43 (2002)
46. S. E. Harris: Appl. Phys. Lett. **9**, 114 (1966)
47. P. St. J. Russell, IEEE J. Quantum Electron. **27**, 830 (1991)
48. Y. J. Ding, S.J. Lee, J. B. Khurgin, Phys. Rev. Lett. **75**, 429 (1995)
49. M. Matsumoto, K. Tanaka, J. Quantum. Electron. **31**, 700 (1995)
50. Y. J. Ding, J. B. Khurgin: Opt. Lett. **21**, 1445 (1996); IEEE J. Quantum Electron. **32**, 1574 (1996)
51. G. D'Alessandro, P. St. J. Russell: A. A. Wheeler, Phys. Rev. A **55**, 3211 (1997)
52. G. D. Landry, T. Maldonado: Opt. Lett. **22**, 1400 (1997); Applied Optics **37**, 7809 (1998); IEEE J. Lightwave Tech. **17**, 316 (1999)
53. P.M. Lushnikov, P. Lodahl, M. Saffman: Opt. Lett. **23**, 1650 (1998)
54. Y. J. Ding, J. U. Kang, J. B. Khurgin: IEEE J. Quantum Electron. **34**, 966 (1998)
55. A. Picozzi, M. Haelterman, Opt. Lett. **23**, 1808 (1998); Phys. Rev. E **59**, 3749 (1999)
56. C. Conti, G. Assanto, S. Trillo: Opt. Lett. **25**, 1134 (1999)
57. K. Gallo, P. Baldi, M. De Micheli, D.B. Ostrowsky, G. Assanto: Opt. Lett. **25**, 966 (2000)
58. C. Conti, S. Trillo, G. Assanto: Phys. Rev. Lett. **85**, 2502 (2000)
59. S. Somekh, A. Yariv: Appl. Phys. Lett. **21**, 140 (1972)
60. M.M. Fejer, G.A. Magel, D.H. Jundt, R.L. Byer: IEEE J. Quantum Electron. **QE-28**, 2631 (1992)
61. J. P. Van Der Ziel, L. M. Ilegems: Appl. Phys. Lett. **28**, 437 (1976)

62. S. Janz, C. Fernando, H. Dai, F. Chatenoud, M. Dion, R. Normadin: Opt. Lett. **18**, 589 (1993)
63. J.U. Kang, Y. J. Ding, W.K. Burns, J.S. Mellinger: Opt. Lett. **22**, 862 (1997)
64. X. Gu, R. Y. Korotkov, Y. J. Ding, J. U. Kang, J. B. Khurgin: J. Opt. Soc. Am. B **15**, 1561 (1998)
65. X. Gu, M. Makarov, Y. J. Ding, J. B. Khurgin, W.P. Risk: Opt. Lett. **24**, 127 (1999)
66. X. Mu, I.B. Zotova, Y.J. Ding, W.P. Risk: Opt. Commun. **181**, 153-159 (2000)
67. V. Berger: Phys. Rev. Lett. **81**, 4136 (1998)
68. N.G.R. Broderick, G.W. Ross, H.L. Offerhaus, D.J. Richardson, D.C. Hanna: Phys. Rev. Lett. **84**, 4345 (2000)
69. J. Coste, J. Peyraud: Phys. Rev. B **40**, 12201 (1989)
70. B. Denardo, B. Galvin, A. Greenfield, A. Larraza, S. Putterman, W. Wright: Phys. Rev. Lett. **68**, 1730 (1992)
71. Y. S. Kivshar: Phys. Rev. Lett. **70**, 3055 (1993)
72. G. Huang, Z. Jia: Phys. Rev. B **51**, 613 (1995)
73. S. John, J. Wang: Phys. Rev. Lett. **64**, 2418 (1990)
74. D. Barday, M. Remoissenet: Phys. Rev. B **41**, 10387 (1990); Phys. Rev. B **43**, 7297 (1991)
75. J. M. Bilbault, M. Remoissenet: J. Appl. Phys. **70**, 4544 (1991)
76. N. Aközbek, S. John: Phys. Rev. Lett. **71**, 1178 (1993); Phys. Rev. E **57**, 2287 (1998)
77. R. Grimshaw, B. A. Malomed: Phys. Rev. Lett. **72**, 949 (1994)
78. V.V. Konotop, G.P. Tsironis: Phys. Rev. E **53**, 5393 (1996)
79. A. E. Kozhekin, G. Kurizki: Phys. Rev. Lett. **72**, 949 (1994); A. E. Kozhekin, G. Kurizki, B. A. Malomed: Phys. Rev. Lett. **81**, 3647 (1998); N. Aközbek, S. John: Phys. Rev. E **58**, 3876 (1998)
80. H. G. Winful, V. Perlin: Phys. Rev. Lett. **84**, 3586 (2000)
81. Z. Cheng, G. Kurizki: Phys. Rev. Lett. **75**, 3430 (1995)
82. V.I. Rupasov, M. Singh: Phys. Rev. Lett. **77**, 338 (1996)
83. O. Zobay, S. Pötting, P. Meystre, E.M. Wright: Phys. Rev. A **59** 643 (1999)
84. S. Wabnitz: Opt. Lett. **14**, 1071 (1989)
85. G. Cappellini, S. Trillo, S. Wabnitz, R. Chisari: Opt. Lett. **17**, 637 (1992)
86. P. Byrd, M. Friedman: *Handbook of elliptic integrals for engineers and physicists* (Springer, Berlin, 1971)
87. C.B. Clausen, Yu.S. Kivshar, O. Bang, P.L. Christiansen: Phys. Rev. Lett. **83**, 4740 (1999)
88. P. Di Trapani, A. Bramati, S. Minardi, W. Chinaglia, S. Trillo, C. Conti, J. Kilius, G. Valiulis: Phys. Rev. Lett. **87** 183902 (2001); C. Conti, S. Trillo, P. Di Trapani, J. Kilius, A. Bramati, S. Minardi, W. Chinaglia, G. Valiulis: J. Opt. Soc. Am. B **19** (4), 852 (2002)
89. M. Grillakis, J. Shatah, W. Strauss: J. Funct. Anal. **74**, 160 (1987); **94**, 308 (1990)
90. W.E. Thirring: Ann. Phys. **3**, 91 (1958)
91. E.A. Kuznetsov, A.V. Mikhailov: Teor. Mat. Fiz. **30**, 193 (1977)
92. F.V. Marchevskii, V.L. Strizhevskii, V.P. Feshchenko: Sov. J. Quantum Electron. **14**, 192 (1984)
93. T.V. Makhviladze, M.E. Sarychev: Sov. Phys. JETP **44**, 471 (1975)
94. A.P. Sukhorukov: *Nonlinear Wave Interactions in Optics and Radiophysics* (Nauka, Moscow 1988) (in Russian)
95. J.M. Soto-Crespo, N.N. Akhmediev, V.V. Afanasjev, Opt. Commun. **118**, 587-593 (1995)

96. M. G. Vakhitov, A. A. Kolokolov: Radiophys. Quantum Electron. **16**, 783 (1973)
97. A. V. Buryak, Y. S. Kivshar, S. Trillo: Phys. Rev. Lett. **77**, 5210 (1996)
98. F. V. Kusmartsev: Phys. Rep. **183**, 2 (1989)
99. Bogolubsky: Phys. Lett. 73A, 87 (1979); A. Alvarez, B. Carreras: Phys. Lett. **86A**, 327 (1981); W. A. Strauss, L. Vazques: Phys. Rev. D **34**, 641 (1986); A. Alvarez, M. Soler: Phys. Rev. D **34**, 644 (1986)
100. V.I. Barashenkov, D.E. Pelinovsky, E.V. Zemlyanaya: Phys. Rev. Lett. **80**, 5117 (1998); V.I. Barashenkov, E.V. Zemlyanaya: Computer Phys. Commun. **126**, 23 (2000)
101. A. De Rossi, C. Conti, S. Trillo: Phys. Rev. Lett. **81**, 85 (1998)
102. J. Schöllmann, R. Scheibenzuber, A.S. Kovalev, A.P. Mayer, A.A. Maradunin: Phys. Rev. E **59**, 4618 (1999); J. Schöllmann, A.P. Mayer: Phys. Rev. E **61**, 5830 (2000)
103. R. L. Pego, M. I. Weinstein: Phyl. Trans. R. Soc. Lond. A **340**, 47 (1992)
104. D. Mihalache, D. Mazilu, L. Torner: Phys. Rev. Lett. **81**, 4353 (1998)
105. M. Johansson, Y. K. Kivshar: Phys. Rev. Lett. **82**, 85 (1999)
106. H.G. Winful, R. Zamir, S.F. Feldman: Appl. Phys. Lett. **58**, 1001 (1991)
107. A.B. Aceves, S. Wabnitz, C. De Angelis: Opt. Lett. **17**, 1566 (1992)
108. C. M. de Sterke: J. Opt. Soc. Am. B **15**, 2660 (1998)
109. H. He, A. Arraf, C. M. de Sterke, P. D. Drummond, B. A. Malomed: Phys. Rev. E **59**, 6064 (1999)
110. M. Yu, C.J. McKinstrie, G.P. Agrawal: J. Opt. Soc. Am. B **15**, 607 (1998)
111. A.R. Champneys, B.A. Malomed, M.J. Friedman: Phys. Rev. Lett. **80**, 4169 (1998)

# Impact of Stimulated Raman Scattering
# in High-Speed Long-Distance Transmission Lines

P.T. Dinda, A. Labruyere, and K. Nakkeeran

Laboratoire de Physique de l'Université de Bourgogne, UMR CNRS No. 5027, Av. A. Savary, B.P. 47 870, 21078 Dijon Cédex, France

**Abstract.** We examine the effects of stimulated Raman scattering on ultra-short pulses propagating in optical fiber systems. In particular we demonstrate that the existing theories for the Raman-induced soliton self-frequency shift give consistent results only in a restricted domain of pulse width which excludes important practical applications to high-speed soliton transmission systems. We present a general theory for the soliton self-frequency shift (SSFS), which applies to any pulse whose spectral bandwidth lies within the third-order telecommunication window. We show that the disastrous impact of the SSFS in high-speed long-distance transmission lines can be suppressed by use of filters whose central frequency is appropriately up shifted with respect to the transmission frequency.

## 1   Introduction

In 1986, Mitschke and Mollenauer [1] discovered an important phenomenon in the context of optical communications, called soliton self-frequency shift (SSFS), in which the Stimulated Raman Scattering (SRS) causes a continuous downshift of the mean frequency of short pulses propagating in optical fibers. Subsequent to this discovery, Gordon [2] elaborated the theory of the SSFS, by analytically demonstrating the functional dependence of the SSFS upon the Raman gain of optical fibers. The SSFS is useful for some optical systems, whereas for various other systems it may become a serious drawback. For example, the SSFS is beneficial for developing new optical femtosecond fiber lasers. In recent developments of soliton lasers, the SSFS is exploited to make the pulse wavelength tunable over large ranges, through a variation of the fiber-input power [3,4]. In contrast, in high-speed optical communication lines, the combined effects of SRS and group-velocity dispersion causes a continual temporal shift of the pulse (away from its rest frame). The disastrous impact of SRS in such communication lines comes from the fact the SSFS of a given pulse in a bit pattern is closely related to the amplitude and width of this pulse. As the noise present in the line inevitably induces a jitter in the pulse parameters during the propagation, it becomes clear that the amount of SSFS (and the associated temporal shift) may substantially differ from one pulse to another at the receiver end, even if the pulses have the same input parameters. This loss of synchronism in the bit pattern may cause a strong transmission penalty. An accurate prediction of the impact of this phenomenon is therefore useful in general, and more particularly for the two types of systems mentioned above. In this context, a fundamental result obtained by Gordon relates the shifting rate of the mean frequency of the pulse to its temporal width $\tau$ by the formula $d\omega_0/dz \propto \tau^{-4}$, where $z$ is the propagation coordinate. By convention we will refer this formula as being the "Gordon law". Lucek and Blow [5,6] suggested a modification in the Gordon law by assuming that the pulse width will grow exponentially because of the optical losses and derived a simple formula for calculating the frequency shift from the knowledge of the initial pulse width. The simplicity of these laws [2,5,6] makes them so attractive that, very often, they are used as models for quickly estimating the SSFS without having to go through any lengthy calculation required when solving directly the nonlinear Schrödinger equation (NLSE). We would like to point out that the use of the Gordon law or Lucek-Blow law for all kinds of situation (i.e., without extreme care) may lead to dramatic consequences in predicting the importance of Raman effects in optical systems, and especially their impact in high-speed long-distance communication lines. Indeed, the Gordon law results from an analytical procedure having two levels of approximations: The first is the assumption of an exact balance between dispersion and nonlinearity in the presence of strong perturbing effects such as the SRS or fiber losses. Second, one assumes that the frequency dependence of the Raman susceptibility is linear [2,5–7]. These two levels of approximation lead to strict limitations to the domain of validity of the Gordon law, and can straight forwardly explain most of the results obtained in previous studies showing a strong discrepancy between this law and the experimental measurements [5,6].

Our purpose in the present study is to carry out a careful analysis of the essential features of SRS on ultra-short pulses propagating in optical fiber systems, at a given carrier frequency. In particular, we present a general theory of the SSFS, which revolves the fundamental insufficiencies of the existing theories mentioned above. We show that the general theory can be effectively applied to the problem of SRS management in optical communication systems. Indeed we propose a method for suppressing the SSFS and the associated timing-shift in long distance pulse transmissions. This method is based on the use of filters, called "up-shifted filters" (USFs), whose peak frequency $\omega_f$ is up shifted with respect to the carrier frequency $\omega_0$. The suppression mechanism is based on an appropriate choice of the filter's bandwidth (BW) and peak frequency $\omega_f$. This study is organized as follows: in section 2 we present the general features of SRS, the theoretical model and the procedure of determination of the SSFS. In section 3, we present our method for suppressing the SSFS in optical communication lines, and in section 4, we give some concluding remarks.

## 2     General Features and Theoretical Model

### 2.1     General Features of SRS

From a fundamental point of view, SRS results from the action of a light wave on the atoms of the dielectric media that make up an optical fiber, as schematically represented in Fig. 1 (a). In this two-level process, an atom which is initially in

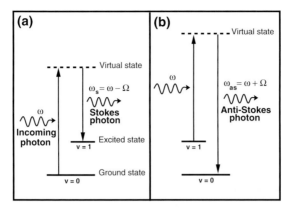

**Fig. 1.** Schematic diagrams of the interactions between incoming photons with frequency $\omega$ and a dielectric material, leading to generation of (a) the Raman Stokes radiation, and (b) the Raman anti-Stokes radiation

the ground state (labeled $v = 0$ in Fig. 1) absorbs all the energy of an incoming photon and moves up to a highly unstable excited state (or virtual state).

As soon as the atom reaches this virtual state, it goes down immediately to a relatively stable excited state labeled $v = 1$. In going down, the atom emits a large fraction (but not the totality) of the photon energy that has been absorbed. This energy is emitted in the form of a new photon whose energy is reduced (compared with the energy of the incoming photon) by an amount that corresponds to the energy difference between the two vibrational states $v = 0$ and $v = 1$. Consequently the frequency of the newly created photon is also reduced compared with the incoming frequency $\omega$. Thus, through SRS, incoming photons are progressively destroyed while new photons, called Stokes photons, are created at a down-shifted frequency $\omega_s = \omega - \Omega$, where $\Omega$ is proportional to the energy difference between the two vibrational states. In other words, this process, which corresponds to an ordinary process of SRS in an optical fiber, will induce an energy transfer from the higher to the lower frequency components of a light wave propagating in the medium. On the other hand, it is quite possible that an atom lying in the excited state $v = 1$ can come back to the ground state. The way of coming back is represented in Fig. 1 (b): the excited atom will absorb an incoming photon and will move up to a highly unstable state before coming back immediately to the ground state. In doing that, the atom will radiate away energy in the form of a photon with an up-shifted frequency $\omega_{as} = \omega + \Omega$, called anti-Stokes photon. But the generation of the Raman anti-Stokes radiation is not a usual process in standard optical fibers. In the present study, we will mainly focus on the ordinary process of SRS in fibers, which can be measured with the help of the Raman susceptibility of $SiO_2$ glasses[8–11], say $\tilde{\chi}_R$, as that obtained in Ref. [8]. In Fig. 2, we have normalized the Raman susceptibility [8] represented in solid curves, to $Re[\tilde{\chi}_R(0)] = 1$.

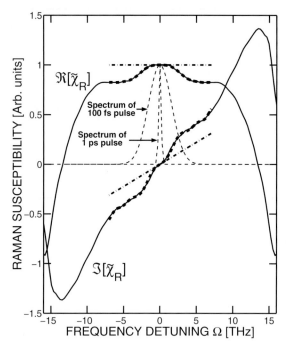

**Fig. 2.** Plot showing the Raman susceptibility $\tilde{\chi}_R$ for fused quartz. The dot-dashed lines represent the linear approximation of the Raman susceptibility, $\tilde{\chi}_{LA}$. Thick dashed curves represent a polynomial fit to $\tilde{\chi}_R(\Omega)$ in the range $-7.3\,THz \leq \Omega \leq 7.3\,THz$, obtained from Eq.(1), with the coefficients $k_l = i^{-l}f_l$ ($l = 0, \cdots, 15$), given by the last lines of Table 1 (a) and 1 (b), respectively.

The energy transfer processes from the incoming wave to the scattered waves are governed by the imaginary part of the Raman susceptibility, that is the Raman gain, whose broad spectrum originates from the amorphous nature of the fused silica. In fact, the susceptibility curves given in Fig. 2 corresponds to the parallel component of the Raman susceptibility, which governs energy transfers between waves with parallel polarizations. In silica fibers, there exists an orthogonal component of the Raman susceptibility, not represented here, which govern energy transfers between waves with orthogonal polarizations [8]. The two components of the Raman susceptibility coincide when the frequency separation $\Omega$ between the incoming and scattered photons is sufficiently small. In other words, in a single channel transmission line operating at low bite-rate (i.e., in a small spectral bandwidth) the Raman effect will be relatively insensitive to the polarization of the incoming photons. But in case of large spectral bandwidth, the two components of the Raman gain will become quite different for large $\Omega$ values, with a much smaller gain for the orthogonal component, which can therefore be neglected.

We would here like to emphasize the following important point. For a sufficiently low bit rate, the spectral bandwidth of the transmission will fall in a small

**Table 1.** Raman Coefficients

(a): Even terms

| $\Omega\,[THz]$ | $f_0$ | $f_2 \times 10^5$ | $f_4 \times 10^5$ | $f_6 \times 10^6$ |
|---|---|---|---|---|
| $\leq 0.3$ | 1 | 0 | 0 | 0 |
| $\leq 1.0$ | 1 | $-4.4089$ | 0 | 0 |
| $\leq 1.75$ | 1 | $-17.323$ | $-6.4026$ | 0 |
| $\leq 2.5$ | 1 | $-22.590$ | $-9.2104$ | $-6.7166$ |
| $\leq 3.5$ | 1 | $-13.246$ | $-8.1168$ | $-7.5048$ |
| $\leq 4.5$ | 1 | $-5.7463$ | $-7.1210$ | $-7.1219$ |
| $\leq 5.5$ | 1 | $6.8572$ | $-5.1455$ | $-4.9375$ |
| $\leq 7.3$ | 1 | $17.502$ | $-3.4126$ | $-2.6922$ |

| $\Omega\,[THz]$ | $f_8 \times 10^7$ | $f_{10} \times 10^9$ | $f_{12} \times 10^{10}$ | $f_{14} \times 10^{12}$ |
|---|---|---|---|---|
| $\leq 0.3$ | 0 | 0 | 0 | 0 |
| $\leq 1.0$ | 0 | 0 | 0 | 0 |
| $\leq 1.75$ | 0 | 0 | 0 | 0 |
| $\leq 2.5$ | 0 | 0 | 0 | 0 |
| $\leq 3.5$ | $-4.0341$ | 0 | 0 | 0 |
| $\leq 4.5$ | $-5.4197$ | $-23.929$ | 0 | 0 |
| $\leq 5.5$ | $-4.0242$ | $-25.901$ | $-9.5831$ | 0 |
| $\leq 7.3$ | $-1.7075$ | $-9.0464$ | $-3.5656$ | $-7.7629$ |

(b): Odd terms

| $\Omega\,[THz]$ | $f_1 \times 10^3$ | $f_3 \times 10^4$ | $f_5 \times 10^5$ | $f_7 \times 10^6$ |
|---|---|---|---|---|
| $\leq 0.3$ | 7.06 | 0 | 0 | 0 |
| $\leq 1.0$ | 7.06 | 9.2893 | 0 | 0 |
| $\leq 1.75$ | 7.06 | 11.593 | $-10.583$ | 0 |
| $\leq 2.5$ | 7.06 | 12.368 | $-15.418$ | 12.794 |
| $\leq 3.5$ | 7.06 | 11.095 | $-12.783$ | 12.178 |
| $\leq 4.5$ | 7.06 | 10.721 | $-12.410$ | 13.080 |
| $\leq 5.5$ | 7.06 | 9.8535 | $-10.657$ | 10.947 |
| $\leq 7.3$ | 7.06 | 7.6294 | $-6.4194$ | 5.0793 |

| $\Omega\,[THz]$ | $f_9 \times 10^7$ | $f_{11} \times 10^8$ | $f_{13} \times 10^{10}$ | $f_{15} \times 10^{11}$ |
|---|---|---|---|---|
| $\leq 0.3$ | 0 | 0 | 0 | 0 |
| $\leq 1.0$ | 0 | 0 | 0 | 0 |
| $\leq 1.75$ | 0 | 0 | 0 | 0 |
| $\leq 2.5$ | 0 | 0 | 0 | 0 |
| $\leq 3.5$ | $-6.999$ | 0 | 0 | 0 |
| $\leq 4.5$ | $-10.917$ | 5.2316 | 0 | 0 |
| $\leq 5.5$ | $-9.9131$ | 6.8353 | $-26.338$ | 0 |
| $\leq 7.3$ | $-3.6584$ | 2.2088 | $-9.789$ | 2.3630 |

region where the Raman susceptibility curves essentially coincide with straight lines, as can be seen in Fig. 2. Previous studies [2,5–7,12] have intensively used this straight line approximation, which corresponds to the first-order Taylor-series expansion of $\tilde{\chi}_R(\omega)$. Hereafter this first-order approximation of $\tilde{\chi}_R(\omega)$ will be simply referred to as being the *linear approximation* of $\tilde{\chi}_R(\omega)$, that we denote hereafter by $\tilde{\chi}_{LA}$. As can be seen in Fig. 2, $\tilde{\chi}_{LA}$ coincides essentially with $\tilde{\chi}_R(\omega)$ within a relatively small spectral bandwidth that corresponds to pulse widths above or $\approx 1\,ps$. Now, subpicosecond pulses cover a large spectral bandwidth in which a clear discrepancy exists between $\tilde{\chi}_R$ and $\tilde{\chi}_{LA}$, as illustrated by a thin dashed curve in Fig. 2 which represents the Fourier transform of a $100\,fs$ pulse. Thus, the smaller the pulse width, the larger will be the magnitude of the discrepancy between $\tilde{\chi}_R$ and $\tilde{\chi}_{LA}$. Consequently, to extend the theory of SRS to ultra-short pulses, one can perform a Taylor series expansion of $\tilde{\chi}_R$ as follows:

$$\tilde{\chi}_R(\Omega) \approx \tilde{\chi}_{PA}(\Omega) = 1 + k_1\Omega + \sum_{l=2}^{N-1} \frac{k_l}{l!}\Omega^l, \tag{1}$$

where $k_l = \left.\frac{\partial^l \tilde{\chi}_R}{\partial \omega^l}\right|_{\omega=\omega_0}$, $\Omega = \omega - \omega_0$, $\omega_0$ is the carrier frequency. The number of terms $N$ required for this expansion of $\tilde{\chi}_R$ is closely related to the size of the spectral bandwidth under consideration. Note that the first two terms in the r.h.s. of Eq. (1) correspond to $\tilde{\chi}_{LA}$. On the other hand, it is useful to note that the curves for $\tilde{\chi}_R$ in Fig. 2 are actually full of small rough portions which are not physically important but which will be systematically taken into account if we perform a Taylor series expansion of $\tilde{\chi}_R$. In other words, such an expansion would involve an extremely large number ($N$) of terms which are simply due to the roughness of $\tilde{\chi}_R$. One can dramatically reduce the number of terms required to describe a given portion of $\tilde{\chi}_R$ by somewhat smoothing the susceptibility curves. Using a polynomial fit of $\tilde{\chi}_{PA}$ in Eq. (1) to $\tilde{\chi}_R$ in Fig. 2, we have obtained the real coefficients

$$f_l = i^l k_l \tag{2}$$

optimized for various spectral domains. The results are displayed in Table 1. The most important point to be noticed in Table 1 is that only sixteen terms [eight for each of $\Re(\tilde{\chi}_R)$ and $\Im(\tilde{\chi}_R)$] are needed to cover a spectral domain of $-7.3\,THz \le \Omega \le 7.3\,THz$, as indicated by thick dashed curves in Fig. 2, which corresponds essentially to the totality of the third-order telecommunication window. Hereafter we will refer to this approximate but highly accurate representation of the Raman susceptibility as being the *polynomial approximation* ($\tilde{\chi}_{PA}$).

We would now like to emphasize the following important point. In a dielectric medium, the SRS will never act alone on a light pulse, but will always act in close connection with the group velocity dispersion (GVD). In the frequency domain, the Raman-induced energy transfer from the higher to the lower frequency components of a light pulse will cause a down frequency shift of the whole pulse's spectrum (SSFS). Through GVD, the SSFS is converted into a temporal shift of the pulse away from the center of its rest frame (or bit slot). The importance and direction of this temporal shift depends on the magnitude and sign of the GVD, respectively. For sake of clarity, we discuss below separately two important situations in the context of optical communications.

*Uniform dispersion line:* Fig. 3 illustrate the combined effects of SRS and GVD in a standard line with uniform GVD. The input pulse corresponds to a chirp-free pulse, that is, a pulse that consists of photons having the same frequency throughout the pulse. When the pulse is injected in the fiber, the GVD immediately enters into play, by continually producing a chirp during the pulse propagation.

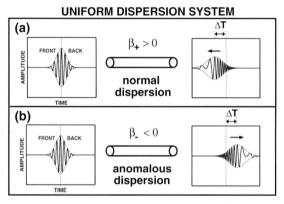

**Fig. 3.** Schematic representation of the combined effects of SRS and group-velocity dispersion on a pulse propagating in (a) a normal-dispersion fiber, and (b) an anomalous-dispersion fiber.

In other words, incoming photons (with frequency $\omega_0$) are continually destroyed in the front and the back of the pulse, while higher (lower) frequency photons are created in the back (front) of the pulse, in the normally dispersive fiber illustrated in Fig. 3 (a). As the newly created photons travel at different velocities, the pulse broadens in the temporal domain. But the most important effect here lies in the action of SRS, which induces an energy transfer from the higher frequency photons (lying in the back of the pulse) to the lower-frequency photons (lying in the front). As result, the pulse is continually pushed away from the center of its rest frame in forward direction of its motion, as indicated by an horizontal arrow in Fig. 3 (a). Note that, in general, the chirp induced by the dispersion is always linear at least in the central part of the pulse. On the other hand, in the anomalous dispersion fiber [Fig. 3 (b)] the higher (lower) frequency photons are created in the front (back) of the pulse, and there, the combined effects of GVD and SRS causes a continual temporal shift in the backward direction of the moving pulse.

*Dispersion-managed (DM) line:* Basically, the dispersion management technique utilizes a transmission line with a periodic dispersion map, such that each period is built up by two types of fibers with generally different lengths ($L_1$, $L_2$) and opposite signs of GVD, as schematically represented in Fig. 4.

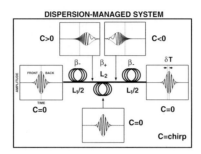

**Fig. 4.** Schematic representation of the combined effects of SRS and group-velocity dispersion on a pulse propagating in a dispersion-managed line.

In this system, an initially unchirped pulse will be continually pushed away from the center of its rest frame, in the backward direction, during propagation in the first half of the anomalous dispersion fiber, like in the case of Fig. 3 (b). As soon as the pulse enters the normal-dispersion fiber, the magnitude of the chirp begins to decrease; which slows down the velocity of the temporal shift of the pulse, until the pulse stops at the free-chirp point ($C = 0$). From there, the sign of the pulse reverses, and the combined effects of SRS and GVD begin to push the pulse in the opposite (or forward) direction, like in the case of 3 (a). In the way back towards the center of its rest frame, the pulse first moves rapidly until it enters the second half of the anomalous-dispersion fiber. From there, the pulse slows down and stops near the center of its rest frame, at the end of the dispersion map. Thus, whereas the pulse will be monotonically and quickly pushed away from its rest frame in either of the two uniform-dispersion fibers ($\beta_+$ or $\beta_-$), by a large amount $\Delta T \approx |\beta_\pm|\Delta x_{5R}$, where $\Delta x_{5R}$ represents the SSFS, the DM system will lead to a dramatic reduction of the temporal shift to a small quantity $\Delta T \approx |\langle\beta\rangle|\Delta x_{5R}$ after each dispersion map. Indeed, DM systems are always designed in such a way to have a relatively high local dispersion $\beta_\pm$, and a very small (anomalous) average dispersion $\langle\beta\rangle = (\beta_+ L_+ + \beta_- L_-)/(L_- + L_+)$, to reinforce the pulse stability. However, the systematic generation of a very small temporal shift $\delta T$ after each dispersion map may lead to a non-negligible cumulated temporal shift for very long (transoceanic) propagation distances.

## 2.2   Generalized Theory of the SSFS

The above qualitative considerations can be made more mathematically precise through a careful examination of the pulse dynamics in a given optical fiber system. The pulse evolution in periodically amplified optical fiber links may be described by the generalized nonlinear Schrödinger equation (GNLSE):

$$\frac{\partial \psi}{\partial z} + \frac{i\beta_2(z)}{2}\frac{\partial^2 \psi}{\partial t^2} + \frac{\beta_3(z)}{6}\frac{\partial^3 \psi}{\partial t^3} - i(1-\rho)\gamma(z)|\psi|^2\psi = -\frac{\alpha(z)}{2}\psi + F[\psi]$$
$$+ R[\psi] + A[\psi]. \qquad (3)$$

Here $\psi$ is the pulse field at position $z$ in the fiber and at time $t$. The parameters $\alpha(z)$, $\beta_2(z)$, $\beta_3(z)$ and $\gamma(z)$ represent the loss, second-order GVD, third-order dispersion, and self-phase modulation (SPM) parameters, respectively. $F[\psi]$ and $A[\psi]$ represent the filtering and amplification actions, respectively. In the following, whenever it will be required, the expressions of $A[\psi]$ and $F[\psi]$ will be mathematically detailed. The Raman contribution in the GNLSE is given by [8]

$$R[\psi] = i\gamma\rho\psi \int\limits_0^\infty \chi_R(s)|\psi|^2(t-s)ds = i\gamma\rho\psi\mathcal{T}^{-1}\left[\tilde{\chi}_R(\omega)\cdot|\tilde{\psi}|^2\right], \qquad (4)$$

where the symbol $\mathcal{T}^{-1}$ represents the inverse Fourier transform and the tilde over a given quantity represents the Fourier transform of that quantity. $\rho = 0.18$ measures the fractional contribution of SRS to the total fiber nonlinearity [8–10].

The procedure for determining the SSFS is fundamentally based on collective variable (CV) theories. To simplify this procedure, it is helpful to make use of the polynomial approximation of $\tilde{\chi}_R$ in Eq. (1) with the coefficients $f_l$ given in Table 1. Then, invoking Eq.(4), the GNLSE (3) takes the following form:

$$\frac{\partial \psi}{\partial z} + \frac{i\beta_2(z)}{2}\frac{\partial^2 \psi}{\partial t^2} + \frac{\alpha}{2}\psi - i\gamma|\psi|^2\psi = i\gamma\rho\psi\sum_{l=1}^N \frac{f_l}{l!}\frac{\partial^l |\psi|^2}{\partial t^l}. \qquad (5)$$

Here we have taken $\beta_3 = 0$, because the effect of the third-order dispersion on the temporal shift associated with SSFS can be neglected in the presence of a sufficiently high second-order GVD. Furthermore, here, we focus on the portions of the line where the pulse freely propagate without any assistance from active or passive elements such as amplifiers or filters, which are localized at discrete points along the line. This is the reason for which amplification and filtering actions do not appear explicitly in the r.h.s of Eq.(5). In previous studies [5–7,12], only the first term in the r.h.s of Eq.(5) were considered, i.e., $N = 1$ (or equivalently, $\tilde{\chi}_R = \tilde{\chi}_{LA}$). To illustrate the general theory of the SSFS, we consider an extremely large spectral bandwidth covering essentially all the third-order telecommunication window, $-7.3\,THz \leq \Omega \leq 7.3\,THz$, which corresponds to $N = 15$ in the GNLSE (5). One can carry out a rigorous collective variable (CV) treatment for the GNLSE (5) by decomposing the soliton field in the following way [13,14]:

$$\psi(z,t) = f(x_1, x_2, \cdots, x_6, t) + q(z,t), \qquad (6)$$

where the ansatz function $f$ is chosen to be the best representation of the pulse configuration, $x_j$'s designate the pulse parameters, and $q$ is the remaining field such that the sum of $f$ and $q$ satisfies the GNLSE (5). This field $q$, that we call as "residual field" accounts for the dressing of the soliton and any radiation coupled to the soliton's motion. The residual field is ignored in deriving the Gordon law. This approximation, (i.e., setting $q = 0$), called bare approximation, yields consistent results only when there is no considerable radiation, i.e., the soliton dressing is negligible. We would here like to raise the following fundamental points. A major approximation made in the existing theories of the SSFS (in addition to the bare approximation and the linear approximation of the Raman susceptibility), is the assumption that the pulse profile corresponds exactly to the hyperbolic secant profile of the first-order soliton throughout the pulse propagation. In other words, the ansatz function used in previous studies assumes an exact balance between the anomalous fiber GVD and SPM [2,5,6,12]. Mathematically, an exact balance between the dispersion length $L_D = x_3^2/|\beta_2|$, and the nonlinear length $L_{NL} = 1/(\gamma x_1^2)$, corresponds to the following relation between the pulse width $x_3$ and its amplitude $x_1$:

$$x_3 = \frac{1}{x_1}\sqrt{\frac{|\beta_2|}{\gamma}}. \tag{7}$$

This relation corresponds to a severe approximation. Indeed, if the NLSE contains any type of strong perturbations (such as losses, spatially varying dispersion, or SRS), the relation between the pulse width and amplitude will be given by a more complicated function than that given by Eq. (7). To resolve this insufficiency one can assume any desired form for the ansatz function but without imposing any prior relation between the pulse parameters. For example one can assume a gaussian profile given by

$$f(z,t) = x_1 \exp\left[-\frac{(t-x_2)^2}{x_3^2} + i\frac{x_4}{2}(t-x_2)^2 - ix_5(t-x_2) + ix_6\right], \tag{8}$$

where $x_1, x_2, x_3, x_4/(2\pi), x_5/(2\pi)$ and $x_6$ represent the pulse amplitude, temporal position, width, chirp, frequency and phase respectively. Using this ansatz, and neglecting the residual field, the GNLSE (5) can be expressed in terms of a set of ordinary differential equations for the dynamics of the CVs [13,14], which describe the spatial evolution of the pulse parameters:

$$\dot{x}_1 = -\frac{\alpha x_1}{2} + \frac{\beta_2 x_1 x_4}{2}, \tag{9}$$

$$\dot{x}_2 = \beta_2 x_5, \tag{10}$$

$$\dot{x}_3 = -\beta_2 x_3 x_4, \tag{11}$$

$$\dot{x}_4 = -\beta_2\left(\frac{4}{x_3^4} - x_4^2\right) - \frac{\sqrt{2}\gamma x_1^2}{x_3^2} + \frac{3\sqrt{2}\gamma_{r2}x_1^2}{x_3^4} - \frac{5\sqrt{2}\gamma_{r4}x_1^2}{x_3^6} + \frac{7\gamma_{r6}x_1^2}{3\sqrt{2}x_3^8}$$
$$- \frac{3\gamma_{r8}x_1^2}{4\sqrt{2}x_3^{10}} + \frac{11\gamma_{r10}x_1^2}{60\sqrt{2}x_3^{12}} - \frac{13\gamma_{r12}x_1^2}{360\sqrt{2}x_3^{14}} + \frac{\gamma_{r14}x_1^2}{168\sqrt{2}x_3^{16}}, \tag{12}$$

$$\dot{x}_5 = -\frac{\sqrt{2}\gamma_{r1}x_1^2}{x_3^2} + \frac{\sqrt{2}\gamma_{r3}x_1^2}{x_3^4} - \frac{\gamma_{r5}x_1^2}{\sqrt{2}x_3^6} + \frac{\gamma_{r7}x_1^2}{3\sqrt{2}x_3^8} - \frac{\gamma_{r9}x_1^2}{12\sqrt{2}x_3^{10}} + \frac{\gamma_{r11}x_1^2}{60\sqrt{2}x_3^{12}}$$
$$- \frac{\gamma_{r13}x_1^2}{360\sqrt{2}x_3^{14}} + \frac{\gamma_{r15}x_1^2}{2520\sqrt{2}x_3^{16}}, \tag{13}$$

$$\dot{x}_6 = \beta_2\left(\frac{1}{x_3^2} - \frac{x_5^2}{2}\right) + \frac{5\gamma x_1^2}{4\sqrt{2}} - \frac{7\gamma_{r2}x_1^2}{4\sqrt{2}x_3^2} + \frac{9\gamma_{r4}x_1^2}{8\sqrt{2}x_3^4} - \frac{11\gamma_{r6}x_1^2}{24\sqrt{2}x_3^6} + \frac{13\gamma_{r8}x_1^2}{96\sqrt{2}x_3^8}$$
$$- \frac{\gamma_{r10}x_1^2}{32\sqrt{2}x_3^{10}} + \frac{17\gamma_{r12}x_1^2}{2880\sqrt{2}x_3^{12}} + \frac{19\gamma_{r14}x_1^2}{20160\sqrt{2}x_3^{14}}, \tag{14}$$

where $\gamma_{rl} = \rho f_l$. Equations (9)–(14) reveal the following features: All the terms coming from the imaginary part of $\tilde{\chi}_{PA}$ (i.e., proportional to $\gamma_{rj}$, with an odd $j$ value) contribute to the SSFS, as the r.h.s. of Eq. (13) shows, whereas the terms coming from the real part of $\tilde{\chi}_{PA}$ i.e., proportional to $\gamma_{rj}$, with an even $j$ value) will produce a chirp as can be seen in the r.h.s. of Eq. (12). More importantly, the first term in the r.h.s. of Eq. (13), which corresponds to the linear approximation of the Raman susceptibility, reveals that the lowest-order approximation of the SSFS rate is given by

$$\frac{d\omega_0}{dz} \propto \frac{x_1^2}{x_3^2}, \tag{15}$$

when no relation is imposed between the width $x_3$ and the amplitude $x_1$. Note that imposing the relation $x_3 \propto 1/x_1$ from Eq. (7), leads directly to the Gordon law $d\omega_0/dz \propto x_3^{-4}$. There is a major qualitative difference between these two SSFS laws. Indeed, according to the Gordon law, the SSFS rate should change for any change in the pulse width $x_3$. Now, the above formula (15) reveals, against general intuition, that any change of the pulse width will have no effect on the SSFS rate if meanwhile the pulse amplitude is changed by the same amount in the same direction. When higher-order Raman effects come into play (i.e., for subpicosecond pulses), the higher-order terms in the r.h.s. of Eq.(13) begin to contribute significantly to the SSFS. The $n$th-order contribution of the Raman gain to the SSFS rate is given by

$$\left.\frac{d\omega_0}{dz}\right|_n \propto \gamma_{rm}\frac{x_1^2}{x_3^{2n}}, \qquad m = 2n - 1. \tag{16}$$

This relation reveals that, at any order higher than the lowest order, the pulse width and amplitude will still act in opposite directions if they are varied in the same direction, but with a major action for the variation of the pulse width.

It is worth noting that high-speed transmission lines with lumped amplifiers are made up of a large number of small fiber sections in which the pulse will freely propagate without any assistance from in-line control elements, and there, the pulse will inevitably undergo a Raman-induced frequency downshift. Fig. 5 illustrate the behavior a $300\,fs$ pulse propagating in such a fiber section with uniform GVD ($\beta_2$): $-4.5\,ps^2/Km$, SPM ($\gamma$): $0.002\,m^{-1}W^{-1}$, losses ($\alpha$): $0.22\,dB/km$, and length $60m$. In Fig. 5, the dashed curves represent the

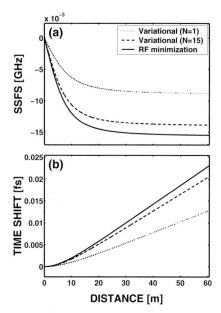

**Fig. 5.** Evolution of (a) the SSFS, and (b) the associated temporal shift for a femtosecond pulse propagating in a uniform-dispersion fiber. GVD ($\beta_2$): $-4.5\,ps^2/Km$. SPM ($\gamma$): $0.002\,m^{-1}W^{-1}$, losses ($\alpha$): $0.22\,dB/km$, Dotted and dashed curves correspond to solutions of the variational equations at first-order ($N = 1$), and the full variational equations ($N = 15$). The solid curve corresponds to the residual field minimization procedure.

solution of the full variational equations (9)–(14) with all the fifteen Raman terms, whereas the dotted curves correspond to the solution of the same variational equations (9)–(14) but at first order ($\gamma_{rj} = 0, j = 2, \cdots, 15$). The solid curves represent the result obtained by means of a rigorous CV approach based on the direct resolution of the GNLSE (5) and subsequent minimization of the residual field [14]. Thus, the dotted curves, which corresponds to the linear approximation of the Raman susceptibility, lead to a large discrepancy with respect to the residual-field minimization procedure (solid curve). Including the necessary higher-order Raman terms in the variational equations substantially improves the analytical prediction of the SSFS (as the dashed curve shows) but not perfectly. The small remaining discrepancy comes from the approximation of neglecting the residual field in the variational equations, which in fact is not zero because the Gaussian function is not the exact solution of the pulse propagation in the line under consideration. However, as the amount of calculations required by the variational equations to obtain the evolution of the pulse parameters represents only an extremely small fraction of the calculations which are required when using the numerical solution of the GNLSE and subsequent residual-field minimization procedure, our variational equations (9)–(14) appear as helpful tools for quickly evaluating the SSFS. Thus, the above results clearly illustrate that a careful management of the SSFS should be done to achieve high performances in transmission of ultra-short pulses. We demonstrate below that the disastrous impact of SRS can be effectively suppressed in high-speed transmission lines.

## 3    Suppression of the Soliton Self-frequency Shift

In this section we show that the disastrous impact of SRS can be effectively suppressed in high-speed transmission systems, by use of filters whose peak frequency is appropriately up-shifted with respect to the transmission frequency. Figure 6 provides a clear qualitative insight into the filtering action (in the spectral domain) leading to suppression of the SSFS.

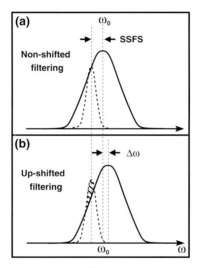

**Fig. 6.** Schematic representation of the filtering action on the pulse spectrum (dashed curve). The solid curve represents the transfer function of the filter.

Between two consecutive filters, the Raman effect inevitably induces an amount of SSFS, say $\Delta x_{5R}$. Then on passing through each filter, a small quantity of energy is extracted from the front of the pulse's spectrum, thus causing a reflection back of the whole spectrum. With non-shifted filters (NSFs), the reflection back of the pulse spectrum is relatively weak, and not sufficient to completely overcome the Raman-induced SSFS, unless one uses excessively narrow filters. But those kind of filters are highly detrimental to the pulse stability, as we will show below. In contrast, up-shifted filters (USFs) with moderate strength can induce a sufficiently strong reflection back of the spectrum, to completely cancel the SSFS, with the pulse stability preserved. Here we examine lumped filters with Gaussian transfer functions defined by:

$$T(\omega) = \exp[-(2\ln 2)(\omega - \omega_f)^2/B_f^2], \tag{17}$$

where $B_f$ is the filter's bandwidth (BW). The central frequency of the filters, $\omega_f$, is up-shifted by $\Delta\omega = \omega_f - \omega_0$ with respect to the transmission frequency $\omega_0$. One can obtain a more precise qualitative picture of the filter action by adopting the usual practice of approximating the lumped filtering by an equivalent continuous distributed filtering action along the line, defined by:

$$F[\psi] = \xi\psi_{tt} + i\eta\psi_t - \delta\psi = \xi[\psi_{tt} + 2i\Delta\omega\psi_t - \Delta\omega^2\psi], \qquad (18)$$

where $\xi = 2\ln 2/(B_f^2 z_f)$ represents the filter strength parameter and $z_f$ is the filtering period (the same as the amplification period). The first term in Eq. (18), which does not depend on $\Delta\omega$, represents the filtering action that is always present whether or not the central frequency of the filter is shifted. Now, shifting the filter up gives rise to two additional terms in Eq. (18), with $\eta \equiv 2\Delta\omega\xi$ and $\delta \equiv \Delta\omega^2\xi$, which will be carefully exploited to suppress the SSFS. One can obtain the mode of action of the filters on the pulse frequency, by means of the collective-variable approach. In doing so we obtain the following expression for the SSFS rate:

$$\frac{dx_5}{dz} = \mathcal{R} + \mathcal{F}_0 + \mathcal{F}(\Delta\omega), \qquad (19)$$

where $\mathcal{R}$, represents the Raman-induced SSFS rate given in the r.h.s of Eq.(13). The term $\mathcal{F}_0 \equiv -P\xi x_5$, where $P = (x_3^4 x_4^2 + 4)/x_3^2$, and which does not depend on $\Delta\omega$, represents the filtering action that is present whether or not the central frequency of the filter is shifted. The last term in the r.h.s. of Eq. (19), $\mathcal{F} \equiv P\xi\Delta\omega$ arises only when USFs are used.

Thus if NSFs are used (i.e., if $\Delta\omega = 0$, or equivalently $\mathcal{F} = 0$), the filtering term $\mathcal{F}_0$, which acts in the opposite direction to the Raman-induced SSFS rate $\mathcal{R}$, will cause the pulse spectrum to be trapped near the transmission frequency $\omega_0$. However NSFs will not completely cancel the average SSFS without an excessively strong filter strength (i.e., a large $\xi$, or a small filter's BW) that can easily destroy the desired dynamics.

When USFs are used, the additional term $\mathcal{F}$ in the r.h.s. of Eq. (19) enters into play, by acting in the direction opposite to the Raman-induced SSFS. The presence of this term make it possible to completely cancel the SSFS by simply tuning the central frequency of the filter, even with a moderate filter strength. Here lies the basic principle of SSFS suppression by up shifted filtering. The value of $\Delta\omega$ for suppressing the SSFS increases as the filter's BW $B_f$ increases. For sufficiently large $B_f$, we will have $|\Delta\omega| \gg |\Delta x_{5R}|$, and $\mathcal{F}$ will become prominent over $\mathcal{F}_0$ (which can therefore be neglected). Moreover, one can neglect all the terms proportional to $x_5$ in $\mathcal{F}$. Furthermore, as the pulse's spectral BW will execute only small variations about its initial value during the dynamics, one can integrate Eq. (19) with $P(z) \simeq P(0)$. In doing so we obtain

$$\delta x_5 \approx \Delta x_{5R} + z_f P(0)\xi\Delta\omega, \qquad (20)$$

where $\Delta x_{5R} = \int_0^{z_f} \mathcal{R}dz$ is the SSFS between two consecutive filters. Setting $\delta x_5 = 0$ gives us the following formula for SSFS suppression:

$$\Delta\omega = -N_f^2 \Delta x_{5R}, \qquad (21)$$

where $N_f \equiv B_f/B_p$ represents the ratio between the filter's BW and the spectral BW of the input pulse, defined by $B_p = \sqrt{2P(0)\ln 2}$. In short, we achieve SSFS suppression as follows: First, we evaluate $\Delta x_{5R}$ by simply letting the stationary pulse propagate without filtering over the distance $z_f$, and we deduce $\Delta x_{5R}$. Then we choose the desired value for $N_f$ (or equivalently $B_f$), and use Eq.(21) to finally calculate the value of $\Delta\omega$ for suppressing the SSFS.

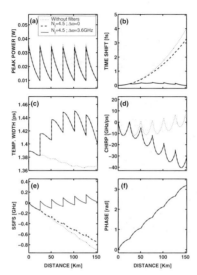

**Fig. 7.** Evolution of the pulse parameters in a periodically amplified DM line, and demonstration of suppression of the SSFS and the associated time shift, over a relatively short propagation distance $z_{max} = 150\,km$. Solid and dashed curves correspond to USFs and NSFs, respectively. The dotted curves correspond to a line without filters.

Figure 7, which represent the pulse parameters (after each dispersion map) obtained by the numerical procedure of minimization of the residual field [14], illustrates the action of NSFs and USFs over a relatively short propagation distance, in a periodically amplified dispersion-managed (DM) fiber line with 36 dispersion maps in one amplification period of $z_A = 25\,km$, and the following parameters. GVD: $d_\pm = \pm 3.5\,ps/nm/km$. SPM: $\gamma_\pm = 0.002\,m^{-1}\,W^{-1}$. Losses: $\alpha_\pm = 0.22\,dB/km$. Here, the subscript $+(-)$ refers to normal (anomalous) fiber section. Lumped amplifiers, which are modelled in the GNLSE (3) by

$$A[\psi] \equiv (\sqrt{G} - 1)\psi \sum_{n}^{M} \delta(z - nz_A),\qquad(22)$$

are used to compensate exactly for the fiber losses. Here we denote the gain of each amplifier by $G$ and the total number of amplifiers by $M$. The input pulses correspond to stationary solutions of the DM line (fixed point) in absence of SRS and filters. The dashed curves in Figs. 7 (b) and (e), illustrate clearly that NSFs with moderate bandwidth (e.g., $N_f = 4.5$) have essentially no effect on the Raman-induced SSFS over relatively short propagation distances such as in the present case where the pulse propagate over $z_{max} = 6 z_A$. Quite in contrast, one can clearly observe [solid curves in Figs. 7 (b) and (e)] that USFs with parameters $N_f = 4.5$ and $\Delta\omega = 3.6\,GHz$, cause a strong reflection back of the pulse spectrum at each filtering site, leading to complete cancelling of the average SSFS over the distance $z_{max}$, and total cancelling of the associated temporal shift [as shown by the solid curve in Fig. 7 (b)]. Thus, for a given value of $N_f$, an appropriate choice of $\Delta\omega$ will permit total suppression of the average SSFS for any propagation distance $z_{max}$.

On the other hand, Fig. 8, which represent the pulse parameters after each amplifier, illustrates the action of NSFs and USFs over a relatively long propagation distance, in the same DM line as for the Fig. 7. Here we observe that in the absence of filters, the pulse undergoes continually a Raman-induced frequency downshift throughout the line, as illustrated by the dotted curves in Fig. 8 (a1) or (a2).

In the presence of NSFs the pulse dynamics exhibit two outstanding regimes [Fig. 8 (a1)]. The first is the *untrapped regime*, which starts as soon as the pulse begins to propagate, in which the pulse executes a continual frequency downshift. As the pulse propagates farther in this regime the action of NSFs progressively

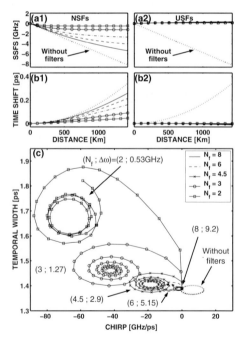

**Fig. 8.** Evolution of the pulse parameters, and demonstration of suppression of the SSFS and the associated time shift, over a relatively long propagation distance, in same DM line as in Fig. 7.

enters into play, by progressively reducing the sliding velocity of the pulse frequency. Then, the pulse ultimately enters the *trapped regime*, in which the sliding velocity falls to zero and remains there during the remaining dynamics. Thus, NSFs cause the soliton ultimately to be trapped near a frequency $\omega_1(N_f) < \omega_0$. This permanent frequency shift $\omega_1 - \omega_0$, which appears clearly in Fig. 8 (a1) for $N_f = 2$ (squares) and $N_f = 3$ (circles), can be reduced by a further decrease in the value of $N_f$, but it cannot be cancelled completely without destabilizing the pulse propagation.

Indeed, Fig. 8 (c), which is a phase diagram of the slow dynamics of the pulse in the phase plane $(x_4, x_3)$, shows that for narrow filters (i.e., small $N_f$) the slow dynamics moves toward increasingly broadened and chirped pulses that do not necessarily correspond to the desired dynamics. In contrast we observe the existence of a critical value of $N_f$, which we denote hereafter by $N_c$, for which the slow dynamics the input pulse parameters are essentially suppressed. In Fig. 8 (c) the critical value corresponds to approximately $N_f = 8$ ($B_f \approx 2.5\,THz$). Note that Fig. 8 (c) is the phase diagram for USFs, which we found to be essentially the same as for NSFs.

We can also clearly illustrate the effectiveness of USFs with the simulation of a bit pattern $\langle 011110101100100 \rangle$ propagation in a high-speed single-channel transmission line operating at 160 Gb/s, with a periodic dispersion management using two types of fiber (perfect cable), with GVD: $d_\pm = \pm 3.5\,ps/nm/km$, third-order dispersion: $0.06\,ps/nm^2/km$, SPM: $\gamma_\pm = 0.002$ m$^{-1}$W$^{-1}$, $\alpha_\pm = 0.22\,dB/km$, amplifier noise figure 4.5 dB, and an amplification period of $z_A = z_f = 25\,km$ that corresponds to 36 dispersion maps. The input pulses (fixed point) correspond to unchirped Gaussian pulses of duration 1.39 ps (FWHM) and energy 0.0526 pJ.

After solving the GNLSE (3), we have evaluated the transmission performance by means of the amplitude and timing $Q$-factors, $Q_A$ and $Q_T$, respectively. Figure 9, demonstrate the effectiveness of USFs in high-speed long-distance transmissions, with excellent performance of stability of the pulse bit pattern over several thousands of kilometers. The best performance is obtained for the following filter parameters: $N_f = 4.5$ ($B_f = 1.42THz$) and $\Delta\omega = 2.9GHz$, as Fig. 9 (e) shows. This figure demonstrate that the optimum filter's BW ($N_f = 4.5$) does not correspond necessarily to the critical filter's BW ($N_c \approx 8$). More importantly, Fig. 10 which shows the eye diagram at $7500\,km$, demonstrates that the disastrous impact of SRS can be effectively suppressed in high-speed long-distance transmission lines by use of appropriate USFs, without any compromise in the system performances.

## 4    Conclusion

In conclusion we have presented a generalized theory for SSFS arising from SRS, which is valid for any duration of pulse, within a spectral domain covering the entire third-telecommunication window. Using a CV approach, we have derived the analytical equations describing the generalized theory for SSFS. The main virtue

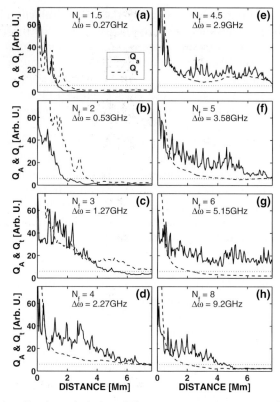

**Fig. 9.** Amplitude and timing $Q$ factors versus propagation distance.

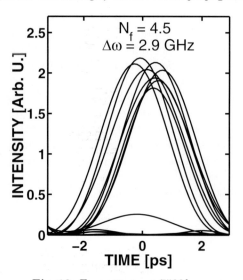

**Fig. 10.** Eye pattern at $7500km$.

of our analytical approach lies in its wide domain of validity compared with that corresponding to previous theories. Recent theoretical results on high-bit-rate optical fiber communication have proved the feasibility of pulse transmission at bit rates higher than 160 Gbit/s per channel [15,16]. For this kind of high speed transmission, one requires pulse with duration less than $1\,ps$. Hence, we believe that our general theory for SSFS can play a vital role in designing such high-speed optical communication systems. In particular, we have demonstrated the feasibility of using up-shifted filters for efficiently suppressing the SSFS. We have also compared the performance of the NSFs and USFs, which proved that an USF with weaker strength than its counterpart NSF, can totally suppress the SSFS while preserving the stability of the pulse propagation. Although here we have considered the effectiveness of up-shifting the central frequency of Gaussian filters, the same method can be utilized for other kind of filters like, super-Gaussian, Fabry-Perot or sliding filters. Hence for ultra-high speed optical communication, the advantages of USFs can be properly utilized for improving the performance of the transmission systems.

### Acknowledgments

The Centre National de la Recherche Scientifique, and the Ministère de l'Education Nationale de la Recherche et de la Technologie (contract ACI Jeunes No. 2015) are gratefully acknowledged for their financial support of this work. K. Nakkeeran wishes to thank the Centre National de la Recherche Scientifique (CNRS) for offering the Research Associate fellowship.

# References

1. F. M. Mitschke and L. F. Mollenauer: Opt. Lett. **11**, 659 (1986).
2. J. P. Gordon: Opt. Lett. **11**, 662 (1986).
3. N. Nishizawa and T. Goto: IEEE Photon. Technol. Lett.**11**, 325 1999.
4. N. Nishizawa, R. Okamura and T. Goto: IEEE Photon. Technol. Lett. **11**, 421 (1999).
5. J. K. Lucek and K. J. Blow: Electron. Lett. **27**, 882 (1991).
6. J. K. Lucek and K. J. Blow: Phys. Rev. A **45**,6666 (1992).
7. T. I. Lakoba, and D. J. Kaup: Opt. Lett. **24**,808 (1999).
8. R. Hellwarth: Prog. Quantum Electron. **5**, 1 (1977).
9. C. Lin: Opt. Comm. **4**, 2 (1983).
10. R. H. Stolen, J. P. Gordon, W. J. Tomlinson and H. A. Haus. J. Opt. Soc. Am. B **6**, 1159 (1989).
11. D. J. Dougherty F. X. Kärtner, H. A. Haus and E. Ippen: Opt. Lett. **20**, 31 (1995).
12. K. J. Blow, N. J. Doran and D. Wood: J. Opt. Soc. Am. B **5**, 1301 (1988).
13. P. Tchofo Dinda, A. B. Moubissi and K. Nakkeeran: newblock j. Phys. A **34**, (2001).
14. P. Tchofo Dinda, A. B. Moubissi and K. Nakkeeran: Phys. Rev. E **63**, 016608, (2001).
15. T. Hirooka, T. Nakada and A. Hasegawa: EEE Photonics Technol. Lett. **12**, 633 (2000).
16. L. J. Richardson, and W. Forysiak: EE Proc.-Optoelectron. **147**, 417 (2000).

# Quasi-linear Optical Pulses in Dispersion Managed Fibers: Propagation and Interaction

M.J. Ablowitz and T. Hirooka

Department of Applied Mathematics, University of Colorado at Boulder
Boulder, CO 80309-0526, USA

**Abstract.** An analytical description of the evolution and interaction of quasi-linear optical pulses in strongly dispersion managed transmission systems is presented. Long-scale pulse dynamics in dispersion-managed optical fibers is governed by a nonlinear nonlocal equation, referred to as the dispersion managed nonlinear Schrödinger (DMNLS) equation, which is obtained by employing an appropriate multiscale expansion of the perturbed nonlinear Schrödinger equation. The role of nonlinearity in quasi-linear pulse transmission is elucidated by asymptotic analysis of the DMNLS equation. In the framework of the DMNLS equation, quasi-linear pulse transmission, where nonlinearity is mitigated, can be viewed as degenerate limits of dispersion managed soliton transmission, where nonlinearity balances dispersion. The DMNLS equation also provides an analytical model that describes nonlinear intrachannel interactions between adjacent pulses in quasi-linear transmission, such as their energy exchange and frequency modulation.

## 1 Introduction

In recent years, considerable research has been devoted to the study of fiber-optic communication systems employing dispersion management for both soliton and non-soliton transmission. In a dispersion-managed system, the fiber is made up of alternating sections of positive and negative group-velocity dispersion (GVD) in such a manner as to create a transmission line with high local and low average dispersion.

In soliton transmission, dispersion management was first introduced to overcome the Gordon-Haus jitter, i.e., timing jitter of solitons arising from their interaction with amplifier noise [1], without degrading the signal-to-noise ratio (SNR) [2]. The motivation was to reduce the Gordon-Haus effect by compensating for dispersion accumulation periodically, since Gordon-Haus jitter is proportional to the accumulated dispersion. Later it was shown numerically that there exists a nonlinear localized solution in a dispersion-managed system, now referred to as a dispersion-managed (DM) soliton, which propagates with periodically changing its shape [3]. The DM soliton is found to have enhanced energy compared to the energy of the soliton in a fiber with constant dispersion equal to the average dispersion. This allows one to reduce the Gordon-Haus effect by lowering the average dispersion without sacrificing the degradation of SNR.

In non-soliton transmission, dispersion management is found to manage fiber nonlinearity and suppress certain nonlinear effects such as self-phase modulation [4,5] and inter-channel crosstalk in wavelength division multiplexed (WDM) systems [6–9]. In a strongly dispersion-managed transmission line, because of high local dispersion, the pulse width expands considerably and thus the peak power is suppressed locally, which in turn mitigates the nonlinearity. Such a system is hence commonly referred to as quasi-linear. In contrast to DM solitons, in quasi-linear system the effective nonlinearity is mitigated for large map strength, and the spectral intensity is found to be an invariant of the propagation in the lossless case [4].

We note the difference between DM soliton transmission where nonlinearity balances dispersion, as compared to quasi-linear transmission in which nonlinearity is managed. Nevertheless, independent of transmission format, the pulse dynamics in dispersion-managed optical fibers is described by the perturbed nonlinear Schrödinger (NLS) equation, where the dispersion coefficient is now varying rapidly as a periodic function of distance. By introducing multiple scales and separating fast scale dynamics due to the large and periodic perturbation from the perturbed NLS equation, one obtains an averaged evolution equation which governs the pulse dynamics over a slow scale characterized by the nonlinear length [10]. The obtained model, referred to as the dispersion-managed NLS (DMNLS) equation, elucidates the role of nonlinearity in dispersion-managed transmission independent of transmission format, and thus provides a unified analytical description of DM soliton and quasi-linear transmission.

Dispersion management considerably modifies the dynamics of pulse evolution because of the periodic variation of GVD. The alternating GVD sign within one period brings about large pulse width oscillation. This leads to strong overlap of neighboring bits in a pulse train, resulting in significant nonlinear interactions [11,12]. Dispersion management also alters the sign of the velocity of the pulse within each map period, and thus the trajectory of the central pulse position draws large zigzags. This zigzag motion leads to repeated collisions among pulses in different wavelength channels in WDM transmission [13,14].

In DM soliton transmission, WDM interactions due to the repeated collisions is found to be a major transmission penalty [15,16]. They are responsible for timing jitter and may even result in collapse. On the other hand, in quasi-linear transmission, because of their strong overlap, residual nonlinearity induces intra-channel crosstalk between adjacent pulses, which lead to serious transmission penalties such as timing and amplitude jitter of the main signals (the 1's) [17–20] and ghost pulse generation at 0 bits [21].

In this article, we present an analytical framework that describes the propagation and interaction of quasi-linear optical pulses in strongly dispersion-managed transmission systems. Preliminary remarks on the normalizations and units of the perturbed NLS equation are summarized in sec.2. We introduce multiple scale analysis and derive the DMNLS equation from the perturbed NLS equation in sec.3. In sec.4 we investigate the pulse evolution and elucidate the role of nonlinearity in quasi-linear transmission using the DMNLS equation. In sec.5, we develop an analytical model to study intra-channel pulse interactions in quasi-linear transmission based on the DMNLS equation.

## 2    Perturbed Nonlinear Schrödinger Equation

Propagation of optical pulses in dispersion-managed fibers in the presence of loss and amplification is described by the perturbed NLS equation

$$i\frac{\partial u}{\partial z} + \frac{D(z)}{2}\frac{\partial^2 u}{\partial t^2} + g(z)|u|^2 u = 0, \tag{1}$$

where all the quantities are expressed in dimensionless units: $t = t_{\text{ret}}/t_*$, $z = z_{\text{lab}}/z_*$, $u = E/\sqrt{gP_*}$, $D = k''/k''_*$ with the characteristic parameters denoted by the subscript $*$, where $t_{\text{ret}}$ and $z_{\text{lab}}$ are the retarded time and the propagation distance, respectively, and $E$ denotes the slowly varying envelope of the optical field. The normalizing variables are determined so that $z_* = z_{\text{NL}} \equiv 1/\nu P_*$ and $k''_* = -t_*^2/z_{\text{NL}}$ where $\nu$ is the nonlinear coefficient. The functions $D(z)$ and $g(z)$ describe the local GVD of the fiber and the variation of power due to loss and amplification, respectively, which are both periodic in $z$ with period $z_a$. The nonlinear coefficient $g(z)$ for lumped amplification based on EDFA's is given by

$$g(z) = g_e \exp[-2\Gamma(z - nz_a)], \quad nz_a \le z < (n+1)z_a, \tag{2}$$

where $g_e = 4G/[1 - \exp(-4G)]$ so that $\langle g \rangle = 1$ ($\langle \cdot \rangle$ denotes the path-average over $z_a$), $\Gamma$ is the dimensionless loss coefficient, and $G = \Gamma z_a/2$. We write the accumulated dispersion in the form

$$\bar{D}(z) = \int_0^z D(z')dz' = C(z) + \langle D \rangle z, \tag{3}$$

where $\langle D \rangle$ is the average dispersion over a period and $C(z)$ is a periodic function with period $z_a$ having zero average $\langle C \rangle = 0$.

## 3    Dispersion Managed Nonlinear Schrödinger Equation

In order to model strong dispersion management, we decompose the GVD $D(z)$ into two parts: a path-average constant $\langle D \rangle$ and the rapidly varying function $\Delta$ corresponding to local GVD:

$$D(z) = \langle D \rangle + \frac{1}{z_a}\Delta(z/z_a), \tag{4}$$

where $z_a (\ll 1)$ is the map period. Note that $\Delta/z_a$ represents a large variation about the average due to strong dispersion management and thus the proportionality factor $1/z_a$ is required in front of $\Delta(z/z_a)$ so that both $\langle D \rangle$ and $\Delta$ are quantities of order one. Since the perturbed NLS equation with $D(z)$ given by Eq.(4) contains both slowly and rapidly varying terms, it is convenient to introduce the fast and slow scales as $\zeta = z/z_a$ and $z$ respectively. We also expand the field $u$ in powers of $z_a$:

$$u(\zeta, z, t) = u^{(0)}(\zeta, z, t) + z_a u^{(1)}(\zeta, z, t) + \cdots. \tag{5}$$

The perturbed NLS equation is now broken into a series of equations corresponding to the different powers of $z_a$. At the leading order in the expansion $O(1/z_a)$, we have

$$i\frac{\partial u^{(0)}}{\partial \zeta} + \frac{\Delta(\zeta)}{2}\frac{\partial^2 u^{(0)}}{\partial t^2} = 0, \tag{6}$$

namely the evolution of the pulse is determined solely by the large variations of $D(z)$ about the average, and nonlinearity and residual dispersion represent only a small perturbation to the linear solution. Eq.(6) can be solved by the Fourier transform

$$\hat{u}^{(0)}(\zeta, z, \omega) = \mathcal{F}\left[u^{(0)}\right] = \int_{-\infty}^{\infty} u^{(0)}(\zeta, z, t)\exp(-i\omega t)dt,$$

and the solution is given by

$$\hat{u}^{(0)}(\zeta, z, \omega) = \hat{U}(z, \omega)\exp[-iC(\zeta)\omega^2/2], \quad C(\zeta) = \int_0^\zeta \Delta(\zeta')d\zeta', \tag{7}$$

where $\hat{U}(z, \omega)$ is the integration constant in terms of $\zeta$ and represents the slowly evolving amplitude of $\hat{u}^{(0)}$, whose exact form is determined from the higher order in the expansion.

At the next order in the expansion $O(1)$, we have (in the frequency domain)

$$i\frac{\partial \hat{u}^{(1)}}{\partial \zeta} - \frac{\Delta(\zeta)}{2}\omega^2\hat{u}^{(1)} = -\hat{P}^{(1)}, \quad \hat{P}^{(1)} = i\frac{\partial \hat{u}^{(0)}}{\partial z} - \frac{\langle D\rangle}{2}\omega^2\hat{u}^{(0)} + g(z)\mathcal{F}\left[|u^{(0)}|^2 u^{(0)}\right], \tag{8}$$

which is written in a more convenient form as

$$i\frac{\partial}{\partial \zeta}\left[\hat{u}^{(1)}\exp(iC\omega^2/2)\right] = -\hat{P}^{(1)}\exp(iC\omega^2/2). \tag{9}$$

In order to remove secularities, namely to avoid resonant growth of $u^{(1)}$ so that the expansion of $u$ in powers of $z_a$ in Eq.(5) is remained to be well ordered, we require the following condition:

$$\int_0^1 \hat{P}^{(1)}\exp(iC\omega^2/2)d\zeta = 0. \tag{10}$$

This condition yields the following equation for $\hat{U}$:

$$i\frac{\partial \hat{U}}{\partial z} - \frac{\langle D\rangle}{2}\omega^2\hat{U} + \int_{-\infty}^{\infty}\int_{-\infty}^{\infty} r(\omega_1\omega_2)\hat{U}(z, \omega + \omega_1)\hat{U}(z, \omega + \omega_2)$$
$$\times \hat{U}^*(z, \omega + \omega_1 + \omega_2)d\omega_1 d\omega_2 = 0, \tag{11}$$

where the kernel

$$r(x) = \frac{1}{(2\pi)^2}\int_0^1 g(\zeta)\exp(iC(\zeta)x)d\zeta \tag{12}$$

represents the structure of GVD profile.

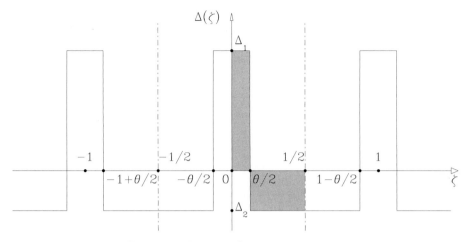

**Fig. 1.** Schematic diagram of a two-step dispersion map.

Let us consider two special cases. First, when there is no loss and amplification $(g(z) = 1)$, the kernel $r(x)$ of a piecewise constant dispersion map (see Fig. 1) is given by $r(x) = (1/(2\pi)^2) \sin sx/sx$, where $s = [\theta\Delta_1 - (1-\theta)\Delta_2]/4$ is a measure of dispersion map strength. Secondly, if the dispersion has no periodic variations $(\Delta = 0$ and thus $s = 0)$, then $r(x) = 1/(2\pi)^2$ and the DMNLS equation reduces to the standard NLS equation, i.e., Eq.(1) with $D(z) = g(z) = 1$.

Special stationary solutions of the DMNLS equation are obtained by looking for solutions of the form $\hat{U}(z, \omega) = F(\omega)\exp(i\lambda^2 z/2)$ which yields the following nonlinear integral equation for $F(\omega)$:

$$(\lambda^2 + \langle D\rangle\omega^2)F(\omega) = 2\int_{-\infty}^{\infty}\int_{-\infty}^{\infty} r(\omega_1\omega_2)F(\omega+\omega_1)F(\omega+\omega_2)F^*(\omega+\omega_1+\omega_2)d\omega_1 d\omega_2.$$

(13)

A rapidly convergent procedure to solve this numerically is described in [22]. The obtained solution is the stationary DM soliton pulses. Some typical pulses are depicted in Fig. 2. We note that Eq.(13) contains a parameter $\lambda$ which characterizes the mode, corresponding to the energy. The main features of the DM solitons for large $s$ are a Gaussian like center with exponentially decaying and oscillating tails. For small $s$, the mode approaches the classical soliton profile.

The fast scale evolution (i.e., with respect to $\zeta$) can be reconstructed by multiplying $\exp(-iC\omega^2/2)$ with the slowly varying Fourier amplitude $\hat{U}$. When observed in this scale the pulse shape is changing periodically because of the local GVD variation, whereas the pulse propagates without changing its shape on a slow scale which is characterized by nonlinearity and the average dispersion. Thus DM soliton is considered to be a stationary pulse where nonlinearity balances the average dispersion.

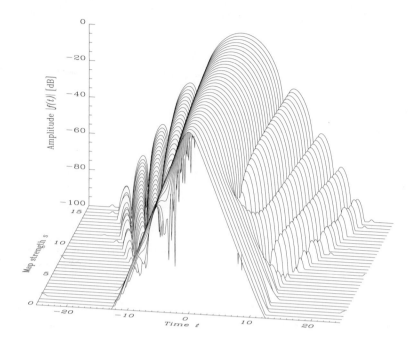

**Fig. 2.** The shape of the stationary pulses for $\langle D \rangle = 1$, $\lambda = 1$ and various values of $s$.

## 4    Quasi-linear Pulse Propagation

The evolution of pulses in the quasi-linear regime ($s \gg 1$) can also be studied based on the DMNLS equation [4,5]. To analyze the behavior of quasi-linear pulses, we assume that $\hat{U}(z, \omega)$ depends only weakly on $s$ and compute an asymptotic expansion of the nonlinear nonlocal term in Eq.(11) for $s \gg 1$.

### 4.1    Lossless Case

When $g(z) = 1$ (i.e., in a lossless system), we have

$$i\frac{\partial \hat{U}}{\partial z} - \frac{\langle D \rangle}{2}\omega^2 \hat{U} + \Psi\left[|\hat{U}|^2\right]\hat{U} = 0, \tag{14}$$

$$\Psi\left[|\hat{U}(z,\omega)|^2\right] = \frac{1}{2\pi s}\left[(\log s - \gamma)|\hat{U}(z,\omega)|^2 - \int_{-\infty}^{\infty} H(\omega' - \omega)|\hat{U}(z,\omega')|^2 d\omega'\right], \tag{15}$$

where $\gamma = 0.57722$ is Euler's constant and $H(\omega) = (1/\pi)\int_{-\infty}^{\infty} \log|t| \exp(-i\omega t)dt$. Eq.(14) can be solved explicitly as

$$\hat{U}(z,\omega) = \hat{U}(0,\omega)\exp\left\{-i\langle D \rangle \omega^2 z/2 + i\Psi\left[|\hat{U}(0,\omega)|^2\right]z\right\}. \tag{16}$$

From these results, we note the following important observations in terms of pulse dynamics in the quasi-linear regime. First, nonlinearity is mitigated by $O(\log s/s)$ and vanishes in the limit $s \to \infty$. In other words, the quasi-linear system is in the regime where nonlinearity is managed. Secondly, nonlinearity is responsible only for phase shift $\phi_{\mathrm{NL}}(z,\omega) = \Psi[|\hat{U}(0,\omega)|^2]z$ in frequency domain, and thus the spectral intensity $|\hat{U}(z,\omega)|$ is preserved during propagation, as opposed to the self-phase modulation in standard fibers. Finally we also note that the quasi-linear pulse can be viewed as degenerate limit of a DM soliton. The DM soliton satisfies the DMNLS equation with $\hat{U}(z,\omega) = F(\omega)\exp(i\lambda^2 z/2)$, where $F(\omega)$ is the solution of Eq.(14) and can be approximated by a Gaussian $F(\omega) \sim \alpha(\lambda)\exp(-\beta(\lambda)\omega^2/2)$. A quasi-linear Gaussian pulse has the same structure but is independent of the "eigenvalue" $\lambda$. The quasi-linear pulse can thus be looked at as a limit $\lambda \to 0$ in DM solitons. Importantly, both types of transmission format can be described with the same analytical framework.

To test our model, we compared the analytical results with direct numerical simulations of Eq.(1) with $\langle D \rangle = 0$ and $z_a = 0.1$. The incident pulse is given by a Gaussian $u(0,t) = (1/\sqrt{\beta})\exp(-t^2/\beta)$. With the choice of $t_* = 2.2$ ps, $\nu = 2.2$ $\mathrm{W}^{-1}\mathrm{km}^{-1}$, and $P_* = 1$ mW (i.e. $z_{\mathrm{NL}} = 450$ km and $k_*'' = -0.01\ \mathrm{ps}^2/\mathrm{km}$), $\beta = 1$ corresponds to the peak power of 1 mW and FWHM of 3.7 ps (at minimum). The fiber loss is 0.2 dB/km and the amplifier period is 45 km. For this initial profile, the nonlinear phase shift at $\omega = 0$ is given by $\phi_{\mathrm{NL}}(z,0) = (1/s)\log(s/\beta)z$.

In Fig. 3, we plot the shape of the quasi-linear pulse when $s = 100$ after a propagation of $z = 20$ (9000 km), as well as the initial profile with $\beta = 1$. As predicted from the model, the spectral intensity is an invariant of the propagation, whereas small pulse broadening induced by the nonlinearity is observed in the time domain as a result of the acquired nonlinear chirp $\partial\phi_{\mathrm{NL}}/\partial\omega$ in the frequency domain (which is similar to the self-phase modulation in the time domain).

Figure 4 shows a plot of the phase shift $\phi_{\mathrm{NL}}(z,\omega = 0)$ at $z = 1$ as a function of $s$ for $\beta = 0.5$ and 1. The nonlinear phase shift is indeed decreasing for large $s$ as $O(\log s/s)$, which further confirms the validity of the asymptotic expansion of Eq.(14) and (15). In this figure, we also plot $\phi_{\mathrm{NL}}(\omega = 0)$ in the presence of loss ($G = 0.5$), which also follows similar lines as the lossless case. Detailed asymptotic analysis in the system with loss and amplification is presented in the next section.

## 4.2   Lossy Case

When $g(z) \neq 1$ (i.e., in the presence of loss and amplification), the kernel $r(x)$ depends not only on $s$ but also on the relative location of the amplifier within one dispersion map period, and the asymptotic analysis must be modified accordingly. In all cases, the nonlinearity is again mitigated by $O(\log s/s)$. However, the spectral characteristics of the pulse evolution depends strongly on the amplifier locations. We define $\zeta_a$ to represent the position of the amplifier within the dispersion map $(-1/2 \leq \zeta < 1/2)$. For instance, $\zeta_a = 0$ means that the amplifier is located at the midpoint of the anomalous GVD segment, and $\zeta_a = 1/4$

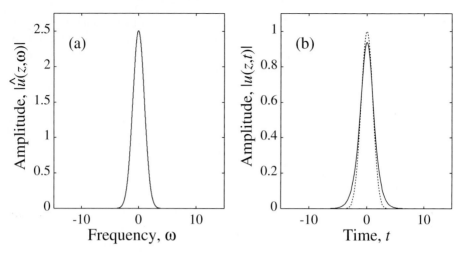

**Fig. 3.** Shape of the quasi-linear Gaussian pulse after a propagation of $z = 20$ (solid curves) and the initial profile (dotted curves) for $s = 100$, $\langle D \rangle = 0$, and $G = 0$: (a) frequency domain, (b) time domain. In (a), the dotted curve is indistinguishable from the solid curve.

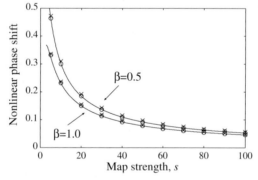

**Fig. 4.** Nonlinear phase shift $\phi_{\mathrm{NL}}(\omega = 0)$ of the quasi-linear Gaussian pulse with $\beta = 0.5, 1$. The solid curve is the analytical result (see text), and the circles and crosses are the numerical results obtained from the direct simulation of Eq.(1) with $G = 0$ and $G = 0.5$, respectively.

corresponds to the boundary between normal and anomalous GVD segment (see also Fig. 1).

When the amplifiers are placed at the locations where the sign of GVD changes (i.e., $\zeta_a = \pm 1/4$), Eq.(11) is reduced to Eq.(14) but with

$$\Psi\left[|\hat{U}(z,\omega)|^2\right] = \frac{1}{s}\left[(L_0 \log s - L_1)|\hat{U}(z,\omega)|^2 - L_0 \int_{-\infty}^{\infty} H(\omega' - \omega)|\hat{U}(z,\omega')|^2 d\omega'\right],$$

$$(17)$$

and the constants $L_0$ and $L_1$ defined as

$$L_0 = \frac{G}{2\pi} \exp(-2G)\operatorname{csch}G, \tag{18a}$$

$$L_1 = \frac{G}{4\pi} \exp(-G)\operatorname{csch}G[\exp(G)I_{G1} + \exp(-G)(3\gamma + \log G - I_{G2})]$$
$$- \frac{G}{2\pi} \exp(G)I_{G1}, \tag{18b}$$

$I_{G1} = \int_G^\infty dx \exp(-x)/x$, and $I_{G2} = \int_0^G dx(\exp(x) - 1)/x$. Neglecting $O(1/s^2)$ terms, the spectral intensity $|\hat{U}(z,\omega)|^2$ is still preserved during pulse propagation, and the solution is Eq.(16) with $\Psi[|\hat{U}|^2]$ in Eq.(17). After the linear phase shift $\exp(-i\langle D\rangle\omega^2 z/2)$ is removed by means of pre- or post-transmission compensation, the averaged dynamics of the quasi-linear pulse transmission is characterized only by the nonlinear phase shift $\phi_{\mathrm{NL}}(z,\omega) = \Psi[|\hat{U}(0,\omega)|^2]z$ as in the lossless limit $G \to 0$ ($L_0 \to 1/2\pi$ and $L_1 \to \gamma/2\pi$).

When the amplifiers are placed in the middle of the normal or anomalous GVD segment ($\zeta_a = -1/4 \pm 1/4$, i.e., $\zeta_a = 0$ and $-1/2$), the spectral intensity varies as $O(1/s)$ and spectral compression or broadening is observed respectively. In this case Eq.(11) is written as

$$i\frac{\partial\hat{U}}{\partial z} - \frac{\langle D\rangle}{2}\omega^2\hat{U} + \Psi\left[|\hat{U}|^2\right]\hat{U} \mp i\frac{K_2}{s}\hat{J}(z,\omega) = 0, \tag{19}$$

where $\Psi[|\hat{U}|^2]$ is now given by

$$\Psi\left[|\hat{U}(z,\omega)|^2\right] = \frac{1}{s}\left[(K_0\log s - K_1)|\hat{U}(z,\omega)|^2 - K_0\int_{-\infty}^\infty H(\omega'-\omega)|\hat{U}(z,\omega')|^2 d\omega'\right], \tag{20}$$

with the constants $K_0$, $K_1$, and $K_2$ defined as

$$K_0 = \frac{G}{2\pi}(\exp(-G)\operatorname{csch}G + 1), \tag{21a}$$

$$K_1 = \frac{G}{4\pi}\operatorname{csch}G[\exp(G)I_{G1} + \exp(-G)(3\gamma + \log G - I_{G2})] + \frac{G}{2\pi}(2\gamma + \log G), \tag{21b}$$

$$K_2 = \frac{G}{4}, \tag{21c}$$

and

$$\hat{J}(z,\omega) = \int_{-\infty}^\infty\int_{-\infty}^\infty \frac{d\omega_1 d\omega_2}{\omega_1\omega_2}\hat{U}(z,\omega+\omega_1)\hat{U}(z,\omega+\omega_2)\hat{U}^*(z,\omega+\omega_1+\omega_2). \tag{22}$$

In the limit of $G \to 0$, Eq.(19) reduces to Eq.(14) with Eq.(15) ($K_0 \to 1/2\pi$, $K_1 \to \gamma/2\pi$, and $K_2 \to 0$). From Eq.(19), the evolution of the intensity is described as

$$\frac{\partial|\hat{U}|^2}{\partial z} = \pm\frac{K_2}{s}f(z,\omega), \tag{23}$$

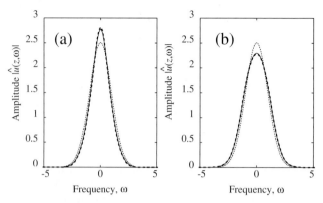

**Fig. 5.** The spectrum of the quasi-linear Gaussian pulse for $s = 50$: (a) $\zeta_a = 0$; (b) $\zeta_a = -1/2$. The solid curve is the pulse at $z = 20$ (i.e., $z_{\text{lab}} = 9000$ km) obtained from the direct simulation, and the dashed curve shows the analytical result. The dotted curve is the initial profile.

where $f(z,\omega) = \hat{J}(z,\omega)\hat{U}^*(z,\omega) + \hat{J}^*(z,\omega)\hat{U}(z,\omega)$. Since $s$ is large, the solution of Eq.(23) can be approximated by

$$|\hat{U}(z,\omega)|^2 \approx |\hat{U}(0,\omega)|^2 \pm \frac{K_2 z}{s} f(0,\omega) \tag{24}$$

for moderate values of $z$.

In Fig. 5 we show a comparison of the spectral profile of a quasi-linear Gaussian pulse with $\beta = 1$ when $s = 50$, $\langle D \rangle = 0$, $z_a = 0.1$, and $G = 0.5$ (namely $\Gamma = 10$), after a propagation of $z = 20$ as well as the initial profile, with (a) $\zeta_a = 0$ (when amplifiers are located in the middle of anomalous GVD segments) and (b) $-1/2$ (in the middle of normal GVD segments). We see that the nonlinearity yields spectral compression and broadening when $\zeta_a = 0$ and $-1/2$ respectively. Remarkable agreement between the analysis and numerical results is obtained, which further confirms the validity of our model. Since the spectral evolution is a consequence of the nonlinearity, spectral reshaping may be observed by increasing the launch power. Note, however, that the nonlinearity is mitigated for large $s$, by $O(\log s/s)$ and the spectral reshaping effect is $O(1/s)$.

We also note that when $\zeta_a = 0$ or $-1/2$, since the spectral reshaping is not associated with phase modulation, the observed spectral compression/broadening accompanies pulse broadening/compression respectively in the time domain. This implies the possibility of transform-limited temporal or spectral compression of laser pulses. The analysis presented here is consistent with the recent experimental and numerical observations of spectral compression of RZ pulses in strongly dispersion-managed lines [23–25].

Fig. 6 shows the evolution of the spectral peak amplitude $|\hat{u}(z,\omega = 0)|$ for $s = 50$, with $\zeta_a = 0, -1/2, -1/4$ (when amplifiers are positioned at the boundary between anomalous and normal GVD segments), and $1/4$ (at the boundary between normal and anomalous GVD segments), obtained from di-

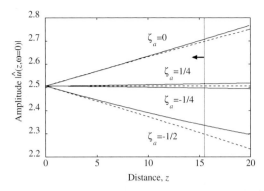

**Fig. 6.** The evolution of the spectral peak amplitude $|\hat{u}(z, \omega = 0)|$ for several values of $\zeta_a$. Solid lines are the results obtained from the direct simulation, and the dashed lines show the analytical approximation obtained from Eq.(24). Dotted line shows a reasonable upper limit of the validity of Eq.(24).

rect numerical simulation and the analytical approximation (24). Since energy is conserved during propagation, the increase and decrease of the peak amplitude when $\zeta_a = 0$ and $-1/2$ demonstrates the spectral compression and broadening respectively. On the other hand, with $\zeta_a = -1/4$ and $1/4$ the spectrum is still conserved, namely nonlinearity is responsible only for a phase shift. The deviation of the analytical approximation from the numerics for large $z$ is a consequence of the growth of the term $(K_2 z/s) f(0, \omega)$ in Eq.(24). To be valid, $(K_2 z/s) f(0, \omega)/|\hat{U}(0,0)|^2$ should be small. We also plot a reasonable upper limit for the validity of the asymptotic [namely $z < (K_2/s) f(0, \omega) = 15.5$ (7000 km)].

## 5    Intra-channel Quasi-linear Pulse Interactions

We have seen in the previous section that nonlinearity is mitigated by the factor $O(\log s/s)$ in the quasi-linear regime. This suggests the employment of strong dispersion management with large map strength $s$.

As $s$ increases, however, different forms of nonlinear interactions take place and result in the main source of signal deformation. In a strongly dispersion managed system, quasi-linear pulses in neighboring bit slots interact with each other because of their large overlap which is as a result of large pulse width broadening associated with high local GVD. This overlap induces nonlinear mixing (crosstalk) between pulses such as their frequency modulation and energy exchange, and leads to the fluctuation of the temporal position and amplitude of the pulses in "1"bits and the generation of "ghost" pulses in "0" bits [19–21]. The timing and amplitude jitter and the ghost pulse growth impose a major limitation to system performance in the quasi-linear regime with large $s$.

In order to study nonlinear intra-channel interactions between the main signal $u_0$ and the adjacent pulses $u_l$, $u_m$, $u_n$ ($u_k$ represents the signal pulse centered at $t = kT$, where $T$ is the bit interval and $k$ is the integer representing the location of the bit slot), we write $u = u_0 + u_l + u_m + u_n$ and substitute this into Eq.(1) to find the evolution of $u_0$ perturbed by one of the nonlinear terms $u_l^* u_m u_n$:

$$i\frac{\partial u_0}{\partial z} + \frac{D(z)}{2}\frac{\partial^2 u_0}{\partial t^2} + g(z)|u_0|^2 u_0 = -g(z)u_l^* u_m u_n. \tag{25}$$

The perturbation terms on the right hand side of Eq.(25) takes two forms, which are either phase-dependent $u_l^* u_m u_n$ ($l \neq m$, $: n$; $m$, $: n \neq 0$) or phase-independent $|u_n|^2 u_0$ ($n \neq 0$). Phase-dependent forcing terms $u_l^* u_m u_n$ ($l \neq m$, $:$ $n$; $m$, $: n \neq 0$) yield ghost pulse generation when the bit is zero and amplitude fluctuation of $u_0$ when the bit slot is occupied by a signal, as a result of nonlinear interactions with nonzero bits $u_l$, $u_m$ and $u_n$. They are commonly referred to as intra-channel FWM. Phase-independent terms $|u_n|^2 u_0$ ($n \neq 0$) bring about a timing shift of $u_0$ due to interaction with another nonzero bit $u_n$. It should be noted that the phase-independent terms do not contribute to the ghost pulse growth or energy exchange, since they maintain energy. Also, the contribution of the phase-dependent terms to the frequency and timing shifts is negligible [26]. Note that the integers $l$, $m$ and $n$ must satisfy the phase-matching condition $l = m + n$ [21,26].

## 5.1   Perturbed DMNLS Equation

In this section, using the multiple scale method, we develop an analytical framework to study nonlinear intra-channel interactions. The key idea behind multiple scale approach is the same as in sec.3, i.e., to introduce fast and slow scales and thus eliminate fast pulse dynamics due to large and periodically varying dispersion. The obtained equation is the DMNLS equation perturbed by interaction terms. The leading order multiple scale approximation is found to agree with the analytical model presented in [19,20]

We start the analysis with the perturbed NLS equation of the form

$$i\frac{\partial u_0}{\partial z} + \frac{D(z)}{2}\frac{\partial^2 u_0}{\partial t^2} = -g(z)|u_0|^2 u_0 - 2g(z)|u_j|^2 u_0 - g(z)u_l^* u_m u_n, \tag{26}$$

where the second and third term of the right hand side is intra-channel XPM and FWM term respectively, where $j \neq 0$ and $l = m + n$ ($m, n \neq 0$) from the phase matching condition.

We carry out the same multiple scale expansion as in sec.3. We expand the field $u_k$ ($k = 0, l, m, n$) in powers of $z_a$

$$u_k(\zeta, z, t) = u_k^{(0)}(\zeta, z, t) + z_a u_k^{(1)}(\zeta, z, t) + \cdots, \tag{27}$$

to find the leading order solution at $O(1/z_a)$:

$$\hat{u}_k^{(0)}(\zeta, z, \omega) = \hat{U}_k(z, \omega)\exp[-iC(\zeta)\omega^2/2] \tag{28}$$

(note $\hat{U}_k(z, \omega) = \hat{U}_0(z, \omega)\exp(-i\omega kT)$). At $O(1)$ in the expansion, we have (in the frequency domain)

$$i\frac{\partial \hat{u}_0^{(1)}}{\partial \zeta} - \frac{\Delta(\zeta)}{2}\omega^2 \hat{u}_0^{(1)} = -\hat{P}^{(1)},$$

$$\hat{P}^{(1)} = i\frac{\partial \hat{u}_0^{(0)}}{\partial z} - \frac{\langle D \rangle}{2}\omega^2 \hat{u}_0^{(0)} + g(z)\mathcal{F}\left[|u_0^{(0)}|^2 u_0^{(0)} + 2|u_j^{(0)}|^2 u_0^{(0)} + u_l^{(0)*} u_m^{(0)} u_n^{(0)}\right].$$

(29)

The nonsecular condition (10) yields the following equation for $\hat{U}_0$:

$$i\frac{\partial \hat{U}_0}{\partial z} - \frac{\langle D \rangle}{2}\omega^2 \hat{U}_0 = -\left\langle g(\zeta)\exp(iC\omega^2/2)\left(\mathcal{F}\left[|u_0^{(0)}|^2 u_0^{(0)}\right] + \mathcal{F}\left[2|u_j^{(0)}|^2 u_0^{(0)}\right]\right.\right.$$
$$\left.\left. +\mathcal{F}\left[u_l^{(0)*} u_m^{(0)} u_n^{(0)}\right]\right)\right\rangle$$
$$\equiv -(\hat{\mathcal{I}}_{\text{SPM}} + \hat{\mathcal{I}}_{\text{XPM}} + \hat{\mathcal{I}}_{\text{FWM}}),$$

(30)

where

$$\hat{\mathcal{I}}_{\text{SPM}} = \int_{-\infty}^{\infty}\int_{-\infty}^{\infty} r(\omega_1\omega_2)\hat{U}_0(z,\omega+\omega_1)\hat{U}_0(z,\omega+\omega_2)\hat{U}_0^*(z,\omega+\omega_1+\omega_2)$$
$$\times d\omega_1 d\omega_2,$$

(31a)

$$\hat{\mathcal{I}}_{\text{XPM}} = 2\int_{-\infty}^{\infty}\int_{-\infty}^{\infty} r(\omega_1\omega_2)\hat{U}_0(z,\omega+\omega_1)\hat{U}_j(z,\omega+\omega_2)\hat{U}_j^*(z,\omega+\omega_1+\omega_2)$$
$$\times d\omega_1 d\omega_2,$$

(31b)

$$\hat{\mathcal{I}}_{\text{FWM}} = \int_{-\infty}^{\infty}\int_{-\infty}^{\infty} r(\omega_1\omega_2)\hat{U}_m(z,\omega+\omega_1)\hat{U}_n(z,\omega+\omega_2)\hat{U}_l^*(z,\omega+\omega_1+\omega_2)$$
$$\times d\omega_1 d\omega_2,$$

(31c)

If we keep only the term $\hat{\mathcal{I}}_{\text{SPM}}$ in Eq.(30), this is reduced to the DMNLS equation obtained previously in Eq.(11).

## 5.2   Energy Transfer

We now apply Eq.(30) to study nonlinear intra-channel interactions. Since the energy of $u_0$ is computed by $W_0 = (1/2\pi)\int_{-\infty}^{\infty}|\hat{U}_0(z,\omega)|^2 d\omega$ in the frequency domain, the energy change is obtained from

$$\frac{dW_0}{dz} = \frac{1}{2\pi}\int_{-\infty}^{\infty}\left(\hat{U}_0^*\frac{\partial \hat{U}_0}{\partial z} + \hat{U}_0\frac{\partial \hat{U}_0^*}{\partial z}\right)d\omega = \frac{i}{2\pi}\int_{-\infty}^{\infty}\left(\hat{U}_0^*\hat{\mathcal{I}}_{\text{FWM}} - \hat{U}_0\hat{\mathcal{I}}_{\text{FWM}}^*\right)d\omega.$$

(32)

By using Eq.(12) and Eq.(28) and employing the inverse Fourier transform, we find

$$\frac{dW_0}{dz} = \frac{i}{(2\pi)^3}\int_0^1 d\zeta g(\zeta)\int_{-\infty}^{\infty} d\omega_1 d\omega_2 d\omega\,[\hat{u}_0^*(\zeta,z,\omega)\hat{u}_m(\zeta,z,\omega+\omega_1)$$
$$\times \hat{u}_n(\zeta,z,\omega+\omega_2)\hat{u}_l^*(\zeta,z,\omega+\omega_1+\omega_2) - \text{c.c.}]$$
$$= \frac{i}{(2\pi)^3}\int_0^1 d\zeta g(\zeta)\int_{-\infty}^{\infty} dt_1 dt_2 dt_3 dt_4 d\omega_1 d\omega_2 d\omega\,\{u_0^*(\zeta,z,t_1)u_m(\zeta,z,t_2)$$
$$\times u_n(\zeta,z,t_3)u_l^*(\zeta,z,t_4)\exp[-i\omega(-t_1+t_2+t_3-t_4)]\exp[-i\omega_1(t_2-t_4)]$$
$$\times \exp[-i\omega_2(t_3-t_4)] - \text{c.c.}\}$$
$$= i\int_0^1 d\zeta g(\zeta)\int_{-\infty}^{\infty} dt\,[u_0^*(\zeta,z,t)u_m(\zeta,z,t)u_n(\zeta,z,t)u_l^*(\zeta,z,t) - \text{c.c.}],$$

(33)

where we used the identity $\int_{-\infty}^{\infty} d\omega \exp(i\omega t) = 2\pi\delta(t)$, and $t = t_1$. Energy change $\Delta W_0(z) = W_0(z) - W_0(0)$ is computed by integration of Eq.(33).

When $\langle D \rangle = 0$, $\hat{u}_k$ is independent of $z$, so that the right-hand side of Eq.(33) is constant in terms of $z$, and thus $\Delta W_0$ grows linearly with respect to distance

$$\Delta W_0(z) = \bar{Q}_{m,n}z,$$

$$\bar{Q}_{m,n} = i \int_0^1 d\zeta g(\zeta) \int_{-\infty}^{\infty} dt \, [u_0^*(\zeta,t)u_m(\zeta,t)u_n(\zeta,t)u_l^*(\zeta,t) - \text{c.c.}] \quad (34)$$

representing the growth rate. Thus we have a resonant growth situation.

So far no particular restrictions on the form of the pulse shape have been imposed in the analysis. In order to obtain explicit formulae to compute the energy exchange, in the following we assume that a signal is given by a Gaussian pulse

$$u_k(\zeta,t) = \frac{\alpha}{\sqrt{2\pi\xi(\zeta)}} \exp\left[-\frac{(t - kT)^2}{2\xi(\zeta)}\right], \quad \xi(\zeta) = \beta + iC(\zeta). \quad (35)$$

Substituting Eq.(35) to Eq.(34), we find

$$\bar{Q}_{m,n} = \frac{i\alpha^4}{4\pi^2}\sqrt{\frac{\pi}{2\beta}} \int_0^1 \left[\frac{g(\zeta)}{|\xi(\zeta)|} \exp\left(-\frac{(m^2 + n^2)\beta + 2imnC(\zeta)}{2|\xi(\zeta)|^2}T^2\right) - \text{c.c.}\right] d\zeta, \quad (36)$$

where we used the condition $l = m + n$. These formulae agree with the results obtained directly from Eq.(26) [20].

Equation (32) provides a more general result when $\langle D \rangle \neq 0$ and the pulse is pre-chirped by the cumulative dispersion $C_0$. We compute $W_0$ by rewriting $\hat{U}(z,\omega)$ as

$$\tilde{U}(z,\omega) = \hat{U}(z,\omega)\exp[-i(C_0 + \langle D \rangle z)\omega^2/2] \quad (37)$$

and substituting to $\hat{\mathcal{I}}_{\text{FWM}}$ Eq.(31c) and $dW_0/dz$ Eq.(32). Furthermore, when we take a limit $s \to 0$, this equation allows one to describe the evolution of energy change in a highly dispersed system [27,28] i.e., large and constant dispersion fibers.

Figure 7 shows plots of the energy change of the bits $u_0$, $u_1$, and $u_2$ for the bit pattern '111' ($k = 0, 1, 2$; see the inset) and 0 elsewhere, obtained from Eq.(34) with Eq.(36) calculated numerically and from direct simulation of Eq.(1). The parameters that are used in the calculation are $\alpha = \sqrt{2\pi}$, $\beta = 1.0$, $T = 8.3$, $\Gamma = 10$, $z_a = 0.1$, and $s = 22$ in dimensionless units. With the choice of $t_* = 3$ ps, $\nu = 2.5$ W$^{-1}$km$^{-1}$, $P_* = 1$ mW, i.e. $z_{\text{NL}} = 400$ km and $k''(= -t_*^2/z_{\text{NL}}) = -2.25 \times 10^{-2}$ ps$^2$/km, they correspond to the transmission of the pulses with the path-average peak power 1 mW, the full-width at half-maximum (FWHM) $\tau_{\text{FWHM}} = 5$ ps (minimum), and the bit interval $t_{\text{bit}} = 25$ ps (corresponding to the bit rate $B = 40$ Gbit/s), in a dispersion-managed fiber with the period 40 km, $k'' = \pm 20$ ps$^2$/km, and the loss 0.22 dB/km. We note that $\Delta W_2 = \Delta W_0$ and $\Delta W_1 = -2\Delta W_0$ from symmetry and the conservation of total energy. In this case, the only combination of integers satisfying the phase-matching condition is $(l, m, n) = (2, 1, 1)$. Good agreement between the

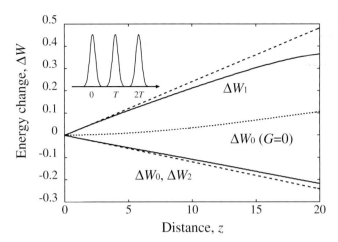

**Fig. 7.** Growth of energy change of the signals in a 111 bit pattern in a lossy case ($\Gamma = 10$). The solid curves are results of direct numerical simulation of Eq.(1) and the dashed curves are the results obtained from Eq.(34) with Eq.(36). The dotted curve shows the energy change $\Delta W_0$ when $G = 0$ obtained from direct numerical simulation.

analytical and the numerical results can be seen. It should be noted that over longer distance, as the energy transfer increases, other effects become important and the assumptions in the model must be modified. We also note that, in the limit of $G \to 0$ (i.e., $g(z) = 1$), we have $\bar{Q}_{m,n} \to 0$ [20]. This implies that in a lossless system the energy change is reduced substantially. This suppression is attributed to the absence of amplification, which introduces periodicity into the system and in turn brings about strong resonance. In Fig. 7 we also show the energy change of $u_0$ in a lossless case obtained from direct simulation. As expected, when $G \to 0$ suppression of the energy change is observed even with the same value of path-average signal power as in the lossy case, which further confirms the analysis.

## 5.3   Frequency and Timing Shifts

The frequency and timing shift of $u_0$ is also computed from Eq.(30). Noting that the mean frequency is given by $\Omega_0 = \int_{-\infty}^{\infty} \omega |\hat{U}_0|^2 d\omega / \int_{-\infty}^{\infty} |\hat{U}_0|^2 d\omega$, we find the frequency shift due to the XPM term to be calculated from

$$\frac{d\Omega_0}{dz} = \frac{\int_{-\infty}^{\infty} \omega \left( \hat{U}_0^* \frac{\partial \hat{U}_0}{\partial z} + \hat{U}_0 \frac{\partial \hat{U}_0^*}{\partial z} \right) d\omega}{\int_{-\infty}^{\infty} |\hat{U}_0|^2 d\omega} = \frac{\int_{-\infty}^{\infty} i\omega \left( \hat{U}_0^* \hat{\mathcal{I}}_{\text{XPM}} - \hat{U}_0 \hat{\mathcal{I}}_{\text{XPM}}^* \right) d\omega}{\int_{-\infty}^{\infty} |\hat{U}_0|^2 d\omega}.$$

$$(38)$$

Using Eq.(12) and Eq.(28) and employing the inverse Fourier transform, we have

$$
\int_{-\infty}^{\infty} i\omega \left( \hat{U}_0^* \hat{\mathcal{I}}_{\mathrm{XPM}} - \hat{U}_0 \hat{\mathcal{I}}_{\mathrm{XPM}}^* \right) d\omega
$$

$$
= \frac{2}{(2\pi)^2} \int_0^1 d\zeta g(\zeta) \int_{-\infty}^{\infty} d\omega_1 d\omega_2 d\omega\, i\omega \left[ \hat{u}_0^*(\zeta, z, \omega) \hat{u}_0(\zeta, z, \omega + \omega_1) \right.
$$
$$
\left. \times \hat{u}_j(\zeta, z, \omega + \omega_2) \hat{u}_j^*(\zeta, z, \omega + \omega_1 + \omega_2) - \text{c.c.} \right]
$$

$$
= \frac{2}{(2\pi)^2} \int_0^1 d\zeta g(\zeta) \int_{-\infty}^{\infty} dt_1 dt_2 dt_3 dt_4 d\omega_1 d\omega_2 d\omega\, i\omega \left\{ u_0^*(\zeta, z, t_1) u_0(\zeta, z, t_2) \right.
$$
$$
\times u_j(\zeta, z, t_3) u_j^*(\zeta, z, t_4) \exp[-i\omega(-t_1 + t_2 + t_3 - t_4)] \exp[-i\omega_1(t_2 - t_4)]
$$
$$
\left. \times \exp[-i\omega_2(t_3 - t_4)] - \text{c.c.} \right\}
$$

$$
= 2 \int_0^1 d\zeta g(\zeta) \int_{-\infty}^{\infty} dt_1 dt_4 \left[ u_0^*(t_1) u_0(t_4) |u_j(t_4)|^2 \int_{-\infty}^{\infty} d\omega\, i\omega \exp[i\omega(t_1 - t_4)] \right.
$$
$$
\left. - u_0(t_1) u_0^*(t_4) |u_j(t_4)|^2 \int_{-\infty}^{\infty} d\omega\, i\omega \exp[-i\omega(t_1 - t_4)] \right]
$$

$$
= -2(2\pi) \int_0^1 d\zeta g(\zeta) \int_{-\infty}^{\infty} dt_4 \left( \frac{\partial u_0^*}{\partial t_4} u_0(t_4) + \frac{\partial u_0}{\partial t_4} u_0^*(t_4) \right) |u_j(t_4)|^2
$$

$$
= -2(2\pi) \int_0^1 d\zeta g(\zeta) \int_{-\infty}^{\infty} dt \frac{\partial |u_0|^2}{\partial t} |u_j|^2
$$

$$
= 2(2\pi) \int_0^1 d\zeta g(\zeta) \int_{-\infty}^{\infty} dt |u_0|^2 \frac{\partial |u_j|^2}{\partial t}, \tag{39}
$$

where we have used the relations $\int_{-\infty}^{\infty} d\omega \exp(i\omega t) = 2\pi\delta(t)$, $\int_{-\infty}^{\infty} dt' f(t') \int_{-\infty}^{\infty} d\omega$ $\times i\omega \exp[\pm i\omega(t' - t)] = \mp 2\pi \int_{-\infty}^{\infty} dt' (df/dt')\delta(t' - t) = \mp 2\pi (df/dt)$, and $t = t_4$. Thus $d\Omega_0/dz$ is obtained from Eq.(38) as

$$
\frac{d\Omega_0}{dz} = 2 \int_0^1 d\zeta g(\zeta) \frac{\displaystyle\int_{-\infty}^{\infty} |u_0(\zeta, z, t)|^2 \frac{\partial}{\partial t} |u_j(\zeta, z, t)|^2 dt}{\displaystyle\int_{-\infty}^{\infty} |u_0(\zeta, z, t)|^2 dt}, \tag{40}
$$

where we used $\int_{-\infty}^{\infty} |\hat{U}_0(z, \omega)|^2 d\omega = \int_{-\infty}^{\infty} |\hat{u}_0(\zeta, z, \omega)|^2 d\omega = 2\pi \int_{-\infty}^{\infty} |u_0(\zeta, z, t)|^2 dt$. Note that, when $\langle D \rangle = 0$, $\hat{u}_k$ is independent of $z$, so that the right-hand side of Eq.(40) is constant in terms of $z$.

The frequency shift yields the shift of the central temporal position of the pulse $t_0 = \int_{-\infty}^{\infty} t|u_0|^2 dt/W_0$ through GVD:

$$
\frac{dt_0}{dz} = D(z)\Omega_0(z). \tag{41}
$$

The timing shift $\Delta t_0(z) = t_0(z) - t_0(0)$ is obtained by integration of Eq.(40) with $\Omega_0(0) = 0$ and using Eq.(41). Interchanging the order of integration yields

$$\Delta t_0(z) = \int_0^z D(z') \left[ \int_0^{z'} \frac{d\Omega_0}{dz''} dz'' \right] dz' = \int_0^z \frac{d\Omega_0}{dz''} \left[ \int_{z''}^z D(z') dz' \right] dz''$$

$$= \int_0^z \frac{d\Omega_0}{dz''} \left( \bar{D}(z) - \bar{D}(z'') \right) dz'' = \delta t_0^{(1)}(z) + \delta t_0^{(2)}(z), \tag{42}$$

where

$$\delta t_0^{(1)}(z) = \bar{D}(z) \Omega_0(z), \tag{43}$$

$$\delta t_0^{(2)}(z) = \int_0^z \left( -\frac{d\Omega_0}{dz''} \right) \bar{D}(z'') dz''. \tag{44}$$

Thus the timing shift is composed of two terms: $\delta t_0^{(1)}$ and $\delta t_0^{(2)}$. We note that when $\bar{D}(z) = 0$, the timing shift is given by $\delta t_0^{(2)}$ alone in Eq.(42). In the case of zero average dispersion, this condition corresponds to $C(z) = 0$, namely at chirp-free points.

When $\langle D \rangle = 0$, $d\Omega_0/dz$ is constant in terms of $z$ and $\bar{D}(z) = C(\zeta)$, from Eq.(42) and Eq.(44) the timing shift $\delta t_0^{(2)}$ grows linearly with respect to distance

$$\delta t_0^{(2)}(z) = \bar{P}_n z,$$

$$\bar{P}_n = -2 \int_0^1 d\zeta C(\zeta) g(\zeta) \frac{\displaystyle\int_{-\infty}^{\infty} |u_0(\zeta, t)|^2 \frac{\partial}{\partial t} |u_j(\zeta, t)|^2 dt}{\displaystyle\int_{-\infty}^{\infty} |u_0(\zeta, t)|^2 dt}. \tag{45}$$

In the following, we assume a signal given by a Gaussian pulse (35). Substituting Eq.(35) to Eq.(45), we have

$$\bar{P}_n = -\frac{\alpha^2 \beta n T}{\sqrt{2\pi}} \int_0^1 \frac{C(\zeta) g(\zeta)}{|\xi(\zeta)|^3} \exp\left( -\frac{\beta n^2 T^2}{2|\xi(\zeta)|^2} \right) d\zeta. \tag{46}$$

These formulae agree with the results obtained directly from Eq.(26) [19].

Figure 8 shows plots of the timing shift of the bits $u_{-2}$, $u_{-1}$, and $u_1$ for the bit pattern '1101' ($k = -2, -1, 0, 1$), obtained from Eq.(45) with Eq.(46) calculated numerically and from direct simulation of Eq.(1). In order to compute the timing shift of $u_k$, $k \neq 0$ using Eq.(45), we need to shift the bit pattern by $-kT$ so that the central position of $u_k$ is moved to $t = 0$, and relabel all the bit slots correspondingly. $\Delta t_{-2}$, for instance, is given by $\Delta t_0(n = 1) + \Delta t_0(n = 3)$. Once again good agreement between the analytical and the numerical results can be seen. It should be noted that, in the limit of $G \to 0$ (i.e., $g(z) = 1$), we have $\bar{P}_n \to 0$ [19]. This implies substantial reduction of timing shifts, like energy change. We also show in Fig. 8 the timing shift of $u_{-2}$ in a lossless case obtained from direct simulation. As predicted, a significant suppression of the timing shift is observed even with the same value of path-average power as in the lossy case.

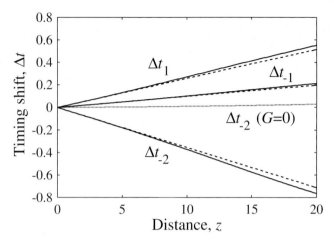

**Fig. 8.** Growth of timing shifts of the signals centered at $t = -2T$ ($\Delta t_{-2}$), $t = -T$ ($\Delta t_{-1}$) and $t = T$ ($\Delta t_1$) in a 1101 bit pattern in a lossy case ($\Gamma = 10$). The solid curves are results of direct numerical simulation of Eq.(1) and the dashed curves are the results obtained from Eq.(45). The dotted curve shows the timing shift $\Delta t_{-2}$ when $G = 0$ obtained from direct numerical simulation.

## 6    Conclusion

An analytical model has been developed which clarifies fundamental properties of quasi-linear dispersion-managed transmission. Quasi-linear pulses can be viewed as degenerate limit of a DM soliton. It was found that dispersion management mitigates nonlinearity by $O(\log s/s)$ in quasi-linear regime. The periodic perturbation due to loss and lumped amplification modifies the averaged dynamics of quasi-linear transmission depending on the relative position of the lumped amplifiers. The amount of spectral reshaping and the nonlinear chirp in the frequency domain due to residual nonlinearity was quantified. Periodic dispersion management results in a resonant intra-channel pulse interactions because of periodic forcing due to lumped amplification supported by phase matching in the time domain. The DMNLS equation obtained from the perturbed NLS equation by means of a multiple scale approach was found to provide a unified analytical framework to study long-scale dynamics of the propagation and interaction of quasi-linear pulses and DM solitons.

### Acknowledgments

The authors would like to acknowledge Gino Biondini and Andrew Docherty for providing us with useful remarks. This work was partially supported by NSF grants ECS-9800152, DMS-0070792.

# References

1. J. P. Gordon and H. A. Haus: Opt. Lett. **11**, 665 (1986).
2. M. Suzuki, I. Morita, N. Edagawa, S. Yamamoto, H. Taga, and S. Akiba: Electron. Lett. **31**, 2027 (1995).
3. N. J. Smith, F. M. Knox, N. J. Doran, K. J. Blow, and I. Bennion: Electron. Lett. **32**, 54 (1996).
4. M. J. Ablowitz, T. Hirooka, and G. Biondini: Opt. Lett. **26**, 459 (2001).
5. M. J. Ablowitz and T. Hirooka: J. Opt. Soc. Am. B **11**, 425 (2002).
6. C. Kurtzke: IEEE Photon. Technol. Lett. **5**, 1250 (1993).
7. R. W. Tkach, A. R. Chraplyvy, F. Forghieri, A. H. Gnauck, and R. M. Derosier: IEEE J. Lightwave Technol. **13**, 841 (1995).
8. N. S. Bergano, C. R. Davidson, M. Ma, A. Pillipetskii, S. G. Evangelides, H. D. Kidorf, J. M. Darcie, E. Golovchenko, K. Rottwitt, P. C. Corbett, R. Menges, M. A. Mills, B. Pedersen, D. Peckham, A. A. Abramov, and A. M. Vengsarkar, in *Digest of Optical Fiber Communication Conference* (Optical Society of America, Washington, D.C., 1998), postdeadline paper PD12.
9. R. Horne: Ph. D dissertation, Department of Applied Mathematics, University of Colorado at Boulder (2001).
10. M. J. Ablowitz and G. Biondini: Opt. Lett. **23**, 1668 (1998).
11. R. J. Essiambre, B. Mikkelsen, and G. Raybon: Electron. Lett. **35**, 1576 (1999).
12. P. V. Mamyshev and N. A. Mamysheva: Opt. Lett. **24**, 1454 (1999).
13. H. Sugahara, H. Kato, and Y. Kodama: Electron. Lett. **33**, 1065 (1997).
14. T. Hirooka and A. Hasegawa: Opt. Lett. **23**, 768 (1998).
15. P. V. Mamyshev and L. F. Mollenauer: Opt. Lett. **24**, 448 (1999).
16. M. J. Ablowitz, G. Biondini, and E. S. Olson: J. Opt. Soc. Am. B **18**, 577 (2001).
17. J. Martensson, A. Berntson, M. Westlund, A. Danielsson, P. Johannisson, D. Anderson, and M. Lisak: Opt. Lett. **26**, 55 (2001).
18. S. Kumar: IEEE Photon. Technol. Lett. **13**, 800 (2001).
19. M. J. Ablowitz and T. Hirooka: Opt. Lett. **26**, 1846 (2001).
20. M. J. Ablowitz and T. Hirooka: Opt. Lett. **27**, 203 (2002).
21. M. J. Ablowitz and T. Hirooka: Opt. Lett. **25**, 1750 (2000).
22. M. J. Ablowitz, G. Biondini, and E. Olson: *Massive WDM and TDM Soliton Transmission Systems*, (Ed. A. Hasegawa, Kluwer, Dordrecht, 2000).
23. S. Shen, C. -C. Chang, H. P. Sardesai, V. Binjrajka, and A. M. Weiner: J. Lightwave Technol. **17**, 452 (1999).
24. S. T. Cundiff, B. C. Collings, L. Boivin, M. C. Nuss, K. Bergman, W. H. Knox, and S. G. Evangelides: J. Lightwave Technol. **17**, 811 (1999).
25. R. -M. Mu, T. Yu, V. S. Grigoryan, and C. R. Menyuk, 'Convergence of the CRZ and DMS formats in WDM systems using dispersion management,' in *Digest of Optical Fiber Communication Conference* (Optical Society of America, Washington, D.C., 2000), paper FC1.
26. M. J. Ablowitz and T. Hirooka: submitted to *IEEE J. Sel. Topics Quantum Electron.* (2002).
27. A. Mecozzi, C. B. Claussen, and M. Shtaif: IEEE Photon. Technol. Lett. **12**, 392 (2000).
28. A. Mecozzi, C. B. Claussen, and M. Shtaif: IEEE Photon. Technol. Lett. **12**, 1633 (2000).

# Bi-Soliton Propagating in Dispersion-Managed System and Its Application to High Speed and Long Haul Optical Transmission

A. Maruta, T. Inoue, Y. Nonaka, and Y. Yoshika

Graduate School of Engineering, Osaka University, 2-1 Yamada-oka, Suita, Osaka, 565-0871, Japan

**Abstract.** We show that, in addition to a stable single soliton, a dispersion-managed system can also support stable bi-solitons. The system parameters in which the bi-solitons can exist are also studied. In addition, we propose novel error preventable line-coding schemes in which binary data are assigned to bi-solitons and single dispersion-managed solitons. By using the schemes, impairments arising from intra-channel interactions can be drastically reduced compared with the conventional scheme for the same bit rate.

## 1 Introduction

Dispersion-management, in which fiber's own properties such as group-velocity dispersion (GVD) and nonlinearity are intentionally changed along the line, is an indispensable technique to achieve high speed and long haul optical transmission [1]. Dispersion-managed (DM) soliton is a periodically stationary pulse propagating in a DM line, and owing to its various advantages it is a promising candidate for achieving a transoceanic high speed optical transmission system over 10,000km with more than 40Gbit/s/ch. DM soliton in more strongly DM line has larger peak power than the ideal soliton under the same averaged dispersion, thereby the signal-to-noise ratio is improved [2]. It is also less influenced by Gordon-Haus timing jitter induced by amplifier noise [3,4] and, four-wave-mixing (FWM) and collision induced frequency shift in wavelength-division-multiplexed (WDM) system [5,6]. However, for a high speed system in which the bit rate per channel is more than 40Gbit/s, intra-channel interactions between neighboring DM solitons play a detrimental role to extend the transmissible distance even in a single channel case. They cause timing jitter and the pulses finally collide [7–9].

In this Paper, we firstly study the dependency of the nonlinear interactions between neighboring DM solitons on various parameters of dispersion map and initial pulse-to-pulse spacing in detail and show that the timing shift due to interactions is negligible for specific system parameters. We then show Bi-soliton solutions [10] by using the numerical averaging scheme [11–13] in the parameter range. It has been reported that a couple of nonlinear pulses propagate for long distance with maintaining their pulse-to-pulse spacing in the cases of anti-symmetric soliton which is a couple of anti-phase pulses [14] and multi-soliton propagating in a DM system in which fiber Bragg gratings are used to compensate the accumulated dispersion [15,16]. However, their reports are not complete because they have only suggested the existence of a stationary state in a specific situation and not shown any shape of stationary solution. The bi-soliton which we have shown in Ref. [10] is a new kind of DM soliton. It is a periodically stationary pulse having two main-lobes and universally exists in a relatively wide parameter range of the system.

By using a preferable feature of the bi-soliton for high bit-rate system, i.e., it is not affected with time position shifts arising from intra-channel interactions, we also propose novel error preventable line-coding schemes [17] in which binary data are assigned to bi-solitons and single DM solitons. The impairments arising from intra-channel interactions can be drastically reduced by using these schemes. Moreover, we show that anti-phase bi-soliton can be more densely packed in the time domain than in-phase one.

## 2    Dispersion-Managed Transmission System

The fundamental equation which governs behavior of optical pulse propagation in a DM line is given by [1]:

$$i\frac{\partial E}{\partial z} - \frac{\beta(z)}{2}\frac{\partial^2 E}{\partial t^2} + S(z)|E|^2 E = ig(z)E , \qquad (1)$$

where $E(z,t)(|E|^2[\mathrm{W}])$, $z[\mathrm{m}]$, and $t[\mathrm{s}]$ represent the complex envelope of electric field, the propagation distance, and the retarded time, respectively. $\beta(z)[\mathrm{s}^2/\mathrm{m}]$, $S(z)[1/(\mathrm{m}{\cdot}\mathrm{W})]$, and $g(z)[1/\mathrm{m}]$ are fiber's GVD, nonlinearity, and loss for $g < 0$ or gain for $g > 0$, respectively. Introducing a new variable $u(z,t)(|u|^2[\mathrm{W}])$ as

$$E(z,t) = a(z)u(z,t), \qquad a(z) = a_0 \exp\left[\int_0^z g(\zeta)d\zeta\right] , \qquad (2)$$

with a dimensionless constant $a_0$ which is determined below, $u(z,t)$ then satisfies

$$i\frac{\partial u}{\partial z} - \frac{\beta(z)}{2}\frac{\partial^2 u}{\partial t^2} + s(z)|u|^2 u = 0 . \qquad (3)$$

Here $s(z) = S(z)a^2(z)[1/(\mathrm{m}{\cdot}\mathrm{W})]$ and represents fiber's effective nonlinearity including the variations both of fiber's own nonlinearity and optical power due to fiber's loss or gain along the transmission line simultaneously. In general, $s(z) > 0$ for any $z$, we then introduce a new coordinate $z'[1/\mathrm{W}]$ defined by

$$z'(z) = \int_0^z s(\zeta)d\zeta , \qquad (4)$$

and Eq.(3) is transformed into

$$i\frac{\partial u}{\partial z'} - \frac{\beta'(z')}{2}\frac{\partial^2 u}{\partial t^2} + |u|^2 u = 0 , \qquad (5)$$

where $\beta'(z') = \beta(z')/s(z')[\text{W}\cdot\text{s}^2]$ and represents fiber's effective dispersion including the variations both of fiber's GVD and effective nonlinearity. Equation (5) permits us to describe the variations of dispersion, nonlinearity, and optical power due to loss or gain by a single variable $\beta'(z')$ on the distance $z'$ which is measured with the accumulation of nonlinearity. For a negative constant $\beta'$, Eq.(5) is called the NLS equation, can be analytically solved for any initial input by using the inverse scattering transformation, and has soliton solution [18].

Here we consider a system in which $\beta(z)$, $S(z)$, and $g(z)$ in Eq.(1) are periodic functions of $z$ with their common period of $L$. This means that $a(z)$ in Eq.(2) and $s(z)$ in Eq.(3) are also periodic functions of $z$ with their period of $L$ and $\beta'(z')$ in Eq.(5) is a periodic function of $z'$ with its period of $L' = \int_0^L s(\zeta)d\zeta\,[1/\text{W}]$. With using the period $L'$ and the pulse's minimum full width at half maximum (FWHM) in a period, $\tau_m[\text{s}]$, we introduce new dimensionless parameters to normalize Eq.(5) defined by

$$U(Z,T) = u(z',t)\sqrt{L'}\ ,\quad Z = \frac{z'}{L'}\ ,\quad T = \frac{t}{\tau_m}\ . \tag{6}$$

Substituting Eqs.(6) into Eq.(5), we obtain

$$i\frac{\partial U}{\partial Z} - \frac{b(Z)}{2}\frac{\partial^2 U}{\partial T^2} + |U|^2 U = 0\ , \tag{7}$$

where $b(Z) = \beta'(Z)L'/\tau_m^2$ and represents fiber's dimensionless effective dispersion.

We here assume a solution of Eq.(7) to have the following Gaussian form with linear chirp [1],

$$U(Z,T) = \sqrt{\frac{E_0 p}{\sqrt{\pi}}}\exp\left[-\frac{p^2}{2}(1 - iC)T^2 + i\theta\right]\ , \tag{8}$$

where $p(Z)$, $C(Z)$, and $\theta(Z)$ represent the reciprocal of pulse width, the chirp, and the phase of the pulse, respectively. $E_0$ is a constant and represents the pulse's dimensionless energy defined by

$$E_0 = \int_{-\infty}^{\infty} |U|^2\ \mathrm{d}T = \frac{L'}{\tau_m}\mathcal{E}_0\ . \tag{9}$$

Here $\mathcal{E}_0 = \int_{-\infty}^{\infty} |u|^2\ \mathrm{d}t\ [\text{J}]$ represents the pulse's energy averaged in a period when $a_0$ in Eq.(2) is determined by

$$a_0 = \sqrt{\frac{L}{\int_0^L \exp\left[2\int_0^z g(\zeta)\ \mathrm{d}\zeta\right]\ \mathrm{d}z}}\ . \tag{10}$$

Note here that the pulse's FWHM of $U(Z,T)$ in Eq.(8) is given by

$$T_m(Z) = \frac{2\sqrt{\ln 2}}{p(Z)} \ . \tag{11}$$

Applying the variational method [6] to Eq.(7) together with its assumed solution, Eq.(8), one obtain the following system of ordinary differential equations,

$$\begin{cases} \dfrac{\mathrm{d}p}{\mathrm{d}Z} = b(Z)p^3 C \ , \\[2mm] \dfrac{\mathrm{d}C}{\mathrm{d}Z} = -b(Z)p^2 \left(1 + C^2\right) - \dfrac{E_0 p}{\sqrt{2\pi}} \ , \end{cases} \tag{12}$$

which give the variations of the pulse's width and chirp along the propagation. Equations (12) will be used to find the DM soliton's energy for a given dispersion map.

## 3   Dispersion Map Parameters

We introduce the following two dimensionless system parameters, the accumulated dispersion $B$ and the map strength $S$ [7], which characterize the dispersion map.

$$B = \int_0^1 b(Z)\mathrm{d}Z = \frac{1}{\tau_m^2}\int_0^{L'}\beta'(z')\mathrm{d}z' = \frac{1}{\tau_m^2}\int_0^{L}\beta(z)\mathrm{d}z \ , \tag{13}$$

$$S = \int_0^1 |b(Z)|\mathrm{d}Z = \frac{1}{\tau_m^2}\int_0^{L'}|\beta'(z')|\mathrm{d}z' = \frac{1}{\tau_m^2}\int_0^{L}|\beta(z)|\mathrm{d}z \ . \tag{14}$$

We here focus on a system in which $\beta(z)$ and $s(z)$ in a period $L$ are given by

$$\beta(z) = \begin{cases} \beta_1(< 0) \left(0 \le z < \dfrac{L_1}{2}\right) \\[2mm] \beta_2(> 0) \left(\dfrac{L_1}{2} \le z < \dfrac{L_1}{2} + L_2\right) \ , \\[2mm] \beta_1(< 0) \left(\dfrac{L_1}{2} + L_2 \le z < L\right) \end{cases} \tag{15}$$

$$s(z) = \begin{cases} s_1 \left(0 \le z < \dfrac{L_1}{2}\right) \\[2mm] s_2 \left(\dfrac{L_1}{2} \le z < \dfrac{L_1}{2} + L_2\right) \ , \\[2mm] s_1 \left(\dfrac{L_1}{2} + L_2 \le z < L\right) \end{cases} \tag{16}$$

where $L_1$ and $L_2$ are the lengths of anomalous and normal dispersion fibers

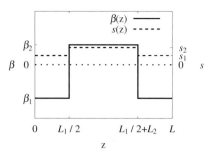

**Fig. 1.** Dispersion-managed transmission line.

respectively, and $L_1 + L_2 = L$. For a DM soliton propagating in such a symmetric dispersion map, $\tau_m$ can be observed at the mid-point of the anomalous dispersion fiber, i.e., at $z = nL$ with integer $n$. $\beta(z)$ and $s(z)$ are shown in Fig. 1. Applying the coordinate transformation of Eq.(4) to $\beta(z)$ and $s(z)$, these parameters are represented by a single parameter $\beta'(z')$ as

$$\beta'(z') = \begin{cases} \beta'_1 = \dfrac{\beta_1}{s_1}(<0) & \left(0 \le z' < \dfrac{L'_1}{2}\right) \\[2ex] \beta'_2 = \dfrac{\beta_2}{s_2}(>0) & \left(\dfrac{L'_1}{2} \le z' < \dfrac{L'_1}{2} + L'_2\right) \\[2ex] \beta'_1 = \dfrac{\beta_1}{s_1}(<0) & \left(\dfrac{L'_1}{2} + L'_2 \le z' < L'\right) \end{cases} , \qquad (17)$$

where $L'_n = s_n L_n$ ; $(n = 1, 2)$ and $L' = L'_1 + L'_2$. The transformed $\beta'(z')$ is shown in Fig. 2. Moreover, applying the coordinate transformation of Eq.(6) to $\beta'(z')$,

$$b(Z) = \begin{cases} b_1 = \beta'_1 \dfrac{L'}{\tau_m^2}(<0) & \left(0 \le Z < \dfrac{\ell_1}{2}\right) \\[2ex] b_2 = \beta'_2 \dfrac{L'}{\tau_m^2}(>0) & \left(\dfrac{\ell_1}{2} \le Z < \dfrac{\ell_1}{2} + \ell_2\right) \\[2ex] b_1 = \beta'_1 \dfrac{L'}{\tau_m^2}(<0) & \left(\dfrac{\ell_1}{2} + \ell_2 \le Z < 1\right) \end{cases} , \qquad (18)$$

where $\ell_n = L'_n/L'$ ; $(n = 1, 2)$ and $\ell_1 + \ell_2 = 1$. The transformed $b(Z)$ is shown in Fig. 3.

For these parameters, $B$ and $S$ defined in Eqs.(13) and (14) are calculated as

$$B = b_1 \ell_1 + b_2 \ell_2 = \frac{\beta'_1 L'_1 + \beta'_2 L'_2}{\tau_m^2} = \frac{\beta_1 L_1 + \beta_2 L_2}{\tau_m^2} , \qquad (19)$$

$$S = -b_1 \ell_1 + b_2 \ell_2 = \frac{-\beta'_1 L'_1 + \beta'_2 L'_2}{\tau_m^2} = \frac{-\beta_1 L_1 + \beta_2 L_2}{\tau_m^2} . \qquad (20)$$

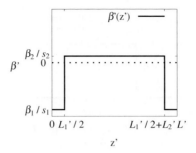

**Fig. 2.** Dispersion-managed transmission line represented in the converted coordinates.

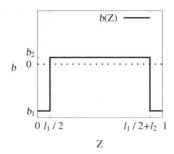

**Fig. 3.** Dispersion-managed line represented in the normalized coordinates.

For characterizing the dispersion map completely, we additionally introduce the following new dimensionless parameter $R$ defined by

$$R = \ell_1 = \frac{L_1'}{L'} = \frac{L_1'}{L_1' + L_2'} = \frac{s_1 L_1}{s_1 L_1 + s_2 L_2} , \tag{21}$$

which represents the ratio of the accumulated nonlinearity in the anomalous dispersion fiber of $\beta_1$ to the total accumulated nonlinearity over a period of $L$.

The three parameters defined in Eqs.(19) - (21) completely characterize a single DM soliton propagating in the dispersion maps shown in Figs. 1-3 [10] and are invariable for the transformations given by Eqs.(4) and (6). Solving Eqs.(19) - (21), $b_1$, $b_2$ and $\ell_1$ are represented by using $B$, $S$, and $R$ as

$$b_1 = \frac{B - S}{2R} , \quad b_2 = \frac{B + S}{2(1 - R)} , \quad \ell_1 = R . \tag{22}$$

Note that the mapping of Fig. 1 to Fig. 2 is not unique. For example, a dispersion map considering fiber's loss shown in Fig. 4 is also mapped to Fig. 2 with using Eqs.(2) and (4). For $S(z) = S_0$ and $g(z) = -\gamma_0$, $s(z)$ and $\beta(z)$ in Fig. 4 are given by

$$s(z) = S_0 \frac{2\gamma_0 L}{1 - \exp(-2\gamma_0 L)} \exp(-2\gamma_0 z) , \tag{23}$$

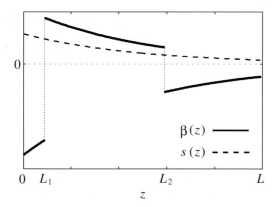

**Fig. 4.** Dispersion-managed lossy transmission line.

and

$$\beta(z) = \begin{cases} \beta_1' s(z) \ (<0) \ (0 \le z < L_1) \\ \beta_2' s(z) \ (>0) \ (L_1 \le z < L_2) \ , \\ \beta_1' s(z) \ (<0) \ (L_2 \le z < L) \end{cases} \qquad (24)$$

where $L_1 = z(z' = L_1'/2)$ and $L_2 = z(z' = L_1'/2 + L_2')$ with

$$z(z') = \frac{1}{2\gamma_0} \log_e \frac{S_0 L}{S_0 L - \{1 - \exp(-2\gamma_0 L)\} z'} \ . \qquad (25)$$

We calculate single DM soliton's energy when the parameters shown in Eqs.(19) - (21) are varied. For the transmission line shown in Fig. 3, a chirp-free Gaussian pulse whose pulse width $T_m(0) = 1$ is used as the input. Solving Eqs.(12) numerically for the initial values, $p(0) = 2\sqrt{\ln 2}$ and $C(0) = 0$, and a given constant $E_0$, we can determine the single DM soliton's energy $E_0$ with which $p(Z)$ and $C(Z)$ are periodic functions with their period of 1 [1]. Figure 5 shows $E_0$ for various $S$ and $R$ with $B = -0.1$. One can see that larger $S$ and/or smaller $R$ result in larger $E_0$.

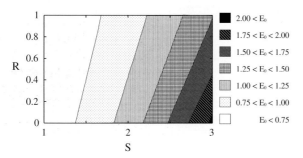

**Fig. 5.** Single DM soliton's energy $E_0$ for various $S$ and $R$ with $B = -0.1$.

## 4    Interactions Between Neighboring DM Solitons

In this section, we study the nonlinear interactions between in-phase neighboring DM solitons with their initial spacing of $T_s$ by solving Eq.(7) numerically. At the input, a couple of chirp-free Gaussian pulses are launched into the DM line. Each initial pulse's energy is adjusted to have a value shown in Fig. 5 and pulses' widths are the same. We firstly calculate the interaction distance of a couple of in-phase DM solitons. The definition of the interaction distance $Z_I$ is the shortest propagation distance at which the pulse-to-pulse spacing increases or decreases more than the pulse's FWHM [9]. Figure 6 show the interaction distance $Z_I$ for various $S$ and $R$ with $B = -0.1, T_s = 3$ or 4. For $T_s = 4$, the longest interaction distance is achieved around $S = 1.65$ for any $R$ as shown in Fig. 6(a). It exhibits the same result as reported in [7,8,19]. On the other hand, narrower pulse-to-pulse spacing of $T_s = 3$ usually results in shorter interaction distance as shown in Fig. 6(b) because of stronger nonlinear interactions. But surprisingly, a strange parameter range is found around $2 < S < 2.7$ and $0.75 < R < 1$ in which a couple of neighboring pulses can propagate over more than 1000 periods with slightly few time position shift. We will call the specific range the "non-collision range".

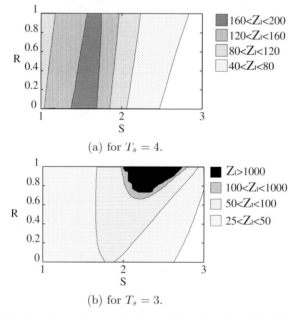

(a) for $T_s = 4$.

(b) for $T_s = 3$.

**Fig. 6.** Interaction distance $Z_I$ for various $S$ and $R$ with $B = -0.1$.

# 5    Bi-Soliton Solution

In the previous section, we have shown the non-collision parameter range in which the interaction distance of neighboring DM solitons is slightly long, i.e., $Z_I > 1000$. We may expect the existence of a periodically stable solution which consists of a couple of DM solitons in the non-collision range. To verify the presumption, we apply the numerical averaging method [11–13] for digging out the fine structure of the periodically stable solution's waveform. Two adjacent Gaussian pulses are used for the input. In the averaging method, the initial pulse and the pulse after propagating for a DM period are numerically averaged after adjusting their phases. Repeating the averaging processes successively, one can judge whether a periodically stable solution exists or not.

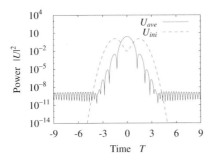

**Fig. 7.** Waveforms of periodically stationary pulse, $U_{ave}$ (solid curve), and initial input, $U_{ini}$ (dashed curve), with initial parameters out of the non-collision range.

In Fig. 7, the solid curve shows the waveform of averaged pulse, $U_{ave}$, and the dashed curve the initial input, $U_{ini}$, for the initial parameters, $(B, S, R, T_s) = (-0.1, 2.3, 0.5, 3)$, out of the non-collision range. On the other hand, Fig. 8 shows the waveforms for $(B, S, R, T_s) = (-0.1, 2.3, 0.8, 3)$ within the non-collision range. While the averaged pulse for the parameters out of the non-collision range has a single main-lobe as shown in Fig. 7, that for the parameters within the non-collision range has two main-lobes as shown in Fig. 8. We then re-define the non-collision range as the range where the periodically stationary solution having two main-lobes exists. The periodically stationary solution shown in Fig. 8 has oscillating tails as the same as the periodically stationary solution having a single main-lobe [11]. Since the solution having a single main-lobe has been called DM soliton, we call the solution having two main-lobes "bi-soliton". Figure 9 show the waveforms of the bi-soliton, in which the solid curve in Fig. 8 is used for the input, observed in a DM period (a) and at every $Z = 100n$ for a long-haul propagation (b). Stable long distance propagation is confirmed in Fig. 9(b). This is the newly found in-phase bi-soliton solution in a DM system and is a different type soliton from either the antisymmetric (anti-phase) soliton [14] or the multi-soliton propagating in a DM system in which fiber Bragg gratings are

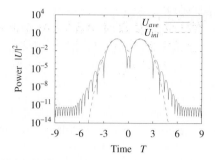

**Fig. 8.** Waveforms of periodically stationary pulse, $U_{ave}$ (solid curve), and initial input, $U_{ini}$ (dashed curve), with initial parameters in the non-collision range.

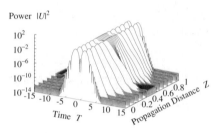

(a) observed in a DM period.

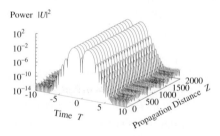

(b) observed at every $Z = 100n$ for a long-haul propagation.

**Fig. 9.** Waveforms of an in-phase bi-soliton.

used [15,16]. In-phase bi-solitons are also found for the other parameters in the non-collision range in Fig. 6(b).

Figure 10 shows the fluctuations of pulse-to-pulse spacing in averaging processes when the initial spacing is varied from $T_s = 2$ to 6 for $(B, S, R) = (-0.1, 2.3, 0.8)$. While the solid curves show the variations of the spacing for the case when a stationary solution is obtained after the averaging, the dashed curves for the case when no stationary solution is obtained. The solid and dashed curves in Fig. 11 show the fluctuations of pulse-to-pulse spacing in the averaging process when the initial spacing are set to $T_s = 2.9$ and 3.2, respectively. We can see in Fig. 10 that periodically stationary pulses exist around $T_s = 3$ and 5. In the vicinity of $T_s = 3$, the fluctuation of the spacing shown in Fig. 11 becomes smaller for larger number of averaging and the spacing finally converges. Even though a little difference exists between the initial spacing and the converged spacing, the spacing converges to a fixed value.

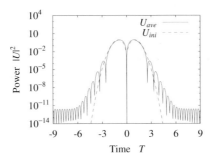

**Fig. 10.** Fluctuations of pulse-to-pulse spacing in averaging process.

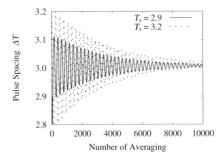

**Fig. 11.** Fluctuations of pulse-to-pulse spacing in averaging process.

In Fig. 12, the solid curve shows the waveform of stationary pulse, $U_{ave}$, and the dashed curve the initial input, $U_{ini}$, for the initial pulse-to-pulse spacing of $T_s = 5$. Comparing Figs. 8 and 12, the number of small peaks observed between two main-lobes of the stationary pulse, $P_n$, is different, i.e., while $P_n = 1$ for Fig. 8, $P_n = 3$ for Fig. 12. This tells us that stationary pulse having the discrete eigennumber of $P_n$ exists according to the pulse-to-pulse spacing at the input and the characteristic parameters of transmission line. For the in-phase stationary pulse in which two main-lobes have the same phase,

$$\frac{\partial |U(Z,T)|}{\partial T}\bigg|_{T=0} = 0 , \tag{26}$$

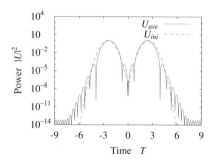

**Fig. 12.** Waveforms of periodically stationary pulse, $U_{\text{ave}}$ (solid curve), and initial input, $U_{\text{ini}}$ (dashed curve), for $T_s = 5$.

because of the waveform's symmetry and continuity at $T = 0$. Therefore $P_n$ should be odd number for the in-phase bi-soliton.

On the other hand, for the anti-phase stationary pulse in which two main-lobes have the opposite phase,

$$|U(Z, T = 0)| = 0 , \tag{27}$$

because of the waveform's anti-symmetry and continuity at $T = 0$. Therefore $P_n$ should be even number including 0 for the anti-phase bi-soliton. As the same as the in-phase bi-soliton, anti-phase bi-solitons having different $P_n$ are obtained for different pulse-to-pulse spacing at the input. In Figs. 13 and 14, the solid curves show the waveforms of stationary pulses, $U_{\text{ave}}$, and the dashed curves the initial inputs, $U_{\text{ini}}$, for $(B, S, R) = (-0.1, 2.3, 0.8)$ and the initial pulse spacings, $T_s = 2$ and 4 respectively. These figures correspond to $P_n = 0$ and 2, respectively. Since the least $P_n = 0$ for anti-phase and 1 for in-phase, the anti-phase configuration with $P_n = 0$ allows bi-soliton to keep narrower peak-to-peak spacing than in-phase one.

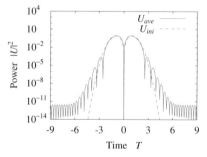

**Fig. 13.** Waveforms of periodically stationary pulse, $U_{\text{ave}}$ (solid curve), and initial input, $U_{\text{ini}}$ (dashed curve), for $T_s = 2$ in the case of anti-phase.

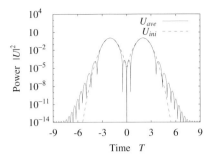

**Fig. 14.** Waveforms of periodically stationary pulse, $U_{\mathrm{ave}}$ (solid curve), and initial input, $U_{\mathrm{ini}}$ (dashed curve), for $T_s = 4$ in the case of anti-phase.

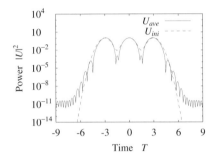

**Fig. 15.** Waveforms of periodically stationary pulse, $U_{\mathrm{ave}}$ (solid curve), and initial input, $U_{\mathrm{ini}}$ (dashed curve), for $T_s = 3$.

In addition to bi-soliton solution, we may expect to find a multi-soliton solution which consists of more than two pulses in a specific parameter range. In Fig. 15, the solid curve shows the waveform of stationary pulse, $U_{\mathrm{ave}}$, and the dashed curve the initial input, $U_{\mathrm{ini}}$, for $(B, S, R, T_s) = (-0.1, 2.1, 0.5, 3)$. This is an in-phase "tri-soliton", which have three main-lobes and oscillating tails. This result encourages us to find multi-solitons having arbitrary number of main-lobes.

## 6    Error Preventable Line-Coding Scheme

By using the bi-soliton together with the single DM soliton, we can originate novel error preventable line-coding schemes which are robust against the impairments arising from intra-channel interactions. Figure 16 show the evolution of waveforms along the transmission line when the amount of input consecutive Gaussian pulses is increased step by step from one to four. System parameters for (a) and (b) are set to $(B, S, R, T_s) = (-0.1, 2.3, 0.8, 3)$ and $(-0.1, 2.3, 0.8, 2)$ with which the in-phase and anti-phase bi-solitons can coexist with single DM soliton, respectively. Note here that we use Gaussian pulses for inputs intentionally in order to show the stability of bi-soliton and single DM soliton simultaneously. As one can see in Figs. 16, more than two consecutive pulses are affected with time position shifts and finally collide because these consecutive pulses do not form any periodically stationary state in the systems. Therefore line-coding schemes in which consecutive pulses more than two are excluded should be designed to apply the bi-soliton to high bit-rate transmission system.

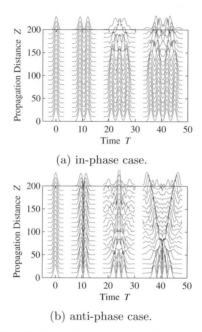

(a) in-phase case.

(b) anti-phase case.

**Fig. 16.** Evolution of consecutive pulses.

An intuitive way to construct such a code is inserting spaces (0's) among more than two consecutive pulses (1's) in given binary data. Considering binary data which is a pseudo-random binary sequence (PRBS) generated by maximum-length-sequence (M-sequence) [20] code with infinite degree, the redundancy due to the inserted spaces is $1/12 = 8.3\%$. This is comparable with conventional forward error correcting (FEC) code such as Reed-Solomon (RS) 255/239 which redundancy is 6.7% [21]. Unfortunately, we cannot use the scheme because it is impossible to distinguish the added spaces from the original data at the receiver. In the following section, we will propose a practical scheme.

# 7    3 Out of 4 Encoding Scheme

The 4 out of 5 (4B/5B) encoding scheme in which groups of four data bits are encoded and transmitted in five bits in order to guarantee that no more than three consecutive spaces (0's) ever occur ; has been used in the fiber distributed data interface (FDDI) network [22]. Improving the scheme, we here propose the 3 out of 4 (3B/4B) encoding as an practical scheme.

We firstly define the code I in which binary data, '1' and '0' are directly assigned to a pulse and a space respectively. This corresponds to the conventional encoding scheme. In the code II, the binary data stream is firstly converted to another stream which consists of three kinds of symbols, '00', '01', and '10', '1' and '0' in each symbol are then assigned to a pulse and a space respectively. By using the code II, it is guaranteed that no more than three consecutive pulses ever appear. We then compare the coding efficiency of these codes. Using the code I with the time slot's width $t_s = 4\tau_m$[s], the resultant bit rate is $0.25/\tau_m$[bit/s]. On the other hand, since the code II consists of three kinds of symbols, the information capacity per symbol is $\log_2 3 = 1.585$[bit/symbol]. Adopting the in-phase bi-soliton of $T_s = t_s/\tau_m = 3$ and the single DM soliton in the code II, the symbol rate is $1/6\tau_m$[symbol/s] and the bit rate is then $0.264/\tau_m$[bit/s]. This means that the bit rate of the proposed code II is 1.05-fold higher than that of the code I.

Let us compare these coding schemes with a concrete symbol assignment shown in Table 1. The code II in Table 1 is corresponding to the 3B/4B encoding scheme which we propose here. Since binary data of three bits are assigned to two symbols of the code II, the information capacity per symbol is 1.5bit/symbol. Using the time slot's width $T_s = 4$ in the code I and 3 in the code II, both of the bit rates are the same. Figure 17(a) shows the evolution of the waveform using the code I for the system parameters of $(B, S, R) = (-0.1, 1.65, 0.5)$, with which in-phase bi-soliton does not exist. For the initial input, binary data of 15 bits, '010110010001111', which is a PRBS generated by a M-sequence, are directly converted to chirp-free Gaussian pulses and spaces. Pulses' time position shifts induce bit error after 200 periods transmission. Figure 17(b) shows the case using the code II for the system parameters used in Fig. 16(a). In this case, the same binary data are replaced by the symbols, '00101000001000011001', according to Table 1, the symbols are then converted to chirp-free Gaussian pulses and spaces. The observed time shift is very small and stable pulse transmission is achieved for long distances. In general, the strength of interactions between neighboring

**Table 1.** Symbol assignment for binary data of three bits.

| Binary data | Code I | Code II |
|:---:|:---:|:---:|
| 000 | '0' '0' '0' | '00' '00' |
| 001 | '0' '0' '1' | '00' '01' |
| 010 | '0' '1' '0' | '00' '10' |
| 011 | '0' '1' '1' | '01' '00' |
| 100 | '1' '0' '0' | '01' '01' |
| 101 | '1' '0' '1' | '01' '10' |
| 110 | '1' '1' '0' | '10' '00' |
| 111 | '1' '1' '1' | '10' '01' |
| ——— | ——— | '10' '10' |

solitons increase exponentially for narrower pulse spacing [23]. While the interactions occur between neighboring pulses of $T_s = 4$ for the code I, 6 for the code II because the interaction induced time shift does not occur in a bi-soliton. This is the reason why the interactions can be drastically reduced by using the proposed code II. For reducing the interactions in the code I, $B$ and/or the peak power of input pulse should be reduced. On the other hand, $B$ and the peak power of input pulse can be kept large value for the code II, and the signal-to-noise ratio can be then improved.

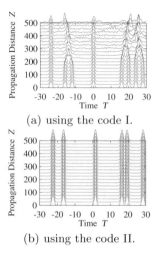

(a) using the code I.

(b) using the code II.

**Fig. 17.** Evolution of bit sequence in a long haul transmission.

For packing the data more densely in time domain, the anti-phase bi-soliton can be used as discussed in Section V. For the anti-phase case, the pulse-to-pulse spacing can be reduced to $T_s = 2$ for the system parameters used in Fig. 16(b). Figure 18(a) shows the evolution of the waveform using the code III whose phases are modulated as 'ss0s0sssss0sssss$\pi$0ss$\pi$'. In the sequence, 's' means a space. This is corresponding to carrier suppressed return-to-zero (CS-RZ) signal format because '0' and '$\pi$' regularly alternate in the neighboring time slots. Interactions between neighboring in-phase pulses of $T_s = 4$ induce the collision in this case as the same as we have seen in Fig. 17(a). On the other hand, Fig. 18(b) shows the case using the improved code IV whose phases are 'ss0s$\pi$sssss0sssss$\pi$0ss$\pi$'. In this code, not only the neighboring pulses of $T_s = 2$ but also pulses of $T_s = 4$ are anti-phase. This is corresponding to duobinary carrier suppressed return-to-zero (DCS-RZ) signal format because '0' and '$\pi$' regularly alternate between neighboring pulses. Since the observed time shift is very small, densely packed transmission is achieved for long distances. This is because the neighboring pulses of $T_s = 4$ also form an anti-phase bi-soliton. Small fluctuations of the spacing between neighboring pulses of $T_s = 2$ are observed in Figs. 18 because the input is just chirp-free Gaussian pulses and it does not have the exact form of anti-

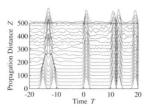

(a) using the phase modulated code III.

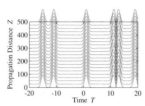

(b) using the phase modulated code IV.

**Fig. 18.** Evolution of bit sequence using anti-phase bi-soliton.

phase bi-soliton. Note that the bit rate of the proposed code IV is 1.5-fold higher than that of the code I. This means that the net capacity increases, even though the redundancy of 33.3% for the codes II-IV is slightly large.

## 8   Conclusion

We have shown that, in addition to a stable single soliton, a DM system can also support stable bi-solitons. The system parameters in which the bi-solitons can exist have been also studied. In addition, we have proposed novel error preventable line-coding schemes using the advantage of bi-soliton. It may be effective for the system whose bit rate is over 40 Gbit/s in which the intra-channel interactions play a detrimental role to extend the transmissible distance. We may also expect to originate a more efficient coding scheme.

## References

1. A. Hasegawa, Y. Kodama, A. Maruta: Opt. Fiber Technol. **3**, 197 (1997)
2. N. J. Smith, F. M. Knox, N. J Doran, K. J. Blow, I. Bennion: Electron. Lett. **32**, 54 (1996)
3. M. Suzuki, I. Morita, N. Edagawa, S. Yamamoto, H. Taga, S. Akiba: Electron. Lett. **31**, 2027 (1995)
4. T. Okamawari, A. Maruta, Y. Kodama: Opt. Commun. **149**, 261 (1998)
5. L. F. Mollenauer, S. G. Evangelides, J. P. Gordon: IEEE/OSA J. Lightwave Technol. **9**, 362 (1991)
6. H. Sugahara, H. Kato, T. Inoue, A. Maruta, Y. Kodama: IEEE/OSA J. Lightwave Technol. **17**, 1547 (1999)
7. T. Yu, E. A. Golovchenko, A. N. Pilipetskii, C. R. Menyuk: Opt. Lett. **22**, 793 (1997)

8. N. J. Smith , N. J. Doran, W. Forysiak, F. M. Knox: IEEE/OSA J. Lightwave Technol. **15**, 1808 (1997)
9. T. Inoue, H. Sugahara, A. Maruta, Y. Kodama: IEEE Photon. Technol. Lett. **12**, 299 (2000)
10. A. Maruta, Y. Nonaka, T. Inoue: 'Symmetric bi-soliton solution in a dispersion-managed system'. In: *Topical Meeting on Nonlinear Guided Waves and Their Applications 2001 (NLGW2001), Clearwater, Florida, March 26-28 2001*, Paper PD4, Electron. Lett., **37**, 1357 (2001)
11. J. H. B. Nijhof, N. J. Doran, W. Forysiak, F. M. Knox: Electron. Lett. **33**, 1726 (1997)
12. J. H. B. Nijhof, W. Forysiak, N. J. Doran: IEEE J. Selected Topics in Quantum Electron. **6**, 330 (2000)
13. V. Cautaerts, A. Maruta, Y. Kodama: Chaos, **10**, 515 (2000)
14. C. Paré, P. -A. Bélanger: Opt. Commun. **168**, 103 (1999)
15. J. D. Ania-Castañón, P. Garcia-Fernández, J. M. Soto-Crespo: Opt. Lett. **25**, 159 (2000)
16. J. D. Ania-Castañón, P. Garcia-Fernández, J. M. Soto-Crespo: J. Opt. Soc. Am. B **18**, 1252 (2001)
17. A. Maruta, Y. Nonaka, T. Inoue: 'Error preventable line coding schemes using bi-soliton to suppress intra-channel interactions in dispersion-managed system,' submitted to IEEE Photon. Technol. Lett.
18. V. E. Zakharov, A. B. Shabat: Sov. Phys. JETP, **34**, 62 (1972)
19. T. Hirooka, T. Nakada, A. Hasegawa: IEEE Photon. Technol. Lett. **12**, 633 (2000)
20. W. W. Peterson, E. J. Weldon, Jr.: *Error-Correcting Codes*, 2nd edn. (MIT Press, Cambridge, Massachusetts 1972) pp.222-223
21. S. Yamamoto, H. Takahira, M. Tanaka: Electron. Lett. **30**, 254 (1994)
22. A. S. Tanenbaum: *Computer Networks*, 3rd edn. (Prentice-Hall International, London 1996) p.320
23. A. Hasegawa, Y. Kodama: *Solitons in optical communications*, (Oxford University Press, Oxford 1995) p.153

# Optical Fiber Soliton Lasers

G.E. Town[1], N.N. Akhmediev[2], and J.M. Soto-Crespo[3]

[1] Department of Electronics, Macquarie University, NSW 2109, Australia
[2] Optical Sciences Centre, Research School of Physical Sciences and Engineering,
    Institute of Advanced Studies, Australian National University, ACT 0200,
    Australia.
[3] Instituto de Optica, Consejo Superior de Investigaciones Scientific, Serrano 121,
    28006 Madrid, Spain.

**Abstract.** Experimental and theoretical developments in optical fiber soliton lasers
are reviewed. We first review the fundamentals of optical fiber soliton lasers from
an experimental viewpoint, including soliton generation and control mechanisms, and
highlight the similarities between developments in soliton fiber lasers and soliton transmission systems. We then review the mathematical theory of soliton lasers based on a
master equation description, highlighting some interesting solutions and recent results
concerning soliton stability.

## 1  Introduction

Optical fiber lasers are ideal systems in which to generate and observe solitons.
Whilst the optical pulses generated in what are widely called soliton lasers are
rarely solitons in the narrow mathematical sense (i.e. in which they may collide
and pass through each other with no change other than a phase shift), they nevertheless can demonstrate most of the properties of Hamiltonian solitons, such
as energy quantization, phase-sensitive interactions, robustness to perturbations,
etc. Consequently in this tutorial review "optical fiber soliton laser" is defined
as any nonlinear optical oscillator containing optical fibers, and which generates
optical pulses due to a balance between various sources of linear and nonlinear
gain, loss, self phase modulation, and dispersion. The variety of such lasers is
very large, and so for the most part we shall further restrict our attention to
lasers in which the gain medium is an optical fiber amplifier.

The next section of this chapter contains a review of fundamental matters
relating to solitons and optical fiber lasers, including typical laser construction and component characteristics, the effects of nonlinearity and dispersion
on pulse propagation, etc. We also highlight parallels between the development
of optical fiber soliton lasers and soliton transmission systems. The following
section reviews experimental developments in optical fiber soliton lasers since
1984, demonstrating how the fundamental ideas are applied in practice. The
final section contains a review of theoretical methods and results of modelling
such lasers. At the time of writing it is almost 20 years since demonstration of
the first "soliton fiber laser", however the field continues to reveal surprises in
both theory and experiment. New optical fibers and fiber components together
with ongoing discoveries of new types of soliton and soliton behavior are likely

to see continued development in the field of optical fiber soliton lasers for some
time yet.

## 2    Fundamentals of Optical Fiber Soliton Lasers

A laser is an optical oscillator in which the presence of optical gain and optical
feedback result in the sustained generation of one or more wavelengths within
the laser cavity. In a linear cavity (i.e. in which the intensity of light does not
change the optical properties of the cavity), the wavelength of each oscillating
mode is set by the boundary conditions, and depends primarily on the optical
path length of the cavity. Optical feedback may be applied by using mirrors
(e.g. as in the Fabry-Perot cavity), or by making a recirculating loop (i.e. ring
cavity). Optical gain may be provided using one or more of the different types
of optical amplifier available. A single lasing mode produces a continuous wave
(CW) single frequency output, however in practice several wavelengths may be
present in the laser output. To obtain pulses, the phase of several lasing modes
must remain locked together (i.e. mode-locking). This is generally achieved by
placing a nonlinear and/or time-varying component in the cavity, e.g. with time
or intensity-dependent loss, or time-varying delay. Other components are also
commonly added to laser cavities for spectral control, coupling of the optical
pump and signal in and out of the laser cavity, etc.

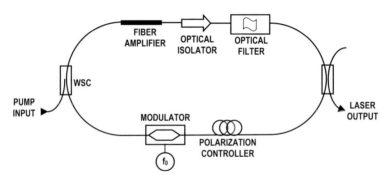

**Fig. 1.** Typical configuration of an actively mode-locked fiber laser

### 2.1    Optical Fiber Laser Components and Construction

A typical optical fiber laser containing the basic elements mentioned above is
shown schematically in Fig. 1. Single mode optical fiber may be chosen to provide
the desired gain, dispersion, and Kerr nonlinearity. The wavelength selective
coupler (WSC) is required to couple the pump into the fiber amplifier without
coupling the signal out. The optical isolator is a nonreciprocal element which may
be used to force unidirectional operation in a ring laser. An optical filter may

be used to help define or tune the lasing wavelength, reject noise, or modify the cavity dispersion. An output coupler is required to out-couple a proportion of the signal in the laser cavity for external use. A polarizatio controller is often required to compensate for unwanted birefringence, e.g. due to bends in the fibers). The mode-locking element, shown in Fig. 1 as an electro-optic amplitude modulator, is required to initiate and sustain periodic pulse generation. For active mode-locking the modulator must be driven at, or very close to, a harmonic of the fundamental cavity frequency (i.e. the inverse of the round-trip time for light within the cavity).

## Optical Fiber Amplifiers

Amplification in optical fibers may be obtained by either stimulated scattering in the silica comprising the fiber, or stimulated emission from rare-earth dopants within the fiber. The main differences between amplifiers based on stimulated scattering and rare-earth-doped amplifiers relate to the pump and gain wavelengths, and to the pump power and length of fiber required. In optical amplifiers based on stimulated Raman scattering in silica fiber the gain is generated at wavelengths approximately 100nm longer than the pump wavelength (or more exactly, 13THz lower in frequency), and may be anywhere within the wide range of wavelengths over which the optical fiber can be used. In rare-earth-doped fiber amplifiers, optical gain is available only in fixed wavelength bands determined by the rare-earth excited-state energy levels, and typically there are only one or two fixed wavelength bands suitable for pumping. For example, erbium-doped fiber amplifiers (EDFAs) can provide large and broadband gain (e.g. 30dB gain over >40nm bandwidth) around $1.55\mu$m in the anomalous dispersion regime of standard step index fibers, and are conveniently pumped by InGaAs 980nm semiconductor diode lasers. The bandwidth of Raman fiber amplifiers is similar to EDFA's, but Raman amplifiers generally require higher pump power and longer lengths of fiber for efficient amplification. An optical fiber Raman amplifier was used in one of the first all-fiber soliton lasers reported [1], however rare-earth-doped fiber amplifiers are now more commonly used. For more details, the interested reader may refer to one of a number of texts on optical fiber amplifiers and lasers [2–4].

The characteristics of fiber amplifiers have can have a significant impact on the performance of optical fiber soliton lasers. For example, optical fiber amplifiers usually have a relatively small pump absorption coefficient, and so several meters of fiber (or even more in Raman amplifiers) are commonly required to absorb most of the pump and produce large gain. Consequently optical fiber lasers are relatively long (typically between 10cm and 100m), and the spacing between modes is correspondingly small (typically between 1GHz and 1MHz). To generate ultrashort optical pulses in such lasers at high repetition rates therefore requires mode-locking of many harmonically-related modes. Secondly, the metastable level lifetime of rare-earth dopants in glass is relatively long (typically milliseconds), hence a large inversion is readily achieved, resulting in low noise amplification with noise figure close to the quantum limit of 3dB. Furthermore, the energy stored in rare-earth doped fiber amplifiers is relatively large (i.e. hundreds of microjoules), especially compared to the energy of solitons in optical fibers. Consequently optical fiber amplifiers generally do not suffer gain depletion from the passage of a single pulse, but rather the gain saturates in response to the average signal power in the laser cavity, which adjusts itself so that the round-trip gain and loss in the cavity are equal. Lastly, whilst relatively broadband, the finite gain-bandwidth of optical fiber amplifiers usually limits the minimum possible pulse duration that can be generated in soliton fiber lasers to tens of femtoseconds.

## Optical Feedback

Broadband optical feedback is readily achieved using either using an optical fiber loop mirror [5] (see also Sect. 3.2), or by splicing the fibers into a loop, as in Fig. 1. In ring cavities the isolator forces unidirectional lasing, thus avoiding complications which can arise due to interactions between counter-propagating waves within the gain medium and laser cavity.

Another convenient form of mirror for providing optical feedback is the Bragg grating, i.e. an optical filter usually used in reflection, and formed by a periodic variation in refractive index along the core of the fiber. Fiber Bragg gratings are also commonly used for spectral and dispersion control in fiber lasers and transmission systems. For further details on fiber Bragg gratings the interested reader may refer to various review articles and texts on the subject, e.g. [6,7].

## Additional Components

Because fiber lasers are relatively long they can be susceptible to environmental perturbations, e.g. changes in temperature, stress, vibration, etc. Consequently it is often necessary in practice to add more components than shown in Fig. 1 to stabilize the laser against such influences, e.g. to stabilize the pulse repetition rate [8], or lock the cavity modes to a stable reference frequency [9].

## 2.2   Nonlinearity and Dispersion

The nonlinearity and dispersion in optical fibers often play a key role in determining the properties of optical fiber soliton lasers. "Nonlinearity" here refers to the Kerr nonlinearity, i.e. the third order nonlinear susceptibility, on result of which is an intensity-dependent refractive index, which causes spectral broadening and self-phase modulation (SPM) of light propagating in the fiber. "Dispersion" refers to group-velocity dispersion (GVD), associated with a wavelength-dependent refractive index, the primary result of which is temporal broadening of optical pulses. These effects are described in the following subsections, however for more detail the interested reader is referred to one of the excellent texts available on nonlinear fiber optics, e.g. [10].

## Self-phase Modulation

The small guiding cross-section and low loss of optical fibers can result in significant nonlinear effects. One of the main nonlinear effects in soliton fiber lasers is the intensity dependent refractive index,

$$n = n_0 + n_2 I \tag{1}$$

where $n_2$ is the nonlinear index coefficient (typically $3 \times 10^{-20} \mathrm{m}^2/\mathrm{W}$ in silica fibers). The nonlinear index results in the accumulation of an intensity-dependent phase during propagation, i.e. self-phase modulation, $\phi_{NL} = (2\pi/\lambda)n_2 Iz$. In the absence of dispersion the effect of the intensity-dependent refractive index on optical pulse propagation may be calculated analytically:

$$u(z,t) = u(0,t)\exp\{i\gamma|u(0,t)|^2 z\}, \tag{2}$$

in which $u$ is the complex amplitude of the pulse envelope (in the slowly varying envelope approximation), $z$ is the distance of propagation, and $\gamma = (n_2\omega_0)/(cA_{\mathrm{eff}})$ is the nonlinearity coefficient, where $\omega_0$ is the optical carrier frequency, and $A_{\mathrm{eff}}$ is the effective area of the core (typically $\gamma \approx 10$ W$^{-1}$km$^{-1}$ in single mode silica fibers). The length scale on which nonlinearity becomes significant is $L_{\mathrm{NL}} = 1/(\gamma P_0)$, in which $P_0$ is the peak optical power. Equation (2) indicates that the output does not change shape during propagation, but broadens spectrally, developing a *chirp* (i.e. time-dependent optical carrier frequency) with instantaneous frequency offset from the carrier $\delta\omega = -\partial\phi_{\mathrm{NL}}(z,t)/\partial t$. For example, Fig. 2 shows the spectral broadening and chirp development due to self-phase modulation in a gaussian input pulse with RMS width $T_0 = 1$ps.

## Group Velocity Dispersion

In the absence of nonlinearity, the effect of GVD on optical pulse propagation may be calculated analytically:

$$U(z,\omega) = U(0,\omega)\exp\{\frac{i}{2}\beta_2\omega^2 z\} \tag{3}$$

in which $U(z,\omega)$ is the Fourier transform of the complex pulse envelope, and $\beta_2$ is the group velocity dispersion coefficient (typically $\beta_2 \approx -20$ps$^2$/km at 1.55 $\mu$m in single mode silica fibers). The length scale on which dispersion may be regarded as significant for an initially unchirped (i.e. transform limited) pulse is $L_{\mathrm{D}} = T_0/|\beta_2|$. In this case the power spectral density does not change with propagation, however the pulse broadens temporally, and develops a chirp. For example, Fig. 3 shows the temporal broadening and chirp development due to anomalous GVD on a gaussian input pulse with RMS width $T_0 = 1$ps.

## 2.3    Typical Characteristics of Optical Fiber Soliton Lasers

The main performance characteristics of interest in soliton fiber lasers are; pulse duration (typically ps or less), pulse bandwidth (typically nm), pulse energy

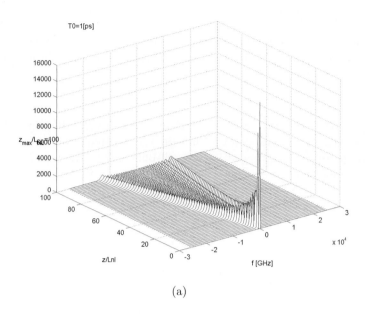

(a)

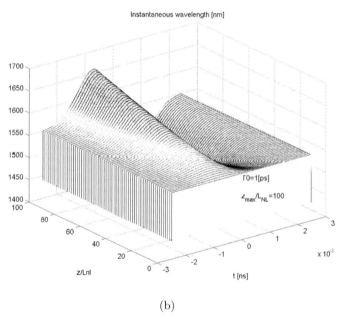

(b)

**Fig. 2.** The effect of self-phase modulation on (**a**) pulse spectral broadening, and (**b**) chirp development

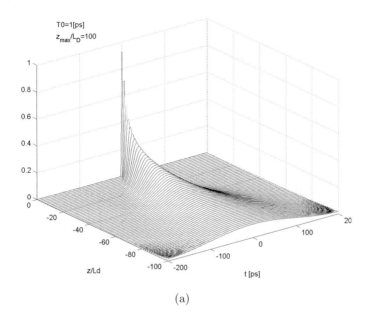

(a)

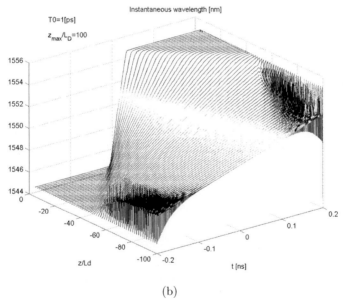

(b)

**Fig. 3.** The effect of group velocity dispersion on (**a**) pulse temporal broadening, and (**b**) chirp development

(typically nJ), repetition rate (typically MHz to many GHz), output power (typically mW), laser stability and self-starting (depends on the laser configuration), and noise and jitter (potentially quantum-limited).

Similar performance tradeoffs apply in soliton lasers as in soliton transmission systems. For example, to generate very short pulses with a given peak power generally requires low average dispersion, however this reduces the energy per pulse due to the energy quantization of solitons, which has also been observed in lasers, [11]. Also, for a given pulse duration, the maximum repetition rate achievable in soliton lasers can be limited by soliton-soliton interactions [12].

Furthermore, in optical fiber lasers and transmission systems the nonlinear effects (which vary with pulse amplitude) and dispersion are often not distributed smoothly throughout the cavity, but lumped into discrete sections, just as gain is localized in the fiber amplifier. The rapid changes perturb the solitons propagating in the cavity, and cause losses to dispersive (i.e. low intensity) waves. The perturbations often have little effect on soliton generation, as provided the cavity length is much shorter than the soliton period, $z_0 = \pi L_D/2$, solitons do not change significantly as they propagate around the cavity, and mainly respond to the average dispersion and nonlinearity in the system [13,14]. In the latter case one may imagine that the chirps induced by the localized nonlinearity and anomalous dispersion in Figs. 2(b) and 3(b) roughly cancel for specific values of pulse duration and peak power. However, if the soliton period is less than several times the cavity length (or the distance between amplifiers in a transmission system) the dispersive waves can build up and interact with the solitons, causing pulse jitter and instability [15–17].

Whilst the description of soliton dynamics outlined above explains the effects of lumped nonlinearity and dispersion, and how solitons might still form in a laser where the components are lumped rather than distributed, it is not the whole picture. For example, pulses may form in optical fiber lasers even with normal average dispersion. The underlying principle is that there must be a balance between the various sources of gain and loss, and between the causes of temporal shortening and broadening, though all these elements may interact. Consequently the dynamics of nonlinear fiber lasers, and the variety of solitons which may be produced, are considerably richer than occurs in simple lossless nonlinear propagation. The soliton control mechanisms mentioned in the following section may be used to alleviate or overcome practical limitations, and to improve the performance of soliton lasers, just as in transmission systems.

## 2.4   Soliton Generation Versus Soliton Transmission

In linear laser cavities, the boundary conditions restrict any stable temporal output of the laser to periodic functions, usually regarded as a superposition of phaselocked modes. This situation is shown schematically in Fig 4.

The latter view of pulse generation presents difficulties in the case of nonlinear soliton lasers. In the presence of nonlinearity the optical path length of the cavity depends on the instantaneous intensity, and hence on the shape of the pulse in time, or on the phase relationships between its Fourier components.

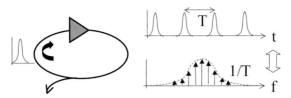

**Fig. 4.** Periodic pulse generation

Consequently the temporal output of a pulsed fiber laser is not exactly a superposition of the cavity's linear modes; the modes depend on the pulse shape and intensity, and vice-versa.

An alternative approach to understanding pulse generation in nonlinear laser cavities is to regard each pulse as an *eigenfunction of the nonlinear operator describing the cavity*. In this sense the pulses are regarded as independent particles which must self-replicate every round-trip for stable pulse generation to continue. This view is also compatible with soliton transmission systems, particularly if each section of the transmission link is identical. For example, if any one section of the periodic transmission system shown schematically in Fig. 5 was looped back on itself, it could form a laser cavity (albeit a relatively long one) like that in Fig. 4.

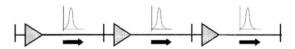

**Fig. 5.** Periodic pulse transmission

The analogy between soliton lasers and soliton transmission systems may be extended by noting the almost parallel development and application of soliton control mechanisms in each. In practical soliton lasers and transmission systems soliton control mechanisms are usually required to limit the effects of soliton interactions [18], soliton self-frequency shift [19], noise and Gordon-Haus jitter [20], and interactions with dispersive waves [21].

The main soliton transmission control mechanisms developed to date are temporal regeneration (e.g. either by active modulation [22], or passively, using a fast saturable absorber [23]), spectral control (e.g. in fixed frequency [24] and sliding-frequency [25] systems), and dispersion (or nonlinearity) management (e.g. by alternating sections of fiber with high and low dispersion, but with low net dispersion [26,27]). Similarities between soliton pulse formation and control mechanisms in fiber lasers and transmission systems will be highlighted in subsequent sections. Table 1 summarizes the development of soliton control mechanisms in fiber lasers and transmission systems.

**Table 1.** Development of soliton control methods in optical fiber lasers and transmission systems

| Soliton Control | Soliton Transmission Systems | Soliton Fiber Lasers |
|---|---|---|
| Temporal (active) | Active temporal regeneration [22] | Actively mode-locked [28], [29] |
| Temporal (passive) | Passive temporal regeneration [23] | Passively mode-locked, with fast or slow saturable absorber [30,31], [32,33], [34,35] |
| Spectral | Spectrally stabilized [24] | Spectrally stabilized [36,37] |
| Sliding | Sliding-filter/frequency [25] | Sliding-frequency [38], [39] |
| Dispersion | Dispersion-managed [26], [27] | Stretched-pulse [40] |

# 3   Experimental Developments in Optical Fiber Soliton Lasers

Ultrashort pulse generation in optical fiber soliton lasers is typically achieved using one of the following mode-locking techniques;

**Active mode-locking:** (slow) active modulation of the cavity loss or length, together with short pulse formation due to the combined effects of nonlinearity and dispersion.

**Passive mode-locking:** fast passive modulation of the cavity loss (e.g. by a fast saturable absorber).

**Soliton mode-locking:** slow passive modulation of the cavity loss (e.g. by a slow saturable absorber), initiating short pulse formation due to the combined effects of nonlinearity and dispersion.

**Hybrid mode-locking:** any combination of the above (i.e. aiming to combine the advantages of each, and to avoid their respective disadvantages).

In the following sections we concentrate on short pulse lasers, i.e. in which the effective cavity dispersion is anomalous. Nevertheless, it should be noted that bright pulse generation is often possible even when the intracavity dispersion is effectively normal, however the resulting pulses tend to be broad both spectrally and temporally, i.e. chirped, with a relatively large time-bandwidth product.

## 3.1   Active Mode-Locking

By actively modulating the loss or delay in the laser cavity at a frequency close to the mode spacing, the laser modes become injection locked to each other, resulting in a periodic pulse output.

According to linear mode-locking theory, actively mode-locked lasers cannot generate pulses as short as passively mode-locked lasers [41], however it was discovered very early in the development of fiber lasers that in the presence of nonlinearity and anomalous dispersion, spectral broadening and temporal shortening of the pulses occurred, resulting in soliton generation with much shorter pulsewidth than otherwise possible [28,29].

Actively mode-locked fiber lasers are attractive options for generating ultrashort optical pulses at very high repetition rates; environmentally stabilized lasers generating picosecond pulses at $>100$GHz repetition rates have been demonstrated, e.g. [42].

## 3.2    Passively Mode-Locked Soliton Fiber Lasers

Passively mode-locked optical fiber soliton lasers are very simple sources of ultrashort (typically sub-picosecond) optical pulses. To promote pulse generation and shortening they use a fast saturable absorber, the function of which is to reject low intensity light from the cavity, but to pass high intensities. The fast saturable absorber function is readily implemented in optical fiber using one of two methods; a nonlinear loop mirror [43], [44], or nonlinear polarization rotation and a polarizer [45]. Both implementations may approximately be regarded as linear interferometry of two pulses derived from the same source, one of which undergoes SPM due to nonlinear propagation, whilst the other does not. It is not difficult to show that such a process leads to pulse shortening, as shown in Fig. 6, independent of the sign of the dispersion.

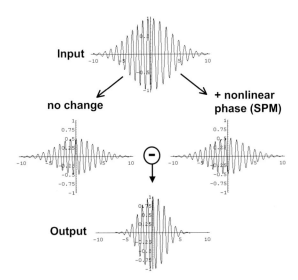

**Fig. 6.** Pulse shortening by linear interferometry of pulses following linear and nonlinear propagation

If the shortened output pulse is fed back into the laser cavity from which it originated, then the pulse shortening process continues until it is balanced by temporal broadening mechanisms, such as bandwidth limiting and group velocity dispersion. The first soliton laser relying on this pulse forming mechanism was reported by Mollenauer and Stolen [46]. When used to promote pulse formation in lasers, the technique is commonly called *additive pulse mode-locking* (APM)

[47]. Passively mode-locked fiber lasers employing APM are described in the next two sections.

## Figure-8 Fiber Laser

The figure-8 laser, which derives its name from its cavity geometry, shown schematically in Fig. 7, is a passively mode-locked optical fiber laser which readily produces subspicosecond optical pulses [30,31]. With reference to Fig. 7 it can be regarded as a ring laser (to the left of the 50:50 coupler), with a nonlinear amplifying loop mirror (NALM) [44], comprising the 50:50 coupler and amplifier loop on the right, providing fast saturable gain. The operation of the NALM is described in the next section, and causes additive-pulse mode-locking of the laser.

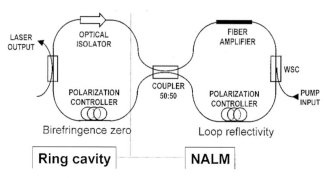

**Fig. 7.** Figure-8 optical fiber soliton laser schematic

**Nonlinear loop mirrors.** The operation of the NALM is similar to its passive predecessor, the nonlinear optical loop mirror (NOLM) [43]; both are antiresonant nonlinear Sagnac interferometers, with intensity-dependent transmission. The key to operation of devices such as the NOLM and NALM is an asymmetry between the clockwise and anticlockwise waves propagating around the loop. In the NOLM the asymmetry is typically provided by using a non-50:50 split ratio in coupler. In the NALM the asymmetry is provided by placing the fiber amplifier closer to one of the coupler arms than the other, as on the right-hand side of in Fig. 7. In either case, light travelling clockwise around the loop has a significantly different intensity to light travelling anticlockwise, and hence a difference in nonlinear phase is accumulated by the counterpropagating waves. On recombining at the coupler, low intensity light (which accumulated no nonlinear phase) is reflected back to the port from which it entered, whilst higher intensity light is transmitted.

It can readily be shown that for CW inputs the NALM with 50:50 coupler has a power-dependent transmission function, given by [44]

$$P_T \approx GP_{\mathrm{IN}} \left\{ 1 - 0.5 \left[ 1 + \cos\left( 0.5\gamma L P_{\mathrm{IN}}(1 - G) \right) \right] \right\}, \tag{4}$$

in which $L$ is the nonlinear loop length, and G the amplifier power gain. The nonlinear transmission is plotted in Fig. 8. The minimum transmission usually occurs at low input power (i.e. zero nonlinear phase difference), however the loop-mirror may be biased to provide any transmissivity at zero-input by adjusting the loop birefringence [48].

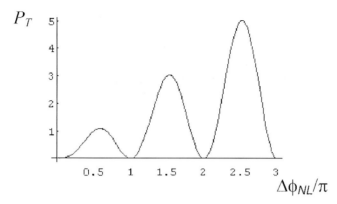

**Fig. 8.** Plot of the transmission of nonlinear optical fiber amplifying loop mirror versus input power, as described by Eq. 4

Apart from promoting short pulse generation in optical fiber lasers, non-linear loop mirrors may be used for ultrafast all-optical switching [49–51], and have been suggested for use as temporal pulse regenerators in soliton transmission systems [23]. Interestingly, solitons can be switched almost completely through nonlinear loop mirrors in the same way as CW [43,50]. It can be shown that maximum energy transmission (i.e. >90%) for fundamental solitons occurs when the peak difference in nonlinear loop phase is $\Delta\phi_{NL} = 0.6\pi$. The excellent switching properties of solitons arise from their uniform soliton phase [52].

**Laser performance and practical considerations.** The typical output from a figure-8 fiber laser in both the time and wavelength domains is shown in Figs. 9(a) and 9(b), respectively.

The results illustrate a number of practical difficulties in passively mode-locked soliton fiber lasers. Firstly, because the pulse peak power and duration are largely determined by the NALM peak transmission power and cavity dispersion, the pulse energy is quantized [11]. Consequently the number of pulses in a passively mode-locked laser cavity varies in proportion to amplifier pump power above threshold. Also, because the pulses circulating in the laser cavity build up from noise, the pulse repetition rate is usually random, with pulses often clumping together in bursts as shown in Fig. 9(a). The latter problem may be alleviated by coupling the laser to a sub-cavity or resonator [8], thus seeding pulse formation at a fixed delay or repetition rate.

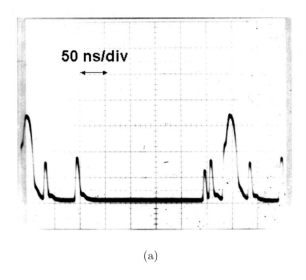

(a)

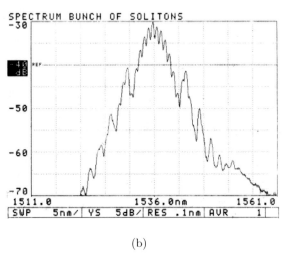

(b)

**Fig. 9.** Typical output from a figure-8 optical fiber soliton laser (**a**) in the time domain (showing pulse energy quantization, random repetition rate, and bunching), and (**b**) in the frequency domain (showing characteristic of chirped solitons, and phasematched sidebands)

For similar reasons as above, the intracavity pulse power and duration are unlikely to correspond to those of fundamental solitons for the fiber used in the laser cavity, hence the output pulses are likely to be slightly chirped. The latter is apparent in the rounded spectrum shown in Fig. 9(b), which would appear more triangular in shape for a transform-limited pulse with $\text{sech}^2$ intensity. Of course, the output pulse chirp, peak power, pulse width, and energy may be modified outside the laser cavity using a variety of techniques involving amplification, and nonlinear and/or dispersive propagation [53].

Furthermore, depending on the NALM characteristics, the figure-8 laser usually has a relatively high self-starting threshold, and the cavity must be perturbed in some way to initiate pulse generation. The latter problem is best solved by using hybrid mode-locking techniques, for example by incorporating a slow saturable absorber into the cavity (see also Sects. 3.3 and 3.4).

Another potential problem is that the laser cavity incorporating the NALM is usually relatively long (i.e. >10m), which can cause a number of difficulties in practice. The first is that the cavity can be susceptible to slow environmental changes. For example, changes in temperature may cause the cavity length to vary by distances comparable to the pulse length, or may cause the net birefringence in the cavity to change and hence change the NALM switching characteristic and the laser operating point. Consequently active control over the cavity length and birefringence is desirable for long-term stability.

Furthermore, if a high repetition rate output is required then many pulses must be circulating within the cavity at a time. Under these circumstances the effects of noise, perturbations, and soliton interactions can cause the pulses in the cavity to move relative to each other and to interact. The introduction of a passband filter into the laser cavity has been observed to reject noise and stabilize the lasing wavelength [36,37], i.e. by reducing noise and Gordon-Haus jitter, preventing soliton frequency shifts, and damping soliton interactions.

Lastly, sudden changes in amplitude and dispersion around the cavity cause the solitons to shed low intensity (i.e. dispersive) radiation [21], which may propagate and interact with the solitons in the cavity [15,16]. The dispersive radiation is phasematched to the solitons at specific wavelengths [17], and appears as sidebands in the laser output, as in Fig. 9(b). Resonances between the soliton period and the periodic perturbations associated with a finite cavity length ultimately limit the minimum pulse duration achievable in any given soliton laser.

## Nonlinear Polarization Rotation Fiber Laser

Nonlinear polarization rotation is the tendency of the polarization ellipse (i.e. describing the polarization state of light) to rotate when propagating in a birefringent nonlinear medium [54,55]. The amount of rotation depends on the instantaneous intensity of light, and results from a combination of self-phase modulation and cross-phase modulation of the two orthogonal polarization components. A polarizer placed after the fiber can be adjusted to reject low intensity components of the signal, whilst passing high intensities; the resulting intensity-dependent transmissivity is similar to the NOLM and NALM, and provides a fast saturable absorber action which can be used to passively mode-lock fiber lasers [32,33].

Some fiber lasers with outstanding properties have been realized using this mode-locking technique, e.g. the shortest (38fs) pulse generated in a fiber laser [56], and high energy (0.5nJ) pulse generation in a dispersion-managed laser [57]. The latter laser is discussed further in Sect. 3.5.

## 3.3   Soliton Mode-Locking

Here we define soliton mode-locking as any pulse generation method in which soliton formation occurs through a combination of nonlinearity and asymmetric confinement in a temporal or spectral window that may be much wider than that of the pulses being generated.

### Mode-Locking with a Slow Saturable Absorber

It was previously shown that short pulse generation was possible in actively mode-locked lasers, despite the weak temporal confinement (i.e. slow loss modulation), due to pulse shortening and soliton formation through a balance between nonlinearity and anomalous dispersion. A similar mechanism can also operate in passively mode-locked lasers, in which the weak temporal confinement is provided by a slow saturable absorber [35].

   The situation is shown schematically in Fig. 10. On arriving at the slow saturable absorber, the energy in the pulse causes a rapid increase in cavity loss. The recovery of the loss is relatively slow (typically picoseconds), however in soliton lasers the recovery time does not directly affect the pulse duration, which is rather determined instead by a balance between anomalous dispersion and nonlinearity in the cavity [58].

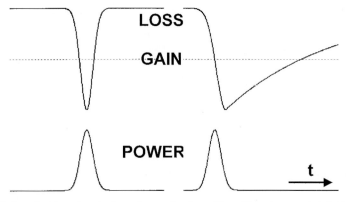

**Fig. 10.** Pulse shaping due to loss dynamics in soliton fiber lasers; (**a**) fast saturable absorber, (**b**) slow saturable absorber and soliton formation

   A variety of semiconductor slow saturable absorber devices have been developed [59], and applied to mode-lock various types of laser, including fiber lasers [34,60–62]. Careful laser design is required to prevent Q-switching instabilities [59,63], particularly in rare-earth doped fiber lasers with long excited-state lifetimes. Some novel soliton states have been observed in these types of lasers [64,65].

## Sliding-Frequency Lasers

The schematic of a sliding-frequency soliton laser is shown in Fig. 11. The principle of operation has much in common with sliding-filter soliton transmission [25,66], the only difference being that the wavelength of the light is shifted every round-trip (e.g. using a CW-driven acousto-optic modulator) rather than changing the wavelength of the filter. In both cases the pulse spectrum is slightly offset from the filter's central wavelength, as shown schematically in Fig. 12. The laser has a relatively low self-starting threshold as CW lasing and noise are suppressed by the frequency shifter and filter; only solitons are able to propagate continuously in the cavity, with the asymmetric spectral loss each round-trip being balanced by the generation of new frequencies through self-phase modulation [38,39,67,68].

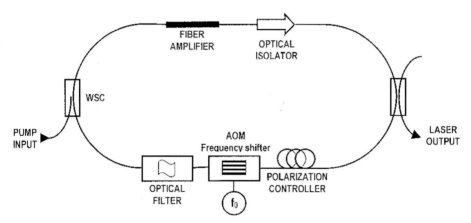

**Fig. 11.** Schematic of sliding-frequency soliton fiber laser cavity

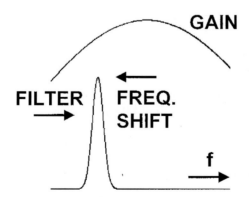

**Fig. 12.** Soliton spectral confinement in the sliding-frequency laser, under the competing effects of the frequency-shifter and self-phase modulation

A characteristic of the sliding-frequency laser is that the spectral width of
the pulses generated is much narrower than the bandwidth of the filter confining them, and consequently it can be difficult to generate pulses much shorter
than 10ps. It can be shown by perturbation analysis that the full-width at half-
maximum duration of solitons generated in the sliding-frequency laser is given
by

$$\Delta\tau = \frac{1.44}{\sqrt[3]{\pi \cdot 10^{-6} \cdot f_s \cdot \Delta\nu^2}}, \tag{5}$$

where $\Delta\tau$ is the pulse width in picoseconds, $f_s$ is the frequency shift per round-
trip in megahertz, and $\Delta\nu$ is the bandpass filter bandwidth in teraradians per
second. Surprisingly, both theoretical and experimental results show that in the
normal dispersion regime the pulse duration is independent of the amount of
cavity dispersion [69].

Apart from the low self-starting threshold, other useful properties of the laser
(and of sliding-frequency transmission systems) which result from the passband
filter and frequency-sliding are; damping of pulse interactions, limited noise and
Gordon-Haus jitter, and suppression of dispersive wave radiation and soliton
resonances.

## 3.4   Hybrid Mode-Locking

Hybrid mode-locking can be used to combine the best properties of the various
mode-locking techniques mentioned in the previous sections. For example, active
mode-locking techniques may be combined with intra-cavity fast saturable gain
or loss to promote ultrashort pulse generation at a stable repetition rate [70].
Similarly, ultrashort pulse generation in a passively mode-locked laser may be
combined with the low threshold of sliding-frequency soliton mode-locking [71].

## 3.5   Other Soliton Lasers

In this section we review two interesting soliton fiber lasers which use mode-
locking techniques discussed in previous sections, but which have significantly
different behaviors. Both have analogues in soliton transmission systems.

### Stretched Pulse Fiber Laser

The laser shown schematically in Fig. 13 contains two equal and opposite sections
of dispersive fiber, such that the total cavity dispersion is close to zero [40]. The
laser is passively mode-locked by a fast saturable absorber based on nonlinear
polarization rotation.

A fiber laser with close to zero dispersion would normally generate very
short pulses, but with very low energy in each pulse, however in the dispersion-
balanced laser the pulse energy can be significantly larger. The reason is that
soliton "breathes" as it propagates around the cavity, undergoing large changes
in temporal width, returning to its minimum temporal width (and maximum

peak power) at the midpoint of each dispersive section. Because the pulse has a high peak power at only two points in the cavity, the average nonlinearity in the cavity is reduced, and the pulse energy is increased accordingly. The same principle is used in dispersion-managed soliton transmission systems to increase the transmitted pulse energy in links with low net dispersion [72]. Other advantages of the stretched-pulse laser in common with dispersion-managed transmission systems are the potential for reduced pulse interactions, and reduced noise and jitter [73–75].

**Multiwavelength Sliding-Frequency Laser**

If it possible to realize a WDM soliton transmission system, then by the analogy used in Sect. 2.4 it should be possible to realize a multiwavelength soliton laser. Such a laser has two particular requirements; a) unless the average dispersion of the cavity is zero, only frequency-domain soliton control methods can used for pulse generation and stabilization, and b) gain cross-saturation between the different lasing wavelengths must be avoided.

Soliton generation at up to nine wavelengths has been observed in a sliding-frequency soliton laser [76,77]. Multiwavelength lasing was achieved by incorporating a broadband Fabry-Perot-like Bragg grating resonator [78] in the cavity to define the lasing wavelengths, and by inhomogeneous broadening the gain of the erbium doped amplifier by cooling it in liquid nitrogen.

The typical output of the laser is shown in Fig. 13. Due to the non-zero cavity dispersion, solitons at different wavelengths had different group velocities, and hence the temporal output of the laser was essentially random, nevertheless some beating between the different wavelength solitons was observable in the autocorrelation.

# 4  Theory of Soliton Lasers

A common method for modelling mode-locked lasers which was pioneered by Haus is the "master equation" approach [79], in which a differential equation is derived to describe the average effect of the various components forming the laser cavity on light propagating in the cavity. In its simplest form, this approach assumes that any pulses generated in the laser change only slightly during each round-trip of the cavity, which can therefore be modelled as a distributed system. In this review we restrict our attention to passively mode-locked lasers, which have often been described by the complex Ginzburg-Landau equation (CGLE) [80,81], or more accurately by the quintic CGLE. Quintic terms in the equation are essential for stable pulse solutions [82]. In the optical context the quintic CGLE has the following form [83]:

$$i\psi_z + \frac{D}{2}\,\psi_{tt} + |\psi|^2\psi + \nu|\psi|^4\psi \; = i\delta\psi + i\epsilon|\psi|^2\psi + i\beta\psi_{tt} + i\mu|\psi|^4\psi, \qquad (6)$$

where $z$ is the cavity round-trip number, $t$ is the retarded time, $\psi$ is the normalized envelope of the field, $D$ is the group velocity dispersion coefficient with

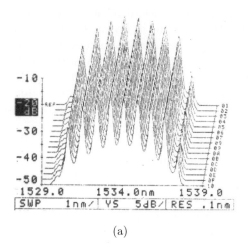

(a)

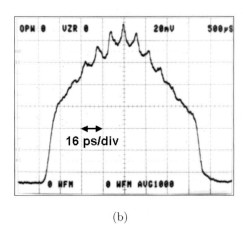

(b)

**Fig. 13.** Multiwavelength sliding-frequency soliton laser (**a**) output spectrum, and (**b**) autocorrelation of the output showing beating between pulses at different wavelengths

$D = \pm 1$ depending on whether the group velocity dispersion (GVD) is anomalous or normal, respectively, $\delta$ is the linear gain-loss coefficient, $i\beta\psi_{tt}$ accounts for spectral filtering or linear parabolic gain ($\beta > 0$), $\epsilon|\psi|^2\psi$ represents the nonlinear gain (which arises, e.g., from saturable absorption), the term with $\mu$ represents, if negative, the saturation of the nonlinear gain, the one with $\nu$ corresponds, also if negative, to the saturation of the nonlinear refractive index.

The above continuous model takes into account the major physical effects occurring in a laser cavity such as dispersion, self-phase modulation, spectral filtering, and gain or loss (both linear and nonlinear). A delicate balance between them gives rise to the majority of the effects observed experimentally.

In the case of slow saturable absorbers, we have to take into account the delayed response of the absorber. Again, the laser is modelled as a distributed system [84] if the pulse shape changes only slightly during each round-trip. The pulse evolution is then governed by a modified CGLE with non-linear non-conservative terms [79,35]:

$$i\psi_z + \frac{D}{2}\psi_{tt} + |\psi|^2\psi = i[g(Q) - \delta_s(|\psi|^2)]\psi + i\beta\psi_{tt}, \tag{7}$$

where notations are the same as above. In addition, $g(Q)$ is the cavity gain $(Q = \int_{-\infty}^{\infty} |\psi|^2\, dt)$, and $\delta_s(|\psi|^2)$ represents the losses in the cavity and in the slow saturable absorber.

The gain term $g(Q)$ in Eq. (7) describes a typical solid-state laser gain medium with a recovery time much longer than the round-trip time of the cavity. For our purposes, the recovery time can be considered infinite. Therefore $g(Q)$, which accounts for gain depletion, depends on the total pulse energy in the following way:

$$g(Q) = \frac{g_0}{1 + \frac{Q}{E_L}}. \tag{8}$$

Here $E_L$ is the saturation energy and $g_0$ is the small signal gain. The value of $g(Q)$ decreases as the energy increases, so that only a limited number of pulses can exist inside the cavity.

The loss modulation in the saturable absorber can be described by the following rate equation [35]:

$$\frac{\partial\delta_s}{\partial t} = -\frac{\delta_s - \delta_0}{T_1} - \frac{|\psi|^2}{E_A}\delta_s, \tag{9}$$

where $T_1$ is the recovery time of the saturable absorber, $\delta_0$ is the loss introduced by the absorber in the absence of pulses, and $E_A$ is the saturation energy of the absorber.

The presence of birefringent elements in the cavity adds new effects. Phase-locking of the two soliton components in the passive birefringent medium has been predicted theoretically [85,86]. Polarization locking effects have been observed experimentally in a laser with a slow saturable absorber [87].

In the latter case the pulse evolution in the laser is governed by a set of two modified nonlinear Schrödinger equations (NLSE) with non-linear and non-conservative terms [64]:

$$i\phi_z + \gamma\phi + \frac{D}{2}\phi_{tt} + |\phi|^2\phi + A|\psi|^2\phi + B\psi^2\phi^* = i[g(Q_1) - \delta_s(|\phi|^2)]\phi + i\beta\phi_{tt},$$

$$i\psi_z - \gamma\psi + \frac{D}{2}\psi_{tt} + |\psi|^2\psi + A|\phi|^2\psi + B\phi^2\psi^* = i[g(Q_2) - \delta_s(|\psi|^2)]\psi + i\beta\psi_{tt}, \tag{10}$$

where $\psi$ and $\phi$ are the normalized envelopes of the two components of the optical field, $\gamma$ is the half-difference between the phase velocities of the two field components, $D$ is the group velocity dispersion coefficient, $A$ is the cross-phase modulation coefficient, $B$ is the coefficient of the energy-exchange term (four-wave-mixing), $\beta$ represents spectral filtering ($\beta > 0$), $g(Q_i)$ is the gain in the cavity which depends on the energy, $Q_i = \int\limits_{-\infty}^{\infty} (|\phi, \psi|^2) \, dt$, in each component of the pulse in one round-trip, and $\delta_s(|\psi|^2)$ is the total loss, including loss in the semiconductor saturable absorber.

In situations where the discrete or lumped nature of the laser cavity cannot be ignored, the above CGL equations can be modelled such that the parameters $D$, $\delta$, $\beta$, $\epsilon$, $\mu$, and $\nu$, vary periodically with $z$, the period corresponding to a cavity round-trip. The goal of this is to capture the essential effects of periodicity and abrupt changes of parameters inside the cavity. This model is similar to that used for dispersion-managed optical transmission lines [88–94]. An example of the map of the parameters is shown in Fig. 14 [95]. Dispersion, nonlinearity, and linear and nonlinear gain act for a certain propagation length, $L_1$ (modelling the evolution in the gain medium). The output coupler is modelled by concentrating losses at the end of each round-trip. For the bulk of the cavity, the pulse propagates a distance $L_2$ under the sole effect of the dispersion term, $D_2$. This model which have distributed parameters. It gives soliton-like solutions which change shape as they propagate within the cavity, and can be called "dispersion-managed solitons".

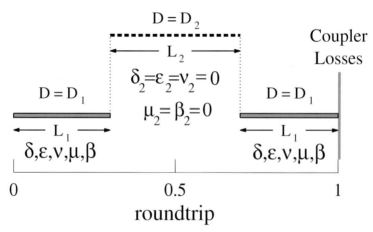

**Fig. 14.** The map of the parameters of the CGLE for a single round-trip in the laser as used in some numerical simulations

In this section the basic equations for modelling the operation of passively mode-locked lasers have been presented. There is a wide variety of such laser systems, and hence the exact model may also vary from one system to another. Nevertheless, the knowledge of these models and their solutions helps to understand the processes of pulse generation in laser systems in general.

## 4.1    Exact Solutions

The CGLE Eq. (6) has no known conserved quantities. This contrasts with the infinite number for the NLSE and the small number for Hamiltonian systems. For an arbitrary initial condition, even the energy is not conserved, as there is energy input exchange with from the external pump. None of the equations (6), (7) and (10) is integrable, and only particular exact solutions can be obtained. In general, initial value problems with arbitrary initial conditions can only be solved numerically. The cubic CGLE, obtained by setting $\mu = \nu = 0$ in Eq. (6), has been studied extensively [96–98]. Exact solutions to this equation can be obtained using a special ansatz [96], the Hirota bilinear method [97], or reduction to systems of linear PDEs [99]. However, it was realized many years ago that the soliton-like solutions of this equation are unstable to perturbations.

We give here the stationary soliton solution of Eq. (6) with zero transverse velocity. This occurs when $\beta \neq 0$:

$$\psi(\tau, z) = A(\tau) \exp(-i\omega z), \tag{11}$$

where $\omega$ is a real constant. The complex function $A(\tau)$ can be written as $A(\tau) = a(\tau) \exp[i\phi(\tau)]$ where $a$ and $\phi$ are real functions of $\tau$ with

$$\phi(\tau) = \phi_0 + d \ln[a(\tau)], \tag{12}$$

where $d$ is the chirp parameter and $\phi_0$ is an arbitrary phase. We suppose $\phi_0 = 0$ for simplicity. For the cubic case, this ansatz covers all pulse-like solutions. In the quintic case, however, Eq. (12) is a restriction imposed on $\phi(\tau)$, because the chirp could have a more general functional dependence on $\tau$.

For the cubic CGLE, that is Eq. (6) with $\nu = \mu = 0$:

$$a(\tau) = BC \operatorname{sech}(B\tau), \quad C = \sqrt{\frac{3d(1 + 4\beta^2)}{2(2\beta - \epsilon)}}, \quad B = \sqrt{\frac{\delta}{d - \beta + \beta d^2}}, \tag{13}$$

and $\omega$ and $d$ are given by

$$\omega = -\frac{\delta(1 - d^2 + 4\beta d)}{2(d - \beta + \beta d^2)}, \tag{14}$$

$$d = d_\pm = \frac{3(1 + 2\epsilon\beta) \pm \sqrt{9(1 + 2\epsilon\beta)^2 + 8(\epsilon - 2\beta)^2}}{2(\epsilon - 2\beta)} \tag{15}$$

where the minus sign is chosen in front of the square root.

For the quintic case, the above formulae also give a soliton solution. However, the solution is only valid for certain relations between the parameters of the equation [83,100]. In general, we need to use some numerical technique to find stationary solutions. One way to do this is to reduce Eqs. (6) to a set of ordinary differential equations (ODEs). We achieve this by seeking solutions in the form:

$$\psi(t, z) = \psi_o(\tau) exp(-i\omega z) = a(\tau) \exp[i\phi(\tau) - i\omega z], \tag{16}$$

where $a$ and $\phi$ are real functions of $\tau = t - vz$, $v$ is the pulse velocity and $\omega$ is the nonlinear shift of the propagation constant. Substituting Eq. (16) into Eq. (6), we obtain an equation for two coupled functions, $a$ and $\phi$. We separate real and imaginary parts, obtain a set of two ODEs, and transform them.

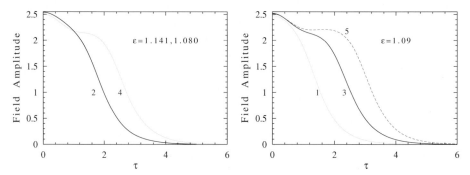

**Fig. 15.** Amplitude profiles of the stationary solutions of the quintic CGLE. The parameter values of CGLE are shown in the figure. Three types of solutions exist for the same $\epsilon = 1.09$.

The resultant set of equations contains all stationary and uniformly translating solutions. The parameters $v$ and $\omega$ are the eigenvalues of this problem. In the $(M, a)$ plane, where $M = a^2\phi'(\tau)$, the solutions corresponding to pulses are closed loops starting and ending at the origin. The latter happens only at certain values of $v$ and $\omega$. If $v$ and $\omega$ differ from these fixed values, the trajectory cannot comprise a closed loop. By properly adjusting the eigenvalue $\omega$, it is possible to find the soliton solution with a "shooting" method.

Numerical simulations show [101] that a multitude of soliton solutions of the CGLE exist. They have a variety of shapes and stability properties. They can even be partly stable and partly unstable. Trajectories on the phase portrait are deflected from their smooth motion near the singular points. As a result, the trajectory can have additional loops and the soliton shape can become multi-peaked [101].

Examples of the amplitude profiles are shown in Fig. 15. Only half of the profiles are plotted as they are even functions of $\tau$. All soliton solutions (or at least those we know about) are interconnected, i.e. continuously changing parameters of CGLE we can transform one type of soliton into another.

Generation of stable pulses is possible in a very narrow range of the laser parameters and requires their careful adjustment. More generally, the pulses may change their shape from one round-trip to another and have complicated dynamics in time, and can have periodic behavior on a time scale larger than the round-trip time [102]. If there are many periods involved in the dynamics, then the pulse shape evolution in time may become chaotic. The system can enter into a chaotic regime in various ways including the classical one through period doubling bifurcations. The most interesting phenomenon is that stationary stable solitons can coexist with the chaotic regime of soliton propagation.

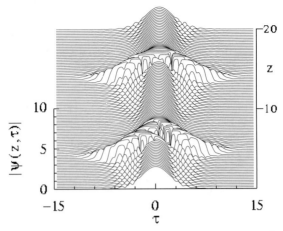

**Fig. 16.** Two periods of the evolution of an exploding soliton. The parameters are $\epsilon = 1.0$, $\delta = -0.1$, $\beta = 0.125$, $\mu = -0.1$, and $\nu = -0.6$. These solutions cannot be found in analytic form. However, they are as common as stationary solutions and exist for a wide range of parameters. The process never repeats itself exactly in successive "periods". However, it always returns to the same shape.

A special class of pulsating solitons can be called "exploding" or "erupting" [102,103] solitons. "Exploding" soliton evolution, shown in Fig. 16, starts from a stationary localized solution which has a perfect soliton shape. After a while, its flanks become covered with small ripples (a form of a small scale instability) which seem to move downwards along the two sides of the soliton, and very soon the pulse is covered with this seemingly chaotic structure. When the ripples increase in size, the soliton cracks into pieces, like a mountain after a strong volcanic eruption or after an earthquake. This can also look like an explosion. This completely chaotic, but well-localized, structure then is filled with "lava" which restores the perfect soliton shape after a "cooling" process. The process repeats forever, although the distance between "explosions" fluctuates, and in each of them the pulse splits into different pieces. The exploding soliton has been observed experimentally [95].

## 4.2   Soliton Solutions for the Case of Slow Saturable Absorber

We consider the limiting case when the pulse amplitude is well below the saturation level. The gain coefficient $g$ is constant if we deal with stationary solutions of Eq. (7) when $Q$ is constant. We also assume that the relaxation time is long in comparison to the pulse width. In this case, $T_1 \to \infty$, and the slow part of the loss variation is given by the approximate formula

$$\delta_s(t) = \delta_0 \, \exp\left[-\int \left(\frac{|\psi|^2}{E_A}\right) \, dt\right] \approx \delta_0 - \alpha \int\limits_{-\infty}^{t} |\psi|^2 \, dt' + \dots \tag{17}$$

where $\alpha = \frac{\delta_0}{E_A}$. Substituting this into Eq. (6), we obtain the equation

$$i\psi_z + \frac{D}{2}\psi_{tt} + |\psi|^2\psi = i\delta\psi + i\beta\psi_{tt} + i\epsilon|\psi|^2\psi + i\alpha\psi \int_{-\infty}^{t} |\psi|^2 dt', \quad (18)$$

where $\delta = g - \delta_0$. This equation is the complex Ginzburg-Landau equation (CGLE) [83], except that it includes a non-conservative nonlinear term on the right-hand-side of Eq. (18), with the term being nonlocal in time.

Eq. (18) has exact pulse-like solutions [104,105] which move with velocity $V$:

$$\psi = A(t - Vz) \, e^{id \, ln[A(t-Vz)]} e^{iKt - i\omega z} \quad (19)$$

where

$$A(x) = \gamma \, C \, sech(\gamma x),$$

$$d = \frac{3(D + 2\epsilon\beta) - \sqrt{9(D + 2\epsilon\beta)^2 + 8(\epsilon D - 2\beta)^2}}{2(\epsilon D - 2\beta)},$$

$$\gamma = \gamma_\pm = \frac{\alpha C^2 \pm \sqrt{\alpha^2 C^4 - 2(2\beta - dD - 2C^2\epsilon)(\beta K^2 - \delta)}}{2\beta - dD - 2C^2\epsilon}, \quad (20)$$

$$C = \sqrt{\frac{3d(D^2 + 4\beta^2)}{2(2\beta - \epsilon D)}}, \qquad K = -\frac{\alpha d C^2}{2\beta(1 + d^2)}, \qquad V = K\left(D - \frac{2\beta}{d}\right),$$

$$\omega = \frac{D}{2}(K^2 + \gamma^2) + \beta d\gamma^2 - (\gamma C)^2$$

We note that the form of $d$ ensures that $C$ is real and positive, because $d$ and $(2\beta - \epsilon D)$ have the same sign. The parameters $(\alpha, \beta, \delta, \epsilon, D)$ must be chosen to ensure that the quantity under the square root sign in $\gamma$ is positive. All the parameters of this solution, including the velocity $V$, are fixed and depend on the parameters of the equation. However, there are two branches of the solution, as specified by the two signs in Eq. (20). An example of the solution for certain values of parameters $\delta$, $\beta$ and $\alpha$ is shown in Fig. 17. Different values of loss/gain on either side of the soliton can cause it to move relative to the reference frame. For $\gamma = \gamma_-$ the solution amplitude is close to zero when $\delta = 0$. For $\gamma = \gamma_+$ it is not close to zero when $\delta = 0$. Note that $\delta$ is equal to the amount of loss (or gain) experienced by the left-hand side of the pulse.

The soliton exists for a certain range of parameters. The limits of existence are defined by the inequality

$$\alpha^2 C^4 - 2(2\beta - dD - 2C^2\epsilon)(\beta K^2 - \delta) > 0. \quad (21)$$

An important parameter for chirped pulses is the amplitude-width product $C$. It does not depend on $\alpha$ or $\delta$ but depends weakly on $\beta$ and $D$. The velocity $V$ of the soliton does not depend directly on $\delta$ but depends linearly on $\alpha$ because $K$ depends linearly on $\alpha$. The velocity varies with $\beta$ and $D$ and can be positive,

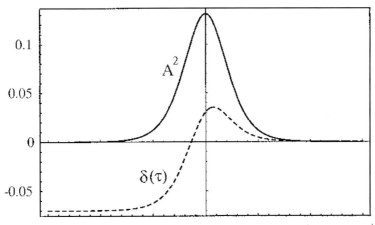

**Fig. 17.** Soliton profile (solid lines) and the loss curve $\delta(t) = \delta + \epsilon|\psi|^2 + \alpha \int\limits_{-\infty}^{t} |\psi|^2 dt$ (dotted line) defined by the exact solution (19) for $\epsilon = 0.1$, $\delta = -0.05$, $\alpha = 0.1$, $\beta = 0.02$, and $D = +1$. The soliton always clings to the gradient of the absorption curve $\delta(t)$.

negative or zero. In the case of $\delta$ positive, the solution exists for both signs in the expression for $\gamma$. Hence, we have simultaneously two solutions for the same set of parameters. Moreover, the solution exists for both normal and anomalous dispersion (negative and positive $D$). This is not surprising [106], because in systems with gain and loss, the pulse is the result of a balance not only of the dispersion and nonlinearity (which is impossible at negative $D$) but also of gain and loss.

### 4.3    Perturbation Approximation

If the coefficients $\delta$, $\beta$, $\epsilon$, $\mu$ and $\nu$ on the right-hand side are all small, then soliton-like solutions of Eq. (6) can be studied by applying perturbative theory to the soliton solutions of the NLSE. Let us consider the right-hand side of Eq. (6), with $D = +1$, as a small perturbation, and write the solution as a soliton of the NLSE, i.e.

$$\psi(\tau, z) = \frac{\eta}{\cosh[\eta(\tau + \Omega z)]} \exp[-i\Omega\tau + i(\eta^2 - \Omega^2)z/2]. \tag{22}$$

In the presence of the perturbation, the parameters of the soliton, i.e. the amplitude $\eta$ and frequency (or velocity) $\Omega$, change adiabatically. The equations for them can be obtained from the balance equations for the energy and momentum [106]. Then we have the equations for the evolution of $\eta(z)$ and $\Omega(z)$:

$$\frac{d\eta}{dz} = 2\eta \left[\delta - \beta\Omega^2 + \frac{1}{3}(2\epsilon - \beta)\eta^2 + \frac{8}{15}\mu\eta^4\right], \quad \frac{d\Omega}{dz} = -\frac{4}{3}\beta\Omega\,\eta^2. \tag{23}$$

The dynamical system of equations, (23), has two real dependent variables and the solutions can be presented on the plane. An example is given in Fig. 18. It has a line of singular points at $\eta = 0$, and, depending on the equation parameters, may have one or two singular points on the semi-axis $\Omega = 0$, $\eta > 0$. The values of $\eta^2$ for singular points are defined by finding the roots of the biquadratic polynomial in the square brackets in Eq. (23). When the roots are negative (hence, $\eta$ is imaginary), there are no singular points and hence no soliton solution. If both roots of the quadratic polynomial (in $\eta^2$) are positive (so that both $\eta$ are real), then there are two fixed points and two corresponding soliton solutions. Both roots are positive when either $\beta < 2\epsilon$, $\mu < 0$ and $\delta < 0$ or $\beta > 2\epsilon$, $\mu > 0$ and $\delta > 0$.

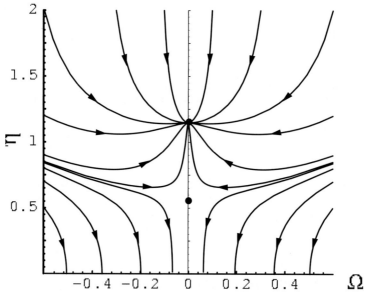

**Fig. 18.** The phase portrait of the dynamical system (23) for $\delta = -0.03$, $\beta = 0.1$, $\epsilon = 0.2$ and $\mu = -0.11$. The upper fixed point is a sink which defines the parameters of a stable approximate soliton-like solution of the quintic CGLE. Any soliton-like initial condition in close proximity to a fixed point will converge to a stable stationary solution. The points on the line $\eta = 0$ are stable when $\delta < 0$ and $\beta > 0$. This condition is needed for the background state $\psi = 0$ to be stable.

The stability of at least one these fixed points requires $\beta > 0$. Moreover, the stability of the background requires $\delta < 0$. In the latter case, we necessarily have $\beta < 2\epsilon$, $\mu < 0$ and the upper fixed point is a sink (as shown in Fig. 18) which defines the parameters of a stable approximate soliton solution of the quintic CGLE. The background $\psi = 0$ is also stable, so that the whole solution (soliton plus background) is stable. Finally, when only one of the roots is positive, there is a singular point in the upper half-plane and there is a corresponding soliton solution. However either the background or the soliton itself is unstable, so that the total solution is unstable. The term with $\nu$ in the CGLE does not influence the location of the sink. It only introduces an additional phase term, $\exp(8i\nu\eta^4 z/15)$, into the solution of Eq. (22).

In cubic CGLE, $\mu = 0$ and $\nu = 0$. The stationary point is then

$$\eta = \sqrt{3\delta/(\beta - 2\epsilon)}, \qquad \Omega = 0. \tag{24}$$

It is stable provided that $\delta > 0$, $\beta > 0$ and $\epsilon < \beta/2$. Clearly, in this perturbative case the soliton and the background cannot be stable simultaneously. Hence, this approach shows that to have both the soliton and the background stable, we need to have quintic terms in the CGLE [82].

This simple approach shows that, in general, the CGLE has stationary soliton-like solutions, and that for the same set of equation parameters two may exist simultaneously (one stable and one unstable). Moreover, this approach results in soliton parameters that are fixed; they do not have free parameters as solitons of the nonlinear Schrödinger equation.

Despite its simplicity and advantages in giving stability and other properties of solitons, the perturbative analysis has some serious limitations. Firstly, it can only be applied if the coefficients on the right-hand-side of Eq. (6) are small, and this is not always the case in practice. Correspondingly, it describes the convergence correctly only for initial conditions which are close to the stationary solution. Secondly, the standard perturbative analysis cannot be applied to the case $D < 1$, when the NLSE itself does not have bright soliton solutions, though the CGLE Eq. (6) has stable soliton solutions for this case as well.

## 4.4   Stability

The vast majority of soliton solutions lack stability. In fact, in some regions of the parameter space, all of them can be unstable. Solving the whole propagation equation only allows us to obtain stable structures [107,108]. However unstable solitons may play an essential role in the overall dynamics when the system starts with an arbitrary initial condition. Therefore, it is important to know all soliton solutions when moving in the parameter space from one point to another. Both types of stationary solitons, stable and unstable, are important and deserve careful study.

The existence of stable and unstable solitons for the CGLE raises an important issue: when we change the parameters of the equation, is it possible that there are connected regions of stable and unstable solitons? If the answer is positive, when does such a transition occur? In other words, what is the stability criterion for solitons in dissipative systems? An attempt to answer the above questions was reported recently [101]. Previous works have dealt with the stability criterion for ground state solitons in Hamiltonian systems [109–111]. Some approaches for higher-order solitons have also been developed [112,113]. However, solitons in dissipative systems are qualitatively different from solitons in Hamiltonian systems. As a result, the stability criterion for Hamiltonian systems cannot easily be generalized to the case of dissipative systems.

Recent works by Kapitula and Sanstede consider the stability of CGLE (dissipative) solitons when they are perturbations of NLSE solitons [114]. This is

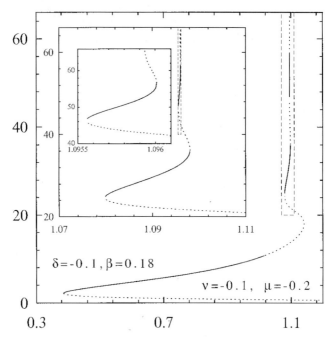

**Fig. 19.** Energy of the soliton $Q$ versus $\epsilon$ for the quintic CGLE solitons. Stable solitons are shown by solid line and unstable solitons by dotted line. Two consecutive magnifications of the small parts of the curve enclosed in dashed rectangles are shown in the insets. Further magnification could reveal more detailed structure, but the accuracy is close to the limit of the numerical method. Local maxima and minima of $\epsilon$ (i.e. the local edges in region of soliton existence) are the points where stability changes.

important for optical transmission lines which are governed by the perturbed NLSE. However, in lasers, for example, dissipative terms are strong and the approach developed in [114] is not sufficiently general.

The main result in [101] is that the points where the stability changes abruptly coincide with the turning points of the branches representing the different families of solitons. These branches are represented by curves which show any soliton parameter (usually its peak amplitude or propagation constant) versus one equation parameter ($\epsilon$). For example, the value of soliton energy, $Q$, is plotted versus $\epsilon$ for CGLE solitons in Fig. 19. The local maxima and minima of $\epsilon$ in the region of soliton existence are the points where the corresponding soliton solutions change their stability.

# References

1. J.D. Kafka, T. Baer: Opt. Lett. **12**, 181 (1987)
2. P.W. France, (ed.): *Optical fiber Lasers and Amplifiers* (CRC Press, 1991)
3. E. Desurvire: *Erbium-Doped Fiber Amplifiers: Principles and Applications* (Wiley, 1994)

4. J.F. Digonnet, (ed.): *Rare-Earth-Doped Fiber Lasers and Amplifiers*, 2nd edn. (Marcel Dekker, 2001)
5. D.B. Mortimore: J. Lightwave Technol. **6**, 1217 (1988)
6. R. Kashyap: *Fiber Bragg Gratings* (Academic Press, 1999)
7. A. Othonos: *Fiber Bragg Gratings: Fundamentals and Applications in Telecommunications and Sensing* (Artech House, 1999)
8. G.T. Harvey, L.F. Mollenauer: Opt. Lett. **18**, 107 (1993)
9. X. Shan, D. Cleland, A. Ellis: Electron. Lett. **28**, 182 (1992)
10. G. Agrawal: *Nonlinear Fiber Optics* 3rd edn. (Academic Press, 1989)
11. A.B. Grudinin, D.J. Richardson, D.N. Payne: Electron. Lett. **28**, 67 (1992)
12. M. Romagnoli, M. Midrio, P. Franco, F. Fontana: J. Opt. Soc. Am. B **12**, 1732 (1995)
13. A. Hasegawa, Y. Kodama: Opt. Lett. **15**, 1443 (1990)
14. K.J. Blow, N.J. Doran NJ: IEEE Photon. Technol. Lett. **3** 369 (1991)
15. S.M.J. Kelly, K. Smith, K.J. Blow, N.J. Doran: Opt. Lett. **16**, 1337 (1991)
16. N. Pandit, D.U. Noske, S.M.J. Kelly, J.R. Taylor: Electron. Lett. **28**, 455 (1992)
17. N.J. Smith, K.J. Blow, I. Andonovic: J. Lightwave Technol. **10**, 1329 (1992)
18. J.P. Gordon: Opt. Lett. **8**, 596 (1983)
19. J.P. Gordon: Opt. Lett. **11**, 662 (1986)
20. J.P. Gordon, H.A. Haus: Opt. Lett. **11**, 665 (1986)
21. J.P. Gordon: J. Opt. Soc. Am. B **9**, 91 (1992)
22. M. Nakazawa, H. Kubota, E. Yamada, K. Suzuki: Electron. Lett. **28**, 1099 (1992)
23. E. Yamada, M. Nakazawa: IEEE J. Quantum Electron. **30**, 1842 (1994)
24. A. Mecozzi, J.D. Moores, H.A. Haus, Y. Lai: Opt. Lett. **16**, 1841 (1991)
25. L.F. Mollenauer, J.P. Gordon, S.G. Evangelides: Opt. Lett. **17**, 1575 (1992)
26. C. Kurtzke: IEEE Photon. Technol. Lett. **5**, 1250 (1993)
27. W. Forysiak, F.M. Knox, N.J. Doran: Opt. Lett. **19**, 174 (1994)
28. J.D. Kafka, T. Baer, D.W. Hall: Opt. Lett. **14**, 1269 (1989)
29. F.X. Kärtner, D. Kopf, U. Keller: J. Opt. Soc. Am. **12**, 486 (1995)
30. D.J. Richardson, R.I. Laming, D.N. Payne, V. Matsas, M.W. Phillips: Electron. Lett. **27**, 542 (1991)
31. I.N. Duling: Electron. Lett. **27**, 544 (1991)
32. M. Hofer, M.E. Fermann, F. Haberl, M.H. Ober, A.J. Schmidt: Opt. Lett. **16**, 502 (1991)
33. V.J. Matsas, T.P. Newson, D.J. Richardson, D.N. Payne: Electron. Lett. **28**, 1391 (1992)
34. E.A. De Souza, C.E. Soccolich, W. Pleibel, R.H. Stolen, J.R. Simpson, D.J. Di-Giovanni: Electron. Lett. **29**, 447 (1993)
35. F.X. Kärtner, U. Keller: Opt. Lett. **20**, 16 (1995)
36. G. Town, M. Sceats, and S. Poole: (Proceedings, 17th Australian Conference on Optical fiber Technology, Hobart, November, 1992) pp. 154
37. K. Tamura, C.R. Doerr, H.A. Haus, E.P Ippen: IEEE Photon. Technol. Lett. **6**, 697 (1994)
38. H. Sabert, E. Brinkmeyer: Electron. Lett. **29**, 2122 (1993)
39. F. Fontana, L. Bossalini, P. Franco, M. Midrio, M. Romagnoli, S. Wabnitz: Electron. Lett. **30**, 321 (1994)
40. K. Tamura, E.P. Ippen, H.A. Haus, L.E. Nelson: Opt. Lett. **18**, 1080 (1993)
41. A.E. Siegman: *Lasers*, (Oxford University Press, 1986)
42. K.S. Abedin, M. Hyodo, N. Onodera: Electron. Lett. **36**, 1185 (2000)
43. N.J. Doran, D. Wood: Opt. Lett. **13**, 56 (1988)

44. M.E. Fermann, F. Haberl, M. Hofer, H. Hochreiter: Opt. Lett. **15**, 752 (1990)
45. R.H. Stolen, J. Botineau, A. Ashkin: Opt. Lett. **7**, 512 (1982)
46. L.F. Mollenauer, R.H. Stolen: Opt. Lett. **9**, 13 (1984)
47. E.P. Ippen, H.A. Haus, L.Y. Liu: J. Opt. Soc. Am. B **6**, 1736 (1989)
48. N. Finlayson, B.K. Nayar, N.J. Doran: Opt. Lett. **17**, 112 (1992)
49. N.J. Doran, D.S. Forrester, B.K. Nayar: Electron. Lett. **25**, 267 (1989)
50. K.J. Blow, N.J. Doran, B.K. Nayar: Opt. Lett. **14**, 754 (1989)
51. B.K. Nayar, K.J. Blow, N.J. Doran: Opt. Comput. & Process. **1**, 81 (1991)
52. K.J. Blow, N.J. Doran, S.J.D Phoenix: Opt. Comm. **88**, 137 (1992)
53. M.E. Fermann, A. Galvanauskas, D. Harter: Appl. Phys. Lett. **64**, 1315 (1994)
54. H.G. Winful: Appl. Phys. Lett. **47**, 213 (1985)
55. C.R. Menyuk: IEEE J. Quantum Electron. **25**, 2674 (1989)
56. M. Hofer, M.H. Ober, F. Haberl, M.E. Fermann: IEEE J. Quantum Electron. **28**, 720 (1992)
57. K. Tamura, C.R. Doerr, L.E. Nelson, H.A. Haus, E.P. Ippen: Opt. Lett. **19**, 46 (1994)
58. F.X. Kärtner, I.D. Jung, U. Keller: IEEE J. Select. Topics Quantum Electron. **2**, 540 (1996)
59. U. Keller, K.J. Weingarten, F.X. Kärtner, D. Kopf, B. Braun, I.D. Jung, R. Fluck, C. Honninger, N. Matuschek, J. Aus der Au: IEEE J. Select. Topics Quantum. Electron. **2**, 435 (1996)
60. W.H. Loh, D. Atkinson, P.R. Morkel, M. Hopkinson, A. Rivers, A.J. Seeds, D.N. Payne: IEEE Photon. Technol. Lett. **5**, 35 (1993)
61. W.H. Loh, D. Atkinson, P.R. Morkel, R. Grey, A.J. Seeds, D.N. Payne: Electron. Lett. **29**, 808 (1993)
62. B.C. Barnett, L. Rahman, M.N. Islam, Y.C. Chen, P. Bhattacharya, W. Riha, K.V. Reddy, A.T. Howe, K.A. Stair, H Iwamura, S.R. Friberg, T. Mukai: Opt. Lett. **20** 471 (1995)
63. R. Paschotta, U. Keller: Appl. Phys. B **B73**,653 (2001)
64. N.N. Akhmediev, J.M. Soto-Crespo, S.T. Cundiff, B.C. Collings, W.H. Knox: Opt. Lett. **23**, 852 (1998)
65. M.J. Lederer, B. Luther-Davies, H.H. Tan, C. Jagadish, N.N. Akhmediev, J.M. Soto-Crespo: J. Opt. Soc. Am. B **16**, 895 (1999)
66. L.F. Mollenauer, E. Lichtman, M.J. Neubelt, G.T. Harvey: Electron. Lett. **29**, 910 (1993)
67. H. Sabert, E. Brinkmeyer: J. Lightwave Technol. **12**, 1360 (1994)
68. M. Romagnoli, S. Wabnitz, P. Franco, M. Midrio, F. Fontana, G.E. Town: J. Opt. Soc. Am. B **12**, 72 (1995)
69. M. Romagnoli, S. Wabnitz, P. Franco, M. Midrio, L. Bossalini, F. Fontana: J. Opt. Soc. Am. B **12**, 938 (1995)
70. T.F. Carruthers, I.N. Duling, M.L. Dennis: Electron. Lett. **30**, 1051 (1994)
71. G. Town, J. Chow, M. Romagnoli: Electron. Lett. **31**, 1452 (1995)
72. N.J. Smith, F.M. Knox, N.J. Doran, K.J. Blow, I. Bennion: Electron. Lett. **32**, 54 (1996)
73. S. Namiki, H.A. Haus: IEEE J. Quantum Electron. **33**, 649 (1997)
74. B. Bakhshi, P.A. Andrekson, X. Zhang: Opt. Fiber Technol. **4**,293 (1998)
75. Waiyapot S, Matsumoto M. Opt. Commun. **188**,167 (2001)
76. G. Town, J. Chow, K. Sugden, I. Bennion, M. Romagnoli: J. Electric. & Electron. Eng. Aust. **15**, 267 (1995)
77. G.E. Town, J. Chow, A. Robertson, and M. Romagnoli: Electro-Optics - Europe (CLEO-Europe '96), Postdeadline paper CPD2.10, Hamburg, September 1996

78. G.E. Town, K. Sugden, J.A.R. Williams, I. Bennion, S.B. Poole: IEEE Photon. Technol. Lett. **7**, 78 (1995)
79. H. Haus: J. Appl. Phys. **46**, 3049 (1975)
80. P.A. Belanger: J. Opt. Soc. Am. B **8**, 2077 (1991)
81. A.I. Chernykh, S.K. Turitsyn: Opt. Lett. **20**, 398 (1995)
82. J.D. Moores: Opt. Comm. **96**, 65 (1993)
83. N.N. Akhmediev, A. Ankiewicz: *Solitons: Nonlinear Pulses and Beams* (Chapman & Hall, 1997)
84. C.-J. Chen, P.K.A. Wai, C.R. Menyuk: Opt. Lett. **19,** 198 (1994)
85. N.N Akhmediev, A.V. Buryak, J.M. Soto - Crespo: Opt. Comm. **112**, 278 (1994)
86. N.N Akhmediev, A.V. Buryak, J.M. Soto-Crespo, D.R. Andersen: J. Opt. Soc. Am. B **12**, 434 (1995)
87. S.T. Cundiff, B.C. Collings, W.H. Knox: Opt. Express **1**, (1997)
88. I. Gabitov, E.G. Shapiro, S.K. Turitsyn: Phys. Rev. E **55**, 3624 (1997)
89. A. Hasegawa: Physica D **123**, 267 (1998)
90. Y. Kodama, S. Kumar, A. Maruta: Opt. Lett. **22**, 1689 (1997)
91. J.N. Kutz, P. Holms, S. Evangelides, J. Gordon: J. Opt. Soc. Am. B **15**, 87 (1998)
92. J. Nijhof, N.J. Doran, W. Forysiak, F.M. Knox: Elect. Lett. **33**, 1726 (1997)
93. M.J. Ablowitz, G. Biondini: Opt. Lett. **23**, 1668 (1998)
94. J.P. Gordon, L.F. Mollenauer: Opt. Lett. **24**, 223 (1999)
95. S.T. Cundiff, J.M. Soto-Crespo, N. Akhmediev: Phys. Rev. Lett. **88**, 073903 (2002)
96. N.R. Pereira, L. Stenflo: Phys. Fluids **20**, 1733 (1977)
97. K. Nozaki, N. Bekki: Phys. Soc. Japan. **53**, 1581 (1984)
98. P.-A. Bélanger, L. Gagnon and C. Paré: Opt. Lett. **14**, 943 (1989)
99. R. Conte, M. Musette: Physica D **69**, 1 (1993)
100. N.N. Akhmediev, V.V. Afanasjev, J.M. Soto-Crespo: Phys. Rev. E **53**, 1190 (1996)
101. J.M. Soto-Crespo, N. Akhmediev, G. Town: Opt. Comm. **199**, 283 (2001)
102. N. Akhmediev, J.M. Soto-Crespo, G. Town: Phys. Rev. E **63**, 056602 (2001)
103. J.M. Soto-Crespo, N. Akhmediev, A. Ankiewicz: Phys. Rev. Lett. **85**, 2937 (2000)
104. N.N. Akhmediev, A. Ankiewicz, M.J. Lederer, B. Luther-Davies: Opt. Lett. **23**, 280 (1998)
105. V.S. Grigoryan, T.C. Muradyan: J. Opt. Soc. Am. B **8**, 1757 (1991)
106. N.N. Akhmediev: *General Theory of Solitons.* In: *Soliton-driven Photonics*, ed. by A.D. Boardman A.P. Sukhorukov (Kluver, 2001) pp. 371
107. N. Akhmediev, J.M. Soto-Crespo: In: *Proceedings of SPIE*, **3666**, 307 (1999)
108. V.V. Afanasjev, N. Akhmediev, J.M. Soto-Crespo: Phys. Rev. E **53**, 1931 (1996)
109. A.A. Kolokolov: Zh. Prikl. Mekh. Tekh. Fiz. **3**, 426 (1973)
110. C.K.R.T. Jones, J.V. Moloney: Phys. Lett. A **117**, 175 (1986)
111. M. Grillakis, J. Shatah, W. Strauss: J. Funct. Analysis **74**, 160 (1987)
112. A.A. Kolokolov, A.I. Sukov: J. Appl. Mech. Tech. Phys. **4**, 519 (1975)
113. N.N. Akhmediev, A. Ankiewicz, H.T. Tran: J. Opt. Soc. Am. B **10**, 230-236 (1993)
114. T. Kapitula, B. Sandstede: Physica D **124**, 58 (1998); Physica D **116**, 95 (1998)

# Nonlinear Phenomena with Ultra-Broadband Optical Radiation in Photonic Crystal Fibers and Hollow Waveguides

A. Husakou, V.P. Kalosha, and J. Herrmann

Max Born Institute for Nonlinear Optics and Short Pulse Spectroscopy,
Max-Born-Str. 2a, D-12489 Berlin, Germany

**Abstract.** In the present work nonlinear optical phenomena with ultra-broadband radiation are studied. For the description of these processes the standard method based on the slowly varying envelope approximation and Taylor expansion for the refractive index can not be applied. Here a generalized theoretical approach without these approximations is used for the study of some interesting physical problems such as the generation of supercontinua and extremely short pulses. It is shown that the recently observed supercontinuum in.photonic crystal fibers with a spectral width exceeding two octaves can not be explained by the effect of.self-phase modulationbut by spectral broadening through fission of higher-order solitons into red-shifted fundamental solitons and blue-shifted non-solitonic radiation. Degenerate four-wave mixing can be achieved in photonic crystal fibers in an extremely broad frequency range reaching from IR to UV. A new method for the generation of extremely short optical pulses by high-order.Stimulated Raman Scatteringis also investigated. Finally, propagation phenomena in resonant media described by the full.Maxwell-Bloch equations are studied and the formation of solitary half-cycle pulses by coherent propagation effects is demonstrated.

## 1 Introduction

The standard theoretical method in nonlinear optics is based on the.slowly varying envelope approximationand the Taylor expansion for the linear refractive index. Both approximations require that the spectrum of the radiation during propagation is much narrower than the central frequency. Most known theoretical predictions in optics and laser physics are restricted to the range of validity of these standard approximations. However there exist interesting physical phenomena which can not be described by this theoretical approach such as nonlinear processes with ultrabroad-band radiation or with extremely short pulses with a duration in the order of one optical cycle. The present contribution is devoted to the study of some of such processes in photonic crystal fibers and hollow waveguides. In Sect. 2 fundamentals for the theoretical description of nonlinear effects without the standard approximations in nonlinear optics are presented. In the following Sections this approach is applied to study the generation of supercontinua and extremely short pulses. In Sect. 3 the nonlinear propagation of femtosecond pulses in photonic crystal fibers (PCF) is investigated. It will be shown that supercontinuum (SC) generation in PCFs for relatively low intensities rests on a new mechanism of spectral broadening which is related to the evolution and fission of higher-order solitons [1]. In contrast to spectral broadening by self-phase modulation, the spectra generated from longer pulses are broader than for shorter pulses with the same intensity. In Sect. 4 degenerate.four wave mixing (FWM) in PCFs is studied. It is shown that phase matching can be realized for FWM in PCFs in a much larger frequency range than for standard fibers with pump wavelengths in the optical range and signal and idler in the IR and the blue region, respectively. The evolution of spectra show that FWM is dominant for short propagation length but SC's can emerge at a later stage only if the spectral part in the anomalous region is large enough to form a higher-order soliton. In Sect. 5 the generation of extremely short pulses by high-order stimulated Raman scattering is studied. The underlying mechanism and temporal and spectral characteristics are examined using analytical and numerical solutions. It is shown that phase- locked pulses as short as 1.7 fs can be generated in impulsively excited media without the necessity of external phase control. Finally, in Sect. 6 coherent propagation effects of few optical-cycle pulses in resonant two-level systems are investigated. It is demonstrated that optical subcycle pulses can be generated in dense media of two-level systems due to pulse splitting and reshaping by coherent effects. The formation of half-cycle full Maxwell-Bloch solitons and two-solitons beyond the limit of SVEA is demonstrated due to this mechanism starting with initial pulses with a duration of 5 fs.

## 2    Theoretical Fundamentals

The standard theoretical method in nonlinear and fiber optics is the slowly-varying envelope approximation (SVEA),in which the rapidly varying part (carrier) of the electric field $\boldsymbol{E}(\boldsymbol{r}, t)$ is separated from the slowly varying envelope $\boldsymbol{A}(\boldsymbol{r}, t)$:

$$\boldsymbol{E}(\boldsymbol{r}, t) = \frac{1}{2} \boldsymbol{A}(\boldsymbol{r}, t) \exp[i\omega_0 t - k(\omega_0)z] + c.c. , \tag{1}$$

where $\boldsymbol{r} = \{x, y, z\}$, $\omega_0$ is the input carrier frequency, $k(\omega) = n(\omega)\omega/c$ and $n(\omega)$ is the frequency-depending refractive index. In the SVEA the slowly varying envelope is assumed to satisfy the condition

$$\left|\frac{\partial A}{\partial t}\right| \ll \omega_0 |A| , \tag{2}$$

which is fulfilled only if the spectral width of a pulse $\Delta\omega$ is much smaller than the carrier frequency $\omega_0$ of the pulse: $\Delta\omega \ll \omega_0$. The latter condition allows to neglect higher-order terms in Taylor expansion of $k(\omega)$ around $\omega_0$ when determining the influence of dispersive effects. Obviously, these approximations are no longer valid for radiation with ultrawide spectra such as pulses with duration approaching one optical cycle.

Pulse propagation without special prerequisites of the SVEA can be studied by the numerical solution of Maxwell equations by the finite-difference time-domain method (see Ref. [2–4] and references therein). However, the large numerical effort in this approach limits the possible propagation lengths to a few mm. In several papers various improved corrected equations has been derived that allow the theoretical description beyond the validity of the standard approximations [5,6,1]. In the following we give a systematic derivation of a first-order unidirectional propagation equation without the use of the SVEA and the Taylor expansion of the linear refraction index. This equation extends previously derived equations into the non-paraxial and extremely nonlinear region from which the basic equations of Ref. [2] or [1] can be derived in a physically transparent manner. The propagation of pulses in nonlinear media is described by the wave equation

$$\left( \frac{\partial^2}{\partial z^2} + \Delta_\perp \right) \boldsymbol{E} - \frac{1}{c^2} \frac{\partial^2}{\partial z^2} \boldsymbol{E} = -\mu_0 \frac{\partial^2}{\partial t^2} \boldsymbol{P} \,, \tag{3}$$

where $\Delta_\perp = \partial^2/\partial x^2 + \partial^2/\partial y^2$ and $\boldsymbol{P}$ is the medium.polarization. Substituting the Fourier transformed field $\boldsymbol{E}(z,\omega,\boldsymbol{k}_\perp) = \int_{-\infty}^\infty e^{i\omega t - i\boldsymbol{k}_\perp \boldsymbol{r}_\perp} \boldsymbol{E}(\boldsymbol{r},t)dtd\boldsymbol{r}_\perp$, $\boldsymbol{r}_\perp = \{x,y\}$, $\boldsymbol{k}_\perp = \{k_x, k_y\}$ into the wave equation (3) we obtain

$$\frac{\partial^2 \boldsymbol{E}(z,\omega,\boldsymbol{k}_\perp)}{\partial z^2} + \beta^2_{\mathrm{NL}}(\omega)\boldsymbol{E}(z,\omega,\boldsymbol{k}_\perp) = 0 \tag{4}$$

with

$$\beta_{\mathrm{NL}}(\omega,\boldsymbol{k}_\perp) = \left\{ \frac{\omega^2}{c^2}[1 + \chi(\omega)] - \boldsymbol{k}_\perp^2 + \mu_0\omega^2 B_{\mathrm{NL}}(z,\omega,\boldsymbol{k}_\perp) \right\}^{1/2} \tag{5}$$

and assuming $\boldsymbol{P_{NL}} || \boldsymbol{E}$

$$B_{\mathrm{NL}} = \frac{P_{\mathrm{NL}}(z,\omega,\boldsymbol{k}_\perp)}{E(z,\omega,\boldsymbol{k}_\perp)} \,. \tag{6}$$

Here we separate the polarization into a linear and a nonlinear part as $\boldsymbol{P} = \boldsymbol{P}_L(\boldsymbol{r},\omega) + \boldsymbol{P}_{\mathrm{NL}}(\boldsymbol{r},\omega)$, where $\boldsymbol{P}_{\mathrm{NL}}(\boldsymbol{r},\omega)$ is the Fourier transform of the nonlinear part of polarization and $\boldsymbol{P}_L(\boldsymbol{r},\omega) = \epsilon_0[n^2(\omega) - 1]\boldsymbol{E}(\boldsymbol{r},\omega)$ its linear part.

In most typical situations the condition $\partial B_{\mathrm{NL}}/\partial z \ll \beta_{\mathrm{NL}}B_{\mathrm{NL}}$ is satisfied with high accuracy, therefore the separation $\partial^2/\partial z^2 + \beta^2_{\mathrm{NL}}(\omega) = [\partial/\partial z - i\beta_{\mathrm{NL}}(\omega)][\partial/\partial z + i\beta_{\mathrm{NL}}(\omega)]$ is possible. The electric field can be separated into a forward $E^+$ and a backward $E^-$ propagating part: $\boldsymbol{E}(z,\boldsymbol{k}_\perp,\omega) = \boldsymbol{E}^+(z,\boldsymbol{k}_\perp,\omega) + \boldsymbol{E}^-(z,\boldsymbol{k}_\perp,\omega)$. Let us consider a pulse propagating into the forward direction along the $z$-axis $\boldsymbol{E}^+(z,\boldsymbol{k}_\perp,\omega) \sim \exp[ik(\omega)z]\boldsymbol{E}_0(z,\boldsymbol{k}_\perp,\omega)$ and neglect waves propagating backward. This requires that the refractive index is constant or a smooth, slowly changing function of the $z$ coordinate. Therefore we have

$$\frac{\partial E^+}{\partial z}(z,\boldsymbol{k}_\perp,\omega) = i\beta_{\mathrm{NL}}(z,\boldsymbol{k}_\perp,\omega)E^+(z,\boldsymbol{k}_\perp,\omega) \,, \tag{7}$$

when the effect of background wave on $B_{\mathrm{NL}}$ can be neglected. For the backward wave

$$\frac{\partial E^-}{\partial z}(z,\boldsymbol{k}_\perp,\omega) = -i\beta_{\mathrm{NL}}(z,\boldsymbol{k}_\perp,\omega)E^-(z,\boldsymbol{k}_\perp,\omega) \,. \tag{8}$$

Equation (7) represents a more general approach than the standard approach and is even more accurate than previously derived evolution equations without

SVEA as presented in [1]. It includes into the theoretical analysis broad bandwidth, sharp temporal features, space-time coupling and higher-order nonlinear dispersive effects. Note that the SVEA with paraxial approximation for the transverse momentum fails to describe.self-focusing in dispersive media accurately long before the temporal structure reaches the time of an optical cycle [7]. This effect is a result of space-time focusing of short pulses leading to a reduced axially projected group velocity of wide-angle rays in the angular spectra. Equation (12) can be numerically solved by the second-order split-step Fourier method.

Expanding the square root in Eq.(5) as

$$\beta(\omega, \boldsymbol{k}_\perp) \simeq k(\omega) - \frac{k_\perp^2}{2k(\omega)} + \frac{\mu_0 \omega^2}{2k(\omega)} B_{\mathrm{NL}} .\tag{9}$$

With the introduction of the moving time coordinates $\xi = z, \eta = t - z/c$ with $\partial/\partial z = \partial/\partial \xi - c^{-1}\partial/\partial \eta$, we obtain the following basic equation in Fourier presentation which we denote in the following as forward Maxwell equation (FME) [1]:

$$\frac{\partial \boldsymbol{E}(\boldsymbol{r}, \omega)}{\partial \xi} = i\frac{\omega}{c}[n(\omega) - 1]\boldsymbol{E}(\boldsymbol{r}, \omega)$$

$$+ \frac{i}{2k(\omega)}\Delta_\perp \boldsymbol{E}(\boldsymbol{r}, \omega) + \frac{i\mu_0 \omega c}{2n(\omega)}\boldsymbol{P}_{\mathrm{NL}}(\boldsymbol{r}, \omega).\tag{10}$$

In a bulk medium the nonuniform beam transverse intensity profile leads to nonuniform phase relations due to diffraction and nonlinear transverse beam reshaping effects. Pulse compression by SPM or HSRS in a nonlinear bulk medium is therefore seriously limited. A possible solution of this problem is using the hollow fibers filled with a gas as nonlinear medium. This technique was first applied for optical pulse compression by self-phase modulation using hollow-core silica waveguides filled with a noble gas [8]. The same technique was also used for impulsive Raman scattering with a Raman-active gas as $SF_6$ or $H_2$ [9,10]. Propagation of higher modes is suppressed in hollow waveguides, making them suitable for guiding pulses with high energies. For the description of wave propagation in hollow fibers the electric field in frequency domain can be separated in the form: $\boldsymbol{E}(x, y, z, \omega) = \boldsymbol{F}(x, y, \omega)\tilde{E}(z, \omega)$, where the transverse fundamental mode distribution $\boldsymbol{F}(x, y, \omega)$ is the solution of the Helmholtz equation

$$\Delta_\perp \boldsymbol{F} + k^2(\omega)\boldsymbol{F} = \beta^2(\omega)\boldsymbol{F}\tag{11}$$

with the eigenvalue $\beta(\omega)$. The longitudinal distribution $\tilde{E}(\xi, \omega)$ satisfies the equation

$$\frac{\partial \tilde{E}(\xi, \omega)}{\partial \xi} = i\frac{[n(\omega) - 1]\omega}{c}\tilde{E}(\xi, \omega) + i\mu_0 c\frac{\omega \alpha(\omega)}{2n(\omega)}P_{\mathrm{NL}}(\xi, \omega) .\tag{12}$$

Here $\alpha(\omega) = \int_S F^4(x, y, \omega)dS/S_1$ is a nonlinearity reduction factor and $S_1$ is the effective mode area $S_1(\omega) = \int_S F^2(x, y, \omega)dS$, $S$ being the cross-section of the fiber.

Equation (12) is a generalization of the so-called *reduced* Maxwell equation [5] which is valid only for a refractive index close to unity and also neglect the back propagating wave. This equation is obtained from Eq.(3) by the substitution $n(\omega) - 1 \simeq [n^2(\omega) - 1]/2$ and back transformation into the time domain:

$$\frac{\partial E}{\partial \xi} = -\frac{1}{2\epsilon_0 c}\frac{\partial P}{\partial \eta} \ . \tag{13}$$

This equation is a useful tool for examining the nonlinear effects of ultrabroadband radiation in gaseous media. It was recently applied for the study of pulse compression and SC generation by the optical.Kerr effect [4,11] or by high-order stimulated Raman scattering [12,13] in hollow waveguides. In Ref. [14] an extended version of this equation was used with inclusion of the diffraction term. However, as one would expect, for solids solutions of this equation differ from the exact Maxwell equation as can be seen by comparison of both in Fig. 1(b).

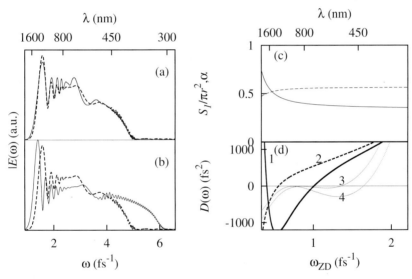

**Fig. 1.** Comparison of spectra by different propagation equation and PCF dispersion parameters. Spectra of 15-fs, 40-TW/cm$^2$ pulse after propagation of 0.5 mm in standard fiber as calculated by FME (a) and reduced Maxwell equation (b) are compared to the results by full wave equation (dashed curves). Effective mode area $S_1$ (solid curve) and nonlinear reduction factor $\alpha(\omega)$ (dashed curve) are shown in (c). Part (d) shows dispersion parameter $D(\omega)$ for PCF with $\Lambda = 1.5\,\mu\text{m}$, $d = 1.3\,\mu\text{m}$ (curve 1) and for bulk silica (curve 2) together with phase-matching for the weakest (curve 3) and the strongest (curve 4) solitons for Fig. 4.

Equation (12) generalizes the standard approximate evolution equation for the envelope in nonlinear optics. Choosing the same prerequisites Eq.(2) and expanding $\beta(\omega)$ as given by Eq.(9) around $\omega_0$ the propagation equation takes the form

$$\frac{\partial A}{\partial z} + \frac{i}{2}\beta''\frac{\partial^2 A}{\partial \eta^2} - \frac{1}{6}\beta'''\frac{\partial A}{\partial \eta^3} = i\gamma\left(|A^2|A + \frac{i}{\omega_0}\frac{\partial|A|^2A}{\partial \eta}\right) , \tag{14}$$

where $\gamma = \alpha(\omega)\omega n_2/(cS_1)$.

The same procedure can be applied if the nonlinear interaction of waves with different carrier frequencies is considered. As an example in degenerate four-wave mixing (FWM) pump, signal and idler pulses with central frequencies $\omega_p, \omega_s, \omega_I = 2\omega_p - \omega_s$ interact by the nonlinearity $\boldsymbol{P}_{\mathrm{K}}$. In this case using SVEA the electric field is presented by $E(z,t) = 1/2\sum_j A_j(z,t)\exp[i(k_jz - \omega_jt)] + \mathrm{c.c.}$ $(j = s,p,I)$ and from Eq.(10) three propagation equations follow for $A_p, A_s$, and $A_I$ describing the process of FWM. The additional advantage of the evolution equation (10) is that all spectral components that can take part in a given process are included automatically in a single equation. This is particularly very favorable if different processes can affect the propagation. The price for this more general approach with a single evolution equation is the necessarily higher resolution of the temporal grid in the scale of the carrier, while in the SVEA the resolution is in the scale of the envelope.

Now we will describe the model for linear and nonlinear dispersion for the systems considered in our work. The Fourier transform of the linear polarization is given by $P_L(z,\omega) = \epsilon_0[\chi_b(\omega) + \chi_g(\omega)]E(z,\omega)$, where $\chi_b$ is the bulk susceptibility $\chi_b(\omega) = \sum_{\ell=1,2} S_\ell/(\omega_\ell^2 - \omega^2)$ with experimentally determined parameters $S_\ell$, $\omega_\ell$. The steady state solution of Eq.(11) for the transverse mode distribution for a dialectic waveguide with mode diameter much larger than the wavelength was found by Marcatili and Schmelzer [15]. For the hollow waveguides with radii $a \ll \lambda$ the waveguide contribution for the $\mathrm{EH}_{11}$ mode is given by $\chi_g(\omega) = (2.405c/wa)^2[1 + i(1 + \nu^2)/(2\pi a\sqrt{1 - \nu^2})]$ [16], where $\nu \simeq 1.45$ is the ratio between the refractive indices of the jacket (fused silica) and the internal gas. For the photonic and tapered fibers, the procedure to obtain this contribution is described in the corresponding section.

In Eq.(12) $P_{\mathrm{NL}}(z,\eta)$ is the medium polarization induced by the electronic Kerr effect $P_{\mathrm{K}}$ and the Raman process $P_{\mathrm{R}}$.

The nonlinear polarization $P_{\mathrm{K}}$ far off medium resonance for not too high intensity in a isotropic medium can be treated as third-order process. Sheik-Bahae et al. [17] have developed a model for the dispersion of the dominate electronic part of $\chi_3$ for semiconductors and wide-gap optical solids. It gives a universal formula for $\chi_3(\omega_1, \omega_2, \omega_3, \omega_4)$ in quite good agreement with measurements. According to this model for $\omega_j \ll \omega_g = E_g/\hbar$, where $j = 1..4$ and $E_g$ is the bandgap energy, up to the first correction term $\chi_3(\omega, \omega, -\omega, \omega)$ has the form

$$\chi_3 = \chi_3^{(0)}\left(1 + 2.8\frac{\omega^2}{\omega_g^2} + \dots\right) . \tag{15}$$

For fused silica $E_g = 9\,\mathrm{eV}$ and therefore the correction terms do not play a significant role even for the extremely broad spectra considered here. With this estimation the nonlinear polarization for linearly polarized waves is $P_{K,x} = \epsilon_0\chi_3^{(0)}E_x^3$.

The Raman polarization $P_R = N\text{Sp}(\hat{\alpha}\hat{\rho})E$ is described by the two-photon two-level system equations (see, e.g., [12]):

$$\left(\frac{\partial}{\partial t} + \frac{1}{T_2} - i\omega_v\right)\rho_{12} = \frac{i}{2\hbar}\left[(\alpha_{11} - \alpha_{22})\rho_{12} + \alpha_{12}w\right]E^2, \tag{16}$$

$$\frac{\partial w}{\partial t} + \frac{w+1}{T_1} = \frac{i}{2\hbar}\alpha_{12}\left(\rho_{12} - \rho_{12}^*\right)E^2, \tag{17}$$

where $\rho_{12} = \frac{1}{2}(u+iv)$, $w = \rho_{22} - \rho_{11}$, $\rho$ is the density matrix, $\hat{\alpha}$ is the two-photon polarizability matrix responsible for the Raman process and Stark shift, $\omega_v$ is the Raman frequency shift, $N$ is the density, $T_1$, $T_2$ are the relaxation times.

The macroscopic nonlinear polarization for a one-photon resonant two-level system $P_B = Ndu$, where $d$ is the dipole moment, is connected with the off-diagonal density matrix element $\rho_{12} = \frac{1}{2}(u + iv)$ and the population difference $w = \rho_{22} - \rho_{11}$, which are determined by the Bloch equations [18]:

$$\left(\frac{\partial}{\partial t} + \gamma_2 - i\omega_{21}\right)\rho_{12} = i\Omega w, \tag{18}$$

$$\frac{\partial w}{\partial t} + \gamma_1(w + 1) = -2\Omega v. \tag{19}$$

Here $\omega_{21} \approx \omega_0$ is the resonant frequency of the two-level system, $\gamma_1$ and $\gamma_2$ are the population and polarization relaxation constants, $\Omega = Ed/\hbar$ is the Rabi frequency.

## 3    Higher-Order Solitons and Supercontinuum Generation in Photonic Crystal Fibers

Photonic crystal fibers (PCF) are currently a topic of high interest because of their unusual optical properties and their potential for important applications. A PCF [19] has a central region of pure silica (core) surrounded by air holes.

The air holes form a low-index cladding around the solid silica core, therefore light is guided similarly as in standard step-index fibers. The refractive index difference between the core and the holey cladding provides a very strong specifically controlled waveguide contribution to dispersion. Therefore the optical properties of a PCF differ remarkably from those of standard fibers. In particular the zero-dispersion wavelength can be shifted into the visible region, and a PCF can show single-mode operation almost over the complete transmission range [20–22]. As a result new features in nonlinear optical effects arise that can not be observed in standard optical fibers. One such phenomenon is the generation of an extremely broadband supercontinuum (SC) covering more than two octaves from low-energy ($\sim 1$ nJ) pulses with an initial duration $\tau_0 = 100$ fs [23]. The analogous effect has also been observed in tapered fibers [24]. In comparison, SC generation in standard fibers requires more than two orders of magnitude higher initial peak intensities $I_0$ [25,26,4]. The dramatic spectral broadening of relatively low-intensity pulses in PCFs is an interesting phenomenon and has already been used in several fascinating applications. Significant progress in frequency metrology [27,28] has been achieved using the generation of an octave-spanning optical frequency "comb" in a PCF by a train of fs laser pulses. SC generated in a PCF is also an excellent source for optical coherence tomography with ultrahigh resolution in biological tissue [29]. Further there is a large potential for numerous applications such as for pulse compression, laser spectroscopy, all-optical telecommunication, dispersion measurements, sensor technique and others. The dramatically increased nonlinear response in PCFs in combination with single-mode behavior over almost the complete transmission range and the engineerable dispersion characteristic suggest that new phenomena could appear also in other nonlinear processes. In the present Section we study theoretically the nonlinear pulse propagation of femtosecond pulses and the physical origin of the above described SC in PCFs. Recently we found that SC generation in PCFs for relatively low intensities rests on a new mechanism of spectral broadening which is related to the evolution and fission of higher-order solitons[1]. In recent work [30] the experimental evidence for this mechanism for SC generation has been provided. In agreement with the theory from Ref. [1] for the same intensity increasing spectral widths of the output radiation with increasing input pulse durations has been observed.

The formation of SC by the interaction of intense pulses with matter has been discovered already in the seventeenth first in condensed matter [31], and later also in single-mode fibers (for an overview see [32]). The physical origin of this effect is a refractive index change by the electric field $\Delta n = n_2 I(t)$, where $I(t)$ is the intensity and $n_2$ the nonlinear refractive index. As a result a time-dependent phase is induced which implies the generation of new spectral components around the input frequency $\omega_0$ with maximum spectral width $\Delta\omega_{SPM}/\omega_0 = 1.39 n_2 I_0 L/\tau_0 c$ [33], $L$ being the propagation length. Normal group-velocity dispersion (GVD) limits this width while in the.anomalous dispersion region the balance of GVD and SPM leads to the formation of solitons. At higher amplitudes a higher-order soliton with soliton number $N$ is formed. Such bounded N-soliton undergoes periodic narrowing and broadening during propagation. Higher-order nonlinear and dispersive effects lead to a break-up of N-solitons into their constituent 1-solitons [9] and the emission of blue-shifted non-solitonic radiation (NSR) [34,35,8,36] (for an overview see [33]).

As will be shown the low-intensity spectral broadening observed in PCF's are caused by a different mechanism and can not be explained by the effect of SPM for pulses in the range of 1 nJ and 100 fs duration. To study pulse propagation in PCFs we need besides the evolution equation presented in Sect. 2 a model for the PCF waveguide contribution to dispersion, which is calculated by the following two-step procedure. First, an infinite photonic crystal fiber without central defect (core) is considered. The fundamental space-filling mode in such system can be analyzed in the scalar approximation, i.e. the transverse dependence of all components of both electric and magnetic fields are given by the same function $\psi(\omega, x, y)$. This function satisfies the Helmhotlz equation. Besides that, the fundamental mode has the same periodical properties as the photonic crystal. We split the photonic crystal into the elementary hexagonal cells so that air holes are in the center of the cells. From the periodicity of $\psi(\omega, x, y)$ follows that the derivative $\partial\psi/\partial n$ vanishes at the cell boundary, where $n$ is the vector normal to boundary. To further simplify the calculations, we consider a circular cell instead of the hexagonal one. The radius of the inner air cylinder in the circular cell equals $d$, the outer radius is defined by the air fraction in the original crystal as $R = \Lambda\sqrt{\sqrt{3}/(2\pi)}$. Now we consider the function $\psi(\omega, x, y)$ in the polar coordinates as $\psi(\omega, \rho, \varphi)$. The condition $\partial\psi/\partial n = 0$ transforms to $\partial\psi/\partial\rho|_R = 0$, and the continuity conditions on the inner air-silica interface have to be satisfied as well. In the inner $(0 < \rho < r)$ area, $\beta^2(\omega) - n_{Air}^2(\omega)\omega^2/c^2 = \kappa^2(\omega) > 0$ and therefore $\psi \sim I_0(\kappa(\omega)\rho)$, $J$, $N$ and $I$ denoting the Bessel functions. In the outer area $(r < \rho < R)$, $\beta^2(\omega) - n_{Silica}^2(\omega)\omega^2/c^2 = -\gamma^2(\omega) < 0$ and $\psi$ is equal to the linear combination of $J_0$ and $K_0$ with coefficients chosen so that the boundary condition at $R$ is satisfied:

$$\psi(\omega, \rho) \sim J_0(\gamma(\omega)\rho) - N_0(\gamma(\omega)\rho)\frac{J_1(\gamma(\omega)R)}{N_1(\gamma(\omega)R)} \,. \tag{20}$$

Substituting $\psi(\omega, \rho)$ into the boundary conditions we obtain the dispersion equation

$$\kappa(\omega)\frac{I_1(\kappa(\omega)r)}{I_0(\kappa(\omega)r)}\left[J_0(\gamma(\omega)r) - N_0(\gamma(\omega)r)\frac{J_1(\gamma(\omega)R)}{N_1(\gamma(\omega)R)}\right] =$$
$$-\gamma(\omega)\left[J_1(\gamma(\omega)r) - N_1(\gamma(\omega)r)\frac{J_1(\gamma(\omega)R)}{N_1(\gamma(\omega)R)}\right]. \tag{21}$$

This equation is solved together with $\kappa^2(\omega) + \gamma^2(\omega) = [n_{Silica}^2(\omega) - n_{Air}^2(\omega)]\omega^2/c^2$ to find subsequently $\kappa(\omega)$, $\beta(\omega) = \sqrt{\kappa^2(\omega) + n_{Air}^2(\omega)\omega^2/c^2}$ and the effective refractive index for the fundamental space-filling mode $n_{eff}(\omega) = \beta(\omega)c/\omega$.

Then we consider the omitted hole as a core of a step-index fiber with diameter $2r = 2\Lambda - d$, where $\Lambda$ is the center-to-center distance between the holes (pitch) and $d$ is the hole diameter, and the surrounding photonic crystal as homogeneous cladding with refractive index $n_{eff}(\omega)$. Note that for certain PCFs the air holes are very large and therefore the central core is supported by the very thin bridges of silica which can be neglected in the calculation of $\beta(\omega)$. Such fiber can be described as isolated strand of silica surrounded by air. Dispersion of tapered fiber, which is a μm-scale silica core surrounded by air, can be described in the same way.

To study the basic mechanism of SC in PCFs we solved the nonlinear evolution equation with the model for the determination of the eigenvalue $\beta(\omega)$ in a PCF as described in Sect. 2 for different input and PCF parameters. First we solve the pulse propagation equation for a PCF with $\Lambda = 1.5$ μm, $d = 1.3$ μm and the zero dispersion frequency $\omega_{ZD} = 2.66$ fs$^{-1}$ ($\lambda_{ZD}=710$ nm) for pulses with large duration (.FWHM) $\tau_0 = 100$ fs and a relatively low intensity $I_0 = \epsilon_0 n(\omega)cE^2/2 = P_0/S_1 = 0.6$ TW/cm$^2$. In order to study the crucial role of the specific PCF dispersion, we consider the propagation of pulses with different initial frequencies presented in Fig. 2.

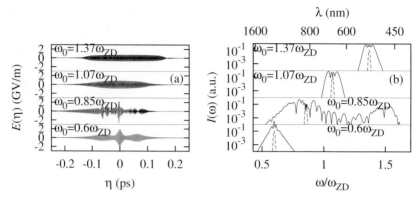

**Fig. 2.** Output pulse shapes (a) and spectra (b) for $L=15\,\text{mm}$, $I_0=0.6\,\text{TW/cm}^2$, $\tau_0=100\,\text{fs}$ and different input frequencies as indicated. Initial spectra (dashed, scaled for clarity here and hereafter) are also presented

For the parameters in Fig. 2 the initial frequency $\omega_0 = 0.6\omega_{ZD}$ is deep in the anomalous region, the magnitude of the negative GVD parameter $D$ is relatively large and the TOD parameter is very small. The soliton number for these parameters is $N=4.3$. The solution presented in Fig. 2 for $\omega = 0.6\omega_{ZD}$ can be identified as a bounded fourth-order soliton which is also well described by the analytical solution of the.nonlinear Schrödinger equation (NSE) with typical periodic evolution with the propagation length and splitting into three distinct peaks which merge again later [33]. At initial frequency $\omega_0 = 1.07\omega_{ZD}$ and $\omega_0 = 1.37\omega_{ZD}$ in.normal dispersion range, spectral broadening typical for SPM can be found, with a spectral width $\Delta\omega/\omega = 0.06$ which agrees approximately with the above given maximum SPM-induced width 0.07. However as can be seen for $\omega_0 = 0.85\omega_{ZD}$ for the same input intensity and pulse duration a qualitative different behavior arise: the spectrum shows a more than one order of magnitude larger width (Fig. 2). The theoretical prediction presented in Fig. 2 for $\omega_0 = 0.85\omega_{ZD}$ is in good agreement with experimental measurements for the same input pulse parameters reported in Ref. [23]. The further evolution of the temporal shape shows that the pulse is successively split into up to finally seven ultrashort peaks moving with different velocities with shapes which do not change their form over a long distance. It is impossible to explain the extremely broad spectrum for the rather long pulses with a relatively small intensity by the effect of SPM. As discussed above the largest spectral broadening by SPM is given by [33] $\Delta\omega_{SPM}/\omega_0 = 1.39n_2I_0L/(\tau_0c) = 0.07$ for the input pulse parameters in Fig. 2 for $\omega_0 = 0.85\omega_{ZD}$ while we obtained a more than one order of magnitude broader spectrum.

Additionally, we find a surprising result if we consider the spectral broadening of a shorter pulse with the same intensity, as shown in Fig. 3. As can be seen in Fig. 3(b), for $\omega_0 = 0.85\omega_{ZD}$ the spectral width of about 50 nm generated by a 17.5 fs pulse is ten times smaller compared with the 100-fs-pulse case. This much narrower spectrum is in direct contrast to the behavior of SPM-induced

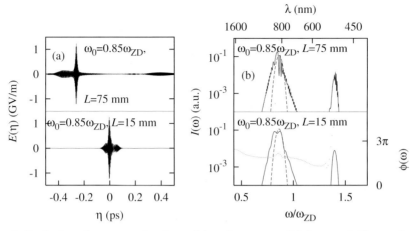

**Fig. 3.** Evolution of output pulse shape (a) and spectrum (b) for $\omega = 0.85\omega_{ZD}$, $I_0=0.6$ TW/cm$^2$, $\tau_0=17.5$ fs.

broadening, where, corresponding the relation for $\Delta\omega_{SPM}/\omega_0$, an about 6 times shorter pulse should yield a correspondingly larger width. The temporal shape presented in Fig. 3(a) top shows the formation of a single short spike together with background radiation. While the spike does not change its form during propagation from 15 mm to 75 mm, the background radiation becomes temporally broadened.

The behavior of SC generation in PCFs described above is qualitatively different from SPM-induced broadening and requires a careful study of its physical origin. Note that the considered input frequency $0.85\omega_{ZD}$ is in the anomalous region and, therefore, soliton dynamics plays a crucial role in the propagation. The input parameters in Fig. 2 for $0.85\omega_{ZD}$ imply the formation of a higher-order soliton [33] with a soliton number $N = \sqrt{n_2 I_0\omega_0\tau_0^2 L/|D|c} = 7.8$. Higher-order solitons of the nonlinear Schrödinger equation (NSE) show periodic changes with propagation and can not explain the effects described above. But in a PCF higher-order dispersion effects are stronger than in standard fibers and play a much more significant role in pulse propagation. Previous studies [9,34,8,35] of the perturbed NSE, taking into account positive TOD, predict the following behavior: A higher-order soliton with number N splits into N pulses with different red-shifted central frequencies and different group velocities [9]. After the fission every pulse emits non-solitonic radiation phase-matched to the corresponding pulse [8,35] while simultaneously moving to IR until the stability is reached[35]. Although the perturbed NSE is not valid for the propagation phenomena and the spectral broadening and shifts in standard fibers are two order of magnitude smaller, the analogous effects in PCFs can be readily identified as the physical origin of the SC generation. The amplitudes and durations of the separated spikes in Fig. 2 for $\omega_0 = 0.85\omega_{ZD}$ satisfy the relation for a fundamental soliton [33]. To corroborate the soliton nature of these spikes, we simulate the propagation of every separated pulse over a distance of 75 mm and do not find any

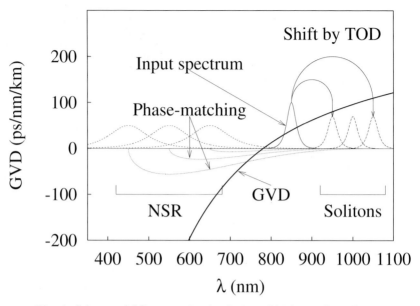

**Fig. 4.** Scheme of SC generation by fission of higher-order solitons.

change in shape and spectrum during propagation. For the same conditions a low-intense pulse would spread by a factor of 200.

The higher-order soliton splits into fundamental red-shifted 1-solitons and loses energy by emitting blue-shifted NSR (see the scheme in Fig. 4).The pulse duration's of the 1-solitons are determined approximately by $\tau_n = \tau_0/\xi_n$ [9] and their frequency-shift by $\Delta\omega_s \approx 7/\tau_n$ [34] where $\xi_n = 2A_0 - n$, $\quad n = 1, 2 \ldots N$ are determined by the eigenvalues of the nonlinear Schrödinger equation. This scenario explains the observed and numerically calculated features in Figures 2 and 3 and the given rough analytical estimates for $\tau_n$ and $\Delta\omega_s$ are supported by our numerical computations, with acceptable deviations. All spectral components of each soliton are phase-locked and the solitons preserve their shape and spectrum in collisions. The calculated spectrum of the three isolated strongest solitons show a red-shift with central frequencies at $0.87\omega_0$, $0.93\omega_0$ and $0.97\omega_0$, and its velocities are close to corresponding group velocities. The phases of a soliton at frequency $\omega_s$ in its moving frame and that of the non-solitonic radiation at $\omega$ in the same frame are given by

$$\phi_s(\omega_s) = n(\omega_s)\omega_s L/c + n_2 I \omega_s L/(2c) - \omega_s L/v_s \qquad (22)$$

and

$$\phi_r(\omega) = n(\omega)\omega L/c - \omega L/v_s \ . \qquad (23)$$

In Fig. 1(d) in the curves 3 and 4 the phase difference $\Delta\phi = \phi_s - \phi_r$ for the strongest and weakest soliton with respect to its corresponding non-solitonic radiation is presented. The strongest soliton is phase-matched with non-solitonic radiation at 400 nm and the weakest at 550 nm. Due to the presence of several solitons with different frequencies, distinct spectral fractions arise and therefore a broad spectrum is generated in the intermediate range between 430 nm to 550 nm. The spectrum in the range between 550 nm and 700 nm arises as a result of nonlinear interactions between the solitons and the blue-shifted continuum. The phase relations discussed above are supported by the numerically calculated spectral phases of the pulses $\tilde{\phi} = \phi(\omega) - \omega L(1/v_s - 1/c)$.

Now the result in Fig. 3 for a shorter input pulse but a narrower output spectrum can be explained. Since for the smaller pulse duration the soliton number with $N = 1.5$ corresponds to one fundamental soliton, no soliton fission can occur and only an isolated blue-shifted side peak is generated.

# 4    Degenerate Four-Wave Mixing and Parametric Amplification

In Sect. 4 we show that phase matching can be realized for FWM in PCFs in a much larger frequency range than for standard fibers with pump wavelengths in the optical range and signal and idler in the IR and the blue, respectively. The design of PCF parameters as well as controlling the pump frequency and intensity enable generation of tunable new frequency components in the IR. The evolution of spectra show that FWM is dominant for short propagation length forming two distinct side peaks in the IR and the blue. SC's can emerge at a later stage only if the spectral part in the anomalous region is large enough to form a higher-order soliton which after fission emits blue-shifted non-solitonic radiation.

The third-order nonlinear polarization leads in general to the interaction of four optical waves with frequencies $\omega_1, \omega_2, \omega_3, \omega_4$ and include such phenomena as FWM and parametric amplification. These processes can be used to generate waves at new frequencies. In the degenerate case $\omega_1 = \omega_2 = \omega_p$ two photons of a strong pump pulse are annihilated with simultaneous creation of two photons at $\omega_s = \omega_p + \Omega$ and $\omega_I = \omega_p - \Omega$. In collinear propagation as in fibers significant FWM occurs only if the phasematching relation $\Delta k = \Delta k_M + \Delta k_{WG} + \Delta k_{NL} = 0$ is nearly satisfied. Here $\Delta k_M, \Delta_{WG}, \Delta k_{NL} = 2\alpha n_2 \omega_p I_p / c$ represent the wavevector mismatch occurring as a result of material dispersion, waveguide dispersion and the nonlinear contribution to the refractive index, respectively. For standard optical fibers FWM has been studied extensively [33]. Phase matching can be achieved in multimode fibers by using different modes or in birefringent fibers by the different effective index for waves propagating with orthogonal polarization. In single-mode nonbirefringent fibers phase matching can be satisfied near the zero-dispersion wavelength around 1.28 μm for very specific conditions for the pump frequency because the material contribution become quite small and can be compensated by the waveguide contribution. When the pump wavelength lies in the anomalous dispersion region the negative material and waveguide contribution can be compensated by the positive nonlinear contribution. In both cases the frequency shift is typically lower than 100THz. At visible frequencies of the pump wave FWM has been demonstrated in weakly birefringent single-mode fibers [37]. In a nonbirefringent single-mode fibers phase matching has been achieved in the visible by a combination of SPM and cross-phase modulation[38]. Recently FWM has been observed in microstructure fibers in the optical region at 753 nm with a frequency shift of about 100 THz [39].

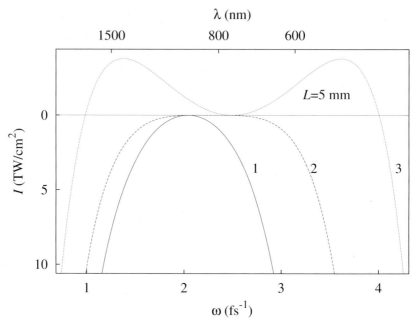

**Fig. 5.** Phase-matching for 2.95-$\mu$m-diameter tapered fiber and different input frequencies as indicated. Phase-matched frequencies are determined as crossing points of the corresponding curve and horizontal line at $I = I_p$.

To study FWM in PCFs or tapered fibers first we consider the phase matching condition. In Fig. 5 the wave vector mismatch

$$\Delta k_m + \Delta k_{WG} = \beta(\omega_s)\omega_s + \beta(\omega_I)\omega_I - 2\beta(\omega_p)\omega_p \qquad (24)$$

is presented. Here $\beta(\omega)$ is the eigenvalue of the Helmholz equation including both the waveguide and the material contribution which was found numerically as described in Sect. 2. Phasematched idler and signal frequencies can be found as common points of horizontal line drawn at pump intensity and the curve corresponding to the pump frequency. As can be seen, phasematching in a PCF shows a rather distinct behavior from standard fibers (compare [33]) caused by the large waveguide contribution $\Delta k_{WG}$ to the wavevector mismatch. In Fig. 5 PCF with large hole radius (or tapered fiber) with parameters as given in the caption and a zero-dispersion wavelength of 830 nm and three different pump frequencies are considered. All considered pump frequencies are within the tuning range of a Ti:Sapphire laser. As can be seen the possible frequency shifts to the idler and signal wave are now in the range of the pump carrier frequency and more than one order of magnitude larger than in standard fibers. In particular for the input frequency $2.5\,\text{fs}^{-1}$ the idler wavelength is about 1600 nm and the signal at 400 nm. Small changing of the pump frequency leads to a huge range of detuning covering the range of IR up to the blue. Since additional the dispersion parameters of the PCF can be adjusted during the fabrication process WFM in PCFs offer the potential for the use of broadband parametric amplifier, frequency shifters and other interesting devices.

We start the numerical study of FWM by examining the propagation of a 200-fs-long pulse with $I_0 = 2.0\,\text{TW/cm}^2$ in a 2.95-$\mu$m-thick tapered fiber or in a PCF structure with very thin silica bridges. The input central frequency $\omega_0 = 2.27\,\text{fs}^{-1}$ coincides with the zero-dispersion frequency of this fiber.

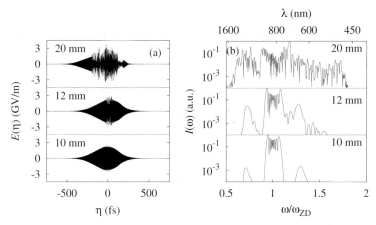

**Fig. 6.** Evolution of output pulse shape (a) and spectrum (b) for $\omega_0 = \omega_{ZD}$, $I_0=2\,\text{TW/cm}^2$, $\tau_0=200\,\text{fs}$ in 2.95-$\mu$m-diameter tapered fiber.

The evolution of the pulse shapes and spectra are illustrated in Fig. 6(a) and Fig. 6(b). During the initial stage of propagation, SPM causes slight broadening of the spectrum around the input frequency. With longer propagation length, two additional peaks arise at wavelengths $\sim 1150$ nm and 640 nm, which agree rather well with those predicted by the phase-matching condition in Fig. 5. At around 12 mm propagation length, seen in the middle section of Fig. 6, additional spectral components are generated. The temporal shape shows in the central part besides fast oscillating also low oscillating components, arising by the superposition of the different side peaks in the spectrum seen on the left side. With longer propagation to 20 mm the part of radiation transferred to the anomalous dispersion region by FWM and SPM is sufficient to form solitons. As can be seen in the upper section now a wide-band SC reaching from 500 nm to 1600 nm is generated and several peaks arise in the temporal shape. With further propagation these peaks move with different velocities and form separated pulses which do not change their form during further propagation (not shown here). This allows the interpretation that the SC in the upper part of Fig. 6 is caused by fission of higher-order solitons analogous as in Sect. 3. However, FWM and SPM is the dominant process on the early stages of propagation, which is responsible for generating new frequency components.

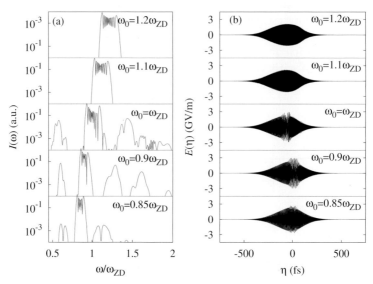

**Fig. 7.** Output spectra (a) and pulse shapes (b) for $L$=3 mm, $I_0$=8.7 TW/cm², $\tau_0$=200 fs and different input frequencies as indicated.

As discussed above and seen in Fig. 5 the dispersion properties of the medium have a large influence on the 4-wave-mixing processes. In Fig. 7, we compare the spectra of pulses with different input frequencies in anomalous ($\omega_0 < \omega_{ZD}$) and normal ($\omega_0 > \omega_{ZD}$) dispersion regions but the same intensity $I_0 = 8.7$ TW/cm² and pulse duration $\tau_0 = 200$ fs with fiber length $L = 3$ mm. In the anomalous dispersion regime, similar to the previous case new spectral components arise on both sides of the input peak. As predicted by comparison of curves 1 and 2 in Fig. 5, the frequency difference to the side peak is larger for the frequencies closer to the zero-dispersion point. This allows to control the wavelength of the generated radiation by slight shift of the input frequency. For $\omega_0 = 0.9\omega_{ZD}$ an additional spectral peak on the blue side is seen; this peaks is generated by cascaded 4-wave mixing. However, for frequencies in the normal dispersion region, no side peaks caused by FWM are generated and only spectral broadening by SPM can be seen. The temporal shapes for the input frequencies in the anomalous dispersion region and at the zero-dispersion frequency shows low and high-frequency components as expected from the spectrum, in contrast to the shapes for $\omega_0 > \omega_{ZD}$ where the low-frequency components to not arise.

# 5    Pulse Compression Without Chirp Control by High-Order Coherent Raman Scattering

Optical pulse compression based on combination of spectral broadening by self-phase modulation (SPM) induced by the optical Kerr effect and subsequent chirp compensation forms the physical basis for ultrashort pulse generation in the few-optical-cycle regime [40]. The key element of pulse compression by this method is the necessary phase compensation technique with negative group delay dispersion (GDD) based on gratings, prisms, chirped mirrors, liquid-crystal modulators and deformable mirrors. However, all methods for phase control are rather complex and several of their drawbacks cause difficulties for complete phase compensation for pulses below 5 fs.

In the present chapter we study an alternative method for pulse compression. In this method an intense pump pulse with duration smaller than the period of the molecular oscillations, $\tau_{\mathrm{pu}} < T_{\mathrm{v}} = 2\pi/\omega_{\mathrm{v}}$ ($\omega_{\mathrm{v}}$ is the Stokes shift), effectively excites the Raman-active medium and a temporally delayed, weak probe pulse with duration $\tau_{\mathrm{pr}} < T_{\mathrm{v}}$ is coherently scattered by the long-living Raman excitation. Then the spectrum of the probe pulse contains an infinite combination of frequency components satisfying the Raman resonance condition. Therefore the spectrum remains continuous and experiences a large spectral broadening and additionally a large blue or red shift depending on the delay time [13]. As it is shown for optimum conditions in molecular $SF_4$, a single pulse can be shortened from 20 fs down to 1.7 fs after the Raman medium. A modification of this method where the Raman-active medium is pumped by two ps-pulses with the frequency difference $\omega_{\mathrm{v}}$ is studied in Ref. [12]. Impulsive excitation considered here is experimentally simpler to realize.

Impulsive stimulated Raman scattering (SRS) and coherent scattering of a delayed probe pulse were first demonstrated and analyzed in Ref. [41]. Recently, using such regime, the transformation of a 160-fs probe pulse into a train of 6-fs pulses following with period $T_{\mathrm{v}}$ was observed [10,42]. However, possible applications of such high repetition trains are limited. In contrast, in the short-pulse regime with $\tau_{\mathrm{pr}} < T_{\mathrm{v}}$, the compression of single pulses with continuous blue- or red-shifted spectra is achieved and the physical mechanism differs significantly from that realized in Ref. [10,42]. In other related work on high-order SRS the generation of a discrete broad rotational spectrum in $H_2$ was demonstrated [43] and the use of electromagnetically induced transparency to produce trains of sub-fs pulses was studied [44].

Spectral and temporal effects of coherent scattering using impulsive SRS were previously analyzed [41,42] in the frame of the standard slowly varying envelope approximation. For the investigation of significant spectral and temporal changes this approximation fails and here we solve the basic equations ( 13, 16, 17) without using it.

First we analyze an approximate analytical solution of Eqs. (13) and (16) given for the first time in Ref. [12] for the case of two-color excitation. Neglecting the influence of dispersion, Stark shift, Kerr effect, relaxation and population changing, for a delay much larger than the pump pulse duration, the solution of

Eq.(16) is given by $\rho_{12} = \rho_0 \exp(i\omega_v\eta)$, where

$$\rho_0 = -\frac{i\alpha_{12}}{2\hbar} \int\limits_{-\infty}^{\infty} E_{\mathrm{pu}}^2(z,\eta')\exp(-i\omega_v\eta')d\eta'$$

is determined by the pump field and depends weakly on distance $z$. For arbitrary weak probe pulses and $\rho_0 = \mathrm{const}$, the solution of Eq.(13) can be found as follows:

$$E_{\mathrm{pr}}(z,\eta) = E_0(s)\frac{\sin(\omega_v s)}{\sin(\omega_v\eta)} , \qquad (25)$$

where $s = s(z,\eta)$ satisfies the relation $\mathrm{tg}(\omega_v s/2) = \mathrm{tg}(\omega_v\eta/2)\exp(-\omega_v\gamma z)$, $E_0(\eta)$ is the field strength of the input probe pulse, $\gamma = \frac{1}{2}c\mu_0 N\alpha_{12}u_0$. For $\omega_v\gamma z \ll 1$ we have $s \approx \eta - \gamma z\sin(\omega_v\eta)$ and $E_{\mathrm{pr}} = A(\eta - \eta_0)\cos\{\omega_{\mathrm{pr}}[\eta - \eta_0 - \gamma z\sin(\omega_v\eta)]\}$, where $A(\eta)$, $\omega_{\mathrm{pr}}$ and $\eta_0$ are the envelope, the carrier frequency and the delay of the initial probe pulse, respectively. This is a frequency-modulated pulse which is not shortened during propagation, with the Fourier-transformed field

$$E_{\mathrm{pr}}(z,\omega) = \sum_{m=-\infty}^{\infty} J_m(\omega_{\mathrm{pr}}\gamma z)A(\omega - \omega_{\mathrm{pr}} + m\omega_v)e^{im\omega_v\eta_0},$$

where $J_m$ are the Bessel functions and the different terms describe high-order Raman components with a continuous spectrum for $\tau_{\mathrm{pr}} < T_v$. If the probe pulse is in-phase with the molecular oscillations, with delay $\eta_0^{(\mathrm{in})} = nT_v$ $(n = 0, \pm 1, ...)$, then $E_{\mathrm{pr}}(z,\omega)$ is real and changes its sign only on the anti-Stokes side ($\omega_{\mathrm{pr}}\gamma z < 2.405$ is supposed), where the spectral phase $\phi(\omega)$ jumps between 0 to $\pi$. Therefore, the spectrum is shifted to the Stokes side. For an out-of-phase probe pulse with delay $\eta_0^{(\mathrm{out})} = (n + \frac{1}{2})T_v$ only the Stokes components show sign jumps and therefore the spectrum is shifted to the anti-Stokes side. This means that, in general, phase-synchronization in high-order SRS does occur neither in the short nor in the long pulse limit.

For significant pulse shortening a large spectral broadening with $\omega_v\gamma z > 1$ is required. For $\Gamma = \exp(\omega_v\gamma z) \gg 1$ we obtain $s = \Gamma\eta + (1 - \Gamma)\eta_0^{(\mathrm{out})}$ and $s = \Gamma^{-1}\eta + (1 - \Gamma^{-1})\eta_0^{(\mathrm{in})}$ in the vicinity of out-of-phase and in-phase points, respectively. Then for a pulse with out-of-phase delay Eq.(25) yields the asymptotic solution for a single compressed probe pulse of the form $E_{\mathrm{pr}}(z,\eta) = \Gamma E_0(\Gamma\eta)$ with a duration $\tau_{\mathrm{pr}}/\Gamma$ and a frequency $\Gamma\omega_{\mathrm{pr}}$ of the spectral maximum. On the other hand, for the in-phase delay the pulse is split into two parts with field maxima at two adjacent out-of-phase points.

In Fig. 8 the solution (25) is represented for a 20-fs Gaussian input probe pulse and $\Gamma = 3.3$. For an in-phase delay the pulse splits into two parts and its oscillating Fourier components are extended into both directions [Fig. 8(a) and 8(c)]. In contrast, the out-of-phase probe is compressed, its Fourier-transformed field is significantly broadened and blue-shifted with sign changes on both the Stokes and anti-Stokes side [Figs. 8(b) and 8(d)]. The specific type of amplitude-phase modulation for in- or out-of-phase delay shown in Figs. 8(c), 8(d) with

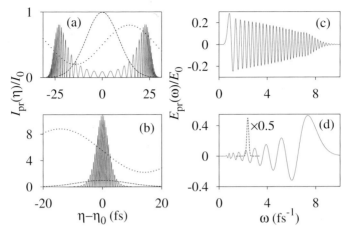

**Fig. 8.** Solution (25) for 20-fs in-phase (a,c) and out-of-phase (b,d) pulse and $\omega_v \gamma z = 1.2$, $\omega_{pr} = 2.39\,\text{fs}^{-1}$ ($\lambda_{pr} = 790\,\text{nm}$), $\omega_v = 587\,\text{cm}^{-1}$ ($T_v = 57\,\text{fs}$): (a,b) pulse intensity (solid curve), initial pulse envelope (dashed), molecular variable $u(\eta)$ in arbitrary scale (dotted); (c,d) Fourier-transformed output (solid) and initial (dashed) field.

real $E(\omega)$ and phase jumps from 0 to $\pi$ is rather unusual in ultrafast optics. However, such pulses can be compressed by standard methods. For instance, the pulse in Fig. 8(b) can be shortened from 5 fs to 1.2 fs by external linear GDD with parameter $\pm 1.4\,\text{fs}^2$, but contrary to SPM by Kerr effect, with both positive and negative sign because $E_{pr}(\omega)$ is real.

Equation (25) describes unlimited compression with increasing length, but in real systems dispersion, pump pulse phase modulation and depletion cause limitations and additionally Kerr effect and Stark shift influence the pulse propagation. To study all these effects we solved the complete basic Eqs. (13, 16, 17) for the vibrational Raman transitions in gaseous $SF_6$. The main limitation for compression arises due to the influence of dispersion. With increasing spectral broadening the central frequency is shifted to the blue-side and therefore the delay of the probe pulse shifts during propagation from the out-of-phase point to the in-phase point, where pulse compression is finished. Therefore in order to reach the region of shorter pulses up to the sub-cycle range one has to realize a high Raman conversion over a short propagation length or to minimize the influence of dispersion. Due to the waveguide contribution to the linear polarization in hollow fibers filled by Raman gaseous medium, the GVD can be controlled by the change of pressure and waveguide radius.

In Fig. 9 the temporal and spectral transformation of the pump pulse is shown. As one can see, the pump pulse is temporally broadened and its spectrum experiences a large red-shift up to the wavelength 1450 nm. In Fig. 10 the shape (a) and the spectrum (b) of the probe pulse is presented for different pressure. For the low pressure the pulse compression and the spectral broadening is small because of the reduced Raman polarization. Increasing the pressure to an intermediate value results to significant pulse compression up to a FWHM of 1.7 fs determined from fitted Gaussian envelope. The spectrum shows a continuous broadening and a continuous shift to the anti-Stokes side up to 250 nm with weak side maxima. With a higher pressure of 1 atm the bulk contribution to dispersion increases, while the waveguide contribution to dispersion remains constant. Due to the frequency shift and larger value of GVD parameter the probe pulse delay now is shifted from the out-of-phase point to the maximum of Raman polarization during a shorter propagation length, therefore the pulse duration becomes larger and the spectral width smaller.

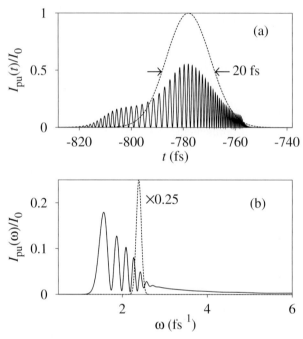

**Fig. 9.** Temporal shape (a) and spectrum (b) of the pump pulse at output of 30-cm-long hollow waveguide with radius 80 $\mu$m and $SF_6$ pressure 0.5 atm for a 790-nm, 20-fs initial pulse (dashed lines) with intensity $I_0 = 50$ TW/cm$^2$.

In conclusion, though the method of spectral broadening by SPM can be extended to achieve ultra-wide spectra with a continuous phase delay sufficient for pulse shortening up to 0.5 fs [4], a simpler method without the necessity of external phase control is of high interest. The results demonstrate that chirp-free pulses up to 1.7 fs can be generated by coherent scattering in impulsively excited Raman media without external chirp compensation. We expect that by further optimization of this method the milestone in pulse compression in the range of 1 fs could be achieved, this study is presently in work. This method has the potential to be developed to a simple to handle, highly reproducible and stable short-pulse source with tunable frequencies and pulse durations, high pulse energies and shortest durations.

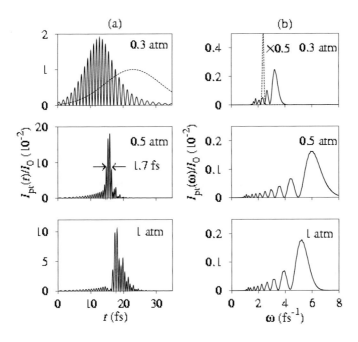

**Fig. 10.** Probe pulse shapes (a) and spectra (b) for different gas pressures (as indicated) in a hollow waveguide and a 0.5 TW/cm$^2$, 20-fs initial pulse (dashed line). Other parameters are the same as in Fig. 9.

## 6    Formation of Optical Sub-cycle Pulses and Full Maxwell-Bloch Solitary Waves in Resonant Two-Level Medium

Recent advances in ultrafast laser technology have made possible the generation of extremely short and intense pulses with only two optical periods or less than 5 fs in duration in the visible region [45,46]. Efforts for generation of still shorter pulses are motivated by the possibility to study light-matter interactions under unique conditions as the pulse duration approaches the duration of a single optical cycle. The most promising conditions for the generation of sub-fs pulses are provided by strong nonlinear processes with high degree of the spatio-temporal coherence. We investigated such a process in which a layer of a resonant dense medium described by a two-level model is driven by a strong and extremely short optical pulse. For pulses with a duration of many cycles, the nonlinear properties of two-level media have been extensively studied. In particular, the resonant interaction of pulses shorter than all relevant relaxation times of the medium gives rise to the effect of .self-induced transparency (SIT) and pulse splitting [18]. Here we explore the propagation of several optical cycle pulses [47–49] without the use of SVEA or rotating-wave approximation (RWA) and solve the full Maxwell equations and the full Bloch equations.

We consider the propagation of an extremely short pulse along the $z$ axis normally to an input interface of a resonant two-level medium at $z=0$. Initially the pulse moves in the free space; then it partially penetrates into the medium and partially reflects backwards; the penetrating part propagates through the medium and finally exits again into the free space through the output interface at $z = L$. With the constitutive relation for the electric displacement for linear polarization along the $x$ axis, $D = \epsilon_0 E + P_B$, the Rabi frequency $\Omega = dE/\hbar$ and $\Psi = \sqrt{\mu_0/\epsilon_0}dH_y/\hbar$ Maxwell's equations for the medium take the form

$$\frac{\partial \Psi}{\partial t} = -\frac{\partial \Omega}{\partial \zeta}, \quad \frac{\partial}{\partial t}(\Omega + \omega_c u) = -\frac{\partial \Psi}{\partial \zeta}, \tag{26}$$

where $E$, $H$ are the electric and magnetic fields, respectively, $d$ is the dipole moment, $\zeta = z/c$, $\omega_c = Nd^2/\epsilon_0\hbar$ is the characteristic coherent time of the medium. In Eqs. (26) the macroscopic nonlinear polarization $P_R$ is defined by the Bloch Eqs. (18,19). The initial condition is $\Omega(t = 0, \zeta) = \Psi(t = 0, \zeta) = \Omega_0 \cos[\omega_p(\zeta - \zeta_0)]\mathrm{sech}[1.76(\zeta - \zeta_0)/\tau_p]$, where $\Omega_0$ is the peak Rabi frequency of the incident pulse, $\omega_p$ is the carrier frequency, $\tau_p$ is the FWHM of the pulse intensity envelope.

Assuming the resonance condition $\omega_p = \omega_0$ and slow relaxation times, the solution of the problem depends only on three normalized parameters $\Omega_0\tau_p$, $\omega_c\tau_p$ and $\omega_0\tau_p$ which permit a rescaling of the five atomic and pulse parameters. Several gaseous atoms have well isolated resonances in the optical region (as, e.g., Rubidium with $\omega_p = 2.4\,\mathrm{fs}^{-1}$), but the necessary high density or high pressure for the required parameters as given below is hard to realize. However, the two-level model can be still applied for the description of coherent effects for materials with a broad distribution of transitions such as inhomogeneously broadened resonance lines in gases and solids [18], but also as an approximation for inhomogeneous quasi-continuous energy bands as, e.g., in semiconductors [50,51]. In these more complex systems the coherent superposition of a band of continuous transitions with approximately the same dipole moment manifests itself as a unified single level with an inhomogeneous linewidth $(T_2^*)^{-1}$. In the present Letter the inhomogeneous broadening is neglected. In the following we consider the dimensionless parameters of the problem in the ranges $\omega_0\tau_p = 11.5$, $1 \leq \omega_c\tau_p \leq 10$, $7 \leq \Omega_0\tau_p \leq 22$. It is instructive to indicate concrete initial pulse and material parameters that meet these conditions: $\tau_p=5\,\mathrm{fs}$, $\omega_p = \omega_0=2.3\,\mathrm{fs}^{-1}$ ($\lambda=830\,\mathrm{nm}$), $d = 2 \times 10^{-29}$ Asm, $\gamma_1^{-1}=1$ ps, $\gamma_2^{-1}=0.5$ ps. For these parameters the density $N = 4.4 \times 10^{20}$ cm$^{-3}$ gives $\omega_c=0.2\,\mathrm{fs}^{-1}$ and the Rabi frequency $\Omega_0=1\,\mathrm{fs}^{-1}$ corresponds to the electric field of $E_x=5 \times 10^9$ V/m or an intensity of $I = 6.6 \times 10^{12}$ W/cm$^2$.

Equations (18), (19) and (26) were integrated by use of Yee's finite-difference time-domain (FDTD) discretization scheme [2] for the fields and the predictor-corrector method for the material variables [47]. Nonreflecting boundary conditions [52] were incorporated with FDTD discretization to avoid the influence of the boundaries. The performance of the numerical scheme was monitored at each time update by calculation of the pulse energy, the energy stored in the medium and the energy flux through the computational-domain boundaries in comparison with the energy of the initial pulse. In all our simulations the total energy was conserved with an accuracy better than 0.001%.

We have obtained solutions of Eqs.(18), (19) and (26) for different input pulse parameters and medium densities. In general, with an input envelope area $\mathcal{A} = 1.76\Omega_0\tau_p$ smaller than $\pi$ and chosen medium densities, the main part of the pulse is reflected, but with increasing area splitting of the penetrating part accompanied by pulse shortening occurs, in agreement with results within the SVEA and RWA [18]. However, details of the pulse evolution dramatically change for input pulses with a duration of 5 to 10 fs, where novel transient features arise which qualitatively differ from standard results. As an example, in Fig. 11 the reflected, penetrating and transmitted pulses are shown for different medium densities and a fixed input pulse maximum $\Omega_0=1.4\,\mathrm{fs}^{-1}$ corresponding to the envelope area $\mathcal{A} = 4\pi$. For the case of lower density with $\omega_c=0.2\,\mathrm{fs}^{-1}$, the reflection is weak and the penetrating part is strong enough to split into two pulses, the stronger of which moves faster than the weaker. In the case of a more dense medium (Fig. 11b), a significant part of radiation with low-frequency oscillations on the back front is reflected from the boundary. With further propagation the pulses in Fig. 11 behave in different manner. For the less dense medium the weaker pulse in Fig. 11(a) decays quickly, while the first pulse propagates over much longer distances. For the high density with $\omega_c=1\,\mathrm{fs}^{-1}$ the electric field, polarization and population at $z=60\,\mu\mathrm{m}$ are shown in Fig. 12. In this case we observe a few-cycle field oscillating with higher frequency as compared with the initial pulse oscillations. The polarization follows the field quite accurately in time, but with opposite direction. As a consequence the phase velocity of this

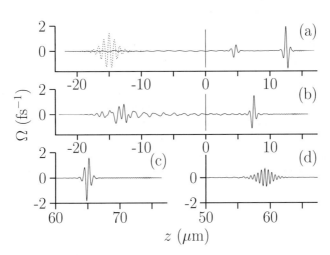

**Fig. 11.** Electric field profiles for different time moments and medium densities near the input face (a,b) and outside the exit face (c,d). Parameters: $\Omega_0=1.4\,\mathrm{fs}^{-1}$, $L=45\,\mu\mathrm{m}$; (a) $t=0.1\,\mathrm{ps}$, $\omega_c=0.2\,\mathrm{fs}^{-1}$; (b) $t=0.1\,\mathrm{ps}$, $\omega_c=1\,\mathrm{fs}^{-1}$; (c) $t=0.3$ ps, $\omega_c=0.2\,\mathrm{fs}^{-1}$; (d) $t=0.4\,\mathrm{ps}$, $\omega_c=1\,\mathrm{fs}^{-1}$. Dotted line: incident pulse at $t=0$.

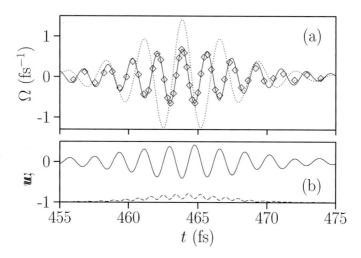

**Fig. 12.** (a) Temporal shape of the electric field (solid line), (b) polarization $u$ (solid) and population $w$ (dashed) at $z$=60 µm inside the medium for $\omega_c$=1 fs$^{-1}$, $\Omega_0$=1.4 fs$^{-1}$, $L$=150 µm. The incident pulse shifted to the same peak position is shown by the dotted line. Diamonds represent the solution (28).

pulse is larger than the light velocity in free space. It is remarkable that in this case the pulse propagation and the two-level atom dynamics show evidence of a soliton propagation regime beyond SVEA and RWA which will be discussed later.

In Fig. 13 we present the propagation of a more intense incident pulse with the Rabi frequency $\Omega_0$=4.4 fs$^{-1}$ and the envelope area $\mathcal{A} = 12.5\pi$ in a more dense medium with $\omega_c$=2 fs$^{-1}$. In this case the reflected pulse is rather weak and the penetrating part splits mainly into three pulses with different field strengths and velocities. With further propagation these pulses evolve into a higher-frequency oscillating pulse and two separated half-cycle pulses of opposite polarity. Both unipolar half-cycle pulses with a duration <1 fs are not attenuated during long-term propagation and show a clear solitary behavior. As remarked above the existence of SIT solutions of the envelope equations is restricted by the validity of SVEA and RWA, which is violated for the conditions considered here. However, there exists an exact nonoscillating solitary solution for field strength of the full.Maxwell-Bloch equations (18), (19) and (26) (with $\gamma_1 = \gamma_2 = 0$), which is given by the expressions

$$\Omega = A\mathrm{sech}[A(t - m\zeta)],$$
$$u = 2\omega_0\beta\Omega, \tag{27}$$
$$w = -1 + 2\beta\Omega^2,$$

where $m = \sqrt{1 + \beta\omega_0\omega_c}$ is the pulse refractive index and $\beta = (A^2 + \omega_0^2)^{-1}$ [53]. The solution (27) describes an unipolar half-cycle pulse with the amplitude $A$, which is the only free parameter determining the pulse duration. As shown in

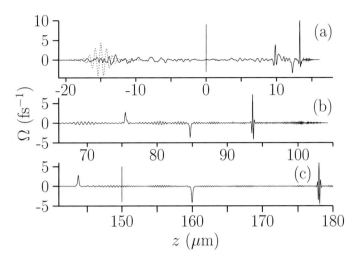

**Fig. 13.** Electric field profiles near the input face located at $z=0$ (a), inside the medium (b) and near output face located at $z=150\,\mu$m (c) for $\omega_c=2\,\text{fs}^{-1}$, $\Omega_0=4.4\,\text{fs}^{-1}$. (a) $t=0.1$ ps; (b) $t=0.4$ ps; (c) $t=0.7$ ps. The incident pulse at $t=0$ is shown by the dotted line.

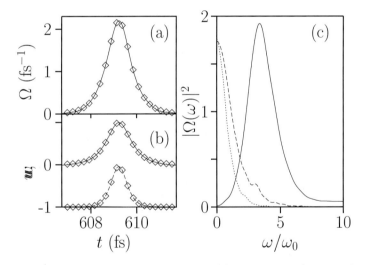

**Fig. 14.** (a) Temporal shape of the electric field, (b) polarization (solid line) and population (dashed) at $z=120\,\mu$m for $\omega_c=2\,\text{fs}^{-1}$, $\Omega_0=4.4\,\text{fs}^{-1}$, $L=150\,\mu$m. Diamonds represent the solution (27). (c) Spectrum of the first high-frequency pulse (solid line), second (long-dashed) and third (short-dashed) unipolar pulses at $z=120\,\mu$m.

Fig. 14 for the third pulse of Fig. 13, the numerically obtained solutions and the solution (27), in which only the amplitude $A$ is determined by a fit, can be clearly identified with each other even though the pulse propagation in Fig. 13

at $z=120\,\mu m$ is still not completely in a steady-state regime because the three pulses still interact. In Fig. 14c the spectra of the three pulses are presented. As expected both half-cycle pulses show extended spectra up to zero frequency, while the first pulse has a large blue shift of the carrier frequency with a mean frequency $\bar\omega \approx 4\omega_0$.

Note that the fastest blue-shifted pulse in Fig. 13 shows also a solitary behavior but, in difference to the half-cycle pulses, that of a.multi-soliton solution. Far away from the input interface, Maxwell's equations can be approximated by a reduced version without the use of SVEA and RWA, but with the neglect of back-reflection. These reduced Maxwell-Bloch equations have exact multi-soliton solutions [54]. Their two-soliton solution describing a localized pulse with "internal" oscillations is given by

$$\Omega = A\mathrm{sech}\vartheta_e \frac{\cos\vartheta - \alpha\sin\vartheta\mathrm{tanh}\vartheta_e}{1 + \alpha^2\sin^2\vartheta\mathrm{sech}^2\vartheta_e}, \tag{28}$$

where the amplitude $A$ and carrier frequency $\bar\omega$ are free parameters, $\vartheta_e = A(t - m_e\zeta)/2$, $\vartheta = \bar\omega(t - m\zeta)$, $\alpha = A/2\bar\omega$. The refractive indices $m_e$, $m$ and the atomic variables are given in Ref. [54]. The solution (28) with an appropriate choice of the amplitude and $\bar\omega = \int_0^\infty \omega|\Omega(\omega)|^2 d\omega / \int_0^\infty |\Omega(\omega)|^2 d\omega$ coincides well with the blue-shifted pulse fields found numerically for both cases of high medium density in Figs. 12 and 13. The direct comparison of analytical and numerical solutions for $\omega_c=1\,\mathrm{fs}^{-1}$ is shown by diamonds in Fig. 12.

### Acknowledgement

We acknowledge financial support from the Deutsche Forschungsgemeinschaft.

# References

1. A. V. Husakou and J. Herrmann: Phys. Rev. Lett. **87**, 203901 (2001).
2. K. S. Yee: IEEE Trans. Antennas Propag. **14**, 302 (1966).
3. V. P. Kalosha, J. Herrmann: Phys. Rev. Lett. **83**, 544 (1999).
4. V. P. Kalosha and J. Herrmann: Phys. Rev. A **62**, 11804 (2000).
5. R. K. Bulough et al.: Physica Scripta **20**, 364 (1979).
6. T. Brabec, F. Krausz: Phys. Rev. Lett. **78**, 3282 (1997).
7. J. E. Rottenberg: Opt. Let. **17**, 1340 (1992).
8. J. N. Elgin, T. Brabec, and S. M. J. Kelly: Opt. Comm. **114**, 321 (1995).
9. Y. Kodama and A. Hasegawa: IEEE J. Quantum Electron. **23**, 510 (1987).
10. W. Wittmann, A. Natarkin, G. Korn: Opt. Lett. **26**, 298 (2001).
11. A. V. Husakou, V. P. Kalosha: and J. Herrmann, Opt. Lett. **26**, 1022 (2001).
12. V. P. Kalosha and J. Herrmann: Phys. Rev. Lett. **85**, 1226 (2000).
13. V. P. Kalosha and J. Herrmann: Opt. Lett. **26**, 456 (2001).
14. M. Geissler et al.: Phys. Rev. Lett. **83**, 2930 (1999).
15. E. A. Marcatili and R. A. Schmeltzer: Bell Syst. Tech. J. **43**, 1783 (1964).
16. M. Nisoli, S. De Silvestri, and O. Svelto: Appl. Phys. Lett. **68**, 2793 (1996).
17. M. Sheik-Bahae et al.: J. Quantum Electron. **27**, 1296 (1991).

18. L. Allen and J. H. Eberly, *Optical Resonance and Two-Level Atoms* (Wiley, New York, 1975).
19. J. C. Knight *et al.*: Opt. Lett. **21**, 1547 (1996).
20. T. A. Birks, J. C. Knight, P. St. J. Russel: Opt. Lett. **22**, 961 (1997).
21. D. Mogilevtsev, T. A. Birks and P. St. J. Russel: Opt. Lett. **23**, 1662 (1998).
22. J. C. Knight *et al.*: IEEE Photonic Technology Letters **12**, 807 (2000).
23. J. K. Ranka, R. S. Windeler, and A. J. Steinz: Opt. Lett. **25**, 25 (2000).
24. T. A. Birks, W. J. Wadsworth, and P. St. J. Russel: Opt. Lett. **25**, 1415 (2000).
25. R. C. Fork *et al*: Opt. Lett. bf 12, 483 (1987).
26. M. Pshenichnikov, W. de Boeij, and D. Wiersma: Opt. Lett. **19**, 572 (1994).
27. D. A. Jones *et al.*: Science **288**, 635 (2000).
28. R. Holzwarth *et al.*: Phys. Rev. Lett. **85**, 2264 (2000).
29. J. Hartl *et al.*: Opt. Lett. **26**, 608 (2001).
30. J. Herrmann, U. Griebner, N. Zhavoronkov, A. Husakou, D. Nickel, J. C. Knight, W. J. Wadsworth, P. St. J. Russel, G. Korn, "Experimental evidence for supercontinuum generation by fission of higher-order solitons in photonic crystal fibers", submitted for publication.
31. R. R. Alfano and S. L. Shapiro: Phys. Rev. Lett. **24**, 592 (1970).
32. R. Alfano, Ed., *The supercontinuum laser source*, (Springer Verlag, New York,1989).
33. G. P. Agrawal, *Nonlinear Fiber Optics* (Academic Press, New York, 1994).
34. P. K. A. Wai *et al.*: Opt. Lett. **12**, 628 (1987).
35. N. Akhmediev and M. Karlsson: Phys. Rev. A **51**, 2602 (1955).
36. A. S. Gouveiv-Neto, M. E. Faldon, and R. J. Taylor: Opt. Lett. **13**, 1029 (1988).
37. S. G. Murdoch, R. Leonard and J. D. Harvey: Opt. Lett. **20**, 866 (1955).
38. J. Zhang *et al.*: Opt. Lett. **26**, 214 (2001).
39. J. E. Sharping *et al.*: Opt. Lett. **26**, 1048 (2001).
40. For a recent review see, T. Brabec and F. Krausz: Rev. Mod. Phys. **72**, 545 (2000).
41. Y.-X. Yan, E. B. Gamble, and K. A. Nelson: J. Chem. Phys. **83**, 5391 (1985).
42. M. Wittmann, A. Nazarkin, and G. Korn: Phys. Rev. Lett. **84**, 5508 (2000).
43. H. Kawano, T. Mori, Y. Hirakawa, and T. Imasaka: Phys. Rev. A **59**, 4703 (1999).
44. S. E. Harris and A. V. Sokolov: Phys. Rev. Lett. **81**, 2894 (1998).
45. J. Zhou *et al.*: Opt. Lett. **19**, 1149 (1994); A.Stingel *et al.*: Opt. Lett. **20**, 602 (1995).
46. M. Nisoli *et al.*: Opt. Lett. **22**, 522 (1997).
47. R. W. Ziolkowski, J. M. Arnold, and D. M. Gogny: Phys. Rev. A **52**, 3082 (1995).
48. M. Müller, V. P. Kalosha, and J. Herrmann: Phys. Rev. A **58**, 1372 (1998).
49. S. Hughes: Phys. Rev. Lett. **81**, 3363 (1998).
50. H. Haug and S. W. Koch, *Quantum Theory of the Optical and Electronic Properties of Semiconductors*, 2nd ed. (World Scientific, Singapore, 1993).
51. V. P. Kalosha, M. Müller, and J. Herrmann: J. Opt. Soc. Am. B **16**, 323 (1999).
52. E. L. Lindman: J. Comp. Phys. **18**, 66 (1975).
53. R. K. Bullough and F. Ahmad: Phys. Rev. Lett. **27**, 330 (1971).
54. J. C. Eilbeck *et al.*: J. Phys. A **6**, 1337 (1973).

# Experimental Study of Modulational Instability and Vector Solitons in Optical Fibers

G. Millot[1], S. Pitois[1], J.M. Dudley[2], and M. Haelterman[3]

[1] Laboratoire de Physique de l'Université de Bourgogne, 21078 Dijon, France
[2] Laboratoire d'Optique P.M. Duffieux, Université de Franche-Comté, 25030 Besançon, France
[3] Service d'Optique et Acoustique, Université Libre de Bruxelles, B-1050 Bruxelles, Belgium.

**Abstract.** This chapter brings forth the experimental study of modulational instability and vector solitons in optical fibers.

## 1 Introduction

The nonlinear coupling between two intense laser beams in normally dispersive optical fibers can lead to fascinating physical phenomena such as modulational instability (MI) [1], whereby a continuous or quasi-continuous wave undergoes a modulation of its amplitude or phase in the presence of noise or any other weak perturbation. The perturbation can originate from quantum noise (spontaneous-MI) or from a frequency shifted signal wave (induced-MI). MI was observed for the first time for a single pump wave propagating in a standard non birefringent fiber (scalar MI) [2]. In this framework, it was shown that scalar MI only occurs when the group velocity dispersion (GVD) is negative (anomalous dispersion regime). As a matter of fact, in the normal GVD regime the propagation of an individual intense beam in a single-mode fiber is not subject to MI. However, as it was first pointed out by Berkhoer and Zakharov, the nonlinear coupling between two different (e.g., polarization) modes by cross-phase modulation(XPM) may extend the domain of MI to the normal GVD regime [3]. This XPM-induced MI is also called vector modulational instability. Several experiments with two polarization modes were performed in order to observe spontaneous-MI in the normal GVD regime either in standard high-birefringence (hibi) fiber [4], [5], [6], or more recently, in polarization-preserving microstructured fibers (MF) [7]. Modulational instability may be detrimental in multichannel communications [1] and, in this context, to avoid the generation of unwanted MI sidebands, one may consider a configuration in which all frequency components of a pump field are linearly polarized in the normal dispersion regime of a single-mode fiber. Note that in a hi-bi fiber, whenever the frequency components of the pump are orthogonally polarized, the pump frequency spacing should be chosen in a region known as the *critical gap*, in which the MI gain vanishes [8].

In the time domain, induced-MI leads to the break-up of the quasi-cw pump wave into a train of ultrashort pulses [9]. The repetition rate of the pulses is determined by the frequency detuning between the signal and the pump (i.e. the modulational frequency). Moreover, families of vector dark solitons solutions were found for the coupled NLS equations that apply to hi-bi fibers [10]. Thus the use of hi-bi fibers in the normal dispersion regime has led to the generation of vector dark soliton trains at repetition rates of 2.5 THz [11]. A major experimental difficulty, however, is that the sub-picosecond structure of the THz optical pulse train cannot be directly measured using photodiodes or streak cameras. Nowadays, autocorrelation and spectral analysis are routinely replaced by the technique of frequency-resolved optical gating (FROG), which can provide complete intensity and phase characterization on sub-picosecond time-scales [12]. Although the FROG technique is typically applied to ultrashort pulses [12], it was recently demonstrated that this method can also be exploited for periodic optical pulse trains by using an adapted pulse retrieval algorithm [13],[14],[15].

Another interesting application of MI is the frequency conversion of a pump wave into a pair of Stokes and anti-Stokes parametric sidebands. Unfortunately the MI gain spectrum is generally narrow (< 1 THz), and consequently, the frequency detuning between the pump and the sidebands should be carefully chosen for the occurrence of frequency conversion. However it has been shown recently that the frequency detuning bandwidth of MI can be significantly extended by overcoming the spectral limitations imposed by phase-matching conditions [16]. The basic idea of the technique is to choose a pump-signal frequency detuning in such a way that one of its harmonics falls within the linear MI-gain spectrum. Modulational instability is then induced by multiple four-wave mixing.

Modulational instability has been also studied in nearly resonant static gratings [17]. However, experimental demonstrations of MI in periodic structures have been limited because of technological difficulties inherent to the resonant nature of the device. Indeed, MI occurs when light propagates into the fiber Bragg grating at a wavelength close to the Bragg resonance, a condition in which the grating becomes highly reflective. A novel scheme has been recently proposed that allows to operate arbitrarily close to the photonic band gap. The periodic modulation of the refractive index is ensured by the temporal beating of two intense waves of different frequencies propagating in a fiber, leading to the generation of a dynamical grating [18],[19].

## 2    Experimental Setup

Figure 1 shows the experimental set-up that we employed for the study of MI in optical fibers. In most of the experiments, we simultaneously injected into an optical fiber nanosecond pulses of different frequencies from two synchronized laser sources. A Q-switched, frequency-doubled and injection-seeded Nd:YAG laser emitting at $\lambda_p = 532.26$ nm, operating at a repetition frequency of 25 Hz, was used to generate the pulses from both laser sources [6]. The Nd:YAG laser beam was subsequently split into two different intensity beams. The less powerful beam was sent into a multipass carbon dioxide cell, which shifts its frequency by means of self-Stimulated Raman Scattering [6]. The resulting first Stokes wave $P_1$, shifted by 1388 cm$^{-1}$ or 41.64 THz from the input laser, was filtered by means

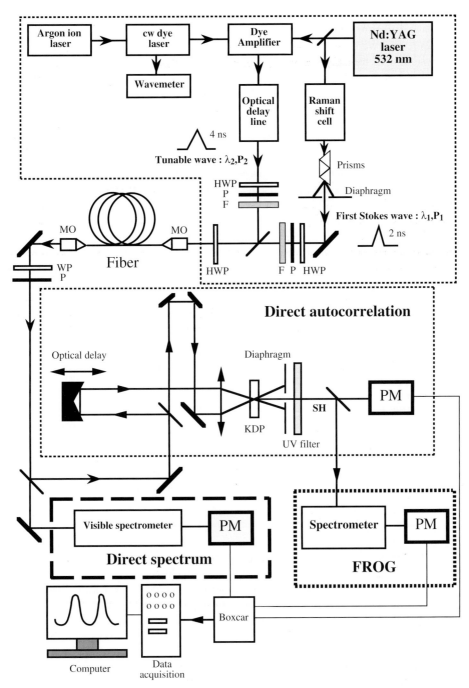

**Fig. 1.** Schematic of the experimental setup used for the recording of direct spectra and autocorrelations and SHG-FROG traces: P's, glan polarizers; BS, beam splitter; MO's 20x microscope objectives; F's, filters; HWP, half-wave plate; WP, wave plate; PM, photomultiplier.

of a direct vision prism. The pulses obtained from the first Stokes beam had a temporal duration of 2 ns measured at the full width at half maximum (FWHM) of power, and were spectrally centered at $\lambda_1 = 574.72$ nm. The most powerful beam served to amplify a cw single-mode tunable ring dye laser pumped by a 5-W cw argon laser. The dye laser was sent into a three-stage dye cell amplifier $P_2$ pumped by the Nd:YAG beam. The pulses at the output of the amplifier had a temporal duration of 4 ns (FWHM of power). The two pulses of wavelengths $\lambda_1$ and $\lambda_2$ were synchronized by sending the dye beam into an optical delay line. The pump and signal beams were finally combined by a beam splitter, and focused into an optical fiber of fixed length. After propagation through the fiber, light was characterized by means of a visible 50 cm-spectrometer with a resolution better than 0.04 nm, a background-free second-harmonic generation (SHG) autocorrelator, and a FROG set-up based on the spectral analysis of the SHG autocorrelation signal using a UV spectrometer. The FROG measurements involved recording the spectrum at each delay with a UV-spectrometer. FROG spectrograms were made up on a $128 \times 128$ grid. The spectral, autocorrelation and SHG-FROG signals were finally sampled and averaged over 30 pulses by means of a boxcar integrator.

## 3     Vector Modulational Instability in Highly Birefringent Fibers

### 3.1     Coupled Nonlinear Schrödinger Equations

We consider a lossless fiber with a strong intrinsic birefringence. Two pump waves polarized along the fast (x) and slow (y) axis, respectively, copropagate along the $z$ axis, in the normal dispersion regime. Hereafter, the two pump waves will be referred to as $P_1$ and $P_2$, with different (angular) frequencies $\omega_1$ and $\omega_2$. Whenever it will be convenient, the single-frequency copropagation can be obtained by just making $\omega_1 = \omega_2$. The total electric field reads as

$$E_j = \frac{1}{2} A_j(z,t) \exp[i(k_j z - \omega_j t)] + c.c., \quad j = 1, 2 \qquad (1)$$

where c.c. denotes complex conjugation, subscripts $j = 1, 2$ refer to the fast and slow axes, and $A_j$ are the slowly-varying field envelopes.

By expanding $k_j$ ($j = 1, 2$) in Taylor series around the two input wave frequencies, the amplitudes $A_j$ of the electric fields are found to satisfy the following set of coupled nonlinear Schrödinger equations (NLSE's):

$$\frac{\partial A_j}{\partial z} + (-1)^j \, \delta \, \frac{\partial A_j}{\partial t} + \frac{1}{2} i \, \beta_j \frac{\partial^2 A_j}{\partial t^2} = i \gamma_j \left( |A_j|^2 + \frac{2}{3} |A_{3-j}|^2 \right) A_j. \qquad (2)$$

In Eqs.(2), $\beta_j$ is the GVD at angular frequency $\omega_j$, $\gamma_j \equiv n_2 \omega_j/(cA_{ej})$ is the nonlinear coefficient, where $n_2 = 3.2 \times 10^{-16}$ cm$^2$/W is the fiber nonlinear refractive index, and $A_{ej}$ is the effective core area. Throughout this paper we will consider the normal dispersion regime of propagation, that is, $\beta_1 > 0$ and $\beta_2 > 0$. The group-velocity mismatch (GVM) between the pumps is defined as

$$\delta \equiv \frac{1}{2}(1/V_g(\omega_2) - 1/V_g(\omega_1)) \approx \frac{1}{2}(\Delta n/c - \Delta\omega(\beta_1 + \beta_2)/2), \qquad (3)$$

where $\Delta n$ is the linear birefringence and $\Delta\omega = \omega_1 - \omega_2$ designates the angular frequency spacing between the pumps.

Note that in the above model [Eq.(2)] we omitted coherent polarization coupling terms $\frac{1}{3}A_j^* A_{3-j}^2 \exp\left[(-1)^j 2i(t(\omega_2 - \omega_1) - z(k_2 - k_1))\right]$. Indeed these coherent terms play a significant role only in a very narrow parameter range close to the condition of equal pump frequencies [6], which is outside the operating conditions of interest in the present paper.

## 3.2    Linear Stability Analysis of Modulational Instability

In this section, we outline the main results of the linear stability analysis (LSA) of MI in the normal dispersion regime of a birefringent fiber. Modulational instability in Eq.(2) is revealed by a LSA of its cw solution,

$$E_j = \sqrt{P_j}\exp[i\gamma_j(P_j + 2P_{3-j}/3)z], \quad j = 1,2, \qquad (4)$$

where $P_j$ is the power for the wave $\omega_j$. Equations 4 show that the total power on each axis remains constant through propagation along the fiber. Introducing first-order perturbations $u$ and $v$,

$$E_1 = (\sqrt{P_1}+u)\exp[i\gamma_1(P_1+2P_2/3)z], \qquad E_2 = (\sqrt{P_2}+v)\exp[i\gamma_2(P_2+2P_1/3)z], \qquad (5)$$

with the modulational ansatz

$$u = u_s(z)\exp\left[i(\Omega t)\right] + u_a(z)\exp\left[i(-\Omega t)\right],$$

$$v = v_s(z)\exp\left[i(\Omega t)\right] + v_a(z)\exp\left[i(-\Omega t)\right], \qquad (6)$$

one obtains a set of four linear ordinary differential equations in $u_s$, $u_a$, $v_s$ and $v_a$:

$$d[Y]/dz = i[M][Y], \quad [Y]^T \equiv [u_a, u_s^*, v_a, v_s^*]. \qquad (7)$$

In Eq.(7), [M] designates the stability matrix of the system.

The eigenvalues of the matrix [M] determine the wave number $K$ of the perturbation. Modulational instability occurs when $K$ possesses a nonzero imaginary part. The importance of MI is measured by a power gain defined by

$$G(\Omega) = 2\left|\text{Im}(K)\right|, \qquad (8)$$

where $K$ is the eigenvalue of $M$ that possesses the largest imaginary part. It is worth noting that an eigenvector $[u_a, u_s^*, v_a, v_s^*] = [a, 0, 0, d]$ corresponds to a anti-Stokes sideband polarized on the fast axis and a Stokes sideband polarized on the slow axis. On the other hand, an eigenvector of the type $[a, b, c, d]$ means that the Stokes and anti-Stokes sidebands possess a component on each axis. When the pump frequency spacing is small (i.e., $(|\Delta\omega| << min(\omega_1, \omega_2))$, then the nonlinear and dispersive parameters of the two pumps can be assumed identical: $\gamma_1 = \gamma_2 \equiv \gamma$ and $\beta_1 = \beta_2 \equiv \beta$. Moreover, for identical pump powers, that is, for $P_1 = P_2 = P$, the absolute frequency shift (with respect to the pump) of the unstable sidebands extends from the low-frequency value $\Omega_- = \sqrt{(2\delta/\beta)^2 - 10\gamma P/3\beta}$ up to the high-frequency value $\Omega_+ = \sqrt{(2\delta/\beta)^2 - 2\gamma P/3\beta}$. This is the spectral band for which both the nonlinear phase-shift [see Eq. 5] and the dispersive mismatch between the pump and the MI sidebands are compensated by the group-velocity mismatch between the two orthogonal modes.

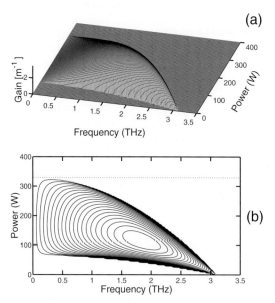

**Fig. 2.** (a) MI gain $G = G(\Omega, P)$ and (b) contour lines as a function of the frequency detuning $\Omega$ and input power $P = P_1 = P_2$. The dotted line in (b) corresponds to the critical power $P_c$.

The outcome of the LSA is summarized in Fig. 2, which we obtained for a single-frequency pump propagating through a hi-bi fiber with a birefringence of $\Delta n = 3.5 \times 10^{-4}$. Here we have: $\gamma_1 \equiv \gamma_2 = \gamma = 0.052$ m$^{-1}$W$^{-1}$, $\beta_1 \equiv \beta_2 = \beta = 0.06$ ps$^2$m$^{-1}$, and $\delta = 0.585$ ps/m. Fig. 2 (a), which represents the power gain $G$ versus the modulational angular frequency $\Omega$ and the pump power $P = P_1 = P_2$, shows that there exists a critical power, $P_c = \frac{3\delta^2}{\gamma\beta} = 329$ W, beyond which MI disappears. This critical power is represented by the dotted line in Fig. 2 (b). For a given power below $P_c$, MI occurs in a frequency region $\Omega_- \leq \Omega \leq \Omega_+$ whose size progressively shrinks as the power decreases. Note in Fig. 2 (a) that both low and high cutoff frequencies, $\Omega_-$ and $\Omega_+$, grow smaller as the pump power $P$ is reduced. In the limit case of low powers, the two cutoff frequencies reduce to the single optimum modulation frequency $\Omega_{opt} \approx \Omega_- \approx \Omega_+ \approx \frac{2\delta}{\beta} = 3.1$ THz.

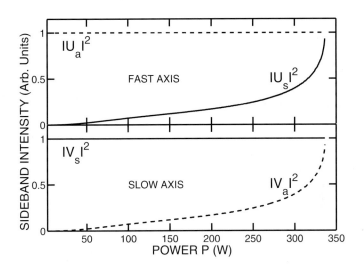

**Fig. 3.** Normalized intensity for Stokes and Anti-Stokes sidebands on the fast and slow axes.

### 3.3 Influence of the Non-phase-matched Waves for a Single-Frequency Pump Field

We now briefly discuss a particularly interesting case that occurs when the input wave power is equally distributed along the two birefringence axes. For a single-frequency, copropagation regime, we have $\omega_1 = \omega_2 = \omega$ and $P_1 = P_2 = P$. We emphasize that, in general, the exponentially growing mode associated with MI is composed of *all the four* sideband components.

Figure 3 shows the evolution of the sideband intensities as a function of power $P$ for $\delta = 0.585$ ps/m. When $P << P_c$ the sideband intensities are such that: $|u_a|^2 \approx |v_s|^2 \approx 1$ and $|u_s|^2 \approx |v_a|^2 \approx 0$. That is, MI generates only one sideband on each fiber axis: an anti-Stokes sideband $u_a$ on the fast axis and a Stokes sideband $v_s$ on the slow axis. These two sidebands are the so-called "phase-matched waves" whereas the two other sidebands, $u_s$ and $v_a$, are called "non phase-matched waves". As $P$ tends to the critical power $P_c$, the non phase-matched sidebands become progressively phase matched. The injection into the fiber of a single-frequency pump beam allows us to record the spectra of perimetrically amplified noise. A typical result obtained for $P_{tot} = 2P = 56$ W is shown in Fig. 4(a). To verify the fact that the frequency of the peak

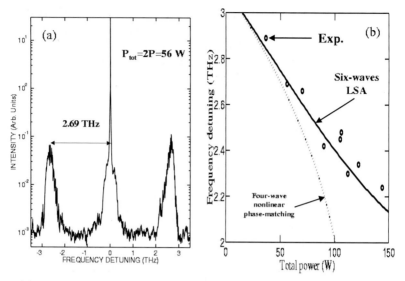

**Fig. 4.** (a) Spontaneous MI spectrum as measured for a fixed total peak power $P_{tot} = P_1 + P_2 = 2P = 56$ W. (b) Measured values of frequency detunings (open circles) of the peak spontaneous MI versus total peak power $P_{tot}$. The solid line is a theoretical fit obtained from the six-wave LSA with a nonlinear coefficient $\gamma = 0.05$. For comparison we report also the four-wave nonlinear phase-matching curve (dashed line).

amplified noise depends on power, we made repeated spectral measurements at several power values up to about $P_{tot} = 150$ W. The results are summarized in Fig. 4(b). The solid curve is obtained from the LSA by adjusting $\gamma$ to fit the data. These results clearly demonstrate the nonlinear nature of the MI process: the peak MI frequency decreases with power in good agreement with the results of the LSA. For the sake of comparison, the dashed line shows that the nonlinear phase-matching frequency obtained from the four-wave mixing model (including the two frequency-degenerated pump waves and the two phase-matched waves) deviates from the observed behavior. These results confirm the influence of the "non-phase-matched" waves in the nonlinear mixing process in agreement with the results of the LSA shown in Fig. 3.

As Eq.(3) shows, an appropriate choice of the frequency separation between the pumps allows for an arbitrary reduction of the magnitude of the GVM, thus leading to an arbitrary reduction of the critical power. Figure 5 illustrates that MI in hi-bi fibers is strongly related to the frequency separation between the two components of the pump (here the pump power is fixed to $P = 100$ W).

## 3.4   Suppression of MI for a Dual-Frequency Pump Field

We considered here the following experimental fiber parameters: the group-velocity dispersion was $\beta = 58.8$ ps$^2$ km$^{-1}$, the birefringence was $\Delta n = 5.7 \times 10^{-4}$ ($\delta = 0.95$ ps m$^{-1}$), and the nonlinear coefficient $\gamma = 0.05$ m$^{-1}$ W$^{-1}$. Fig-

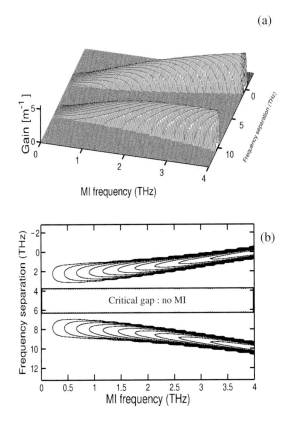

**Fig. 5.** (a) MI gain and (b) contour lines as a function of the frequency detuning $\Omega$ and frequency separation $\Delta\omega$ between the two pumps.

ure 5 (a) shows the power gain versus frequency $\Omega$ and frequency separation $\Delta\omega$. Figure 5 (b) shows the contour lines of the gain versus frequency $\Omega$ and frequency separation $\Delta\omega$. We observe that there exists a critical gap in which MI disappears [8]. The gap center corresponds to $\delta = 0$ and $\Delta\omega = \Delta n / c\beta$, with $\beta \approx \beta_1 \approx \beta_2$. This critical gap gets narrower as the pump power decreases, as shown in Fig. 6(a) for $P = 9$ W. Figure 6(b) show a sequence of experimental observations that exhibit the *critical regime* of MI suppression. The spectra labeled A, B, C were obtained for different values of frequency detuning $\Delta\omega$, and by launching equal power $P = 9$ W, into each birefringence axis. The operating conditions were chosen by making use of the results in Fig. 6(a). The cases of spectra (A) and (C), obtained for a frequency detuning $\Delta\omega = 1.443$ THz and 8.1 THz exhibit, as expected, the MI phenomenon. On the other hand, we may observe in spectrum (B) that when the frequency detuning is equal to $\Delta\omega = 5.15$ THz (which corresponds to an operating condition within the *critical gap* as predicted by Fig. 6(a)), the power of the sidebands is reduced down

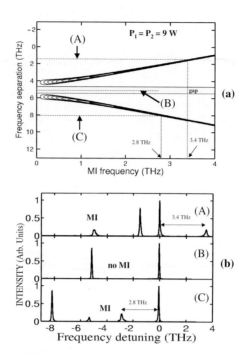

**Fig. 6.** (a) Contour lines of the MI gain as a function of the frequency detuning $\Omega$ and frequency separation $\Delta\omega$ between the two pumps. (b) Output spectra for increasing frequency separation between the pump waves. (A) $\Delta\omega = 1.443$ THz. (B) $\Delta\omega = 5.15$ THz. (C) $\Delta\omega = 8.1$ THz.

to zero. Therefore Fig. 6(b) provides a clear experimental evidence that, for a given power $P$, there is a critical frequency gap where MI is suppressed: this gap corresponds to a finite range of the frequency separations between the pumps.

## 4  Generation of Terahertz Vector Dark-Soliton Trains from Induced Modulational Instability

Let us consider the case of a single frequency pump ($\omega_1 = \omega_2 = \omega_p$) polarized at 45° from the birefringence axes ($P_1 = P_2 = P$). On the other hand, a small Stokes signal $P_s$ with a frequency $\omega_s$ is polarized parallel to the slow axis. We show here, that in the case of normal dispersion, induced-MI in a hi-bi fiber may be exploited to create trains of vector dark solitons. In the normal GVD regime, an exact vector dark-soliton solution of the coupled NLS equations can be found with the help of the change of variables [10],[20],

$$U = E_1 \exp\left( i \left[ -\frac{\delta}{\beta} t + \frac{\delta^2}{2\beta} z \right] \right), \qquad V = E_2 \exp\left( i \left[ \frac{\delta}{\beta} t + \frac{\delta^2}{2\beta} z \right] \right), \quad (9)$$

which leads to the two coupled NLS equations in terms of $U$ and $V$

$$\frac{\partial U}{\partial z} + \frac{1}{2} i \beta \frac{\partial^2 U}{\partial t^2} = i \gamma \left[ \left( |U|^2 + \frac{2}{3} |V|^2 \right) U \right]$$

$$\frac{\partial V}{\partial z} + \frac{1}{2} i \beta \frac{\partial^2 V}{\partial t^2} = i \gamma \left[ \left( |V|^2 + \frac{2}{3} |U|^2 \right) V \right]$$

(10)

Note that the envelope $U$ (and $V$) represents the complex amplitude of a field whose optical carrier is frequency-shifted by $\delta\omega = -\delta/\beta$ (or $\delta/\beta$) with respect to the original field $E_1$ (or $E_2$). This formal frequency shift allows us to eliminate the walk-off term in Eqs. (10). On the other hand, if one injects into the fiber two frequency-shifted ( by $\pm\delta\omega$), orthogonally polarized fields, XPM-induced cross-trapping leads to dark vector solitons [10],[20]. Indeed, such waves have the property of maintaining unchanged along the propagation direction both their intensity profile and their input state of polarization. A symmetric vector dark soliton solution of Eq.(10) reads as [10],[20]

$$U = V = U_o \tanh(\sqrt{\frac{5\gamma}{3\beta}} \, U_o z) \exp(\frac{5}{3} i \gamma U_o^2 z) \qquad (11)$$

The above vector dark soliton solution (11) represents the effect of mutual trapping between the two polarization components. The link between soliton solutions of the coupled NLS equations and MI was analyzed in the case of a weakly birefringent fiber, where the group-velocity walk-off between the two polarizations can be neglected [21]. In a similar way, we can interpret the solution (11) as the vector dark solitons that are associated to the MI effect in hi-bi fibers. Numerical simulations of the coupled NLS equations (Eqs. 2) allows us to determine in particular the initial parameters (frequency modulation and signal power) which yield a vector dark soliton train at the fiber output, for fixed values of the input pump power and fiber length. In the simulations, we took into account the experimentally accessible conditions for the seed, namely, a single-sideband excitation (i.e., a Stokes signal is injected along the slow axis). The corresponding initial conditions were

$$E_1(z = 0, t) = \sqrt{P}, \qquad E_2(z = 0, t) = \sqrt{P} + \sqrt{P_s} \exp(2 i \pi f_{mod} t), \quad (12)$$

where $P_s$ is the input signal power, $f_{mod}$ is the frequency detuning between the pump and the signal Stokes beam. In the numerical simulations, we considered the following experimental fiber parameters: group-velocity dispersion $\beta = 60$ ps$^2$ km$^{-1}$, birefringence $\Delta n = 3.5 \times 10^{-4}$ ($\delta = 0.585$ ps m$^{-1}$), nonlinear coefficient $\gamma = 0.052$ m$^{-1}$ W$^{-1}$, and fiber length $L = 1.8$ m. The initial conditions (fiber input) for the numerical simulations of the coupled NLSE's were chosen to be continuous waves in both fiber axes. Figures 7(a) and (c) show the evolution along the fiber length of the power in the two orthogonal components of the field, for $P = 56$ W, $P_s = 2$ W and $f_{mod} = 2.5$ THz. As can be seen, an initial sinusoidal modulation of the cw wave in the slow axis gradually induces

a full modulation on both fiber axes. Figures 7(b) and (d), which represent the
theoretical power-spectrum evolution on a logarithmic scale along the slow and
fast axis of the fiber, respectively, show that the pumps, the Stokes [in Fig.7(b)]
and anti-Stokes [in Fig.7(d)] sidebands have exactly the same power at the fiber
output. The dark soliton nature of the pulse train generated at the fiber output [see Figs.7 (a) and (c)] is clearly demonstrated by the results presented in
Fig. (8). In the upper part of Fig.(8), we show along with the intensity profile, the
phase profile (dashed curve) of the dark soliton train (solid curve), which shows
a series of abrupt $\pi$ phase-shifts between any two adjacent peaks. Note that
the phase remains constant between any two consecutive peaks. The comparison
between the phase profiles of the two polarization components in the upper part
of Fig. (8) shows that the two coupled dark solitons trains are phase-conjugated
solitons. The power spectra of each polarization component in the lower part
of Fig. (8) also show that the two dark soliton trains along the slow and fast
axes are frequency down-shifted and up-shifted by $f_{mod}/2$, respectively, with respect to the mean frequency of the fields. Indeed, the vertical dashed lines in the
spectra of Fig. (8) indicate the carrier frequency of each individual dark soliton
train. The output spectra are symmetric with respect to the carrier frequency,
which is a characteristic feature of the dark-soliton nature of the trains. The two
orthogonal dark solitons are mutually trapped by the fiber nonlinearity. In fact,
the slow (fast) components of the field speeds up (slows down) in order to attain
the same group velocity for both polarization components, which then permits
the nonlinear cross-trapping effect.

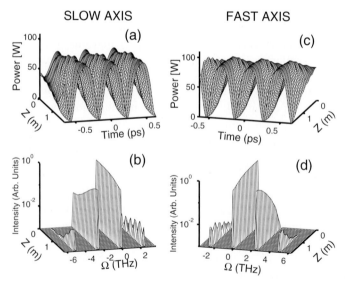

**Fig. 7.** Theoretical evolution with distance of the powers in the (a) slow and (c) fast-
polarization components of the field versus time. Theoretical power spectrum evolution
with distance along the (b) slow and (d) fast axis. The total pump (signal) power is
112 W (2 W), the frequency detuning is 2.5 THz, and the fiber length is 1.8 m.

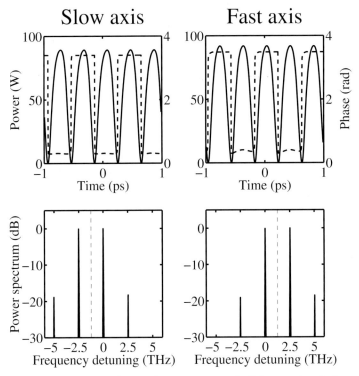

**Fig. 8.** Theoretical time dependence of output powers (solid curves) and phases (dashed curves) of cw dark soliton trains in the slow (left) and fast (right) fiber axis, and corresponding output spectra.

To observe experimentally the generation of vector dark soliton trains we used a 1.8 m hi-bi fiber in which MI is induced by a small Stokes signal polarized parallel to the slow axis. The signal had a 2.1 W power and a wavelength $\lambda_s = 574.72$ nm. On the other hand a 56 W pump with a frequency shift of 2.5 THz from the signal was injected on both fast and slow axes. The strongly birefringent fiber was the HB600 fiber from Fibercore with a linear birefringence $\Delta n = 3.5 \times 10^{-4}$. The linear parameters of the fiber are estimated to be $\beta_2 = 60$ ps$^2$ km$^{-1}$ (GVD at 572 nm) and a group delay $\delta = 0.585$ ps/m. We performed SHG-FROG measurements that demonstrate a direct intensity and phase characterization of the 2.5 THz dark soliton train which is generated along the slow axis, in good agreement with the simulations [13],[14],[15]. Gray scale intensity plots of the measured and retrieved SHG-FROG traces, obtained under the conditions of Fig. 7, are shown in Figs. 9 (a) and (b). As can be seen, the FROG trace shows a complex structure, with spectral components at the SHG wavelengths corresponding to each input beam, as well as sidebands of higher order which are characteristic of a reshaping of the input modulated wave. We also note the presence of spectral bands at wavelengths corresponding to sum frequency mixing. Very good agreement between the two traces is observed, as confirmed

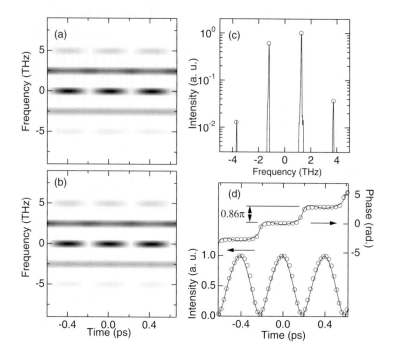

**Fig. 9.** (a) Measured and (b) retrieved SHG-FROG traces of the dark soliton train at 2.5 THz. (c) Measured spectrum (lines) compared with that calculated from the retrieved pulse train (circles). (d) The solid lines show the retrieved intensity (left axis) and phase (right axis), whilst the circles show the expected results from coupled NLSE simulations. With the frequency axis used in the figure, zero frequency corresponds to the mean frequency of the pump and signal waves.

by the FROG retrieval error G=.002 [14],[15]. Retrieved intensity and phase, shown in Fig. 9 (d), yield a spectrum (open circles) in excellent agreement with independent measurements (solid lines) as displayed by Fig. 9 (c). We verified that the autocorrelation function as obtained from the retrieved intensity and phase in Fig.9 (d) was in very good agreement with the independently measured autocorrelation. As can be seen in Fig. 9 (d), the dark solitons possess a reduced modulation depth of 96%, in excellent agreement with NLSE's simulations with a 0.96 blackness parameter [14],[15] as shown by the solid lines in Fig. 9 (d) [1].

## 5    Modulational Instability in Highly Birefringent Air-Silica Microstructure Fiber

There has recently been intense interest in the use of microstructured fibers (MF) because of their unique guidance properties and enhanced nonlinearity. Introducing controlled birefringence in MF yields polarization-preserving characteristics which will become increasingly necessary for numerous applications

[22],[23],[24]. Indeed, birefringence has been previously introduced in a high-delta MF fabricated by Lucent Technologies to ensure polarization preservation during broadband supercontinuum generation [25]. Birefringence or polarization mode dispersion (PMD) can be characterized by the frequency dependence of the difference in propagation constants $\Delta\beta = \beta_x - \beta_y$ between two polarization eigenstates. We measured this experimentally using tunable broadband radiation ($\Delta\lambda = 3nm$) launched at low power equally along the fast and slow axes of a length $L = 3.93m$ of Lucent MF. Fringes on the output spectrum were observed at a modulation frequency $\Delta\nu = (Ld\Delta\beta/d\omega)^{-1}$ (see Fig. 10(a)), allowing the PMD $d\Delta\beta/d\omega$ to be accurately determined. The circles in Fig. 10(b) show these measurements over $550 - 650nm$. These results allow vectorial beam propagation method calculations of the fiber polarization properties to be verified. Scalar modelling of this MF has shown good agreement with experimental chromatic dispersion measurements using a regular hexagonal structure with hole diameter $1.4mm$ and pitch $L = 1.6mm$. Here, however, in order to model the fiber birefringence, the central ring of air-holes was inscribed in an ellipse as shown in Fig. 10(c). The calculations then yield well-defined perpendicular polarization eigenmodes as shown in Fig. 10(b). The mode propagation constants and PMD $d\Delta\beta/d\omega$ were calculated over $550 - 650nm$, and Fig. 10(b) shows excellent agreement between calculations (solid lines) and experiment [7]. The calculations have also been confirmed by the first observation of vector MI in the normal dispersion regime of a MF. The zero dispersion wavelength for the fundamental mode of the MF is around $765nm$. Experiments injected $90W$ peak power transform-limited nanosecond pulses at $625.54nm$ equally along the fast and slow fiber axes. The calculated MI gain curve in Fig. 11(b) (based on the computed PMD at $625.54nm$) yields maximum MI gain at $3.95THz$, in excellent agreement with the measured $3.90THz$ sideband shift on the output spectra shown in Fig. 11(a) [7]. We note that such a large frequency shift at these low peak powers is possible because of the high MF nonlinearity ($A_{eff} = 2.5\mu m^2$).

# 6   Four Wave Mixing-Induced Modulational Instability in Highly Birefringent Fibers

MI in optical fibers can be used for frequency conversion of a pump wave into a single sideband pair. In the regime of small frequency conversion the usual phase-matching condition determines the optimum modulation frequency but, more generally, the gain spectrum is obtained from the usual LSA (see section 3.2). In general, the MI gain bandwidth is small, such that the frequency detuning between the pump and the signal should be specifically chosen to obtain the frequency conversion. This limitation comes from the phase-matching conditions. But a violation of the phase-matching conditions has been previously reported in two experiments. On the one hand, in the strongly depleted regime, the optimum conversion was obtained for a pump-signal frequency detuning just outside the linear MI gain [26]. On the other hand, an efficient frequency conversion was achieved by means of a highly mismatched degenerate FWM assisted

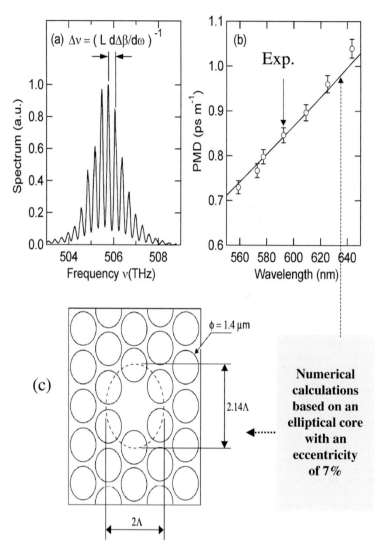

**Fig. 10.** (a) Typical output spectrum from the MF, modulated because of interference between orthogonal polarization components. (b) Measured PMD as a function of wavelength (circles) compared with numerical calculations based on an elliptical core (solid line). (c) Elliptical MF structure used in numerical calculations.

by Raman amplification [27]. The generation of new frequencies by induced-MI in the case of a pump-signal frequency detuning, which is clearly outside the linear bandwidth was reported recently [16]. The basic idea of the technique is to choose a pump-signal frequency detuning in such a way that one of its harmonic falls within the linear MI-gain  spectrum (see Fig. 12). MI is then induced by multiple FWM. The technique leads to new MI gain bandwidths

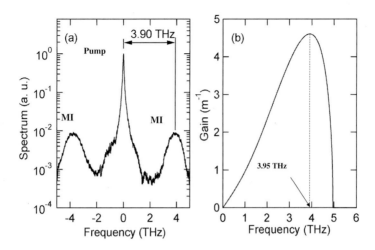

**Fig. 11.** (a) Measured output spectra at $624.54nm$ showing MI sidebands at $3.90THz$. (b) Vector MI gain curve using calculated birefringence parameters.

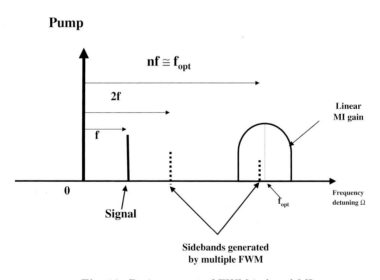

**Fig. 12.** Basic concept of FWM-induced MI.

and permits frequency conversion in both Stokes and anti-Stokes sides of the pump spectrum. We consider here the particular case of a hibi fiber, in which a pump beam is polarized at $45°$ between the fiber axes. We will focus our attention to the case in which MI is induced by an anti-Stokes signal wave polarized along the fast axis. The fiber length is $14m$ and the fiber parameters have the following values: birefringence $\Delta n = 3.5 \times 10^{-4}$, dispersion $\beta = 60 \text{ ps}^2 \text{ km}^{-1}$ and non linearity $\gamma = 0.05 \text{ m}^{-1} \text{ W}^{-1}$. The pump and signal peak powers are

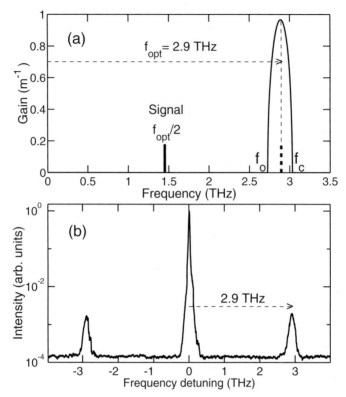

**Fig. 13.** (a) The solid line gives the MI gain versus sideband frequency detuning from a 30 W cw pump equally divided on each axis. The solid stick gives the frequency detuning between the pump and the anti-Stokes signal; the second harmonic of this detuning is shown by the dashed stick. (b) Spontaneous MI spectrum as measured for a total pump power $P_p = 30W$ and a fiber length $L = 25m$.

30W and 0.5W, respectively. In the following, we consider the particular case, in which the second-order harmonic of the pump-signal detuning falls within the MI-gain spectrum. At a pump power of $30W$, the MI gain extends over a narrow band of frequencies ranging between two limits $f_o$ and $f_c$ (see Fig. 13(a)). The optimum frequency at which the gain is maximum is $f_{opt} = 2.9THz$. As shown by Fig. 13(b) this value is in very good agreement with the frequency of the noise-induced sidebands measured on a spectrum recorded with a longer fiber ($L = 25m$). We have investigated the propagation of the pump and signal waves by numerical solutions of the coupled NLS equations. Figure 14 shows the spectrum evolution of the fast and slow components. During the first stage of propagation the pump and signal interact with each other through multiple four-wave mixing. Indeed, a partially degenerate four wave mixing between the pump and the signal leads to the generation of a Stokes wave with a frequency detuning $-f$. Simultaneously a non-degenerate four wave mixing between the pump and the signal generates Stokes and anti-Stokes sidebands at frequency detunings

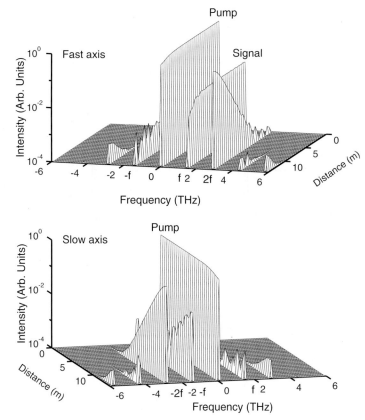

**Fig. 14.** Theoretical power spectrum evolution (in logarithmic scale) with distance in the fast and slow polarization components of the field in the fiber. The total cw pump (signal) power is $30W (0.5W)$, the frequency detuning is $1.45THz$, and the fiber length is $14m$.

$-f$ and $2f$, respectively. The effect of XPM and FWM leads to the generation of a pair of sidebands on the slow axis, located at frequency detunings $-f$ and $+f$. During the first stage of propagation, MI is highly phase-mismatched and do not contribute in the energy-exchange process. On the other hand, in the normal dispersion regime, the FWM interactions are mismatched with relatively small coherence lengths $L_c$ as shown by this fast oscillation. So the intensity of the second-order anti-Stokes sideband remains relatively small, but becomes sufficiently large to induced MI. Thus during the second stage of propagation, MI strongly amplifies the second-order anti-Stokes sideband and leads to the generation of the corresponding Stokes sideband, polarized on the slow axis. Figure 15(a-f) show a set of experimental spectra obtained at the fiber output for different values of the pump-signal frequency detuning $f$. For $f = 1.45THz$, such that $2f = f_{opt}$, we clearly observe an efficient generation of the MI sidebands as predicted by the simulations of Fig. 14. On the other hand, when the

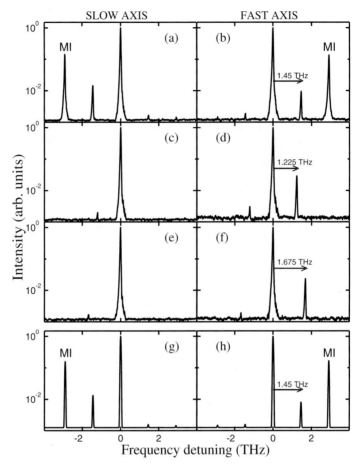

**Fig. 15.** Output experimental spectra for slow and fast axes with the pump-signal detunings: (a,b) $f = 1.45THz$; (c,d) $f = 1.225THz$; (e,f) $f = 1.675THz$. Theoretical pulse averaged spectra from (g) slow and (h) fast axes for $f = 1.45THz$.

pump-signal detuning is decreased down to $f = 1.225THz$, such that $2f < f_o$, or when it is increased up to $f = 1.675THz$, such that $2f > f_c$, the MI sidebands do not appear as usually expected for an initially highly mismatched process. The numerical solutions of the coupled NLS equations, which take into account the pulse nature of the beams and the finite resolution of the spectrometer, are in very good agreement with the experimental spectra (see Figs. 15(g,h)).

# 7    Bragg Modulational Instability Induced by a Dynamic Grating in an Optical Fiber

Nonlinear wave propagation in optical periodic structures has been increasingly studied during the last few years. Indeed, these structures exhibit peculiar dispersion properties characterized by the presence of photonic band gaps. Among all the band gap materials proposed up to now, the fiber Bragg grating plays a particular role owing to its simplicity. For this reason, it has been extensively studied and can now be considered as a paradigmatic system, as well as an ideal experimental test bed, for the investigation of basic photonic band gap phenomenology. Modulational instability has been studied in fiber Bragg gratings [17]. However, experimental demonstrations of this phenomenoun have nonetheless been limited because of technological difficulties inherent to the resonant nature of the device. Indeed, MI occurs when light propagates into the fiber Bragg grating at a wavelength close to the Bragg resonance, a condition in which the grating becomes highly reflective. A novel scheme was proposed recently that allows us to operate arbitrarily close to the photonic band gap. In this scheme the periodic modulation of the refractive index is ensured by the temporal beating of two intense waves of different frequencies propagating along the slow axis of a high-birefringence optical fiber [18],[19]. These waves, which we call "grating waves" in the following, generate a temporal sinusoidal beating that modulates the refractive index along the fast axis through cross-phase modulation (XPM). This index modulation can be expressed as $\Delta n(t) = (2/3) \, n_2 \, (a^2/2) \, cos(2\Omega t)$, where $2\Omega$ represents the frequency separation between the two waves of amplitude $a/2$ and $t$ is the time defined in the reference frame that travels at the average group-velocity of the grating waves. The advantage of such a dynamical grating is that the coupling of light into the grating is very efficient and allows us, in particular, to operate arbitrarily close to the photonic band gap. Let us now consider a pump wave that propagates along the fast axis of the optical fiber in the normal dispersion regime. The evolution of this wave is ruled by the following equation

$$\frac{\partial E}{\partial z} = -\frac{i}{2}\beta_2 \frac{\partial^2 E}{\partial t^2} + i\gamma\Delta n(t)E + i\gamma \mid E \mid^2 E \ , \tag{13}$$

where $E$ is the electric field envelope. To describe the evolution of the electric field $E$ in the presence of the grating, we use the standard coupled-mode theory and express $E$ as the sum of a forward wave $F$ and a backward wave $B$ in the travelling reference frame $(z,t)$[28]: $E(z,t) = F(z,t)exp[i(-\Omega t + Kz)] + B(z,t)exp[i(\Omega t + Kz)]$, where $K = \Omega^2\beta/2$. By direct insertion of this expression into Eq. 13, the following nonlinear coupled-mode equations are obtained

$$\frac{\partial F}{\partial z} = -\frac{i}{2}\beta_2 \frac{\partial^2 F}{\partial t^2} - \Omega\beta_2 \frac{\partial F}{\partial t} + i\kappa B + i\gamma(\mid F \mid^2 +2\mid B \mid^2)F \tag{14}$$

$$\frac{\partial B}{\partial z} = -\frac{i}{2}\beta_2 \frac{\partial^2 B}{\partial t^2} + \Omega\beta_2 \frac{\partial B}{\partial t} + i\kappa F + i\gamma(\mid B \mid^2 +2\mid F \mid^2)B \tag{15}$$

where $\kappa = \gamma a^2/6$. Apart from the dispersion terms, these equations have the form of the standard nonlinear coupled mode equations that describe nonlinear wave propagation in a grating. In particular, the terms proportional to $\kappa$ represent the coupling strength of the Kerr grating while those proportional to $\Omega\beta_2$ represent counterpropagation of the two waves in the reference frame travelling at the grating velocity. More precisely, the factor $\Omega\beta_2$ represents the walk-off velocity that we shall express in the following in $ps/m$. Consequently, by carefully exchanging the time and space variables, the usual nonlinear grating theory can be directly applied to investigate the dynamics of our system. The continuous wave (cw) stationary solutions of Eqs. 15 are looked for under the form [17],[28]

$$F(t,z) = A^{+}exp[i(\beta z - \omega t)], \qquad B(t,z) = A^{-}exp[i(\beta z - \omega t)] \qquad (16)$$

where $\beta$ represents the detuning between the wave number of the field and the Bragg wave number of the grating while $\omega$ is the corresponding frequency detuning. We also define the total pump power $P = (A^{+})^2 + (A^{-})^2$ and the amplitude ratio $r = A^{+}/A^{-}$. Figure 16 shows the dispersion relation $\beta(\omega)$ for $\kappa = 1$ and a total pump power $P = 15$ W considering the following fiber parameters $\beta_2 = 60$ ps$^2$/km and $\gamma = 0.05$ m$^{-1}$W$^{-1}$. Two distinct solutions are associated with each value of $\omega$, which leads to two branches. On the upper (lower) branch, the amplitude ratio $r$ is positive (negative). A standard LSA applied to the cw solutions allows us to calculate the Bragg MI gain spectra and the associated optimal modulation frequency.

The experiment was performed with a highly birefringent optical fiber with intrinsic linear birefringence $\delta n = 3.75 \times 10^{-4}$. The first grating wave is obtained from a frequency-doubled injection-seeded Nd:YAG laser at $\lambda_1 = 532nm$. The second grating wave is produced by using a tunable single-mode dye laser pumped by the Nd:YAG laser and emitting around $575nm$. The two grating waves that have a duration of about $4.5ns$ are mixed by means of a dichroic mirror. The (forward) pump wave is obtained by frequency shifting a part of the Nd:YAG laser beam through stimulated Raman scattering in a multiple-pass chamber of carbon dioxide. We use the first Stokes order corresponding to a frequency shift of $41.64$ $THz$. The resulting pump pulse wavelength is thus fixed to $\lambda_p = 574.78$ $nm$ while its duration is about $2ns$. The wavelength separation between the two grating waves is tuned around 46 nm, which corresponds to a modulation frequency $2\Omega$ around $44.8$ $THz$. The detuning parameter $\omega$ is determined by the wavelength $\lambda_2$ of the second "grating" wave that constitutes

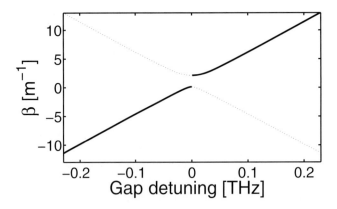

**Fig. 16.** Dispersion relation of the dynamical Bragg grating for the fiber and wave parameters given in the text. Solid curves: $|\ r\ | \geq 1$, dashed curves: $|\ r\ | \leq 1$.

therefore the control parameter in our experiment. Pump and grating waves with orthogonal polarizations are combined through a beam splitter and focused in $11m$ of highly birefringent fiber.

Let us note that the Bragg condition corresponds to a zero walk-off velocity of the pump wave in the grating reference frame. If there was no birefringence in the system, this condition could only be obtained when the forward and backward waves (that compose the pump wave) have the same wavelengths as those of the grating waves. This situation would however lead to complex interactions between the two sets of waves because of four-wave mixing. The introduction of birefringence allows us to avoid such interactions since the coupling between orthogonally polarized waves becomes purely incoherent. In the presence of birefringence, equalization of the velocities of the pump and grating waves (i.e., the Bragg condition) is obtained through dispersion by means of a slight frequency shift given by $\Delta\omega = \delta n/(c\beta_2)$. Note that this situation of velocity matching is precisely the condition needed to avoid XPM-induced modulational instability (see Sect. 3.4). Any MI spectrum observed in our experiment can therefore be strictly attributed to Bragg MI. Figure 17(a) shows typical experimental spectrum resulting from spontaneous Bragg MI obtained for the value of the detuning $\omega = -0.8THz$. The central peak corresponds to the wavelength of the initial forward pump wave. Stokes and anti-Stokes Bragg MI sidebands are clearly visible. Figure 17(b) shows the measured optimal modulation frequencies as a function of the frequency detuning $\omega$. The experimental frequencies (crosses) are in excellent agreement with the LSA (solid line) that predicts an almost linear dependence. These results show a strong detuning-dependence of the modulation frequency over a very broad range of positive and negative detuning values and constitute a remarkable confirmation of the validity of the coupled mode theory of nonlinear Bragg gratings [18],[19].

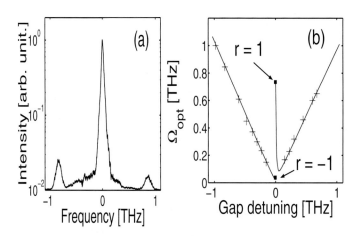

**Fig. 17.** (a) Experimental Bragg MI spectrum obtained for $\omega = -0.8$ THz. (b) Optimal modulation frequency $\Omega_{opt}$ versus the frequency detuning $\omega$. Solid curve is calculated from linear stability analysis and crosses are the experimental measurements.

# 8   Conclusions

In summary, we have presented here an overview of some of the most important effects of MI in normally dispersive hi-bi fibers. We described experiments showing the influence of the non-phase matched waves in the MI process for a single-frequency pump field equally distributed along the two axes of a hi-bi fiber. On the other hand, we analyzed experiments showing that with dual-frequency, orthogonal polarization pumping one may achieve the suppression of MI. We showed that an induced process of MI may be exploited for the generation of THz train of vector solitons. In particular, experimental results were presented for the generation of a $2.5 - THz$ train of dark solitons generated in a hi-bi fiber. The FROG technique was used to completely characterize the intensity and phase of the vector soliton trains. We also reported experiments on MI in a hi-bi air-silica microstructure fiber, leading to $3.9THz$ MI sideband shifts which are in good agreement with theoretical predictions based on the PMD characteristics. In addition, we reported observation of a strong frequency conversion in hi-bi fibers for an initially highly phase-mismatched MI process, thus leading to new MI bandwidths. This frequency conversion results from MI assisted by multiple FWM interactions. Finally we reported the experimental observation of Bragg MI induced by a dynamic grating obtained through XPM with a beating wave in a hi-bi fiber.

**Acknowledgments**

The authors wish to thank P. Grelu, F. Gutty, L. Provino, P. Tchofo Dinda, E. Seve, S. Trillo and S. Wabnitz for their collaboration on many aspects of the material presented in this paper.

# References

1. G.P.Agrawal: *Nonlinear fiber Optics*, 3 rd Edition, (Academic Press Inc,San Diego, 2001).
2. K.Tai, A.Hasegawa, and A.Tomita:Phys. Rev. Lett. **56**, 135 (1986).
3. A.L.Berkhoer and V.E.Zakharov:Sov. Phys. JETP **31**, 486 (1970).
4. J.E.Rothenberg:Phys. Rev. A **42**,682(1990).
5. P.D.Drummond, T.A.B.Kennedy, J.M.Dudley, R.Leonhardt, and J.D.Harvey :Opt. Comm. **78**, 137(1990).
6. E.Seve, P.Tchofo Dinda, G.Millot, M.Remoissenet, J.M.Bilbault, and M.Haelterman :Phys. Rev. A **54**, 3519(1996).
7. G.Millot, A.Sauter, L.Provino, and J.M.Dudley(submitted)
8. P.Tchofo Dinda, G.Millot, E.Seve, and M.Haelterman:Opt. Lett.**21**, 1640(1996).
9. A.Hasegawa:Opt. Lett. **9**, 288 (1994).
10. Y.S.Kivshar and S.K.Turitsyn: Opt. Lett. **18**, 337 (1993).
11. E.Seve, G.Millot, and S.Wabnitz:Opt. Lett.**23**, 1829 (1998).
12. R.Trebino, K.W.DeLong, D.N.Fittinghoff, N.Sweetser, M.A.Krumbogel, and B.A.Richman:Rev. Sci. Instrum. **68** 3277 (1997).

13. F.Gutty, S.Pitois, P.Grelu, G.Millot, M.D.Thomson, and J.M.Dudley :Opt. Lett. **24** 1389 (1999).
14. J.M.Dudley, M.D.Thomson, F.Gutty, S.Pitois, P.Grelu and G.Millot :Electron. Lett. **35** 2042 (1999).
15. J.M.Dudley, F.Gutty, S.Pitois, and G.Millot:IEEE J. Quantum Electron. (2000).
16. G.Millot:Opt. Lett.**26**, 1391 (2001).
17. C.Martijn de Sterke:J.Opt. Soc. Am. B **15**, 2660 (1998).
18. S.Pitois, M.Haelterman, and G.Millot:Opt. Lett.**26**, 780 (2001).
19. S.Pitois, M.Haelterman, and G.Millot:J. Opt. Soc. Am. B. **19** (2002).
20. A.P.Sheppard and Y.S.Kivshar:Phys.Rev. E **55**,4773 (1997).
21. M.Haelterman:Opt. Comm.**111**, 86 (1994).
22. A.Ortigosa-Blanch, J.C.Knight, W.J.Wads worth, J.Arriaga, B.J.Mangan, T.A.Birks, and P.St.Russell:Opt. Lett. **25**, 1325 (2000).
23. T.P.Hansen, J.Broeng, S.E.B.Libori, E.Knudsen, A.Bjarklev, J.R.Jensen, and H.Simonsen:Opt. Lett. **13**, 588 (2001).
24. M.J.Steel, T.P.White, C.Martijn de Sterke:Opt. Lett. **26**, 488 (2001).
25. J.K.Ranka, R.S.Windeler, and A.J.Stentz:Opt.Lett. **25** 25 (2000).
26. E.Seve, G.Millot, and S.Trillo:Phys. Rev. E **61**, 3139 (2000).
27. T.Sylvestre, H.Maillotte, E.Lantz, and P.Tchofo Dinda:Opt. Lett. **24** 1561 (1999).
28. C.Martijn de Sterke and J.E.Sipe:Gap solitons, Progress in Optics **33** 203 (1994).

# Self-structuration of Three-Wave Dissipative Solitons in CW-Pumped Backward Optical Parametric Oscillators

C. Montes

Centre National de la Recherche Scientifique
Laboratoire de Physique de la Matière Condensée
Université de Nice - Sophia Antipolis, Parc Valrose
F-06108 Nice Cedex 2, France

**Abstract.** Generation of ultra-short optical pulses in cw-pumped cavities are mostly associated to mode locking in active media, as doped fibers or solid-state lasers. The cavity contains not only a gain element (atoms or ions) but also a nonlinear element of the host medium, such as self-phase modulation (SPM) or intensity dependent absorption. Our aim here is to present another mechanism for pulse generation in an optical cavity due to the nonlinear three-wave counter streaming interaction. We show that the same mechanism, responsible for symbiotic solitary wave morphogenesis in the Brillouin-fiber-ring laser, may act for picosecond pulse generation in a quadratic optical parametric oscillator (OPO). The resonant condition is automatically satisfied in stimulated Brillouin backscattering (SBS); however, in order to achieve counter-streaming quasi-phase matching (QPM) between the three optical waves in the $\chi^{(2)}$ medium, a grating of sub-$\mu$m period is required. Such a quadratic medium supports solitary waves that result from energy exchanges between dispersionless waves of different velocities. The structure of these temporal localized solitary waves is determined by a balance between the energy exchange rates and the velocity mismatch between the three interacting waves. The backward QPM configuration spontaneously generates tunable picosecond solitary pulses from noise when the quadratic material is placed inside a single resonant OPO. We show, by a stability analysis of the degenerate backward OPO in the QPM decay interaction between a CW-pump and a backward signal, that the inhomogeneous stationary solutions are always unstable, whatever the cavity length and pump power above single OPO threshold. Starting from any initial condition, the nonlinear dynamics exhibits self-pulsing of the backward signal with unlimited amplification and compression. Above a critical steepening, dispersion may saturate this singular behavior leading to a new type of dynamical solitary structures.

## 1 Introduction

Generation of ultra-short optical pulses in CW-pumped cavities are mostly associated with mode locking in active media, such as doped fibers or solid-state (*e.g.* Ti-Sa) lasers. The cavity contains not only a gain element (atoms or ions) but also a nonlinear element permitting self-phase modulation (SPM) or intensity dependent absorption. Spontaneous generation of a pulse train in CW-pumped optical fiber cavities without gain elements can been also obtained through modulation instability caused by the combined action of SPM and group-velocity dispersion

(GVD) on the CW optical beam [1]. Our aim is to present another mechanism for pulse generation in optical cavities due to the nonlinear three-wave counter-propagating interaction. We show that the same mechanism, responsible for symbiotic solitary wave morphogenesis in the Brillouin-fiber-ring laser [2–4], may act for picosecond pulse generation in a quadratic optical parametric oscillator [5] [6]. The resonant condition is automatically satisfied in stimulated Brillouin backscattering (SBS) when the fiber ring laser contains a large number of longitudinal modes beneath the Brillouin gain curve. However, in order to achieve quasi-phase matching between the three optical waves in the $\chi^{(2)}$ medium, a grating of sub-$\mu$m period is required. Recent experiments of backward second-harmonic generation in periodically-poled LiNbO$_3$ [7] [8] and KTP [9] avoid this technical difficulty by using higher-order gratings. Such a quadratic medium supports solitary waves that result from energy exchanges between dispersionless waves of different velocities. The structure of these temporal localized solitary waves is determined by a balance between the energy exchange rates and the velocity mismatch between the three interacting waves. The backward quasi-phase-matching configuration spontaneously generates tunable picosecond solitary pulses from noise when the quadratic material is placed inside a singly resonant OPO.

Three-wave solitons result from the dynamic compensation between the wavefront slope produced by the three-wave parametric instability and the pump depletion due to nonlinear saturation, and belong to the class of symbiotic solitary waves resulting from energy exchanges between dispersionless waves of different velocities [10]-[13]. These structures were introduced in the context of self-induced transparency [11,14]. They have been experimentally investigated in Stimulated Raman Scattering in gases [15] and more recently in Brillouin fiber ring lasers [2–4] but no experiments have been reported to date for symbiotic solitary waves of quadratic optical materials. We have shown [16] that dissipation is required in order to avoid turbulent behavior and to obtain a localized attractor structure. Moreover, dissipation naturally arises in optical cavities through mirror reflection losses.

Here we study self-structuration of three-wave solitary waves in quadratic media with absorption losses. We will consider nondegenerate and degenerate counter-streaming quasi-phase-matching configurations so that both the signal and idler fields propagate backward with respect to the direction of the pump field. These configurations are of interest since the solitary waves can be generated from noise in the presence of a CW-pump when the quadratic material is placed inside a singly resonant optical parametric oscillator (SOPO). We thus consider a quadratic material in which such counter-streaming nondegenerate (or degenerate) three-wave interactions take place [17] [18] [7] [19].

Although a coherent pump has been considered for this self-structuration, it has been shown that, under certain conditions on the group velocities, a coherent structure can be generated and sustained from a highly incoherent source [20]. This phenomenon relies on the advection between the interacting waves and leads to the formation of a novel type of three-wave parametric soliton composed of both coherent and incoherent fields.

## 2    Three Wave Model

The spatiotemporal evolution of the slowly varying envelopes of the three counter-streaming resonant interacting waves $A_i(x,t)$, for nondegenerate three-wave interaction in $\chi^{(2)}$ medium, is given by

$$(\partial_t + v_p\,\partial_x + \gamma_p + i\beta_p\partial_{tt})\,A_p = -\,\sigma_p A_s A_i \tag{1a}$$

$$(\partial_t - v_s\,\partial_x + \gamma_s + i\beta_s\partial_{tt})\,A_s = \sigma_s A_p A_i^* \tag{1b}$$

$$(\partial_t - v_i\,\partial_x + \gamma_i + i\beta_i\partial_{tt})\,A_i = \sigma_i A_p A_s^* \tag{1c}$$

where $A_p(\omega_p, k_p)$ stands for the CW-pump wave, $A_s(\omega_s, k_s)$ for the backward signal wave, and $A_i(\omega_i, k_i)$ for the backward idler wave. The resonant conditions in 1-D space configuration are:

$$\omega_p = \omega_s + \omega_i \;; \quad k_p = -k_s - k_i + K$$

where $K = 2\pi/\Lambda_{QPM}$, and $\Lambda_{QPM}$ is the grating pitch for the backward quasi-phase matching. The group velocities $v_j$ ($i = p, s, i$) as well as the attenuation coefficients $\gamma_j$ are in general different for each wave. Eqs.(1a-c) also hold for standard forward phase-matching configurations in which case the velocities $v_{s,i}$ must be taken negative. The nonlinear coupling coefficients are $\sigma_j = 2\pi d v_j / \lambda_j n_j$, where $n_j$ is the refractive index at frequency $\omega_j$ and $d$ is the effective nonlinear susceptibility. Eqs.(1a-c) account for the effect of chromatic dispersion, which is necessary when the temporal structures generated are sufficient narrow. The effects of group velocity dispersion (GVD) are represented by the second derivaties with respect to time, so that the dispersion parameters are given by $\beta_j = |v_j| k_j''$ where $k_j'' = (\partial^2 k/\partial\omega^2)_j$, $k$ being the wave vector modulus, $k = n(\omega)\omega/c$.

## 3    Solitary Wave Solution

In the absence of dispersion ($\beta_j = 0$) Eqs.(1a-c) have been extensively studied in the literature. Their solitary wave solutions have been first derived in the absence of dissipation ($\gamma_j = 0$) [11] [12]. In the context of stimulated scattering in nonlinear optics, the existence of dissipative solitary waves in the case where one of the velocities $v_{s,i}$ is zero was also shown [2] [21]. More recently, Craik et al. have shown, for the particular case of degenerate three-wave interaction (second harmonic generation or parametric decay), that solitary waves still exist in the presence of dissipation [22]. On the basis of these previous theoretical works, we have calculated the dissipative symbiotic solitary wave of the nondegenerate parametric three-wave interaction described by Eqs.(1a-c). Looking for a solitary wave structure induced by energy transfer from the pump wave to the other two waves, we have to assume zero loss for the pump ($\gamma_p = 0$). It is the only way to keep constant the energy transfer that compensates here for the signal and the idler losses, so as to generate stationary field structures. If $\gamma_p$ was not zero, the pump wave would follow an exponential decay and the three-wave structure would evolve when propagating along this exponential pump profile preventing, in this way, the formation of a stationary solitary wave. When $\gamma_p = 0$ it is easy to find by substitution the following solution to Eqs.(1a-c):

$$A_p = \delta - \beta \tanh\left[\Gamma(x + Vt)\right] \tag{2a}$$

$$A_s = \eta\Gamma \operatorname{sech}\left[\Gamma(x + Vt)\right] \tag{2b}$$

$$A_i = \kappa\Gamma \operatorname{sech}\left[\Gamma(x + Vt)\right] \tag{2c}$$

where $\beta$ is the only free parameter. All other parameters depend on the material properties and on $\beta$. One finds $\delta = [\gamma_s\gamma_i/\sigma_s\sigma_i]^{1/2}$, $\Gamma = \beta[\sigma_i\sigma_s/(V - v_s)(V - v_i)]^{1/2}$, $\eta = [(V + v_p)(V - v_i)/\sigma_i\sigma_p]^{1/2}$, $\kappa = [(V + v_p)(V - v_s)/\sigma_s\sigma_p]^{1/2}$, and $V = (v_s/\gamma_s - v_i/\gamma_i)/(1/\gamma_s - 1/\gamma_i)$. This last expression shows that the velocity $V$ of the solitary wave is fixed by the material parameters, unlike in the nondissipative case where $V$ is undetermined [11]. Let us point out that, in order to keep $\Gamma$ real, the solitary wave must be either superluminous, $V > max(v_s, v_i)$, or subluminous, $V < min(v_s, v_i)$. Note that the superluminous velocity does not contradict by any means the special theory of relativity [2] even if the velocity $V$ becomes infinite when the signal and idler waves undergo identical losses, $\gamma_s = \gamma_i$. This can be easily explained by remembering that the velocity of this type of symbiotic solitary wave is determined by the energy transfer rate which depends on the shape of the envelope of each component. The infinite velocity is here simply due to the fact that the width of the solitary wave $\Gamma^{-1}$ also becomes infinite for $\gamma_s = \gamma_i$.

The free wave parameter $\beta$ fixes, in combination with the material parameters, the amplitude and the width of the solitary wave. According to Eq.(2a), $\beta$ is determined by the initial pump amplitude $A_p = E_p(x = -\infty)$ through the relation $A_p = \beta + \delta$. In practice, this means that, for a given material, the solitary wave is completely determined by the pump intensity at the input face of the crystal. Note that if the losses are such that $\delta > \beta$ the solitary wave no longer exhibits a $\pi$-phase change [16], unlike in the nondissipative case [11].

Figure 1 shows the dissipative symbiotic solitary wave for the case of a quasi-phase-matched backward three-wave interaction with $\lambda_p = 1$ $\mu m$, $\lambda_s = 1.5$ $\mu m$ $\lambda_i = 3$ $\mu m$, $\Lambda_{QPM} = 2\pi/K = 0.226$ $\mu m$, and with a pump field of amplitude $E_p = 0.25$ MV/m (i.e., a pump intensity of $I_p = 10$ kW/cm$^2$) propagating in a quadratic $\chi^{(2)}$ material with the following typical values of the parameters: $d = 20$ pm/V, $n_p = 2.23$, $n_s = 2.21$, $n_i = 2.21$, $v_p = 1.34 \times 10^8$m/s, $v_s = 1.35 \times 10^8$m/s, $v_i = 1.38 \times 10^8$m/s, and the loss coefficients $\alpha_s = 2\gamma_s/v_s = 0.23$m$^{-1}$ and $\alpha_i = 2\gamma_i/v_i = 11.5$m$^{-1}$. Note that these parameters lead to a pulse width of approximately 10 psec. With such pulse durations one can expect that the zero pump loss approximation ($\gamma_p = 0$) is valid in practice in the neighbourhood of the solitary wave structure. Indeed, if the characteristic absorption length $v_p/\gamma_p$

is much larger than the pulse width $\Gamma^{-1}$, one can anticipate that the solitary wave undergoes adiabatic reshaping during propagation so as to adapt locally its profile to the exponentially decaying pump intensity.

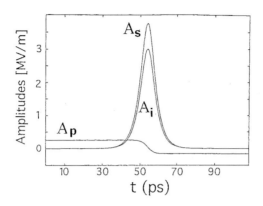

**Fig. 1.** Envelopes of the dissipative three-wave solitary solution.

## 4   Two Wave Adiabatic Approximation

A new solitary wave solution may be obtained when the idler wave is heavily damped and the dynamics may be governed by two coupled equations for the optical pump and signal intensities. If we have $|(\partial_t - v_i \partial_x) A_i| \ll \gamma_i A_i$, $A_i$ can be considered as a slave variable ($A_i = \sigma_i A_p A_s^* / \gamma_i$), and Eqs.(1) after normalization, $[I_{1,2} = \sigma_{p,s}(n\varepsilon_0 c/2)|A_{p,s}|^2/I_{cw}; t\sigma_i I_{cw}/\gamma_i \to t; xn\sigma_i I_{cw}/(c\gamma_i) \to x;$ $a_{1,2} = \gamma_{p,s}\gamma_i/(\sigma_i I_{cw}); nv_{p,s}/c \simeq v]$, yield the intensity equations:

$$(\partial_t + v\partial_x + a_1)I_1 = -I_1 I_2 \tag{3a}$$

$$(\partial_t - v\partial_x + a_2)I_2 = I_1 I_2, \tag{3b}$$

where $I_i \equiv I_3 \propto I_1 I_2/\gamma_i^2$. Performing the change of frame moving in the backward direction ($\xi = x + vt$ ; $\tau = t$), we obtain

$$\left[\partial_\tau + (1+v)\partial_\xi\right]I_1 = -I_1 I_2 - a_1 I_1 \tag{4a}$$

$$\left[\partial_\tau + (v-1)\partial_\xi\right]I_2 = I_1 I_2 - a_2 I_2. \tag{4b}$$

Then, by defining the $J_i$'s intensities as

$$J_1 = |1+v|I_1 ; \quad J_2 = |v-1|I_2 \tag{5}$$

and looking for stationary solutions in the new frame, we have:

$$\partial_X J_1 = -J_1 J_2 - b_1 J_1 \tag{6a}$$

$$\partial_X J_2 = J_1 J_2 - b_2 J_2 \tag{6b}$$

where  $X = \xi/|(v-1)(1+v)|$ ;  $b_1 = a_1|v-1|$ ;  $b_2 = a_2|1+v|$. We shall be only concerned with the solutions for which $v > 1$. Since $v \simeq 1 + \Delta v$ ($|\Delta v| \ll 1$) and $b_1 \ll b_2$ we shall also neglect the pump damping. Setting $b_1 = 0$ and $b_2 = b$, we obtain the following set of coupled equations

$$J_1' = -J_1 J_2 \ ; \quad J_2' = J_1 J_2 - b J_2 \tag{7}$$

yielding for $J_1$ the equation

$$J_1 J_1'' - (J_1')^2 - J_1^2 J_1' + b J_1 J_1' = 0. \tag{8}$$

Introducing the change $J_1 = \exp U$  which satisfies  $dU/dX = J_1'/J_1 = -J_2$, Eq.(8) yields

$$dU/dX = \exp U - b\, U + C, \tag{9}$$

where $C$ is a constant that can be removed with the change $\big[W = U - C/b;$ $X = \zeta \exp(C/b)$ ; $a = a_2 \exp(1/a_2)\big]$

$$\frac{dW}{d\zeta} = \exp W - aW, \tag{10}$$

the constants $C$, $b$ and $a$ being directly related to $v$ and to the dissipation $a_2$ by taking $I_1(\zeta \to -\infty) = 1$. For $a > $ e, (*i.e.* for the localization threshold condition $a_2 < 1$), we obtain two finite values of $W$, say $W_1$ and $W_2$, where all the derivatives of $W$ vanish when $\zeta \to \pm\infty$. Therefore Eq.(10) yields a travelling localized structure of the kink form for $I_1$ and of the pulse form for $I_2$. E.g. for $I_1(\zeta \to -\infty) = 1$ and $a_2 = \gamma_s \gamma_i/(\sigma_i I_{cw}) = 0.625$ we have $a = a_2 \exp(1/a_2) = 3.09564$, $W_2 = 1/a_2 = 1.6$, $W_1 = 0.573$, $I_1(\zeta \to +\infty) = \exp(W_1 - W_2) = 0.358$. The amplitude $I_2$ and the width $\Delta_2$ of the signal structure are related to the velocity $v = 1 + \delta v$ through

$$I_2 = [1 - a_2(1 - ln\ a_2)](v+1)/\delta v, \tag{11}$$

$$\Delta_2 = \delta v/(1 + a_2), \tag{12}$$

giving extremely high amplitude and narrow pulses in the ps range, which should be useful for optical telecommunications. Figure 2 shows the dissipative two-wave adiabatic solution for $a_2 = 0.625$.

However, the intensity model (3) is singular [3] and is not appropriate for describing alone the nonlinear dynamics in a resonator, since no stable saturated regime can be reached. The instantaneous response is responsible for unlimited compression and amplification during the backward evolution, like the degenerate parametric decay case we shall consider below. In a resonator, even for high $\gamma_i$ values, which could justify the adiabatic approximation, only the coherent model (1) lead to a stable periodic pulsed regime. However it is interesting to remark that narrow localized structures are related to high dissipation of one wave. We may use this property to generate short pulses in the cavity.

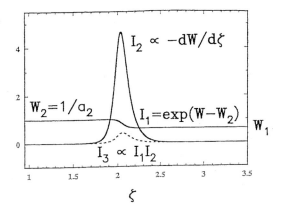

**Fig. 2.** Envelopes of the dissipative two-wave adiabatic solitary solution.

## 5  Self Pulsing in a Cavity

In order to make these solitary waves of practical interest one must find a way to generate them. We have shown in [3] [4] the morphogenesis of nanosecond dissipative Brillouin solitary waves in a fiber ring cavity when $v_i \simeq 0$. We have followed the same idea for the backward-phase-matched quadratic material inside a ring cavity [5] [6], but in order to generate the solitary wave, we have to consider the backward configuration for the idler wave too (Fig. 3).

For the sake of simplicity we first investigate the case of a cavity in which only the signal wave is reinjected. This is known as the singly resonant optical parametric oscillator (SOPO) with a backward quasi-phase-matched parametric amplification process. In order to investigate the dynamics of this system we have to integrate Eqs.(1) with the following boundary conditions

$$A_p(x = 0, t) = E_{p,cw} \tag{13a}$$

$$A_s(x = L, t) = \sqrt{R} \, A_s(x = 0, t) \tag{13b}$$

$$A_i(x = L, t) = 0, \tag{13c}$$

where $R = |\rho_s|^2$ is the intensity feedback coefficient of the input-output coupler and $L$ is the length of the cavity. We assume that the nonlinear medium fills the whole cavity. Note that, for simplicity, we have omitted the phase factor in the boundary condition of the signal field merely because it does not enter into play in the dynamics of the singly resonant OPO [24], similar to the CW-pumped Brillouin fiber ring laser [4].

We integrate Eqs.(1) numerically by following the procedure of Ref. [12] with a fourth order Runge-Kutta algorithm [3] [4] [16]. As initial conditions in $t = 0$ we take a continuous wave (CW) envelope for the pump $A_p(x, t = 0) = E_p$, a random complex noise from $x = 0$ to $x = L$ for the signal envelope $A_s(x, t = 0)$ and zero amplitude for the idler envelope $A_i(x, t = 0) = 0$, because this last

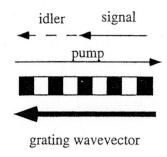

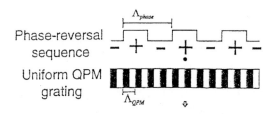

**Fig. 3.** Configuration of the backward optical parametric oscillator. To achieve quasi-phase matching, the spatial period $\Lambda_{QPM}$ of the domains is made by the phase-reversal sequence shown below.

one is automatically generated from the other two waves. At the beginning of the propagation the parametric interaction amplifies $A_s$ and $A_i$ and the system exhibits complex dynamics. As the signal field builds up in the cavity, the system undergoes an instability that results in the slow formation of large oscillations in the envelope of the signal field. This is illustrated in Fig. 4 for the same material parameters as in Fig. 1 (except that now $\gamma_p = \gamma_s$), a reflection coefficient of $R = 0.96$, a cavity length of 3 cm and a pump intensity of 27 kW/cm$^2$ ($E_p$=0.38 MV/m). After 51200 cavity round-trips we see that the oscillations lead to the formation of bright pulses in the signal and idler fields. These pulses narrow progressively until they reach a steady state in which, as shown in Fig. 4d, the envelopes take the form of the symbiotic solitary wave described above. This result is quite general. It has been observed in a wide parameter range including, in particular, reflection coefficients as small as $R = 0.01$, showing that the symbiotic solitary wave is a robust attractor in the system. However, for small values of $R$, the amplitude of the signal pulse depends significantly on $R$ but its shape is still reminiscent of the analytical solution [Eqs.(2)]. Note that, in the example of Fig. 4, the amplitude of the idler wave generated in the cavity is smaller than that of the analytical solitary wave corresponding to the same pump amplitude. This is due to the fact that the idler wave must be regenerated at each round-trip in the cavity and the cavity is not long enough to make its amplitude reach its theoretical asymptotic limit.

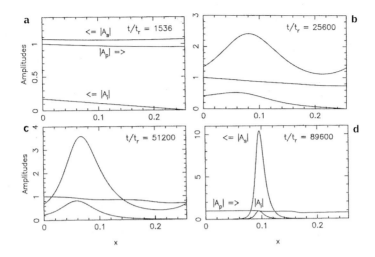

**Fig. 4.** Soliton morphogenesis: Spatial profiles of the three waves interacting in the cavity of length $L = 3$ cm at increasing cavity round trips $t/t_r$ (amplitudes are given in units of $E_p = 0.38$ MV/m.)

Let us point out that this generation process requires backward interaction. The mechanism is similar to the Hopf bifurcation appearing in the counter-streaming Brillouin cavity [3]. Numerical simulations with the more usual forward phase-matching conditions only lead to the steady-state regime. This shows that the distributed feedback nature of the interaction plays a fundamental role in the pulse generation process. This observation corroborates the conclusions of Ref. [19] where complex temporal pattern formation in backward-phase-matched second harmonic generation is studied.

In order to illustrate the potential interest of the proposed device for applications to high-repetition-rate ultrashort pulse train generation, we plotted in Fig. 5 the signal field envelope at the output of the SOPO. As remarkable features, we notice the short duration of the pulses (picosecond range), the absence of background field (the background is actually 7 orders of magnitude less than the pulse amplitude) as well as the absence of phase distortion (transform-limited sech pulses).

As was shown in Ref. [25], chromatic dispersion of the waves interacting in a cavity can also lead to the formation of periodic temporal patterns. Dispersion is thus liable to compete with the mechanism described above and could therefore strongly influence the solitary wave formation process. To investigate this problem we included dispersion in our simulations. With typical values of the dispersion parameters $\beta_{p,s,i}$ in LiNbO$_3$ [26], *i.e.*, 0.1-1 ps$^2$/m, we obtained no significant changes with respect to the case of zero dispersion.

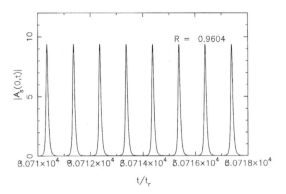

**Fig. 5.** Temporal signal output of the backward SOPO in the asymptotic regime measured in cavity round trip time $t/t_r$ units (amplitude is in units of $E_p = 0.38$ MV/m).

## 6    Three Wave Model

In order to simulate a more realistic experiment we have also to consider the guided optical parametric oscillator in the Fabry-Perot configuration, where all the waves may exhibit natural Fresnel reflections $\rho_{p,s,i} = (n_{p,s,i} - 1)/(n_{p,s,i} + 1)$ at the boundaries of the cavity, due to the difference between air- and material-refractive index. Now we solve a six wave model where the six waves circulate in the resonator:

$$(\partial_t + v_p\,\partial_x + \gamma_p + i\beta_p\partial_{tt})\,A_{p+} = -\,\sigma_p A_{s-} A_{i-} \tag{14a}$$

$$(\partial_t - v_s\,\partial_x + \gamma_s + i\beta_s\partial_{tt})\,A_{s-} = \sigma_s A_{p+} A_{i-}^* \tag{14b}$$

$$(\partial_t - v_i\,\partial_x + \gamma_i + i\beta_i\partial_{tt})\,A_{i-} = \sigma_i A_{p+} A_{s-}^* \tag{14c}$$

$$(\partial_t - v_p\,\partial_x + \gamma_p + i\beta_p\partial_{tt})\,A_{p-} = -\,\sigma_p A_{s+} A_{i+} \tag{14d}$$

$$(\partial_t + v_s\,\partial_x + \gamma_s + i\beta_s\partial_{tt})\,A_{s+} = \sigma_s A_{p-} A_{i+}^* \tag{14e}$$

$$(\partial_t + v_i\,\partial_x + \gamma_i + i\beta_i\partial_{tt})\,A_{i+} = \sigma_i A_{p-} A_{s+}^* \tag{14f}$$

where $A_{p\pm}(\omega_p, k_p)$ stand for the forward (+) or backward (−) CW-pump wave, $A_{s\mp}(\omega_s, k_s)$ for the backward (−) or forward (+) signal wave, and $A_{i\mp}(\omega_i, k_i)$ for the backward (-) or forward (+) idler wave. The resonant conditions for the Fabry-Perot 1-D space configuration are:

$$\omega_{p+} = \omega_{s-} + \omega_{i-}\ ;\qquad k_{p+} = -k_{s-} - k_{i-} + K$$

$$\omega_{p-} = \omega_{s+} + \omega_{i+}\ ;\qquad k_{p-} = -k_{s+} - k_{i+} + K$$

and the boundary conditions are:

$$A_{p+}(x = 0, t) = E_{p,cw} + \rho_p\,A_{p-}(x = 0, t) \tag{15a}$$

$$A_{p-}(x = L, t) = \rho_p\,A_{p+}(x = L, t) \tag{15b}$$

$$A_{s-}(x = L, t) = \rho_s \, A_{s+}(x = L, t) \tag{15c}$$

$$A_{s+}(x = 0, t) = \rho_s \, A_{s-}(x = 0, t) \tag{15d}$$

$$A_{i-}(x = L, t) = \rho_i \, A_{i+}(x = L, t) \tag{15e}$$

$$A_{i+}(x = 0, t) = \rho_i \, A_{i-}(x = 0, t) \tag{15f}$$

The simulation is made for a Type I (eee) polarization configuration in a 3 cm length cavity of LiNbO$_3$ with a couple of quasi-phase-matched backward three-wave interactions with $\lambda_p = 1.06$ μm, $\lambda_s = 1.55$ μm $\lambda_i = 3.35$ μm, corresponding to the $\Lambda_{QPM} = 2\pi/K = 0.247$ μm grating pitch, and with a pump field of amplitude $E_p = 0.725$ MV/m (*i.e.*, a pump intensity of $I_p = 100$ kW/cm$^2$) propagating in the quadratic $\chi^{(2)}$ material with the following values of the parameters: $d = 20$ pm/V, $n_p = 2.16$, $n_s = 2.14$, $n_i = 2.09$, $v_p = 1.353 \times 10^8$ m/s, $v_s = 1.371 \times 10^8$ m/s, $v_i = 1.359 \times 10^8$ m/s, the loss coefficients $\alpha_p = 2\gamma_p/v_p = 6.90$ m$^{-1}$ and $\alpha_s = 2\gamma_s/v_s = 2\gamma_i/v_i = 4.60$ m$^{-1}$, the reflection coefficients $\rho_p = 0.367$, $\rho_s = \sqrt{R}$ (with a reinjection $R = 0.8464$), $\rho_i = 0.352$ and $\beta_{p,s,i} = 1$ ps$^2$/m. It is important to keep the actual value for the temporal walk-off between signal and idler due to their small velocity difference, since it is the cause of dynamics stabilization. The numerical treatment of Eqs.14 takes into account a walk-off of $2(v_s - v_i)/(v_s + v_i) = 1/128$. Figures 6 to 8 show the dynamics of the self-pulsing $\chi^{(2)}$ cavity for the non degenerate Fabry-Perot backward OPO, yielding stable trains of 10 ps width pulses.

## 7 Stability of the Degenerate Backward Optical Parametric Oscillator

Quasi-phase-matched (QPM) second harmonic generation for copropagating waves, in which the wavenumber mismatch between the fundamental and second-harmonic (SH) waves is compensated with the assistance of spatial gratings of

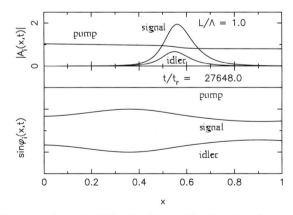

**Fig. 6.** Fabry-Perot backward OPO: Spatial profiles for the three wave amplitudes $A_{p+}$, $A_{s-}$ and $A_{i-}$ and their phases at the 9216 round trip (amplitude is in units of $E_p = 0.725$ MV/m).

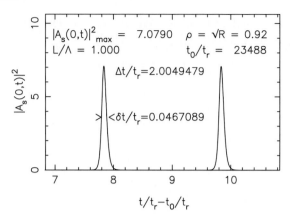

**Fig. 7.** Fabry-Perot backward OPO: Signal output intensity of a pair of dissipative solitons of 10 ps width in the asymptotic regime separated by the round trip time $2t_r = 440$ ps.

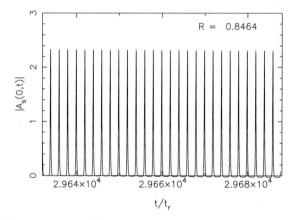

**Fig. 8.** Fabry-Perot backward OPO: Signal output of a train of dissipative solitons showing its stability in the asymptotic regime (amplitude is in units of $E_p = 0.725$ MV/m).

nonlinear or linear material constants, has been extensively studied [27]. If a grating with a sufficient short period is available, the fundamental pump and the SH wave travelling in opposite directions can be phase matched and the SH wave is generated in reflections [28] [17] [18] [7]. Static and dynamical analysis of QPM SH generation by backward propagating interaction have been done and the stability of the counter-propagating interaction in the cavity has been performed [19], showing complex temporal pattern formation. Here we are concerned with the opposite mechanism, *i.e.* parametric down conversion, with the pump at $2\omega$ and where the backward (signal or idler) wave at $\omega$ can spontaneously build up from quantum noise and is then amplified in the OPO.

The equations governing the QPM backward parametric down conversion without optical damping and dispersion, in dimensionless units are:

$$(\partial_t + \partial_x)\, P = -S^2 \tag{16a}$$

$$(\partial_t - \partial_x)\, S = PS^* \tag{16b}$$

with the boundary conditions for the singly resonant cavity

$$P(x = 0, t) = 1 \; ; \quad S(x = L, t) = \sqrt{R}\, S(x = 0, t) \tag{16c}$$

where $P$ stands for the pump amplitude $A_p$ at $2\omega$ and $S$ for the signal $A_s$ (or idler $A_i$) amplitude at $\omega$, $L$ for the cavity length, and $R$ for the intensity feedback parameter of the singly resonant OPO for $S$. At zero phase mismatch and in the absence of dispersion we can look for the stability of the real problem ($S^* = S$). From Eqs.(16a,b) we obtain the closed nonlinear PDE

$$(\partial_{tt} - \partial_{xx})\, \log S = (\partial_t + \partial_x)P = -S^2$$

$$S\partial_{tt}S - (\partial_t S)^2 - S\partial_{xx}S + (\partial_x S)^2 = -S^4. \tag{16d}$$

The stationary equations reading

$$P' = -S^2 \tag{17a}$$

$$S' = -PS \tag{17b}$$

where the prime means space derivative, admit two stationary solutions above threshold $R = exp(-2L)$:
(a) for $P^2 - S^2 = D^2$ and $D = const$:

$$P_{st}(x) = D\, \coth(Dx + \phi_0) = D\, \frac{\coth(Dx) + D}{D\coth(Dx) + 1} \tag{18a}$$

$$S_{st}(x) = \frac{D}{\sinh(Dx + \phi_0)} = \frac{D\sqrt{1 - D^2}}{D\cosh(Dx) + \sinh(Dx)} \tag{19a}$$

$$\coth(\phi_0) = 1/D \; ; \quad \phi_0 = \frac{1}{2}\log(\frac{1 + D}{1 - D})$$

$$R = \frac{S(L)^2}{S(0)^2} = \frac{D^2}{[D\cosh(DL) + \sinh(DL)]^2} \tag{20a}$$

and
(b) for $S^2 - P^2 = D^2$ and $D = const$:

$$P_{st}(x) = -D\, \tan(Dx + \phi_0) = -D\, \frac{\tan(Dx) - 1}{D + \tan(Dx)} \tag{18b}$$

$$S_{st}(x) = \frac{D}{\cos(Dx + \phi_0)} = \frac{D\sqrt{1 + D^2}}{D\cos(Dx) + \sin(Dx)} \tag{19b}$$

$$\tan(\phi_0) = -1/D \;\; ; \;\; \phi_0 = \arg \tan(-\frac{1}{D})$$

$$R = \frac{S(L)^2}{S(0)^2} = \frac{D^2}{[D\cos(DL) + \sin(DL)]^2} \tag{20b}$$

taking into account the boundary conditions (16c).

Let us look for the stability of such solutions, by considering the time perturbative problem like in Ref. [3],

$$P(x,t) = P_{st}(x) + \delta P(x,t) = P_{st}(x) + X(x) \, exp(-i\omega t) \tag{21a}$$

$$S(x,t) = S_{st}(x) + \delta S(x,t) = S_{st}(x) + Y(x) \, exp(-i\omega t) \tag{21b}$$

where $X(x)$ and $Y(x)$ are the space dependent perturbations and $\omega = \omega_r + i\omega_i$ the complex eigenvalue frequency yielding instability when Im $\omega \equiv \omega_i > 0$. From now on we will call $P \equiv P_{st}(x)$ and $S \equiv S_{st}(x)$.

Linearizing Eq.(16d) and taking into account expressions (21a,b), we obtain

$$Y'' - \frac{2S'}{S}Y' + (\omega^2 + \frac{2S''}{S} - 4S^2)Y = 0 \tag{22a}$$

with the boundary condition (16c)

$$Y(L) = \sqrt{R} \, Y(0). \tag{22b}$$

Linearizing Eq.(16b) one obtains

$$X = \frac{1}{S}(-i\omega - P)Y - Y' \tag{23a}$$

with the boundary condition

$$X(0) = 0. \tag{23b}$$

Let us introduce the pertinent variable

$$U = \frac{Y}{S} \; ; \tag{24}$$

Eqs.(22a) and (23a) become

$$U'' + (\omega^2 - 2S^2)U = 0 \tag{25a}$$

$$X = -i\omega U - U' \tag{26a}$$

with the respective boundary conditions

$$U(L) = U(0) \tag{25b}$$

$$X(0) = -U'(0) - i\omega U(0) = 0 \tag{26b}$$

Introducing in (25a) the respective stationary solution (19a,b) for $S$ we may integrate the linear inhomogeneous equation (25a), and conditions (25b)(26b) determine the eigenvalue problem for $\omega = \omega_r + i\omega_i$. Equation (25a) admit solutions in terms of Bessel functions for short enough cavities:

$$(Dx)^2 \ll 1 \quad ; \quad P^2 - S^2 = \pm D^2 \quad ; \quad S^2 = \frac{1 \mp D^2}{(1+x)^2} \tag{27}$$

and for any cavity length when $D = 0 \; [S = P = 1/(1+x)]$, since it takes the form

$$y^2 U'' + [\omega^2 y^2 - 2(1 \pm D^2)]U = 0 \tag{28}$$

with $y = 1 + x$, which solution is [29]

$$U = \sqrt{y} \left[ C_1 J_\nu(\omega y) + C_2 Y_\nu(\omega y) \right] \quad ; \quad \nu^2 - 1/4 = 2(1 \pm D^2). \tag{29}$$

We have performed the analysis for case $D = 0$ ($\nu = 3/2$), for any $x$, and for $D \neq 0$ in the limit $|\omega y| \gg 1$, where an asymptotic expansion of Bessel functions (29) in terms of trigonometric angular functions [29] is available:

$$J_\nu(z) = \sqrt{\frac{2}{\pi z}} \left[ \cos \chi - \frac{4\nu^2 - 1}{8z} \sin \chi \right] \; ; \; Y_\nu(z) = \sqrt{\frac{2}{\pi z}} \left[ \sin \chi + \frac{4\nu^2 - 1}{8z} \cos \chi \right] ; \tag{30}$$

where $z \equiv \omega y$ and $\chi = z - (2\nu + 1)\pi/4$. It allows to analytically treat the interesting problem of the stability of the cavity modes Re $\omega \simeq 2\pi N/L$ ($N$ integer). By introducing (30) into (29) and setting solvability for any couple $(C_1, C_2)$, we obtain the complex dispersion relation for $\omega = \omega_r + i\omega_i$:

$$A + B \; \sin(\omega L) + C \; \cos(\omega L) = 0 \tag{31}$$

with

$$A = (1 + L)(\omega^3 + a(a - 1)\omega)$$
$$B = i(1 + L)\omega^3 - aL\omega^2 + a^2(i\omega - 1)$$
$$C = -(1 + L)\omega^3 - iaL\omega^2 - a\omega(a - 1 - L)$$

where $a = (4\nu^2 - 1)/8 = 1 \pm D^2$. From the real and imaginary parts one obtains a couple of equations for $\omega_r$ and $\omega_i$ which is solved by a two variable Newton method; its solution always presents *instabilty* (Im $\omega > 0$). The overall instability is confirmed in the general case, beyond the asymptotic limit, by numerically solving Eqs.(1a,b). Fig. 1 shows the growth rate Im $\omega$ of the first cavity mode ($N = 1$) *vs.* the feedback rate $R$ of the backward signal intensity for different OPO length $L$ (and same pump input).

This striking result about the instability of the degenerate backward OPO is not so surprising, since the solitary wave which is generated exhibits unlimited amplification and compression above threshold, and therefore is able to stand out whatever the cavity length. We have obtain this behavior by numerically solving the dynamics governed by Eqs.(16a,b) in the cavity. We also know that these equations for the unbounded problem of a dissipative signal wave backward propagating with respect to a CW pump wave does not have a stable attractor solution. Indeed, we may perform an asymptotic Kolmogorov-Petrovskii-Piskunov

(KPP) analysis of the undepleted linear problem [16] [30] in order to prove the non existence of such attractor. Therefore it is necessary to stop this singular behavior by taking into account an additional physical mechanism. Since the solitary wave generated becomes extremely steep, dispersion will play now a basic saturation role. Therefore the dimensionless equations read:

$$(\partial_t + \partial_x + i\tilde{\beta}_p \, \partial_{tt})P = -S^2 \tag{32a}$$

$$(\partial_t - \partial_x + i\tilde{\beta}_s \, \partial_{tt})S = PS^* \tag{32b}$$

where $\tilde{\beta}_{p,s}$ characterize the mode dispersion. The numerical behavior shows that the amplification of the solitary pulses is dynamically saturated by temporal modulation of the envelopes, yielding the dynamical solitary structure shown in Fig. 9, similar to that obtained in Ref.[30] for spatial diffraction. Figure 10 shows the envelope amplitude of the  pulse maxima at the output of the cavity, exhibiting dynamical saturation when the pulse is so steep that dispersion becomes high enough. The values of the dimensionless dispersion parameters $\tilde{\beta}_{p,s}$ of Eqs.(32a,b) correspond to actual (e-e) polarization interaction in LiNbO$_3$ at 1000C, namely $\beta_p = 0.1131$ ps$^2$/m for the pump $P$ at $\lambda_p = 0.775$ $\mu$m, and $\beta_s = 0.8930$ ps$^2$/m for the backward signal wave $S$ at $\lambda_p = 1.55$ $\mu$m. The saturation dynamics has been obtained for a set of different pump intensities and cavity lengths. The power requirement for backward OPO operation depends on the ability to achieve low-order QPM over centimeter lengths. Thus, e.g. for a first order QPM of the above pump and signal wavelengths in LiNbO$_3$ the grating pitch is as small as $\Lambda_{QPM} = 2\pi/K = 0.179$ $\mu$m. For a CW-pump field $E_p = 0.725$ MV/m (i.e., a pump intensity of $I_p = 100$ kW/cm$^2$) propagating in the quadratic $\chi^{(2)}$ material with the following values of the parameters: $d = 20$ pm/V, $n_p = 2.179$, $n_s = 2.141$, $v_p = 1.323 \times 10^8$m/s,

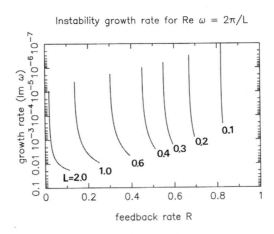

**Fig. 9.** Degenerate backward OPO: Modulated dynamical solitary structure generated when the presence of dispersion saturates the unlimited amplification and compression of the backward propagating pulse if this dispersion is absent (arbitrary start time).

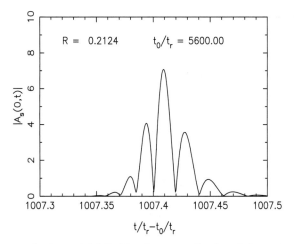

**Fig. 10.** Degenerate backward OPO: Envelope amplitude of the pulse maxima at the output of the cavity, exhibiting dynamical saturation at round trip 6000, when the pulse is so narrow that dispersion is no more perturbative.

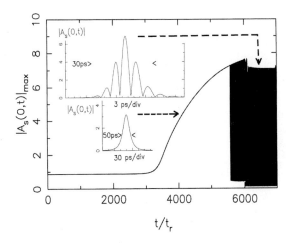

**Fig. 11.** Degenerate backward OPO: Pair of consecutive deeply modulated dynamical solitary structures showing a central peak of 5 ps width (arbitrary start time).

$v_s = 1.371 \times 10^8 \mathrm{m/s}$, $v_i = 1.359 \times 10^8 \mathrm{m/s}$, the nonlinear characteristic time yields $\tau_0 = (\sigma_p E_p/2)^{-1} \simeq 0.37$ ns, where $\sigma_p = 2\pi d v_p/\lambda_p n_p$ is the coupling coefficient, and the nonlinear characteristic length of the cavity is $\Lambda = v_p \tau_0 = 5$ cm. For $L/\Lambda = 1$ the w solitary pulses are compressed until 7.5 ps before dispersion begins the saturation process. The dynamical solitary structure being deeply modulated, the central peak has about 5 ps width (Fig. 11), while the whole pulse spreads over some tens of picoseconds.

# 8    Conclusion

In the first part, we numerically showed the existence of dissipative symbiotic solitary waves sustained by the backward-phase-matched nondegenerate three-wave parametric interaction in optical ring or Fabry-Perot cavities, which are stable structures even in the absence of dispersion due to the finite temporal walk-off between the backward signal and idler waves. These solitary waves constitute a stable and robust attractor for the three waves interacting inside singly or multiple resonant cavities. As a result, background-free high-repetition-rate trains of picosecond transform-limited pulses are generated spontaneously from noise in such cavities, like nanosecond pulses in Brillouin-fiber-ring lasers. These last have been observed in actual experiments [2–4]; the experimental observation of the former would be of great interest because, on the one hand, it would allow for the study of fundamental nonlinear waves and, on the other hand, it could be applied to the realization of a versatile wavelength tunable ultrashort pulse laser source.

In the second part, we have performed an analytical stability analysis of the degenerate backward quasi-phase-matching OPO for the parametric decay interaction in a singly resonant cavity, and show that the stationary solutions are *always unstable* above threshold, whatever the cavity length and pump power. Starting from any initial condition above threshold the backward signal wave evolves to an unlimited compressible pulse. Taking into account dispersion we may saturate this singular behavior and obtain a new type of dynamic solitary wave.

### Acknowledgements

The author wish to thank Professor K. Porsezian for his kind invitation to the International Workshop on Optical Solitons OSTE 2002 in Kochi. He also thanks A. Picozzi, K. Gallo, M. Taki, and C. Durniak for stimulating discussions.

# References

1. Agrawal G. P.: *Nonlinear fiber optics*, (Academic Press, New York, 1989).
2. Picholle E., Montes C., Leycuras C., Legrand O., and Botineau J.:Phys. Rev. Lett. **66**, 1454 (1991).
3. Montes C., Mamhoud A., and Picholle E.:Phys. Rev. A **49**, 1344 (1994).
4. Montes C., Bahloul D., Bongrand I., Botineau J., Cheval G., Mamhoud A., Picholle E., and Picozzi A.:J. Opt. Soc. Am. B **16**, 932 (1999).
5. Picozzi A. and Hælterman M.: Opt. Lett. **23**, 1808 (1998).
6. Montes C., Picozzi A., and Hælterman M.: Optical Solitons: Theoretical Challenges and Industrial Perspectives- Lecture 16, Les Houches Workshop, V.E. Zakharov and S. Wabnitz eds., EDP Springer, 283-292 (1999).
7. Kang J.U., Ding Y.J., Burns W.K., and Melinger J.S.:Opt. Lett. **22**, 862 (1997).
8. Gu X., Korotkov R.Y., Ding Y.J., Kang J.U., and Khurgin J.B.:J. Opt. Soc. Am. B **15**, 1561 (1998).

9. Gu X., Makarov M., Ding Y.J., Khurgin J.B., and Risk W.P.:Opt. Lett. **24**, 127 (1999).
10. Armstrong J.A., Jha S.S., and Shiren N.S.:IEEE J. Quant. Elect. **QE-6**, 123 (1970).
11. Nozaki K. and Taniuti T.:J. Phys. Soc. Jpn. **34**, 796 (1973).
12. Kaup D.J., Reiman A., and Bers A.:Rev. Mod. Phys. **51**, 275 (1979).
13. Trillo S.: Opt. Lett. **21**, 1111 (1996).
14. McCall S.L. and Hahn E.L.:Phys. Rev. Lett. **18**, 908 (1967).
15. Drühl K., Wenzel R.G. and Carlsten J.L.: Phys. Rev. Lett. **51**, 1171 (1983).
16. Montes C., Picozzi A., and Bahloul D.: Phys. Rev. E **55**, 1092 (1997).
17. Matsumoto M. and Tanaka K. IEEE J. Quantum Electron. **31**, 700 (1995).
18. Ding Y. J. and Khurgin J.B.: IEEE J. Quantum Electron. **32**, 1574 (1996).
19. D'Alessandro G., Russell P.St., and Wheeler A.A.:Phys. Rev. A **55**, 3211 (1997).
20. Picozzi A. and M. Hælterman M.: Phys. Rev. Lett. **86**, 2010 (2001).
21. Morozov S.F., Piskunova L.V., Sushchik M.M., and Freidman G.I.: Sov. J. Quant. Electron. **8**, 576 (1978).
22. Craik A.D.D., Nagata M., and Moroz I.M.: Wave Motion **15**, 173 (1992).
23. Botineau J., Leycuras C., Montes C., and Picholle E.: Opt. Comm. **109**, 126 (1994).
24. Yang S.T., Eckaerdt R.C., and Byer R.L.:J. Opt. Soc. Am. B **10**, 1684 (1993).
25. Trillo S. and Haelterman M.: Opt. Lett. **21**, 1114 (1996).
26. Dmitriev V.G., Gurzadyan G.G., Nikogosyan D.N.: *Handbook of Nonlinear Optical Crystals*, (Springer-Verlag 1991).
27. Armstrong J.A., Bloembergen N., Ducuing J., and Perhan P.S.: Phys. Rev. **127**, 1918 (1992).
28. Fejer M.M., Magel G.A., Jundt D.H., and Byer R.L.: IEEE J. Quantum Electron. **28**, 2631 (1992).
29. Abramowitz M. and Stegun I.A.: *Handbook of Mathematical Functions*, 8th ed. (Dover Public., New-York, 1972).
30. Picozzi A. and Haælterman M.: Phys. Rev. Lett. **84**, 5760 (2000).

# Spatial Semiconductor-Resonator Solitons

V.B. Taranenko and C.O. Weiss

Physikalisch-Technische Bundesanstalt, Braunschweig 38116, Germany

**Abstract.** We demonstrate experimentally and numerically the existence of spatial solitons in multiple-quantum-well semiconductor microresonators driven by an external coherent optical field. We discuss stability of the semiconductor-resonator solitons over a wide spectral range around the band edge. We demonstrate the manipulation of such solitons: switching solitons on and off by coherent as well as incoherent light; reducing the light power necessary to sustain and switch a soliton, by optical pumping.

## 1   Introduction

Spatial resonator solitons theoretically predicted in [1–5] can exist in a variety of nonlinear resonators, such as lasers (vortices), lasers with saturable absorber (bright solitons), parametric oscillators (phase solitons), driven nonlinear resonators (bright/dark solitons). Such resonator solitons can be viewed as self-trapped domains of one field state surrounded by another state of the field. The two different field states can be a high and a low field (bright/dark solitons), positive and negative field (phase solitons), right-hand and left-hand polarized field (polarization solitons) or high and zero field (vortices).

Bright solitons in laser resonators with saturable absorber were initially shown to exist in [6], which was limited to single stationary solitons. Existence of moving solitons and simultaneous existence of large numbers of stationary solitons was shown in [7,8]. Phase solitons in degenerate parametric wave mixing resonators were predicted in [9] and demonstrated in [10]. Theoretically it was shown in [11] that not only 2D spatial resonator solitons exist but that also in 3D such structures can be stable, linking the field of optical solitons with elementary particle physics [12]. Vortex solitons differ from other soliton types in that they possess *structural stability* in addition to *dynamical stability*, the only stabilizing mechanism of the other solitons. The existence of vortices in lasers initially shown in [13] and later also the existence of vortex solitons [14].

Since the resonator solitons are bistable and can be moved around they are suited to carry information. Information can be written in the form of a spatial soliton, somewhere, and then transported around at will; finally being read out somewhere else; possibly in conjunction with other solitons. In this respect the spatial resonator solitons have no counterpart in any other kind of information-carrying elements and lend themselves therefore to operations not feasible with conventional electronic means, such as an all-optical pipeline storage register ("photon buffer") or even processing in the form of cellular automata. Experiments on the manipulation of bright solitons as required for such processing

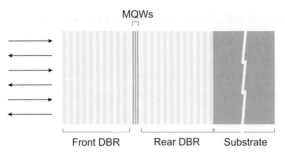

**Fig. 1.** Schematic of semiconductor microresonator consisting of two plane distributed Bragg reflectors (DBR) and multiple quantum wells (MQW).

tasks were first carried out on a slow system: laser with (slow) saturable absorber. In particular it was demonstrated how to write and erase solitons and how to move them or localize them. For reviews see [15,16].

In order to be applicable to technical tasks it is mandatory to operate in fast, miniaturized systems. For compatibility and integrability with other information processing equipment it is desirable to use semiconductor systems. We chose the semiconductor microresonator structure (Fig. 1) as commonly used for Vertical Cavity Surface Emitting Lasers (VCSEL) [17] consisting of multiple quantum well (MQW) structure sandwiched between distributed Bragg reflectors (DBR).

The resonator length of such a structure is $\sim \lambda$ while the transverse size is typically 5 cm. Therefore such short length and wide area microresonator permits only one longitudinal mode (Fig. 2 (a)) and an enormous number of transverse modes that allow a very large number of spatial solitons to coexist. The resonator is obviously of the plane mirror type, implying frequency degeneracy of all transverse modes and thus allowing arbitrary field patterns to be resonant inside the resonator. This is another prerequirement for existence and manipulability of spatial solitons.

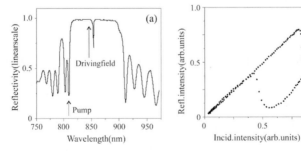

**Fig. 2.** Semiconductor microresonator reflectance spectrum (a) and typical bistability loop in reflection (b). Arrows mark the driving field that is detuned from the resonator resonance and the pump field that is tuned to be coupled into resonator through one of short-wavelength interference notches of the resonator reflectance spectrum.

The resonator soliton existence is closely linked with the plane wave resonator bistability (Fig. 2 (b)) caused by longitudinal nonlinear effects: the nonlinear changes of the resonator length (due to nonlinear refraction changes) and finesse (due to nonlinear absorption changes) [18]. The longitudinal nonlinear effects combined with transverse nonlinear effects (such as self-focusing) can balance diffraction and form resonator soliton. Generally these nonlinear effects can cooperate or act oppositely, with the consequence of reduced soliton stability in the later case.

In the present paper we demonstrate experimentally and numerically existence of bright and dark spatial solitons as well as extended hexagonal patterns in MQW-semiconductor microresonators at room temperature. We discuss stability of the semiconductor-resonator solitons over a wide spectral range around the band edge. We demonstrate the manipulation of such solitons in view of technical application: switching solitons on and off by coherent as well as incoherent light; reducing the light power necessary to sustain and switch a soliton, by optical pumping.

## 2    Model and Numerical Analysis

We consider phenomenological model of a driven wide area MQW-semiconductor microresonator similar to [19,20]. The optical field $E$ inside the resonator is described in mean-field approximation [21]. The driving incident field $E_{\text{in}}$ assumed to be a stationary plane wave. Nonlinear absorption and refractive index changes induced by the intracavity field in the vicinity of the MQW-structure band edge are assumed to be proportional to the carrier density $N$ (normalized to the saturation carrier density). The equation of motion for $N$ involves nonresonant (to avoid saturation effect) pumping $P$, carrier recombination and diffusion. The resulting coupled equations describing the spatio-temporal dynamics of $E$ and $N$ have the form:

$$\partial E/\partial t = E_{\text{in}} - \sqrt{T}E\{[1 + C\text{Im}(\alpha)(1-N)] + i(\theta - C\text{Re}(\alpha)N - \nabla_\perp^2)\},$$

$$\partial N/\partial t = P - \gamma[N - |E|^2(1-N) - d\nabla_\perp^2 N], \tag{1}$$

where $C$ is the saturable absorption scaled to the resonator DBR transmission $T$ ($T$ is assumed to be small since the DBR reflectivity is typically $\geq 0.995$). $\text{Im}(\alpha)(1-N)$ and $\text{Re}(\alpha)N$ describe the absorptive and refractive nonlinearities, respectively, $\theta$ is the detuning of the driving field from the resonator resonance, $\gamma$ is the photon lifetime in the resonator scaled to the carrier recombination time, $d$ is the diffusion coefficient scaled to the diffraction coefficient and $\nabla_\perp^2$ is the transverse Laplacian.

Linear effects in the resonator are spreading of light by diffraction and diffusion (terms with $\nabla_\perp^2$ in (1)). The material nonlinearity that can balance this linear spreading can do this in various ways. It has a real (refractive) and imaginary (dissipative) part and can act longitudinally and transversely. The nonlinear changes of the resonator finesse (due to nonlinear absorption change) and length (due to nonlinear refractive index change) constitute longitudinal nonlinear effects, also known under the name *nonlinear resonance* [22]. The transverse effect of the nonlinear refractive index can be self-focusing (favorable for bright and unfavorable for dark solitons) and self-defocusing (favorable for dark and unfavorable for bright solitons). Absorption (or gain) saturation (bleaching) leading to nonlinear gain guiding in laser parlance, a transverse effect. Longitudinal and transverse effects can work oppositely, or cooperate.

There are two main external control parameters: the driving field intensity $|E|^2$ and the resonator detuning $\theta$. Then for driving intensities not quite sufficient for reaching the resonance condition for the whole resonator area, the system "chooses" to distribute the light intensity in the resonator in isolated spots where the intensity is then high/low enough to reach the resonance condition thus forming bright/dark patterns. Instead of saying "the system chooses" one would more mathematically express this by describing it as a modulational instability (MI). The detuned plane wave field without spatial structure with intensity insufficient to reach the resonance condition is unstable against structured solutions. According to our numerical solutions of (1) a large number of such structured solutions coexist and are stable (see e.g. patterns in Fig. 7).

Figure 3 shows the existence domains (in coordinates $\theta$ - $|E|^2$) of MI, dark spatial solitons and plane-wave bistability calculated for the case of purely dispersive (defocusing) nonlinearity. The dark soliton structure (inset in Fig. 3) can be interpreted as a small circular switching front. A switching front connects two stable states: the high transmission and the low transmission state. Such a front can in 2D surround a domain of one state. When this domain is comparable in diameter to the "thickness" of the front, then each piece of the front interacts

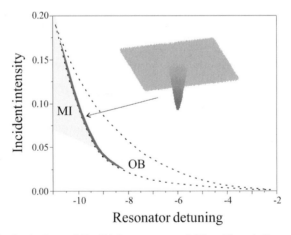

**Fig. 3.** Numerical solutions of Eq.(1) for unpumped ($P = 0$) and dispersive/defocusing ($\mathrm{Im}(\alpha) = 0$) case. Area limited by dashed lines is optical bistability domain for plane waves. Grey shaded area is modulational instability domain. Dark shaded area is domain of stability for dark solitons. Inset is dark soliton in 3D representation. Parameters: $C = 10$, $T = 0.005$, $\mathrm{Re}(\alpha) = -1$, $d = 0.01$.

with the piece on the opposite side of the circular small domain, which can lead, particularly if the system is not far from a modulational instability (see Fig. 3), to a stabilization of the diameter of the small domain. In which case the small domain becomes an isolated self-trapped structure or a dissipative resonator soliton.

## 3    Experimental Technique

Figure 4 shows the optical arrangement for the semiconductor-resonator soliton experiments, and in particular for their switching on or off. The semiconductor microresonators consisting of MQW (GaAs/AlGaAs or GaInAs/GaPAs) structures sandwiched between high-reflectivity ($\geq$ 0.995) DBRs (Fig. 1) operate at room temperature. The microresonator structures were grown on the GaAs substrates by molecular beam epitaxy technique that makes possible high quality MQW structures with small radial layer thickness variation. The best sample tested in our experiments has only $\sim$ 0.3 nm/mm variation of the resonator resonance wavelength over the sample cross section.

The driving light beam was generated by either a tunable (in the range 750-950 nm) Ti:Sa laser or a single-mode laser diode, both operating in continuous wave regime. For experimental convenience to limit thermal effects we perform the whole experiments within a few microseconds, by admitting the light through an acousto-optical modulator. The laser beam of suitable wavelength is focused on the microresonator surface in the light spot of $\sim$ 50 $\mu$m in diameter, thus providing a quite large Fresnel number ($\geq$ 100).

Part of laser light is split away from the driving beam and then superimposed with the main beam by a Mach-Zehnder Interferometer arrangement, to serve as the address beam. The address beam is sharply focused and directed to some particular location in the illuminated. The switching light is opened only for a few nanoseconds using an Electro-optic modulator. For the case of

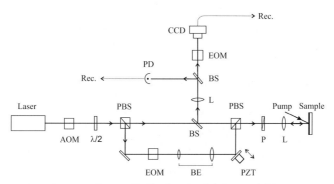

**Fig. 4.** Experimental setup. Laser: Ti:Sa (or diode) laser, AOM: acousto-optic modulator, $\lambda/2$: halfwave plate, PBS: polarization beam splitters, EOM: electro-optical amplitude modulators, BE: beam expander, PZT: Piezo-electric transducer, P: polarizer, L: lenses, BS: beam splitters, PD: photodiode.

incoherent switching the polarization of the address beam is perpendicular to that of the main beam to avoid interference. For the case of coherent switching the polarizations should be parallel and a phase control of the switching field is thus always needed: for switching on as well as switching off a soliton. One of the interferometer mirrors can be moved by a piezo-electric element to control the phase difference between the driving light and the address light.

Optical pumping of the MQW-structures was done by a multi-mode laser diode or a single-mode Ti:Sa laser. To couple the pump light into the microresonator the laser wavelength was tuned into the high transmission spectral window of the microresonator reflectance spectrum as shown in Fig. 2 (a).

The observation is done in reflection by a CCD camera combined with a fast shutter (another electro-optic modulator), which permits to take nanosecond snapshots at a given time of the illuminated area on the resonator sample. Recording movies on this nanosecond time scale is also possible. To follow intensity in time in certain points (e.g. at the location of a soliton) a fast photodiode can be imaged onto arbitrary locations within the illuminated area.

## 4   Experimental Results and Discussions

To find the most stable resonator solitons for applications one can play with the nonlinear (absorptive/dispersive) response by choice of the driving field wavelength, with the resonator detuning, and finally with the carrier population inversion by the pumping. We recall that all nonlinearities change their sign at transparency i.e. at the point where in the valence- and the conduction band populations are equal. Going from below transparency (absorption) to above transparency (population inversion, producing light amplification), nonlinear absorption changes to nonlinear gain, self-focusing changes to self-defocusing and vice versa, and decrease of resonator length with intensity changes to increase (and vice versa). The population of the bands can be controlled by pumping ($P$ in Eq.(1)) i.e. transferring electrons from the valence band to the conduction band. This can be done by optical excitation [23], with radiation of wavelength shorter than the band edge wavelength or - if the structure is suited to support electrical currents (i.e. if it is a real VCSEL-structure) - by electrical excitation.

### 4.1   Below Bandgap Hexagons and Dark Solitons

Working well bellow the bandgap when the driving field wavelength is $\sim 30$ nm longer than the band edge wavelength we observed spontaneous formation of hexagonal patterns (Fig. 5). The hexagon period scales linearly with $\theta^{-1/2}$ [24] indicating that they are formed by the tilted-wave mechanism [25] that is the basic mechanism for resonator hexagon formation [26]. Dark-spot hexagon (Fig. 5 (a)) converts in to bright-spot hexagon (Fig. 5 (b)) when the driving intensity increases. Experiment shows that individual spots of these patterns can not be switched on/off independently from other spots as it is expected for strongly correlated spot structure (or coherent hexagons).

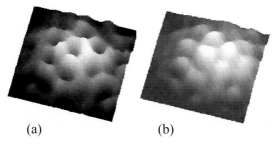

**Fig. 5.** (a) Bright (dark in reflection) and (b) dark (bright in reflection) hexagonal patterns for the dispersive/defocusing case. Driving light intensity increases from (a) to (b).

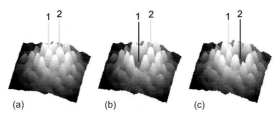

**Fig. 6.** Switching-off of individual spots of incoherent hexagonal structure with addressed pulses located at different places (marked 1 and 2) of the pattern.

However when we further increased the driving intensity we found that the bright spots in such hexagonal patterns can be switched independently by the addressed focused optical (incoherent) pulses [24]. Figure 6 shows the experimental results. Fig. 6 (a) shows the hexagonal pattern formed. The focused light pulse can be aimed at individual bright spots such as the ones marked "1" or "2". Figure 6 (b) shows that after the switching pulse aimed at "1" spot "1" is off. Figure 6 (c) shows the same for spot "2". We remark that in these experiments we speak of true logic switching: the spots remain switched off after the switching pulse (if the energy of the pulse is sufficient, otherwise the bright spot reappears after the switching pulse). These observations of local switching indicate that these hexagonal patterns are not coherent pattern: the individual spots are rather independent, even at this dense packing where the spot distance is about the spot size.

These experimental findings can be understood in the frame of the model (1) [27]. Figure 7 shows the bistable plane wave characteristic of the semiconductor resonator for conditions roughly corresponding to the experimental conditions. At the intensities marked (a) to (d) patterned solutions exist.

The pattern period in Fig. 7 (a) corresponds precisely to the detuning in the following way. When the driving field is detuned, the resonance condition of the resonator cannot be fulfilled by plane waves travelling exactly perpendicularly to the mirror plane. However, the resonance condition can be fulfilled if the wave plane is somewhat inclined with respect to the mirror plane (the tilted wave solution [25]). The nonlinear system chooses therefore to support resonant, tilted

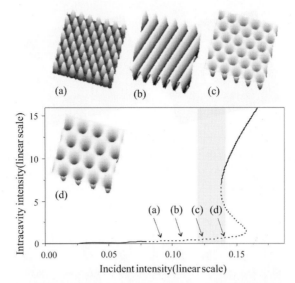

**Fig. 7.** Numerical solution of Eq.(1) for intracavity light intensity as function of in-cident intensity: homogeneous solution (dashed line marks modulations unstable part of the curve) and patterns (a-d). Shaded area marks existence range for dark-spot hexagons. Parameters: $C = 10$, $T = 0.005$, $\mathrm{Im}(\alpha) = 0.1$, $\mathrm{Re}(\alpha) = -1$, $\theta = -10.3$, $d = 0.01$.

waves. Figure 7 (a) is precisely the superposition of 6 tilted waves that support each other by (nonlinear) 4-wave-mixing. The pattern period corresponds to the resonator detuning as in the experiment for structures Fig. 5 (a). Thus the pattern formation in Fig. 7 (a) is mostly a linear process. In this pattern the bright spots are not independent. Individual spots cannot be switched as in the experiment Fig. 5.

On the high intensity pattern Fig. 7 (d) the pattern period is remarkably different from Fig. 7 (a) even though the detuning is the same. This is indication that the internal detuning is smaller and means that the resonator length is nonlinearly changed by the intensity-dependent refractive index (the nonlinear resonance). From the ratio of the pattern periods of Fig. 7 (a) and (d) one sees that the nonlinear change of detuning is about half of the external detuning. That means the nonlinear detuning is by no means a small effect. This in turn indicates that by spatial variation of the resonator field intensity the detuning can vary substantially in the resonator cross section. In other words, the resonator has at the higher intensity a rather wide freedom to (self-consistently) arrange its field structure. One can expect that this would allow a large number of possible stable patterns between which the system can choose.

Figure 8 shows that the high intensity conditions of Fig. 7 (d) allow reproduc-ing the experimental findings on switching individual bright spots. Figure 8 (a) is the regular hexagonal pattern, Fig. 8 (b) shows one bright spot switched off

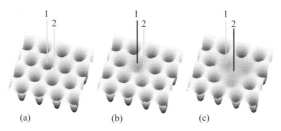

**Fig. 8.** Stable hexagonal arrangements of dark partial solitons: (a) without defects, (b) with single-soliton defect, (c) with triple-soliton defect.

as a stable solution and Fig. 8 (c) shows a triple of bright spots switched off as a stable solution, just as observed in the experiments [27].

Thus while Fig. 7 (a) is a completely coherent space filling pattern, Fig. 7 (d) is really a cluster of (densest packed) individual dark solitons. The increase of intensity from (a) to (d) allows the transition from the extended patterns to the localized structures by the increased nonlinearity, which gives the system an additional internal degree of freedom. We note that the transition from the coherent low intensity pattern to the incoherent higher intensity structure proceeds through stripe-patterns as shown in Fig. 7 (b) [27]. For the intensity of Fig. 7 (c) the individual spots are still not independent as in the experiment Fig. 5 (b).

## 4.2   Near Bandgap Bright and Dark Solitons

Working at wavelengths close to the band edge we found both bright and dark solitons, as well as collections of several of the bright and dark solitons, several solitons existing at the same time (Figs 9, 10). Figure 10 demonstrates that shape and size of bright spots are independent on the shape and intensity of the driving beam that is a distinctive feature of spatial solitons.

Nonlinearity of the MQW structure near the band edge is predominantly absorptive. Therefore in the first approximation we can neglect the refractive part of the complex nonlinearity in the model equations (1) and describe the nonlinear structure as a saturable absorber. Numerical simulations for this case (Fig. 11) confirm existence of both bright and dark resonator solitons as they are observed in the experiment (Figs 9, 10). We can contrast these resonator spatial solitons with the propagating (in a bulk nonlinear material) spatial solitons: the latter can not be supported by a saturable absorber.

Figure 12 (c) and (d) show details of the spontaneous formation of the bright soliton (as in Fig. 9 (c)). As discussed in [28] temperature effects lead in this case to a slow formation of solitons, associated with the shift of the band edge by temperature [29]. Figure 12 (c) gives the incident and reflected intensity at the location of the bright soliton and Fig. 12 (d) gives the reflectivity on a diameter of the illuminated area. At 1.3 $\mu$s (arrow) the resonator switches to high transmission (low reflection). The switched area then contracts relatively slowly to the stable structure Fig. 12 (d), which is existing after $t \approx 3.0$ $\mu$s.

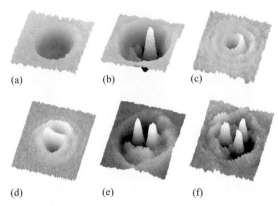

**Fig. 9.** Solitons in the semiconductor microresonator: (a) switched area without solitons (for completeness); (b) dark soliton in switched area; (c) bright soliton on unswitched background; (d) 2 bright solitons; (e), (f) 2,3 dark solitons.

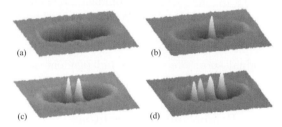

**Fig. 10.** Switched domain (a) and dark solitons (b)-(d) in an illuminating area of elliptical shape.

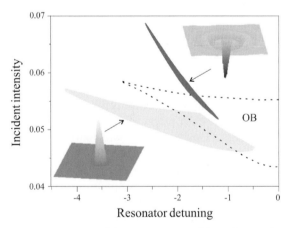

**Fig. 11.** Numerical solutions of Eq.(1) for unpumped ($P = 0$) and absorptive ($\mathrm{Re}(\alpha) = 0$) case. Area limited by dashed lines is optical bistability domain for plane waves. Shaded areas are domains of existence of bright (grey area) and dark (black area) solitons. Insets are bright (left) and dark (right) soliton. Parameters: $C = 20$, $T = 0.005$, $\mathrm{Im}(\alpha) = 1$, $d = 0.01$.

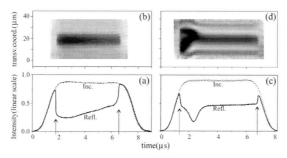

**Fig. 12.** Comparison of bright soliton formation above bandgap (left) and below bandgap (right). Reflectivity on a diameter of the illuminated area as a function of time (b), (d). Intensity of incident (dotted) and reflected (solid) light, at the center of the soliton as a function of time (a), (c). Arrows mark the switch-on and -off.

After the resonator switches to low reflection its internal field and with it the dissipation is high. A rising temperature decreases the band gap energy [29] and therefore shifts the bistable resonator characteristic towards higher intensity. Thus the basin of attraction for solitons which is located near the plane wave switch-off intensity (see locations of the existence domains for the bright solitons and the plane wave bistability in Fig. 11) is shifted to the incident intensity, whereupon a soliton can form. Evidently for different parameters the shift can be substantially larger or smaller than the width of bistability loop, in which case no stable soliton can appear. We note that in absence of thermal effects (good heat-sinking of sample) solitons would not appear spontaneously but would have to be switched on by local pulsed light injection.

Figure 13 (a) shows the incoherent switch-on of bright soliton, where the perpendicularly polarized switching pulse ($\sim$ 10 ns) is applied at $t = 4$ $\mu$s. As apparent, a soliton forms after this incoherent light pulse. The slow formation of the soliton is apparent in Fig. 13 (a) (using roughly the time from $t = 4$ to $t = 4.5$ $\mu$s).

It should be emphasized that this thermal effect is not instrumental for switching a soliton on. However, it allows switching a soliton off incoherently [30]. This is shown in Fig. 13 (b) where the driving light is initially raised to a level at which a soliton forms spontaneously (note again the slow soliton formation due to the thermal effect). The incoherent switching pulse is then applied which leads to disappearance of the soliton.

The soliton can thus be switched on and also off by an incoherent pulse. The reason for the latter is thermal. Initially the material is "cold". A switching pulse leads then to the creation of a soliton. Dissipation in the material raises the temperature and the soliton is slowly formed. At the raised temperature the band edge (and with it the bistability characteristic) and the existence range of solitons is shifted so that a new pulse brings the system out of the range of existence of solitons. Consequently the soliton is switched off.

Thus switching on a soliton is possible incoherently with the "cold" material and switching off with the "heated" material. When the driving intensity

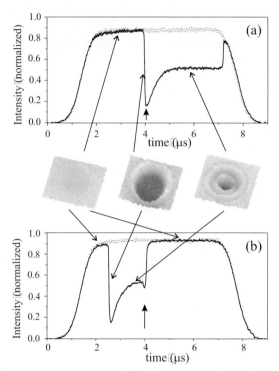

**Fig. 13.** Recording of incoherent switching-on (a) and switching-off (b) of a soliton. Snapshot pictures show unswitched state (left), circular switched domain (center). and a soliton (right). Dotted trace: incident intensity.

is chosen to be slightly below the spontaneous switching threshold the nonlinear resonator is cold. An incoherent pulse increases illumination locally and can switch the soliton on that causes local heating of the resonator. Another incoherent pulse aimed into the heated area can then switch the soliton off and thereby cool the resonator to its initial state, so that the soliton could be switched on/off again.

### 4.3    Above Bandgap Bright Solitons

At excitation above bandgap bright solitons form analogously to the below/near bandgap case [31]. Figure 14 shows the bright solitons as observed in the reflected light with their characteristic concentric rings with the same appearance as the bright solitons below band gap (Fig. 9 (c)).

Figure 12 compares the dynamics of the bright soliton formation for excitation above bandgap (left) and below bandgap (right). Difference between these two cases can be understood from the model. From Eq.(1) we obtain the reflected intensity as a function of incident intensity for wavelengths above ($\mathrm{Re}(\alpha) > 0$), as well as below the bandgap ($\mathrm{Re}(\alpha) < 0$), for plane waves (Fig. 15). One sees that the bistability range is large below and small above the bandgap. Solving

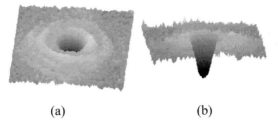

(a)                                    (b)

**Fig. 14.** Bright soliton above bandgap in 3D representation: view from above (a), from below (b).

Eq.(1) numerically the typical bright soliton (top of Fig. 15) is found coexisting with homogeneous intensity solutions in the shaded regions of Fig. 15 (a), (b).

After switching on the resonator below band gap (Fig. 15 (b)), the intensity in the resonator is high and with it the thermal dissipation. The temperature consequently rises, which shifts the band gap [29] and with it the bistability characteristic, so that the switch-off intensity, close to which the stable solitons exist, becomes close to the incident intensity. Then the resonator is in the basin of attraction for the solitons and the soliton forms as observed in Fig. 12 (c), (d).

Above the band gap Fig. 12 (a), (b) show that the soliton is switched on "immediately" without the slow thermal process. Figure 15 (a) shows why. The plane wave characteristic of the resonator above band gap is either bistable but very narrow, or even monostable (due to the contribution of the self-focusing reactive nonlinearity [32]) but still with bistability between the soliton state (not plane wave) and the unswitched state. In this case the electronic switching leads directly into the basis of attraction for solitons and the switch-on of the

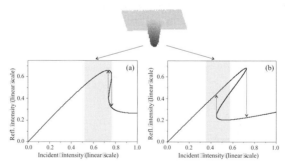

**Fig. 15.** Steady-state plane wave solution of Eq.(1) above bandgap (a): $\text{Re}(\alpha) = 0.05$; and below band gap (b): $\text{Re}(\alpha) = -0.05$. Other parameters: $C = 30$, $\text{Im}(\alpha) = 0.99$, $\theta = -3$, $P = 0$, $d = 0.1$. The soliton solution shown exists for incident intensities corresponding to the shaded areas, in coexistence with homogeneous solutions. For a temperature increase the characteristics together with soliton existence ranges shift to higher incident intensities. Reflected and incident intensities normalized to the same value.

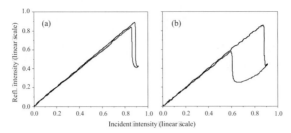

**Fig. 16.** Bistability characteristics (reflected light intensity versus incident light intensity) measured at the center of an illuminated area of 100 $\mu$m diameter, for above band gap (a) and below band gap (b) excitation.

soliton is purely electronic and fast. The width of the bistability characteristics observed experimentally (Fig. 16) scale in agreement with Fig. 15.

Nonetheless, also above bandgap there is strong dissipation after the switch-on. The associated temperature rise influences and can even destabilize the soliton. The effect can be seen in Fig. 12 (a). Over a time of a few $\mu$s after the soliton switch-on the soliton weakens (reflectivity increases slowly) presumably by the rise of temperature and the associated shift of the band gap. At 6.5 $\mu$s the soliton switches off although the illumination has not yet dropped.

Thus, while the dissipation does not hinder the fast switch-on of the soliton, it finally destabilizes the soliton. After the soliton is switched off, the material cools and the band gap shifts back so that the soliton could switch on again.

Figure 17 shows the soliton coherent switching observations [33]. Figure 17 (a) shows switching a soliton on. The driving light intensity is chosen slightly below the spontaneous switching threshold. At t $\approx$ 1.2 $\mu$s the addressing pulse is applied. It is in phase with the driving light, as visible from the constructive interference. A bright soliton results, showing up in the intensity time trace as a strong reduction of the reflected intensity. Figure 17 (b) shows switching a soliton off. The driving light is increased to a level where a soliton is formed spontaneously. The addressing pulse is then applied in counterphase to the driving light, as visible from the destructive interference. The soliton then disappears, showing up in the intensity time trace as reversion of the reflected intensity to the incident intensity value. The Fig. 17 insets show 2D snapshots before and after the switching pulses for clarity.

## 4.4 Optical Pumping

The thermal effects discussed above result from the local heating caused by the high intracavity intensity at the bright soliton location. They limit the switching speed of solitons and they will also limit the speed at which solitons could be moved around, limiting applications. The picture is that a soliton carries with it a temperature profile, so that the temperature becomes a dynamic and spatial variable influencing the soliton stability.

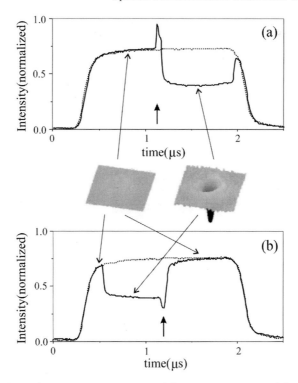

**Fig. 17.** Recording of coherent switching-on (a) and switching-off (b) of a bright soliton. Heavy arrows mark the application of switching pulses. Dotted traces: incident intensity. The insets show intensity snapshots, namely unswitched state (left) and soliton (right).

As opposed, a spatially uniform heating will not cause such problems, as it shifts parameters but does not constitute a variable in the system. The largely unwanted heating effects are directly proportional to the light intensity sustaining a soliton. For this reason and quite generally it is desirable to reduce the light intensities required for sustaining solitons.

Conceptually this can be expected if part of the power sustaining a resonator soliton could be provided incoherently to the driving field, e.g. by means of optical pumping. Optical pumping of MQW structure generates carriers and allows converting from absorption to gain. In the last case the semiconductor microresonator operates as VCSEL [17]. To couple pump light into the microresonator the pump laser was tuned in one of short-wavelength transparency windows in the reflection spectrum of the microresonator as shown in Fig. 2 (a).

When the pumping was below the transparency point and the driving laser wavelength was set near the semiconductor MQW structure band edge the bright and dark solitons formed (Fig. 18 (a), (b)) [34] similar to the unpumped case Figs 9, 10.

Fig. 18. Intensity snapshots of structures observed in reflection from pumped (below transparency) semiconductor microresonator illuminated near resonance showing bright (a) and dark (b) soliton. The illuminating beam from the laser diode has an elliptical shape (c).

Fig. 19. Intensity snapshots of typical beam structures at optical pump intensities slightly above lasing threshold (pump increases from (a) to (c)).

Analysis of Eq.(1) shows that increase of the pump intensity leads to shrinking of the resonator solitons' existence domain and shifting towards low intensity of the light sustaining the solitons. Such reduction of the sustaining light intensity was observed experimentally in [23]. That is why soliton switchings in the pumped case are fast [23] and not mediated by thermal effects as for soliton formation without pumping (Fig. 12 (c), Fig. 13).

When the pump intensity approaches the transparency point of the semiconductor material, the resonator solitons' domain of existence disappears. It reappears above the transparency point. In the experiment we have quite strong contribution of the imaginary part (absorption/gain) of the complex nonlinearity at the working wavelength (near band edge). Therefore the transparency point is very close to the lasing threshold so that inversion without lasing is difficult to realize.

Slightly above threshold we observe in presence of illumination structures (Fig. 19) reminiscent of the solitons in electrically pumped resonators [35]. We note that optical as opposed to electrical pumping allows more homogeneous pumping conditions [36]. This suggests that optically pumped resonators lend themselves more readily for localization and motion control of solitons then electrically pumped ones.

There is difference between below bandgap resonator solitons in pumped and unpumped cases. The nonlinear resonance mechanism of soliton formation [22] requires a defocusing nonlinearity below transparency (Fig. 3) and a focusing nonlinearity above transparency. Transverse nonlinear self-focusing effect is generally furthering soliton formation. Figure 20 shows typical examples of calculated resonator solitons for pumped semiconductor microresonator. Bright solitons have a large existence range in the pumped case (Fig. 20), dark solitons exist, though with smaller range of stability, in the unpumped case (Fig. 3).

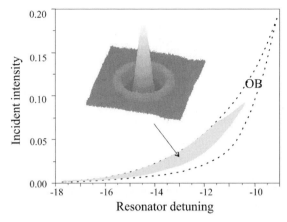

**Fig. 20.** Results of numerical simulations of below bandgap (purely dispersive) solitons using the model (1) for pumped above the transparency point microresonator ($P = 2$). Inset is bright soliton. Shaded area is domain of existence of resonator bright solitons. Area limited by dashed lines is optical bistability domain for plane waves. Other parameters same as in Fig. 3.

Thus optically pumped semiconductor resonators are well suited for sustaining solitons below bandgap: (i) background light intensity necessary to sustain and switch resonator solitons is substantially reduced by the pumping and therefore destabilizing thermal effects are minimized, (ii) above the transparency point only the dispersive part of semiconductor nonlinearity stabilizes a soliton, therefore the domain of existence of below bandgap (purely dispersive) bright solitons can be quite large.

## 5   Conclusion

In conclusion, we have shown experimentally and numerically existence of spatial solitons in driven semiconductor microresonators over a wide spectral range around the bandgap: below bandgap hexagons and dark solitons, near bandgap bright and dark solitons, above bandgap bright solitons, optically pumped below bandgap bright solitons. We have demonstrated the manipulation of such solitons: switching them on and off by coherent as well as incoherent light; reducing the light power necessary to sustain and switch a soliton, by optical pumping.

### Acknowledgment

This work was supported by the Deutsche Forschungsgemeinschaft under Grant We 743/12-1.

# References

1. D.W. Mc Laughlin, J.V. Moloney, A.C. Newell: Phys. Rev. Lett. **51**, 75 (1983)
2. N.N. Rosanov, G.V. Khodova: Opt. Spectrosk. **65**, 1375 (1988)
3. M. Tlidi, P. Mandel, R. Lefever: Phys. Rev. Lett. **73**, 640 (1994)
4. G.S. Donald, W.J. Firth: J. Opt. Soc. Am. B **73**, 1328 (1990)
5. M. Brambilla, L.A. Lugiato, M. Stefani: Europhys. Lett. **34**, 109 (1996)
6. V.Yu Bazhenov, V.B. Taranenko, M.V. Vasnetsov: *Transverse optical effects in bistable active cavity with nonlinear absorber on bacteriorhodopsin*, In: *Proc. SPIE*, **1840**, (1992) pp. 183
7. K. Staliunas, V.B. Taranenko, G. Slekys, R. Viselga, C.O. Weiss: Phys. Rev. A **57**, 599 (1998)
8. G. Slekys, K. Staliunas, C.O. Weiss: Opt. Comm. **149**, 113 (1998)
9. K. Staliunas, V.J. Sanchez-Morcillo: Phys. Rev. A **57**, 1454 (1998)
10. V.B. Taranenko, K. Staliunas, C.O. Weiss: Phys. Rev. Lett. **81**, 2236 (1998)
11. K. Staliunas: Phys. Rev. Lett. **81**, 81 (1998)
12. C. Rebbi, G. Soliani: *Solitons and Particles*, (World Scientific, 1984)
13. C. Tamm: Phys. Phys. Rev. A **38**, 5960 (1988)
14. K. Staliunas, C.O. Weiss, G. Slekys: *Optical vortices in lasers*. In: *Optical Vortices.* ed. by M. Vasnetsov and K. Staliunas (Nova Science Publishers, 1999) pp. 125
15. C.O. Weiss, M. Vaupel, K. Staliunas, G. Slekys, V.B. Taranenko: Appl. Phys. B **68**, 151 (1999)
16. *Soliton Driven Photonics*, ed. by A.D. Boardman, A.P. Sukhorukov (Kluver Academic Publishers, 2001)
17. T.E. Sale: *Vertical Cavity Surface Emitting Lasers*, (Wiley, 1995)
18. H.M. Gibbs: *Optical Bistability - Controlling Light with Light*, (Academic Press, 1985)
19. L. Spinelli, G. Tissoni, M. Brambilla, F. Prati, L.A. Lugiato: Phys. Rev. A **58**, 2542 (1998)
20. D. Michaelis, U. Peschel, F. Lederer: Phys. Rev. A **56**, R3366 (1997)
21. L.A. Lugiato, R. Lefever: Phys. Rev. Lett. **58**, 2209 (1987)
22. K. Staliunas, V.J. Sanchez-Morcillo: Opt. Comm. **139**, 306 (1997)
23. V.B. Taranenko, C.O. Weiss, W. Stolz: Opt. Lett. **26**, 1574 (2001)
24. V.B. Taranenko, I. Ganne, R. Kuszelewicz, C. O. Weiss: Phys. Rev. A **61**, 063818 (2000)
25. P.K. Jacobsen, J.V. Moloney, A.C. Newell, R. Indik: Phys. Rev. A **45**, 8129 (1992)
26. W.J. Firth, A.J. Scroggie: Europhys. Lett. **26**, 521 (1994)
27. V.B. Taranenko, C.O. Weiss, B. Schäpers: Phys. Rev. A **65**, 013812 (2002)
28. V.B. Taranenko, I. Ganne, R. Kuszelewicz, C.O. Weiss: Appl. Phys. B **72**, 377 (2001)
29. T. Rossler, R.A. Indik, G.K. Harkness, J.V. Moloney, C.Z. Ning: Phys. Rev. A **58**, 3279 (1998)
30. V.B. Taranenko, C.O. Weiss: Appl. Phys. B **72**, 893 (2001)
31. V.B. Taranenko, C.O. Weiss, W. Stolz: J. Opt. Soc. Am. B **19**, 8129 (1992)
32. S.H. Park, J.F. Morhange, A.D. Jeffery, R.A. Morgan, A. Chavez-Pirson, H.M. Gibbs, S.W. Koch, N. Peyghambarian, M. Derstine, A.C. Gossard, J.H. English, W. Weidmann: Appl. Phys. Lett. **52**, 1201 (1988)
33. V.B. Taranenko, F.-J. Ahlers, K. Pierz: Appl. Phys. B (2002) in print
34. V.B. Taranenko, C.O. Weiss: nlin.PS/0204048 (2002)
35. Report as given in www.pianos-int.org
36. W.J. Alford, T.D. Raymond, A.A. Allerman: J. Opt. Soc. Am. B **19**, 663 (1992)

# Propagation and Diffraction
# of Picosecond Acoustic Wave Packets
# in the Soliton Regime

O.L. Muskens and J.I. Dijkhuis

Atom Optics and Ultrafast Dynamics, Debye Institute, University of Utrecht, P.O. Box 80 000, 3508 TA, Utrecht, Netherlands

**Abstract.** Recent experiments on propagation of picosecond acoustic wave packets in condensed matter have opened up a new, exciting area of soliton physics. Single cycle strain pulses as short as several picoseconds can be generated in a thin metallic film, yielding local strain fields of the order of $10^{-4}$. The combination of phonon dispersion and anharmonicity of the atomic interaction potential may give rise to strongly nonlinear, but stable propagation of the wave packets over a distance of the order of several millimeters in a single crystalline material. We present new results on nonlinear propagation of acoustic wave packets created by nJ femtosecond optical pulses in a lead molybdate single crystal, employing the Brillouin Scattering technique as a local probe of acoustic strain. Studies of diffraction of narrow discs of acoustic strain show anomalous diffraction of the various Fourier components constituting the wave packet. Propagation of virtually one-dimensional nature is studied by exciting the metal film over a large area using an amplified femtosecond laser. We show that these data can be interpreted by means of the Korteweg-de Vries equation and strongly suggest the development of acoustic solitons.

## 1   Introduction

The discovery of nonlinear optical phenomena in various areas of experimental condensed matter physics has lead to a multitude of potential technological applications. In comparison with the field of nonlinear optics, nonlinear wave phenomena in acoustics have received remarkably little attention. Only recently, acoustic solitons, an archetypical nonlinear wave phenomenon, have been generated in an elastic rod [26], using an exploding metallic flyer foil in water as the generator of high acoustic strain. Here, unipolar (compressional) acoustic strain fields were produced of several microseconds duration. To observe soliton development, a very substantial degree of anomalous dispersion is required. This dispersion was provided in the rod by the modes of an acoustic waveguide.

A completely different regime for stable soliton formation has been explored by Hao and Maris [6,13]. It was already predicted by Breazale and Ford [15] that, given the weak anharmonicity of a realistic solid, for a piezoelectrically generated ultrasonic wave in the MHz regime a discontinuity - or shock wave, another prototype nonlinearity - would only develop after a travelled distance of, say, 500 cm. This explains why in ultrasound experiments in crystalline solids nonlinear acoustical phenomena have remained a relatively unexplored field for a long time. The situation changes dramatically if one replaces the MHz strain field by a wave packet containing much higher frequency components, say, of the order of tens of GHz, as can be generated using high intensity, ultrashort optical excitation in a thin metal transducer [6,13,11]. Under these conditions the discontinuity distance drops to the order of a few millimeters and experiments are well within reach. A fortuitous circumstance is the fact that the metal film is not necessarily destroyed or modified in the optical experiments, which enables doing repetitive experiments. The acoustic strain generated by the method of pulsed optical excitation in a metal film in general is bipolar, therefore both anomalous and normal dispersion may lead to acoustic soliton formation for either polarities. It turns out that the normal phonon dispersion of longitudinal acoustic lattice modes in crystals in combination with the lattice anharmonicity can be sufficient to develop stable propagating solitons [6]. In general, experiments at these high acoustic frequencies require working at liquid helium temperatures, in order to reduce thermal damping of these acoustic fields, that otherwise prevents the buildup of solitary waves.

The most striking difference between picosecond acoustic wave packets and those studied in nonlinear optics is the fact that the initial excitation is a *single cycle* wave packet, rather than an slowly varying modulation of the carrier frequency. This makes the acoustic system a very interesting and new system to study one- and higher-dimensional propagation effects in detail. As the disturbance is propagating with the velocity of sound and can be easily detected by optical means, it is a much more accessible subject to study than an optical disturbance. In particular, it is feasible to obtain information both on the amplitude and the phase of the strain field by means of time domain interferometry(see e.g. Hurley and Wright [12]).

## 2   Theory

In the following we will briefly introduce the ideas and concepts of nonlinear acoustics. A finite local distortion of an anisotropic lattice $\boldsymbol{u}(\boldsymbol{r},t)$ generates a deformation $\Delta$, which is usually defined in terms of the acoustic strain tensor $\eta(\boldsymbol{r},t)$, by [18]

$$\Delta = \eta_{ij}\, dr_i\, dr_j$$
$$\eta_{ij} = \frac{1}{2}\left(q_{ij} + q_{ji} + q_{ik}q_{kj}\right)\,, \tag{1}$$

where

$$q_{ij} = \frac{\partial u_i}{\partial r_j} \tag{2}$$

describes the displacement gradient matrix elements. To obtain the correct nonlinear elastic wave equations, one has to take into account this complete expression for the strain tensor, including the last term on the right hand side (which is neglected in linear elasticity theory). The generated strain yields a stress $T$ in the crystal via the interparticle interactions. This can be found in terms of the derivative of the lattice free energy $\phi$ to the Lagrangian coordinates $q_{ij}$ [25]

$$T_{ij} = \frac{\partial \phi}{\partial q_{ij}} \ . \tag{3}$$

The equation of motion for finite strains can then in turn be formulated in terms of this stress tensor and the mass density $\rho$

$$\rho \boldsymbol{u}_{tt} = \boldsymbol{\nabla} \cdot T \ , \tag{4}$$

where the subscript denotes the partial time derivative. It has been shown that for adiabatic deformation, the internal energy $\phi$ can be expressed as a power series of the first three invariants of the strain tensor, in combination with the appropriate elastic constants of second and third order [25,16]. For isotropic solids there are five of these elastic moduli (the so called 'five-constant' theory, see e.g. Ref. [19]). Expressions for several crystal groups of high symmetry have also been found by Seeger and Buck [16]. For these expressions for the internal energy (consisting of 36 terms for the cubic symmetry group) we refer to literature. In the case of one-dimensional propagation along an axis of high symmetry, the equation of motion however reduces to the simple form [15]

$$\rho u_{tt} = \gamma u_{zz} + \alpha u_z u_{zz} \ . \tag{5}$$

The last term on the right side is the quadratic nonlinearity, due to the geometric nonlinearity of Eq.(1) and the cubic terms in the inter-atomic potential. The nonlinearity coefficient $\alpha$ depends only on the propagation direction in the crystal [3]. For the [001] direction in a cubic crystal the two constants of Eq.(5) take on the form [15]

$$\gamma = C_{33}$$
$$\alpha = (3C_{33} + C_{333}) \ , \tag{6}$$

where the coefficients $C_{33}$ and $C_{333}$ are the second- and third-order elastic moduli in the [001] direction. For most solids, the contribution of the third order modulus is larger than the geometric term and has a negative sign, yielding an $\alpha < 0$.

Up to this point we have not taken into account any dispersion in the equation of motion. In the case of longitudinal acoustic lattice vibrations (LA phonons) in a crystalline solid, the dispersion due to discreteness of the lattice can be written as

$$\omega = \sqrt{\frac{4C_{33}}{M}} \sin|ka/2| \ . \tag{7}$$

As we will be dealing with vibrations of small wavevector, i.e. in the center of the Brillouin zone, it is sufficient to approximate this dispersion relation by its first two nonzero expansion terms

$$\omega = v_0 k + \beta k^3 \ ,$$
$$\beta = -\frac{v_0^3}{6\omega_{max}^2} \ , \tag{8}$$

where $\omega_{max}$ is the LA angular frequency at the edge of the Brillouin zone. This dispersive correction can be put into Eq.(5), leading to a fourth order spatial derivative (see e.g. Hao and Maris [6]). At this point it is further convenient to switch from the displacement coordinate $u$ to the z-component of the acoustic strain $\eta$. This is done by differentiation of Eq.(5) with respect to the z-coordinate. Given the initial wave packet at $t = 0$ of amplitude $\eta_0$ and shape $\psi(z)$, the resulting boundary value problem including dispersion, reads

$$\eta_{tt} - c_0^2 \eta_{zz} - \frac{\alpha}{\rho} \frac{\partial}{\partial z} (\eta \eta_z) - 2c_0 \beta \eta_{zzzz} = 0 \ ,$$

$$\eta(z, t = 0) \qquad\qquad\qquad = \eta_0 \psi(z) \ . \qquad (9)$$

Finally it is convenient to transform to a moving frame coordinate system, defined by the parameters $t' = t$, $y = z - c_0 t$. After substitution of these variables we arrive at terms consisting of only one derivative with respect to the travelling coordinate $y$, except for one term having a double time derivative $\eta_{t't'}$. Neglecting this term will not change the behavior up to first order [19], as this is a 'slow' coordinate with respect to the evolution of the wave packet. Integrating the resulting expression once, we finally obtain the equation

$$\eta_{t'} + \frac{\alpha}{2\rho c_0} \eta \eta_y + \beta \eta_{yyy} = 0$$

$$\eta(y, t' = 0) \qquad\qquad = \eta_0 \psi(y) \ . \qquad (10)$$

This is the well-known Korteweg-de Vries (KdV) equation, describing for example the formation of stable wave packets (solitons) in a narrow water channel. It can be shown that solutions of Eq.(10) also fulfill its spatial derivative, although the reverse it not necessarily true. In the next section we shall describe how to connect this equation to experiments on acoustic wave packets.

## 3  Physical Parameters

To make the step from Eq.(10) to an experimental configuration, the parameter range of interest for nonlinear acoustics will be discussed in this section. A general discussion of the boundary value problem of the Korteweg-de Vries equation for arbitrary $\psi(y, t = 0)$ can be found e.g. in Karpman [17]. The relevant parameter for KdV solitons turns out to be the similarity parameter $\sigma$, defined as

$$\sigma = l_0 \left( \frac{\alpha \eta_0}{2\rho c_0 \beta} \right)^{1/2} \ . \qquad (11)$$

Here $l_0$ and $\eta_0$ denote the spatial width and the amplitude of the initial strain perturbation, $\alpha$ and $\beta$ are the nonlinear and dispersive parameters, respectively. The condition for the generation of a soliton from an initial perturbation depends on the integral - or first moment - of the initial wavepacket:

$$p_0 = \int_{-\infty}^{\infty} \psi(\xi) \, d\xi \ . \qquad (12)$$

**Table 1.** Parameters characterizing the experiments of Hao and Maris compared with those of the present experimental configurations (secs. 5 and 6). ** Values for a typical strain of $10^{-4}$

| Parameter | Units | Hao and Maris [6] | | | | sec.5 | sec.6 |
|---|---|---|---|---|---|---|---|
| $l_0$ | nm | 20 | | | | 180 | 290 |
| $I_{pump}$ | J | $2 \cdot 10^{-9}$ | | | | $10^{-8}$ | $10^{-3}$ |
| $w_{pump}$ | cm | $1.5 \cdot 10^{-3}$ | | | | $1.1 \cdot 10^{-3}$ | 0.15 |
| $P_{max}$ | W cm$^{-2}$ | $10^9$ | | | | $10^{10}$ | $10^{11}$ |
| | | Si [100] | MgO [100] | Al$_2$O$_3$ [0001] | | PbMoO$_4$ [001] | |
| $c_0$ | $10^5$ cm/s | 8.48 | 9.05 | 11.23 | | 3.63 | |
| $\beta$ | $10^{-11}$ cm$^3$s$^{-1}$ | 1.8 | 1.6 | 3.75 | | 13.6 | |
| $\alpha$ | TPa | -0.373 | -4.02 | -1.83 | | -1.0 | |
| $\sigma$ ** | | 4.6 | 12.5 | 4.8 | | 21.7 | 35.0 |
| $N$ ** | | 0.6 | 1.6 | 0.6 | | 2.8 | 4.5 |
| $l_1$ ** | nm | 13.9 | 4.3 | 13.0 | | 21.5 | 21.0 |
| $\Delta c_1/c_0$ ** | $10^{-4}$ | 0.04 | 2.28 | 0.09 | | 2.5 | 2.9 |
| thickness $d$ | mm | 0.315 | 0.495 | 2.01 | | 5.0 | 5.0 |

If $p_0 \neq 0$ then always at least one soliton will form. When, however, $p_0 = 0$, a soliton will only develop if $\sigma^2 > 7$ [17]. With increasing $\sigma$ more solitons will separate from the wave packet. An approximate value for the number of solitons can be found from the inverse scattering transform for large values of $\sigma$ [17]:

$$N = \left(\frac{\sigma}{\pi\sqrt{6}}\right) \int_{\psi(\xi)<0} \sqrt{|\psi(\xi)|} \, d\xi \ . \tag{13}$$

Note that in this limiting case, the number of solitons only depends on the negative part of the initial perturbation.

In Table 1, relevant parameters both of the experiments by Hao et al. and ours, described later (see sec.5 and 6) are shown. Clearly one can notice important differences between both experimental configurations. The most significant difference is the initial pulse width $l_0$, caused by the difference in transducer thickness. The resulting higher value of $\sigma$ is partially compensated by the difference in propagation velocities and, most importantly, by the high dispersion coefficient of lead molybdate. This influences the soliton solutions of the system, both in energy content and in propagation velocity. It is well known that the shape of a Korteweg-de Vries soliton is given (in moving frame coordinates) by

$$\eta_r(y, t') = a_r \, \text{sech}^2 \, l_r^{-1} \, (y - \Delta c_r t') \ , \tag{14}$$

where $\eta_r$ denotes the wave function in strain of the $r$-th soliton developing from an initial wave packet. The soliton width, $l_r$, is given by

$$l_r = \sqrt{\frac{24\rho c_0 \beta}{\alpha a_r}} \; , \tag{15}$$

and the velocity change of the soliton relative to the sound velocity, $\Delta c_r/c_0$, is given by

$$\frac{\Delta c_r}{c_0} = \frac{\alpha a_r}{6\rho c_0^2} \; . \tag{16}$$

For large values of $\sigma$, the strain amplitude of the $r$-th soliton can be approximated by [17]

$$a_r = \frac{3\eta_0}{\sigma^2} \left(1 + \sqrt{1 + 2\sigma^2/3} - 2r\right)^2 . \tag{17}$$

From these expressions it follows that, for large $\sigma$, the amplitude and velocity of the solitons depend mainly on the material parameters and $\eta_0$, and not so much on $l_0$. Due to the high dispersion coefficient of lead molybdate, solitons in the current experimental configuration will not only be broader than those observed in the experiments of Hao and Maris, they will also propagate relatively faster. Theory predicts that in all four configurations of Table 1 one or more acoustic solitons will be generated at an initial strain of $10^{-4}$. It should however be realized that a 290 nm wide pulse contains much more energy than one of 20 nm width of equal strain, meaning that much more pump pulse energy will be needed to achieve such a perturbation.

## 4    Experimental Setup

Inelastic scattering of a single-mode argon-ion laser operating at 514 nm was used to probe the local strain Fourier components in a transparent crystal (see Fig. 1). The scattered radiation, for anti-Stokes scattering shifted upward in frequency and for Stokes shifted downward, is analyzed using a quintuple-pass Fabry-Pérot interferometer (Burleigh model RC-110) and standard photon counting apparatus. For the kHz-excitation experiments of sec.6, additional electronic gating of the photomultiplier signal was required to cope with the small duty cycle because of the short interaction time in the scattering volume and the low pump repetition frequency. Wavevector conservation in the scattering plane ensures sensitivity of the Brillouin signal for *single* Fourier components of the wave packet, although this condition is partly relaxed due to the tight focusing ($w_B = 3.5$ $\mu$m) of the argon-ion laser.

The sample is a lead molybdate (PbMoO$_4$) single crystal of 10x5x6 mm$^3$ dimension, with its c-axis oriented perpendicular to the 10x5 mm$^2$ interface. Lead molybdate was mainly chosen because of its extraordinary high acousto-optic coupling parameter. Further, the large acoustic dispersion parameter, as shown in Table 1, leads to a large $l_{\mathrm{sol}}$ and thus to a larger fraction of the acoustic power in the lower part of the spectrum, that is accessible in Brillouin scattering experiments.

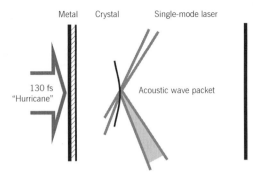

**Fig. 1.** Top view of the Brillouin scattering configuration in the crystal

A 500-nm gold transducer is deposited onto the 10x5 mm$^2$ interface, with a 5-nm chromium interlayer for better adhesion. It is known from experiments by Wright et al. [10] that the width of an acoustic wave packet generated by femtosecond optical absorption in gold is not determined by the optical skin depth of 12 nm, but rather by the ballistic transport length of hot electrons within the electron-phonon relaxation time of about 1 ps. The tails of the acoustic wave packet will therefore be several hundred nanometers wide. The sample is contained in an optical flow cryostat at liquid helium temperatures, which is necessary to eliminate damping of the acoustic wave packet from thermal phonons. Damping of monochromatic, low-amplitude GHz phonon beams has been studied previously in lead molybdate by Damen et al. [2] and is found to correspond at liquid-helium temperatures to mean free paths much larger than the crystal dimensions.

Two different types of femtosecond pulsed laser systems have been used as an excitation source for the acoustic strain. For the diffraction experiments of sec.5, a mode-locked Ti:sapphire laser operating at 800 nm was used, delivering 160 femtosecond pulses of 10 nJ/pulse at 76 MHz (Coherent Mira 900). Previous experiments on mode-locked acoustic wave packets using this setup have been described in Ref. [24]. Recently, an amplified system was incorporated in the setup, delivering 130 femtosecond laser pulses of 1 mJ/pulse at 1 kHz (Spectra Physics Hurricane). Section 6 describes the first results obtained using this excitation source. A Longitudinal acoustic wave packet generated in the transducer by optical absorption generally has two polarities of strain. The initial compressional strain propagates in two opposing directions, of which one is usually reflected from the interface with a lower impedance material (helium), giving an inversion of the strain polarity. The shape of the wave packet can be controlled by changing the acoustic environment of the transducer. We use two different excitation geometries, one is excitation from the helium-gold interface, the other from the crystal-gold interface.

# 5   Propagation and Diffraction of Picosecond Acoustic Wave Packets

Before we describe our experiments we note that for a detectable strain amplitude in case of a modelocked Ti:sapphire laser, the pump beam has to be focused to a waist $w_0$ of about 11 $\mu$m. The limited width of the corresponding acoustic beam leads to significant diffraction of the lower, GHz frequency components. Diffraction of low amplitude monochromatic phonon beams has been studied extensively in lead molybdate by Damen et al. [27] and it was shown that the waist of the strain, $w_{ph}(z)$, follows the Fraunhofer diffraction law corrected with a weak phonon (de)focusing:

$$
w_{ph}(z) = w_0 \left( 1 + \left( \frac{z}{z_0(1 - 2p)} \right)^2 \right)^{1/2} ,
\tag{18}
$$

where $p$ denotes the phonon focusing parameter ($p = 0.173$ for lead molybdate [1]). The corresponding theoretical divergence angle $\theta_{th}$ is given by

$$
\theta_{th} = \frac{w_0}{z_0(1 - 2p)} = \frac{\lambda_{ac}}{(\pi w_0(1 - 2p))} .
\tag{19}
$$

We will measure the diffraction of several frequency components of an acoustic wave packet using the same method as in Ref. [1]: scans of the pump beam transverse to the acoustic beam were taken with micrometer resolution, while keeping the Brillouin scattering condition fixed. The interaction volume (overlap between acoustic beam and argon-ion laser focus) is determined by the focal volume of the argon-ion laser in the sample, which is estimated to be a cylinder of 10 $\mu$m$^2$ by 50 $\mu$m inside the crystal.

First, the input Fourier spectrum of the wave packet was determined at room temperature, by measuring the diffracted intensity directly after the transducer at various scattering angles. We use the crystal-gold side excitation geometry in these experiments, yielding a Brillouin signal in the frequency region of interest, a factor of 50 higher than in the case of the gold-helium geometry. The observed Fourier spectrum is shown in Fig. 2. The spectral resolution in these measurements is limited to 280 MHz due the small size of the interaction volume (and the concomitant spread in diffraction angles)

At the frequencies indicated by the arrows in Fig. 2 (corresponding to maxima and minima in the spectrum), further experiments were performed at $T = 5$ K. At each scattering configuration, the acoustic beam width was measured as function of propagation distance in the crystal for different excitation intensities. Typical traces are shown in Fig. 3 for an acoustic frequency of 7.2 GHz. At low excitation intensities, the observed beam divergence corresponds to the behavior expected for linear propagation, according to Eq.(18). For increasing pump intensities, however, anomalous behavior is observed, leading first to an increase in diffraction and, above 1.5 mJ/cm$^2$, to a steep decrease of the beam divergence.

The measured diffraction angle normalized to the theoretical value of Eq.(19), versus pump intensity, is shown in Fig. 4, for the frequencies indicated in Fig. 2.

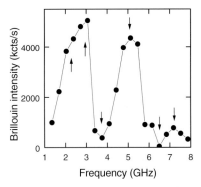

**Fig. 2.** Frequency spectrum of the acoustic wave packet (crystal-gold side excitation). Arrows indicate positions where diffraction experiments have been performed.

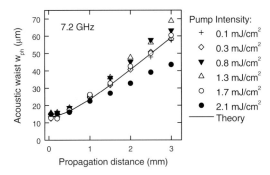

**Fig. 3.** Typical scans of the acoustic beam width as a function of propagation distance at $\nu = 7.2$ GHz, for several pump intensities. The solid line denotes the theoretical divergence Eq.(18) due to Fraunhofer diffraction

Again, the diffraction is seen to converge to the linear theory towards low pump intensities for most of the frequencies under study. Significant deviation occurs however, both in the low-intensity and the general behavior, for the minima in the spectrum of Fig 2 (at 3.8 and 6.5 GHz frequencies). These show a diffraction ratio significantly smaller than unity at all pump intensities. At the other frequencies, the normalized diffraction is seen to change for increasing excitation intensities. At intensities above 1.5 mJ/cm$^2$ clearly a steep decrease in diffraction is found at the maxima of the spectrum (see Fig. 2). The behavior of the 2.2 GHz component differs from that of the other frequencies by a decrease of the diffraction even at lower pump intensities. Unfortunately, our current sensitivity is not sufficient to measure below 0.1 mJ/cm$^2$ pump intensities and does not permit to enter the purely linear diffraction regime (of Ref. [1]). At higher pump intensities the experiment is limited by the output power of the Ti:sapphire laser.

In order to study the propagation of the complete Fourier components, the total acoustic power in the selected modes is computed as a function of $z$ and power by integration of the Brillouin signals over each transverse scan. The re-

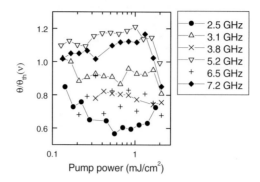

**Fig. 4.** Divergence angle of the different frequency components indicated in Fig. 2 normalized to the theoretical values of Eq.(19), as a function of pump intensity.

sulting traces, normalized to the value measured directly after the transducer, are shown in Fig. 5. The main observation is that the acoustic power is continuously redistributed among the Fourier components constituting the wave packet. It is further remarkable that the behavior is different for the various selected Fourier components and depend on the pump intensity. For example, the acoustic power in the minima of the spectrum of Fig. 2 (at 3.8 GHz and 6.5 GHz) tends to increase during propagation, while the maxima show a significant decrease. This suggests that the spectral minima are the most sensitive to the effects of acoustic nonlinearity, which might be a clue in understanding their anomalous diffraction at low pump intensities in Fig. 4. All our measurements suggest that we are dealing with a regime in which nonlinearity plays an important role and significantly influences the diffraction of the wave packet. Attempts have been made to describe the observations of Fig. 5 by simulations based on the one-dimensional wave equation of section 2 (results not presented in this article). Although exact fits could not be obtained, the qualitative behavior of the different parts of the spectrum could be reproduced within reasonable agreement. These simulations, for the relevant parameters (see Table 1), show significant self-steepening of the wave packet, combined with the onset of dispersion, for strains of the order of $10^{-5}$. Work is in progress to include diffraction into the wave equation in order to explain the nonlinear diffraction of the wave packet in this regime.

Acoustic beam propagation in the nonlinear regime has been studied extensively in the absence of dispersion, giving rise to the formation of shock waves (see e.g. Ref. [19]). It was shown that compressional shock waves tend to defocus while rarefaction shock waves will focus. As we are dealing with bipolar wave packets in which only the compressional wave is stable to dispersion (as we learned from the simulations mentioned above), an increase in diffraction with excitation power would be expected from this theory, which clearly is contradictory to our observations at high pump intensities.

Another mechanism which could lead to a decrease in diffraction is the refraction of the acoustic beam due to a radial thermal gradient in the crystal.

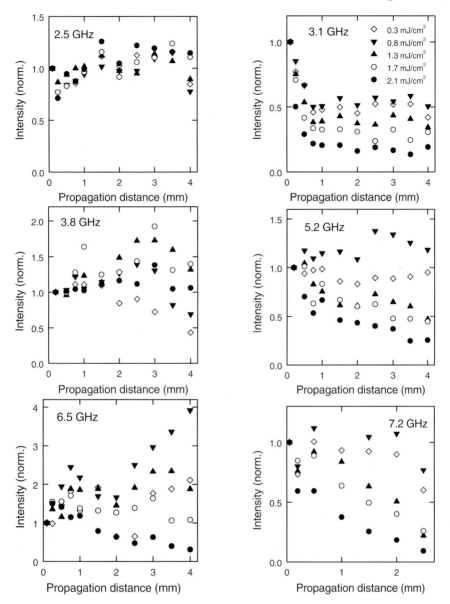

**Fig. 5.** Propagation of the frequency components as indicated in Fig. 2 into the lead molybdate crystal, for different pump intensities.

It is known that an increase in lattice temperature in the center of the beam leads to refraction of an acoustic wave which has a focusing effect [19]. This mechanism however would equally work for all Fourier components, and would strongly depend on the amount of energy deposited in the system and thus on the pump intensity. Further, at low temperatures the phonon mean free paths are so long that significant localized heating of the crystal is absent.

The problem of propagation of a cylindrical symmetric beam and the formation of solitons in a quadratic nonlinear medium *with* dispersion has become a lively subject to study very recently in the case of optical fields [5,4,22,23]. The mechanism leading in Ref. [4] to stable spatial solutions is the continuous up- and down-conversion between coupled modes, which in combination of either normal or anomalous dispersion leads to a wave-front reshaping which compensates the positive curvature due to diffraction. In the optical case of second harmonic generation generally only three modes are present, whereas we are dealing with a large number of modes of the acoustic wave packet (note that one can see a single cycle wave packet as a very broad spectrum of harmonics). It seems however, that the physical principles underlying the coupling between these modes are equal, so similar behavior might be expected. To which extent these mechanisms are indeed present in our current experiment will be a subject of future study.

## 6  One-Dimensional Propagation Experiments

Although the physics of three-dimensional wave packet propagation is an extremely interesting subject of studies, one would certainly also want to be able to perform a more one-dimensional propagation experiment. It is not possible to make a large-area acoustic field which is intense enough for nonlinear acoustics to play a role using a mode-locked Ti:sapphire laser. Therefore, experiments have been initiated employing an amplified ultrafast laser system as described in section 4. The 1 mJ/pulse output of this laser is, without attenuation, weakly focused onto the metal transducer by a lens of $f = 1.6$ m focal distance. The local optical intensity at the transducer is varied by changing the distance between lens and sample (always kept smaller than $f$), in combination with a 1.5 mm pinhole near the sample to maintain a constant illuminated area. It was observed that even transparent materials like quartz show irreversible damage when exposed to a peak power density above $10^{13}$ W/cm$^2$. The damage threshold for a gold film at room temperature was determined to be an order of magnitude lower. The minimal optical waist at the gold film was therefore limited to 1.5 mm. At these high pump intensities significant nonlinear optical effects could be observed in the lead molybdate crystal, therefore experiments were performed in the helium-gold side excitation geometry, preventing the light to enter the crystal.

The incident acoustic frequency spectrum was obtained at $T = 5$ K, which is shown in Fig. 6. We observe a peak at an acoustic frequency of 3 GHz, which corresponds to the gold film thickness of 500 nm. The inset shows the acoustic wave packet that fits the measured power spectrum. The obtained fitting parameters are: $l_0 = 290$ nm for the width of the tails of the packet and $r = 0.4$ for the reflectivity between gold and lead molybdate.

At the frequencies indicated by the arrows in Fig. 6, traces were made at $T = 5$ K as a function of propagation distance into the crystal for different pump intensities. The results of these experiments are shown in Fig. 7. One

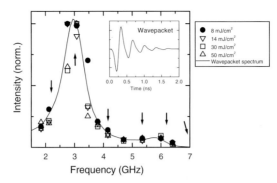

**Fig. 6.** Frequency spectrum of the acoustic wave packet, obtained using a 1kHz, high intensity pump laser (helium-gold side excitation). Inset: acoustic wave packet profile obtained from fitting the power spectrum.

immediately sees that the propagation is strongly anomalous at all frequencies under study, indicating the presence of nonlinear acoustical effects. One is able to discern oscillations in the propagation at frequencies above 4 GHz, which indicate that acoustic power is redistributed over the spectrum in a coherent manner. The initial spectral peak at 3.1 GHz is decreasing rapidly as the wave packet propagates into the crystal, until the intensity is about as high as at the other frequencies in the spectrum. This is also a strong evidence for the redistribution of acoustic power over the spectrum.

Numerical simulations of the one-dimensional KdV equation, with an initial wave packet as found from Fig. 6 yield a propagation of the Fourier components in qualitative agreement with our experimental findings. Typical evolution of the spectral components over the propagation distance is shown in Fig. 8. The spectral maximum at 3.1 GHz decreases within 1 mm propagation distance, while oscillations appear at significantly higher frequencies than generated in the initial wave packet, and start shifting through the spectrum. In these simulations, multiple solitons can be observed at a strain of the order of $10^{-4}$, which is in agreement with the estimate of Table 1. The similarity between simulations and experiments strongly suggests that we are indeed in a regime in which wave packet propagation is dominated by nonlinear and dispersive terms. To which extent dispersion plays a role to form stable packets, i.e. to an acoustic soliton, can however not yet be determined with certainty. Better quantitative agreement between simulations and experiments is needed to determine whether the wave propagation can be described effectively by the one-dimensional KdV equation. Further experiments, also on different crystalline materials, will have to shed more light on these questions in the near future.

## 7   Conclusion

The propagation of an 80 picosecond acoustic wave packet has been studied in a lead molybdate single crystal. Diffraction of individual frequency compo-

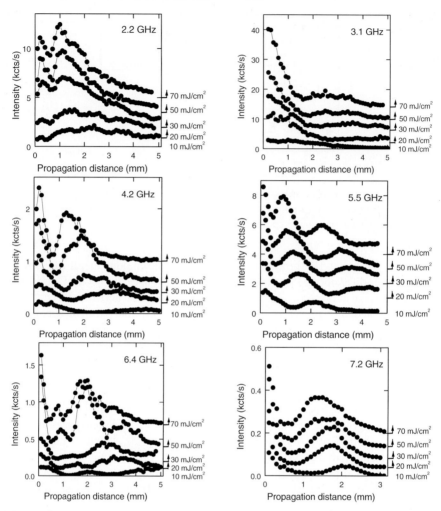

**Fig. 7.** Scans of the acoustic power of the frequency components as indicated in Fig. 6 as a function of propagation distance into the lead molybdate crystal, for different pump intensities. Arrows on the right indicate vertical offsets for the individual traces.

nents has been shown to converge to the linear theory of Fraunhofer diffraction combined with phonon focusing at low excitation intensities, for all frequencies except those at the minima of the spectrum. At higher pump intensities however, anomalous diffraction is observed at all frequencies, its characteristics changing over the spectrum in a nontrivial manner. This anomalous behavior has not yet been explained quantitatively, although we suspect that the explanation might be found in the coupling between acoustic frequencies in combination with dispersion. This mechanism might provide the necessary reshaping of the wave front which can counterbalance diffraction, as has been observed for light pulses in second harmonic generation in quadratically nonlinear optical systems. Further

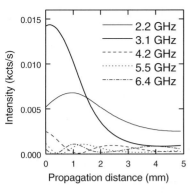
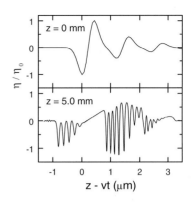

**Fig. 8.** (right) Propagation of the acoustic wave packet obtained from Fig. 6 over a distance of 5 mm through the crystal. (left) Evolution of individual spectral components as a function of propagation distance.

experiments, extending the dynamic range in both the high- and low excitation intensity regimes, will be performed in the near future.

Results on one-dimensional propagation at high peak intensities show strongly nonlinear propagation of frequency components of the wave packet. Oscillations appear which can be explained by a coherent redistribution of acoustic power over the spectrum. Comparison with the theory of nonlinear wave propagation shows that similar features are an indication of self steepening and even soliton formation. This strongly suggests that we are in a regime of strong nonlinear behavior, which is remarkable considering the relatively low frequency range under studies. It is however consistent with the expected parameters for nonlinearity and dispersion in the lead molybdate system. Future experiments, especially in combination with time domain spectroscopy, will certainly provide more insight in the nature of wave packet evolution at these experimental conditions.

The authors would like to thank C.R. de Kok, P. Jurrius and F.J.M. Wollenberg for their technical support. This work was supported by the Netherlands Foundation "Fundamental Onderzoek der Materie (FOM)" and the "Nederlandse Organisatie voor Wetenschappelijk Onderzoek (NWO)."

# References

1. E.P.N. Damen, A.F.M. Arts and H.W. de Wijn: Phys. Rev. Lett. **75**, 4249 (1995)
2. E.P.N. Damen, A.F.M. Arts and H.W. de Wijn: Phys. Rev. B **59**, 349 (1999)
3. T. Bateman, W.P. Mason and H.J. McSkimin:J. Appl. Phys **32**, (1961)
4. W.E. Torruelas, Z. Wang, D.J. Hagan, E.W. VanStryland, G.I. Stegeman, Ll. Torner and C.R. Menyuk, A. Hasegawa: Phys. Rev. Lett. **74**, 5036 (1995)
5. Ll. Torner, D. Mazilu and D. Mihalache: Phys. Rev. Lett. **77**, 2455 (1996)
6. H.-Y. Hao and H.J. Maris: Phys. Rev. B **64**, 4302 (2001)
7. D.A. Reis, M.F. deCamp, P.H. Bucksbaum and R. Clarke, Y. Kodama:Phys. Rev. Lett. **86**, 3072 (2001)

8. V.E. Gusev and O.B. Wright: Phys. Rev. B **57**, 2878 (1998)
9. H.-Y. Hao and H.J. Maris: Phys. Rev. B.**63**, 4301 (2001)
10. O.B. Wright:Phys. Rev. B **49**, 9985 (1994)
11. O.B. Wright and K. Kawashima:Phys. Rev. Lett. **69**, 1668 (1992)
12. D.H. Hurley and O.B. Wright:Opt. Lett. **24**, 1305 (1999)
13. H.-Y. Hao and H.J. Maris:Phys. Rev. Lett. **84**, 5556 (2000)
14. M.F. DeCamp, D.A. Reis, P.H. Bucksbaum, B. Adams and J.M. Caraher: Nature **413**, 825 (2001)
15. M.A. Breazale and J. Ford: J. Appl. Phys **36**, 3486 (1965)
16. A. Seeger and O. Buck: Z. Naturforschg. **15**, 1056 (1960)
17. V.I. Karpman: *Nonlinear waves in dispersive media*, (Pergamon Press 1975)
18. B.A. Auld:*Acoustic fields and waves in solids*, (Robert E. Krieger Publishing Company, 1990)
19. K. Naugolnykh and L.Ostrovsky: *Nonlinear wave processes in acoustics*, (Cambridge University Press, 1998)
20. A.M. Samsonov:Acta Techn. CSAV **41**, 1 (1996)
21. J. H. Ferziger and M Perić: *Computational methods for fluid dynamics*, (Springer-Verlag, 1999)
22. X. Liu, K. Beckwitt and F. Wise: Phys. Rev. E **62**, 1328 (2000)
23. D. Mihalache, D. Mazilu, L.-C. Crasovan, Ll. Torner, B. Malomed and F. Lederer: Phys. Rev. E **62**, 7340 (2000)
24. O.L. Muskens and J.I. Dijkhuis: Physica B (In Press)
25. R.S. Rivlin: Phil. Trans. A **240**, 459 (1948)
26. A.M. Samsonov, G.V. Dreiden, A.V. Porubov and I.V. Semenova: Phys. Rev. B **57**, 5778 (1998)
27. E.P.N. Damen, D.J. Dieleman, A.F.M. Arts and H.W. de Wijn: Phys. Rev. B **64**, 4303 (2001)

# Lecture Notes in Physics

For information about Vols. 1–567
please contact your bookseller or Springer-Verlag

# Monographs
For information about Vols. 1–29
please contact your bookseller or Springer-Verlag